NUCLEAR STRUCTURE AT HIGH SPIN, EXCITATION, AND MOMENTUM TRANSFER

AIP CONFERENCE PROCEEDINGS 142

RITA G. LERNER
SERIES EDITOR

NUCLEAR STRUCTURE AT HIGH SPIN, EXCITATION, AND MOMENTUM TRANSFER

INDIANA UNIVERSITY 1985

EDITOR: HERMANN NANN
INDIANA UNIVERSITY

AMERICAN INSTITUTE OF PHYSICS   NEW YORK 1986

Copy fees: The code at the bottom of the first page of each article in this volume gives the fee for each copy of the article made beyond the free copying permitted under the 1978 US Copyright Law. (See also the statement following "Copyright" below.) This fee can be paid to the American Institute of Physics through the Copyright Clearance Center, Inc., 21 Congress Street, Salem, MA 01970.

Copyright © 1986 American Institute of Physics

Individual readers of this volume and non-profit libraries, acting for them, are permitted to make fair use of the material in it, such as copying an article for use in teaching or research. Permission is granted to quote from this volume in scientific work with the customary acknowledgment of the source. To reprint a figure, table or other excerpt requires the consent of one of the original authors and notification to AIP. Republication or systematic or multiple reproduction of any material in this volume is permitted only under license from AIP. Address inquiries to Series Editor, AIP Conference Proceedings, AIP, 335 E. 45th St., New York, NY 10017.

L.C. Catalog Card No. 86-70837
ISBN 0-88318-341-2
DOE CONF-8510137

Printed in the United States of America

Contents

Session A

Nuclear Structure Seen by Medium Energy Probes

Relativistic Effects in Nuclear Physics 5
  G.E. Brown
Comparison of Electromagnetic and Hadronic Probes of Nuclear Structure 27
  J.J. Kelly

Session B

Collective Modes of Motion at High Energy

Isospin Character of Giant Resonances from Pion Inelastic Scattering 54
  S. Seestrom-Morris
Electroexcitation of the Nuclear Continuum 68
  K.T. Knöpfle
Decay of the Isoscalar Giant Monopole Resonance in ^{208}Pb 95
  A. van der Woude

Session C

Excitation of Stretched Configuration High-Spin States with Various Probes and Reactions

Evidence of Partial Occupancy of Orbits from Electron Scattering 110
  C.N. Papanicolas
Effects of Mesonic Exchange Currents, Unbound Wave Functions, and Knockout Exchange Amplitudes on the Extraction of Particle-Hole Strengths for 1 $\hbar\omega$ Stretched States 133
  R.A. Lindgren, M. Leuschner, B.L. Clausen,
  R.J. Peterson, M.A. Plum, and F. Petrovich
Stretched State Excitation in (p,n) Reactions 155
  B.D. Anderson, J.W. Watson, and R. Madey
High-Spin States with the (p,π) Reaction 181
  W.W. Jacobs
Selective Population of Stretched High-Spin States in the (d,α) Reaction 208
  H. Nann

Session D

Nuclear Structure from Transfer and Knockout Reactions

Towards Absolute Spectroscopic Factors 220
  G.J. Wagner
Complementary Aspects of the Quasi-Free ($e,e'p$) Reaction and the ($d,^3$He) Reaction 233
  P.K.A. de Witt Huberts

Quasi-Free Knockout Reactions Induced by Hadrons 246
    N.S. Chant
High-Spin Inner and Outer Subshells via Transfer Reactions 272
    S. Galès

Session E

Shell Model Calculations

Effects of Particle-Hole Excitations in Light Nuclei 316
    P.W.M. Glaudemans
Large Scale Calculations of the Nuclear Spectrum 327
    K.W. Schmid, E. Hammarén, and F. Grümmer
General Properties of the Residual Interaction as Measured Near Closed Shells 344
    W.W. Daehnick
Analysis of Features in the A = 34–48 Region in Terms of $d_{3/2}$ and $f_{7/2}$ Degrees of Freedom 357
    S.T. Hsieh, X. Ji, R. Mooy, and B.H. Wildenthal
Relativistic Shell-Model Calculations 376
    R. Furnstahl

Session F

Non-Nucleonic Degrees of Freedom

Charge Exchange to the Δ-Region 391
    C. Ellegaard
Models of Nucleon and Pion Substructure 403
    W. Weise
Probing Non-Nucleonic Degrees of Freedom with Strong and Electromagnetic Interactions 416
    B. Frois
Looking for Quark Degrees of Freedom 437
    I. Sick
Strangeness in Nuclei 450
    R. Büttgen, K. Holinde, B. Holzenkamp, and J. Speth

Session G

Outlook

Summary and Outlook 467
    T.W. Donnelly

Conference Participants 479

Proceedings of the Workshop on

Nuclear Structure at High Spin, Excitation, and Momentum Transfer

Indiana University Cyclotron Facility
Bloomington, Indiana 47405
October 21–23, 1985

Preface

As it is now tradition, the fifth annual fall workshop organized by the Indiana University Cyclotron Facility was held at McCormick's Creek State Park. Again, the pastoral setting of the park at the peak of Indiana's beautiful fall colors provided the backdrop for informal discussions and exchange of results and ideas.

The theme of the workshop was "Nuclear Structure at High Spin, Excitation, and Momentum Transfer". The speakers covered an extensive array of topics at the forefront of current research. Experimental research into nuclear structure exploiting different probes and reactions for one common purpose (implied in the workshop title) were discussed. A range of theoretical approaches to nuclear structure were presented. The advancements in the field are clearly being driven by new experimental facilities and large-scale computing techniques.

The scientific program of the workshop was structured into six sessions, each devoted to a particular topic. At the end of each session a critical examination of the presentations was given by expert reviewers. This modus operandi contributed much to a lively discussion.

The workshop was clearly successful in bringing the electromagnetic and hadronic interaction physics and the related communities together. There is vigorous activity going on in both areas, and increasingly, inter-probe comparisons are being undertaken to extract complementary information on nuclear structure.

Among the many people who have contributed their time and effort to the organization of the workshop I would like to specially thank Diana McGovern for her central role in dealing with the nitty-gritty of the workshop arrangements. From the early planning stages, Chuck Foster, Laurie Hicks, Phil Thompson, Becky Westerfield, and Bob Woodley not only dedicated a sizeable fraction of their time to this workshop, but their experience from earlier workshops proved invaluable. I greatly appreciate their effort and commend them for their enthusiasm. I also wish to thank Kent Berglund for the photographs that grace these proceedings. The help of T. Throwe and the graduate students J. Adams, S. Aziz, V. Cupps, M. Fatyga, W. Fox, J. Gering, J. Goodwin, E. Korkmaz, D. Low, J. Miranda, K. Pitts, B. Raue, and J. Templon for driving people to and from the airports and for taping the talks and discussions of the workshop is greatly appreciated.

Thanks are also due to the Harshaw-Filtrol Corporation for its financial contribution toward the expenses of the social functions.

Finally, I want to thank the speakers, reviewers, session chairmen and all participants for their active role that made the workshop worth the effort.

Bloomington, Indiana
January, 1986

Hermann Nann

Sponsors

U.S. National Science Foundation
Indiana University Cyclotron Facility
Physics Department, Indiana University
Off. of Res. and Grad. Development, Indiana University

International Advisory Board

E. Adelberger, University of Washington, Seattle, WA
A. Arima, University of Tokyo, Tokyo, Japan
W. Bertozzi, MIT, Cambridge, MA
G.E. Brown, SUNY, Stony Brook, NY
P. de Witt Huberts, NIKHEF-K, Amsterdam, The Netherlands
H. Ejiri, Osaka University, Osaka, Japan
A. Faessler, University of Tübingen, Tübingen, West Germany
B. Frois, DPhN/HE, Saclay, France
D. Kurath, Argonne National Laboratory, Argonne, IL
S.P. Pandya, Physics Research Laboratory, Ahmedabad, India
J.P. Schiffer, Argonne National Laboratory, Argonne, IL
I. Sick, University of Basel, Basel, Switzerland
J. Speth, IKP-KFA Jülich, Jülich, West Germany
S.Y. van der Werf, KVI, Groningen, The Netherlands
G.J. Wagner, University of Tübingen, Tübingen, West Germany

Scientific Program Committee

B.D. Anderson, Kent State University, Kent, OH
A.D. Bacher, Indiana University, Bloomington, IN
L.C. Bland, Indiana University, Bloomington, IN
B.A. Brown, Michigan State University, East Lansing, MI
N.S. Chant, University of Maryland, College Park, MD
G.T. Emery, Indiana University, Bloomington, IN
D.W. Miller, Indiana University, Bloomington, IN
H. Nann (Chair), Indiana University, Bloomington, IN
P. Schwandt, Indiana University, Bloomington, IN
B. Serot, Indiana University, Bloomington, IN
P.P. Singh, Indiana University, Bloomington, IN
B.H. Wildenthal, Drexel University, Philadelphia, PA

SESSION A

NUCLEAR STRUCTURE SEEN BY MEDIUM ENERGY PROBES

RELATIVISTIC EFFECTS IN NUCLEAR PHYSICS

G.E. Brown*
Department of Physics
State University of New York at Stony Brook
Stony Brook, New York 11794

ABSTRACT

Relativistic corrections to the nonrelativistic nuclear many-body problem are treated in perturbation theory, beginning from a zero-order wave function in which nucleons are in positive energy states in a plane-wave representation. Corrections arise from virtual pairs. Two important relativistic effects arise: (1) A repulsive term in the energy per particle $(\overline{\delta\epsilon})_{rel} = 2.4(\rho/\rho_o)^{8/3}$ MeV, where ρ_o is nuclear matter density, comes from the virtual pair terms. This term can equivalently be viewed as coming from a density-dependent correction to the mass of the exchanged scalar boson. (2) The nucleon-nucleon spin-orbit interaction is modified. Effectively, the nucleon mass which enters into this interaction is changed locally by scalar fields which connect to virtual pair states.

When treated consistently, relativistic effects represent small corrections to the nonrelativistic many-body problem, and can easily be grafted on to the latter. In this way one can exploit the nonrelativistic many-body calculations, which have been carried out with much more detailed and sophisticated treatment of correlations than in the relativistic formulation.

In high-energy reactions, which are treated only briefly here, the modification of the spin-orbit term has important consequences.

*Work supported by U.S. Dept. of Energy Contract
No. DE-AC02-76ER13001.

INTRODUCTION

Considerable effort is presently being expended to derive an effective meson theory from QCD. The nucleon is then obtained as a soliton solution in this (nonlinear) meson theory. From the effective Lagrangians coupling mesons and nucleons, one constructs a field theory for the many-body system. This relativistic field theory is a necessary link between the effective Lagrangian and the highly successful nonrelativistic Fermi liquid theory which is used to describe nuclear structure. In this talk, I wish to discuss this link and what we have learned from relativistic field theory, in particular, where relativity adds new aspects to the description of nuclear structure.

Two excellent and extensive reviews on the role of relativity in nuclear physics are appearing. The first one, by Serot and Walecka[1] summarize efforts in making a relativistic Hartree theory, essentially a mean field theory, which have gone on over

many years. The authors call this theory "Quantum Hadrodynamics."
The second one, a book by Celenza and Shakin on relativistic
nuclear physics, to be published by World Scientific Press,
reviews[2] the developments in giving a theoretical foundation to
the highly successful Dirac phenomenology begun chiefly by
Bunny Clark[3] and collaborators at Ohio State.

Last winter a workshop on Dirac phenomenology was held in
Los Alamos. A senior-diplomat style critique by John Negele[4]
will soon appear in Comments.

Because of considerable success in the scattering field, Dirac
phenomenology and relativistic theories are now being applied in
nuclear structure physics, often in a haphazard fashion,
disregarding what we know about nuclear structure. As an outsider
to the above developments, but with some credentials for having
participated in the formulation of the relativistic many-body
theory in atomic physics, I would like to take stock of the
relativistic developments. Most of the work reported in the
following was done together with Wolfram Weise. We plan to publish
a review later.

2. Relativistic Mean Field Theory

Relativistic mean field theory seems to be the natural
generalization of Hartree-Fock theory to include relativity. Most
of the past work has been carried out in Hartree theory, and I
shall defer comments on the Fock part.

One must be careful in generalizing nonrelativistic many-body
theory to a relativistic one. As Brown and Ravenhall pointed
out[5], and this point has been developed more recently by
Sucher[6], one can easily get into trouble with a relativistic
many-body wave function. Let me illustrate: Suppose one has an
A-particle product wave function of Hartree type, as often used.
This wave function is generally written down, without explicitly
expressing the knowledge that negative-energy nucleon states are
filled. Thus, there exist an infinite number of other A-particle
states of identical energy, in which two of the particles have
interacted, one going to a positive-energy continuum state and the
other to a negative-energy state. Consequently, strictly speaking,
one cannot write down a localized A-particle state, because it will
exist in a continuum of degenerate delocalized states. Through
processes like autoionization, the localized state will dissolve,
spreading over all space.

In some sense, we have erected a straw man, which we will now
proceed to knock down. But in a deeper sense, one sees that
"Quantum Hadrodynamics" can only be a complete theory, if in
further extensions to take into account correlations of two-body
nature, it is accompanied by projection operators specifying that
negative-energy states are full; i.e., it must go over towards a
real field theory. (In field-theoretical equations like the
Bethe-Salpeter one, there is no trouble with the "Brown-Ravenhall
disease" because projection operators correctly define this
situation.)

The full relativistic mean field theory may be all right; it is used extensively in many parts of physics. I prefer to proceed, starting from the nonrelativistic many-body problem, introducing relativistic effects by perturbation theory (in fact, not going beyond first order in corrections to wave functions). In this way, everything is well defined. Furthermore, relativistic effects on valence particles, which live mostly in the nuclear surface where densities and velocities are low, are generally not large, and my procedure will allow a unification of relativistic and many-body corrections. General features will be clarified, and it will be seen that there are no objections to using the full relativistic mean field theory for inner-shell nucleons where the effects may be so large that perturbation theory may be inadequate.

I particularly do not wish to carry spinors and interactions off shell at the same time, as is often done. As I will discuss later, this can bring one into grave difficulties. By my procedure, spinors are kept on shell, corrections being introduced by perturbation theory. Interactions can be carried off shell, although for the schematic calculations I perform this is hardly necessary. Interactions do not, however, change much when carried off shell in a sensible way, as I shall discuss.

3. Perturbation Theory

The developments by Celenza, Shakin[2] and collaborators incorporate much of what I wish to do, but for pedagogical reasons, I'll stick to first-order perturbation theory.

The Dirac equation is

$$(p_o - \underset{\sim}{\alpha} \cdot \underset{\sim}{p} - \beta m) \psi = 0 \qquad (3.1)$$

with

$$\alpha_k = \begin{pmatrix} 0 & \sigma_k \\ \sigma_k & 0 \end{pmatrix} \qquad \beta = \begin{pmatrix} 1 & 0 \\ 0 & -1 \end{pmatrix} \qquad (3.2)$$

I define Casimir projection operators

$$\Lambda_{\pm}(p) = \frac{\sqrt{p^2+m^2} \pm \underset{\sim}{\alpha} \cdot \underset{\sim}{p} \pm \beta m}{2\sqrt{p^2+m^2}} . \qquad (3.3)$$

A relativistic (but not covariant) many-body Hamiltonian which operates only in the space of positive-energy states can be written down[5,6]:

$$\mathcal{H}_D = \sum_{i=1}^{A} \mathcal{H}_D(i) + \sum_{i,j} \Lambda_+(p_i)\Lambda_+(p_j) V(\underset{\sim}{x}_i, \underset{\sim}{x}_j) \Lambda_+(p_i)\Lambda_+(p_j) \qquad (3.4)$$

with
$$\mathcal{H}_D(i) = \underset{\sim}{\alpha}_i \underset{\sim}{p}_i + \beta_i m \qquad (3.5)$$

Now, $\mathcal{H}_D$ possesses a well-defined solution Ψ_{++} which may be approximated by a product (Hartree) wave function $(\Psi_{++})_{Hartree}$.

$$(\Psi_{++})_{Hartree} = \Pi \, \psi_{+i}(\underset{\sim}{x}_i). \qquad (3.6)$$

The relation of small to large components in $\psi_{+i}(\underset{\sim}{x}_i)$ is given by

$$\Lambda_-(p_i) \, \psi_{+i} = 0 \qquad (3.7)$$

so that

$$\psi_{+i} = \sqrt{\frac{p_o + m}{2p_o}} \begin{pmatrix} \chi^m_{1/2} \\ \frac{\sigma \cdot p_i}{p_o + m} \chi^m_{1/2} \end{pmatrix} \cong \begin{pmatrix} \chi^m_{1/2} \\ \frac{\sigma \cdot p_i}{2m} \chi^m_{1/2} \end{pmatrix} \qquad (3.8)$$

and we list the negative-energy solutions

$$\psi_{-i} \cong \begin{pmatrix} -\frac{\sigma \cdot p_i}{2m} \phi^m_{1/2} \\ \phi^m_{1/2} \end{pmatrix} \qquad (3.9)$$

Here $p_o \equiv \sqrt{p^2 + m^2}$, and it will be sufficiently accurate for our work to consider $p_o = m$. Note that p_o does not include any interactions! Although the equations to date hold for wave functions in configuration space, with $p_i = \hbar \nabla_i / i$, for simplicity we will now work with the infinite system in momentum space. In this case,

$$\chi^{1/2}_{1/2} = \begin{pmatrix} 1 \\ 0 \end{pmatrix} = \phi^{1/2}_{1/2}$$

$$\chi^{-1/2}_{1/2} = \begin{pmatrix} 0 \\ 1 \end{pmatrix} = \phi^{-1/2}_{1/2} \qquad (3.10)$$

We begin by making a calculation which removes the tadpole, Fig. 1, to first order in perturbation theory. We

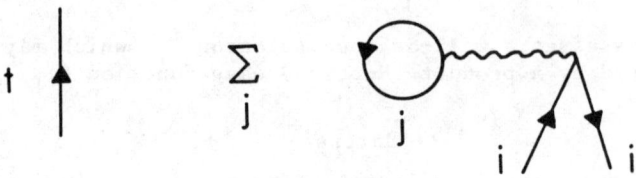

Fig. 1. Tadpole diagram. The upward going line on the right represents a positive-energy wave function; the downgoing one, a negative-energy one.

use Dirac hole theory, which is equivalent to pair theory as long as no interaction loops close on a particle, to the order we are working, at least.

The matrix element shown in fig. 1 is easy to evaluate. At this stage we consider only scalar and vector two-body interactions. The vector matrix elements vanish for the process, fig. 1. In terms of the scalar interaction $U(x-x')$ one has the matrix element

$$M = \rho_s \bar{U} \left(\phi^+ \frac{\sigma \cdot p}{m} \chi \right) \tag{3.11}$$

where

$$\rho_s \bar{U} = \int U(x-x') \rho_s(x') d^3x' \tag{3.12}$$

and ρ_s is the scalar density

$$\rho_s = \sum_n \psi_n^+ \gamma_o \psi_n(x). \tag{3.13}$$

The correction to the wave function due to coupling to negative energy states by the scalar potential is

$$\delta\psi_n(p) \stackrel{\sim}{=} -\sum_{spins} \bar{U}\rho_s \begin{pmatrix} 0 \\ \phi_s \end{pmatrix} \frac{(\phi_s^+ \frac{\sigma \cdot p}{m} \psi_n)}{2m}$$

$$= -\frac{\bar{U}\rho_s}{m} \begin{pmatrix} 0 \\ \frac{\sigma \cdot p}{2m} \end{pmatrix} \chi \tag{3.14}$$

where we have neglected the small component of the negative-energy wave function. The energy denominator is taken as 2m. We can thus express our corrected wave function as

$$\psi_n(p) + \delta\psi_n(p) \stackrel{\sim}{=} \begin{pmatrix} \chi^m_{1/2} \\ \left(1 - \dfrac{\overline{U\rho}_s}{m}\right) \dfrac{\sigma \cdot p}{2m} \chi^m_{1/2} \end{pmatrix} \quad (3.15)$$

The correction to the small component is the same as if

$$\frac{\sigma \cdot p}{2m + 2\overline{U\rho}} \chi^m_{1/2} \quad (3.16)$$

were expanded to the first order, (3.16) being the small component that would result from mean field theory. We can write the denominator of (3.16) in more familiar form by defining*

$$\varepsilon_p = m + \overline{V\rho} + \overline{U\rho}_s, \quad (3.17)$$

in which case

$$\frac{\sigma \cdot p}{2m + 2\overline{U\rho}_s} = \frac{\sigma \cdot p}{\varepsilon_p + m - \overline{V\rho} + \overline{U\rho}_s} \quad (3.18)$$

Note that $\overline{V}$ is positive, $\overline{U}$, negative. Eq. (3.18) begins looking like Dirac, phenomenology.

4. The Many-Body Problem.

Brueckner theory is formulated in the ++ space. (See Chapter 9 of ref. 1 or ref. 2). We, therefore, wish to construct, by perturbation theory, an effective interaction to represent the corrections for coupling to negative energy states. Relativistic kinematics will enter in through the small components of the on-shell spinors (3.8). Much of the latter is subsured in "minimal relativity", and this is not as interesting as the former. We calculate the process, fig. 2.

*This is the nonrelativistic approximation to the form familiar in the Walecka theory

$$\varepsilon_p = \overline{V}\rho + \sqrt{p^2 + \tilde{m}^2} \quad (3.17a)$$

where

$$\tilde{m} = m + \overline{U}\rho_s. \quad (3.17b)$$

The forms (3.17a) and (3.17) differ by terms $\sim p^2/2\tilde{m}$.

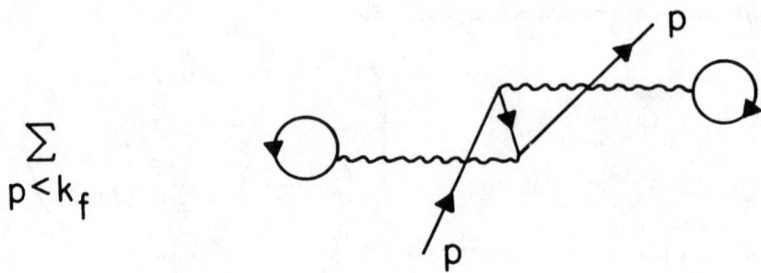

Fig. 2. Correction to the single-particle energy of single-particle k through coupling to negative-energy states.

This calculation is straightforward. Squaring the matrix element (3.11) and dividing by the intermediate-state energy 2m, we find

$$\delta\varepsilon = \left(\frac{\rho_s \bar{U}}{m}\right)^2 \frac{p^2}{2m} . \qquad (4.1)$$

This gives an average relativistic correction of

$$\overline{\delta\varepsilon} = \left(\frac{\rho_s \bar{U}}{m}\right)^2 \bar{T} \qquad (4.2)$$

to the energy per nucleon. Here

$$\bar{T} = \frac{\overline{p^2}}{2m} \simeq 24 \text{ MeV} \qquad (4.3)$$

is the average kinetic energy (of a nucleon of mass m). Choosing

$$U(x) = \frac{g_\sigma^2}{4\pi} \frac{e^{-m_\sigma r}}{r} \qquad (4.4)$$

with $(g_\sigma^2/4\pi) = 7$, $m_\sigma = 4m$, about right for the nucleon-nucleon interaction, we find that

$$\bar{U}\rho_s = 385 \text{ MeV} \qquad (4.5)$$

with $\rho_o = 0.16/\text{fm}^3$ as nuclear-matter density. We find

$$\delta\varepsilon = 4.06 \left(\frac{\rho}{\rho_o}\right)^{8/3} \text{MeV}. \qquad (4.6)$$

This is to be compared with the Celenza-Shakin $3.6(\rho/\rho_o)^{2.4}$ MeV. We prefer our power of 8/3. Celenza and Shakin take into account effects of distortion and exchange in the matrix element, fig. 1. One can see that these do not give large effects.

From fig. 2 one has an alternative interpretation of relativistic effects[8]. They can be viewed as a small density-dependent correction to the mass of the σ-particle. In vacuo, the σ will have self-energy effects, shown in fig. 3, through coupling to virtual nucleon-antinucleon pairs. In the presence of nuclear

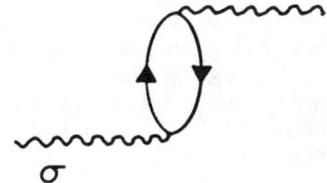

Fig. 3. Self energy of the σ through coupling to virtual nucleon-antinucleon pairs.

matter, processes with momentum $k<k_f$ are blocked. This gives rise to the repulsive contribution, fig. 2.

We see that the relativistic contribution, fig. 2, is really just the difference of lowest-order processes, fig. 4, for bare and modified σ mass.

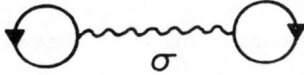

Fig. 4. Lowest-order (Hartree) process in the energy involving σ-exchange.

Since corrections for exchange and correlations are applied to the lowest-order process, they should also be applied to the modified one. From the work of ref. 2 (The corrections are not made there.) one can estimate from the treatment of the lowest-order process, fig. 4, that these corrections give a factor of 0.6. Thus, our final result is obtained by multiplying $\delta\varepsilon_p$ of eq. (4.6) by a factor of 0.6,

$$(\overline{\delta\epsilon})_{rel.} = 2.4 \left(\frac{\rho}{\rho_o}\right)^{8/3} \text{MeV.} \qquad (4.7)$$

This will considerably stiffen the nuclear equation of state. A behavior of $\rho^{8/3}$ would imply an adiabatic index of $\gamma=11/3$ at high densities where (4.7) would dominate. This will be somewhat softened because correlation effects are density dependent. At normal nuclear-matter this correction will contribute

$9 \times \frac{8}{3} \times \frac{5}{3} \times 2.4$ MeV = 96 MeV to the compression modulus of

nuclear matter, nearly all of the modulus if one accepts the estimate of Brown and Osnes[9].

One sees clearly at this stage that simply adding the relativistic correction to the nonrelativistic calculated energies[10] will not solve the nuclear matter problem. Detailed nonrelativistic many-body calculations[10] underbind nuclear matter by several MeV at normal nuclear matter density ρ_o. They give saturation at $\sim 2\rho_o$ with roughly the correct binding energy there. It is clear that <u>several</u> MeV of attraction have to come from somewhere. In chiral models[11,12] one has the three-body term shown in fig. 5. This was

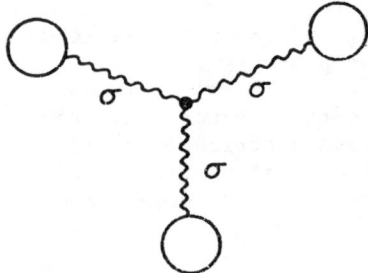

Fig. 5. Three-body interaction from chiral models.

invoked by Jackson, Rho and Krotscheck[8] to give a bit more binding at nuclear matter densities and to partially counterbalance $(\overline{\delta\epsilon})_{rel}$. This problem is now being reconsidered[13] with a better description of the two-body problem.

Consideration of possible processes within the chiral Lagrangian framework[8,11] show the process of fig. 2 to be only one of a number of processes that have to be considered. Indeed, ref. 8 shows that the meson loop correction to the process of fig. 2 is larger than the lowest-order process (and has the $\rho^{8/3}$ dependence). We shall adduce considerable empirical proof to show that the result (4.7), supplemented by a small attractive three-body interaction of the type fig. 5, gives a reasonable description of many-body effects, provided that the density is not too high. The more general considerations of ref. 8 are summarized in an appendix.

If our assessment of the empirical situation is correct, there must be considerable cancellation amongst loop corrections. We see an analogous situation in nuclear physics. Whereas the three-body term of the type fig. 5 in the lowest-order chiral Lagrangian would give an additional binding energy of ~ 11 MeV/nucleon, loop corrections cut it down to the extent that it is hard to identify.

In calculations of the many-body problem an effective mass m* enters in. It gives the relation between momentum and velocity at the Fermi surface,

$$\frac{k_f}{v_f} = m^*. \qquad (4.8)$$

In Fermi-liquid theory, m* is related to m in terms of the Fermi-liquid coefficient F_1,

$$\frac{m^*}{m} = (1 + \frac{F_1}{3}). \qquad (4.9)$$

The coefficient F_1 describes the <u>velocity</u> dependence of the two-body interaction.

In the relativistic Hartree theory an effective mass

$$\tilde{m} = m + \bar{U}\rho_s \simeq 0.6\, m \qquad (4.10)$$

comes in. At first sight it does not appear to be the same as the m* of eq. (4.9). One doesn't see any velocity-dependence in the scalar interaction giving rise to it.

The density of single-particle levels at the Fermi surface is

$$\rho(E_f) = \frac{2k_f\, m^*}{\pi^2}. \qquad (4.11)$$

This can be reconstructed from the empirical position of single-particle levels near the Fermi surface. It may be more convenient to obtain this m* from the volume integral of the real part of the nuclear potential[14]. One sees clearly from fig. 3 of ref. 14 that this volume integral <u>increases</u> with increasing energy, as one moves away from the Fermi surface, showing that m*/m>1. Indeed, detailed analysis[15] shows it to be

$$\frac{m^*}{m} = 1.15 \pm 0.1. \qquad (4.12)$$

The story is an old, but important one.[16] Coupling to collective excitations increases m* from the old Bruecker-theory or k-mass of 0.7 m up to m*/m $\sim$ 1.15 in ^{208}Pb.

Indeed, relativistic Hartree theory gives levels that are a factor of ~ 2 too widely spaced, showing that $\tilde{m}$ acts like the effective mass m* discussed above. In fact, Baym and Chin[17] (eq. (13) of this reference) show that the relativistic generalization of the Landau effective mass relation for excitations above the ground state is

$$v_F = \frac{p_f}{\mu(1 + \frac{1}{3} F_1)} \quad . \tag{4.13}$$

Here μ is the chemical potential, equal to $m-B$ where B is the binding energy of a particle at the Fermi surface, $B \sim 8$ MeV. Thus, to an accuracy of $\sim 1\%$, μ can be replaced by m.

In relativistic Hartree theory,

$$v_F = \frac{\partial \varepsilon_p}{\partial p} = \frac{p_f}{\tilde{E}_{p_f}} \tag{4.14}$$

where we have taken ε_p from eq. (3.17a). Here we use

$$\tilde{E}_{p_f} = \sqrt{p_f^2 + \tilde{m}^2} \stackrel{\sim}{=} \tilde{m}. \tag{4.15}$$

Consistency requires

$$\mu(1 + \frac{1}{3} F_1) = \tilde{E}_{p_f} \tag{4.16}$$

Neglecting binding energy effects and $p^2/2\tilde{m}$ as compared with $\tilde{m}$, this implies

$$m(1 + \frac{1}{3} F_1) = \tilde{m} \tag{4.17}$$

so that, to within these (pretty good) approximations,

$$\tilde{m} \stackrel{\sim}{=} m^*. \tag{4.18}$$

That the important equality is satisfied has been shown by Matsui[18], and by Kurasawa and Suzuki[19]; there is a velocity dependence in the theory, and it is such as to ensure (4.16). This velocity-dependence comes from the three-vector part of the ω, the exchange of which gives an interaction

$$V_\omega = \frac{g_\omega^2}{4\pi} (1 - \underset{\sim}{\alpha}_1 \cdot \underset{\sim}{\alpha}_2) \frac{e^{-m_\omega r}}{r} \quad . \tag{4.19}$$

The $\underset{\sim}{\alpha}_1 \cdot \underset{\sim}{\alpha}_2$ piece of this gives rise to a velocity-dependence, and the F_1 comes from the processes shown in fig. 6.

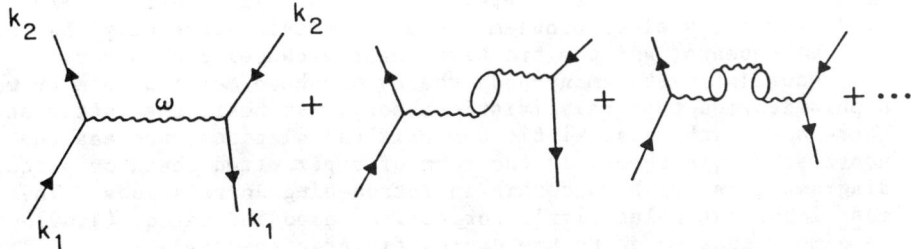

Fig. 6. The Fermi-liquid coefficient F_1 arises from the first-order exchange term plus induced interaction.

The value of F_1 is

$$F_1 = \frac{-\rho(E_F)\left(\frac{g_\omega}{m_\omega}\right)^2 v_f^2}{\left[1 + \frac{1}{3}\rho(E_F)\left(\frac{g_\omega}{m_\omega}\right)^2 v_f^2\right]} \qquad (4.20)$$

where $\rho(E_f) = 2p\,\tilde{E}_f/\pi^2$. The form here is familiar[20]; the numerator is the direct F_1 and the denominator includes effects of the induced interaction. Matsui[18] and Kurasawa and Suzuki[19] verify (4.15) by explicit calculation. We shall essentially repeat their calculation in the next section when we consider magnetic moments.

The interaction shown in fig. 6 giving rise to the F_1 of eq. (4.20) arises from differentiating the Fock term, arising from three-vector ω-exchange, in the self energy. Indeed, in order to obtain consistent results, we were forced to include the self screening of the exchange term. This shows that we cannot work consistently in only Hartree theory (or, indeed, in only Hartree-Fock theory) if one wants to satisfy what is essentially a Ward Identity ensuring eq. (4.16).

We thus see how familiar nonrelativistic Fermi-liquid parameters and relations come out of the relativistic theory. Corrections to the nonrelativistic theory are modest, $\sim B/m_n$, where B is the binding energy, or p_f^2/m_n^2, as we might guess. The relativistic theory introduces considerable formalism and complication to handle these (small) corrections, which are easily treated as small perturbations on nonrelativistic theory.

Note that the near equality of relativistic and nonrelativistic theories follows only because the vector and scalar interactions nearly cancel each other, leaving a net correction to the energy which is small compared with m. We have used this fact in approximating μ by m in going from eq. (4.15) to (4.16). Of course, one can see from my early discussion where the many-body problem was cast in the ++ space that relativistic effects are small for the nuclear problem, because in this space only the sum $V\rho + U\rho$ appears, and the two terms nearly cancel each other.

Nonrelativistic many-body theory has been dealt with much more sophistication than relativistic theory. At best, the latter stops these days with relativistic G-matrix calculations, whereas the nonrelativistic theory in the form of hypernetted chain or parquet diagrams goes far beyond this in introducing correlations. The most important relativistic correction, embodied in eq. (4.6), can be simply grafted on to the nonrelativistic theory.

The empirical effective mass m* is poorly given by the relativistic Hartree theory. It must be nearly doubled by inclusion of Fock terms, correlations and higher-order couplings.

5. Magnetic Moments of Valence Nucleons

It might appear that the isoscalar piece of the magnetic moment of a Dirac proton, which involves the expectation value of $[\underline{\alpha} \times \underline{r}]$, would be enhanced by a factor $m/\tilde{m}$; indeed, if one takes the expectation value of $\underline{\alpha}$ in wave functions of the relativistic Hartree theory, this seems to be so. Were it so, it would wreak havoc with our understanding of Schmidt lines, etc. Screening by the vector interaction[18,19], fig. 7, reduces the correction to one of order B/M_n, i.e., to ~1%. The cancellation of

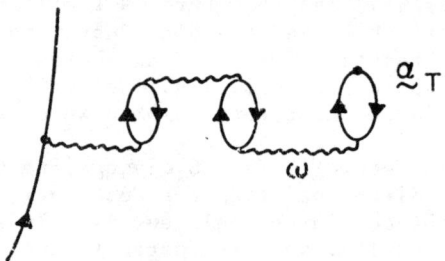

Fig. 7. Screening of the expectation value of the transverse velocity $\underline{\alpha}_T$ by the vector interaction.

the apparent enhancement is accomplished by eq. (4.15). the screening slows the particle down.

This cancellation is quite general[21]. A Ward identity connects the vertex Λ_μ involved in measuring the magnetic moments with the effective mass of the particle. If one corrects the effective mass by medium corrections, then the vertex function must also be corrected.

Magnetic moments will be corrected by exchange currents

connected with Fock terms in the interaction, and by the coupling of nucleons to isovector collective excitations[22]. These corrections can be expressed in terms of a Ward identity connecting $(\delta g_\ell)_{proton}$ with the contribution to the effective mass from these two agencies. The Fock terms from $\pi+\rho-$ exchange decrease m* and increase $(\delta g_\ell)_{proton}$; the coupling to isovector collective excitations works, in each case, in the opposite direction. This can all be summarized through the equation

$$(\delta g_\ell)_{proton} = -\frac{2}{3}\left(\frac{m^*}{m} - 1\right) \quad (5.1)$$

where m*−m is the effective mass depletion (enhancement) <u>from all of these isovector agencies</u>. This δg_ℓ is isovector in nature, so the neutron moment would be changed in the opposite direction. The relation (5.1) should not include changes in the effective mass resulting from isoscalar exchanges, such as that of ω-exchange discussed in the last section. Fock terms from isoscalar exchanges should not be included in eq. (5.1).

Although a very extensive review[16] of the effective mass m* has recently appeared, the contributions to the enhancement from isovector excitations have not been separated from those from isoscalar enhancement to date. However, recent work by Esbensen and Bertsch[23] indicates that it is almost completely the surface isoscalar mode which gives the large effective mass in the neighborhood of the Fermi surface. Indeed, as these authors show, in the limit of a large surface, the effective mass would go to ∞.

The $(\delta g_\ell)_{proton}$ has been connected with the enhancement κ in the giant-dipole sum rule by Fujita and Hirata[24]:

$$\kappa = 2(\delta g_\ell)_{proton}. \quad (5.2)$$

From the Fock terms from $\pi+\rho$-exchange, Brown and Rho[22] found $(\delta g_\ell)_{proton} = 0.22$. A recent accurate measurement[25] gives = 0.46 ± 0.05 for ^{209}Bi. This is the enhancement corresponding to the energy region of the giant-dipole resonance which, Fujita and Hirata point out, is the appropriate region for the relation (5.2). An analysis of mesonic effects on nuclear magnetic moments has been carried out by Yamazaki[26]. After correcting for first-order configuration mixing he arrived at anomalous orbital g-factors of $(\delta g_\ell)_{proton} = 0.15 \pm 0.02$ and $(\delta g_\ell)_{neutron} = -0.05 \pm 0.03$.

In order to extract the meson-exchange part δg_ℓ^{meson} from g_ℓ, the relation

$$\delta g_\ell = \delta g_\ell^{meson} + \delta g_\ell^{high}$$

should be used[26]. Here δg_ℓ^{high} summarizes higher-order corrections due to correlations, especially those tensor in nature, of the nucleons.

From Yamazaki's values, one finds[25]

$$(\delta g_\ell^{meson})_{proton} = 0.28 \pm 0.02$$
$$(\delta g_\ell^{meson})_{neutron} = -0.18 \pm 0.02$$

and

$$\delta g_\ell^{high} = -(0.13 \pm 0.03)\tau_3.$$

This latter value is consistent with calculated values obtained by Shimizu, Ichimura and Arima (Nucl. Phys. **A226**, 282 (1974)).

The Brown-Rho value[22] was calculated for isospin-symmetric matter. However, in ^{209}Bi the effective charge of the proton is N/A, not 1/2. Therefore, this value should be corrected to

$$(\delta g_\ell^{meson})_{proton} = 0.22 \frac{2N}{A} = 0.27$$
$$(\delta g_\ell^{meson})_{neutron} = -0.22 \frac{2Z}{A} = -0.17,$$

in good agreement with the findings of Nolte et al[25].

Since the expectation value of the operator γ_5 which occurs in the timelike component of the axial current brings in negative energy components linearly, it might appear that relativistic theories would give an enhancement $\sim m/m^*$ in this quantity. McNeil and Shepard showed[27] that the correct operator, determined by a PCAC consistency condition, for the axial current has the form

$$J_A^o \sim \bar{\psi} \, \frac{m + \bar{U}\rho_s}{m} \, \gamma^5 \, \psi. \qquad (5.3)$$

Thus, the enhancement obtained for the small components is cancelled by the factor $(m+\bar{U}\rho_s)/m$. One might expect this cancellation, because the timelike component of the axial current behaves like the pion field, which should be unaffected by the introduction of a scalar field.

The axial Ward identity can be written as

$$q^\mu \, {}^5\tilde{\Gamma}^i (p+q, p) = \tilde{g}_A^{-1} [\tilde{S}^{-1}(p+q) \tfrac{1}{2}\tau^i \gamma_5 + \tfrac{1}{2}\tau^i \gamma_5 \tilde{S}^{-1}(p)] \quad (5.4)$$

where the "twiddles" on top of quantities indicate that they are finite, renormalized, medium-dependent quantities. Now $\tilde{S}^{-1}$ involves the mean-field self energy Σ in relativistic Hartree theories (or Hartree-Fock theories or any other mean-field theory). The only term in Σ which does not commute with γ_5 is the scalar mass term. Thus, scalars break the chirality, and should modify β-decays, etc., which depend upon chirality. Since the three-vector part of the axial current is concerned with chirality, the small components of the wave function enter squared, and the effects are not large. Chirality is, however, in principle a relativistic quantity, not something which can be treated by corrections to nonrelativistic theories. Furthermore, axial-vector vertices like

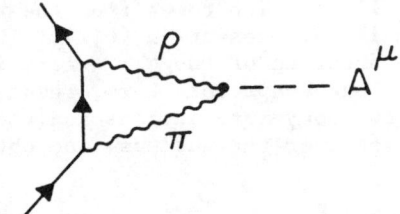

are not related to the self energies like

so one cannot here easily implement the sorts of connections between effective mass and vertex corrections that we did earlier.

Modifications of the spin-orbit interaction, which involves matrix elements of the spin in much the same way as they enter into the chirality in beta decay, by the scalar interaction, are very important. These modifications account for the great success of Dirac phenomenology, so we will discuss them in a separate section.

6. The Spin-Orbit Interaction

The great success of Dirac phenomenology has been in predicting the very detailed structure in the spin-rotation parameter, once the parameters of the theory -- strengths and shapes of potentials -- have been chosen so as to reproduce the elastic scattering and polarization. Since the spin only emerges in a natural fashion from the Dirac theory, it is very reasonable that a relativistic equation is needed to adequately handle it.

One immediately encounters a puzzle when looking at Dirac phenomenology. Here, the nucleon self energy is written, in our notation, as

$$\Sigma = \overline{U}_s + \gamma_o \overline{V}\rho. \qquad (6.1)$$

The spin-orbit interaction then emerges from the Dirac equation as

$$V_{SO} = \frac{\hbar^2}{4m^2 r}\left[\frac{d}{dr}(\overline{U}\rho_s) - \frac{d}{dr}(\overline{V}\rho)\right] \vec{L}\cdot\vec{\sigma} \qquad (6.2)$$

where we have employed a local-density approximation for ρ, which is now a function of r. Because $\overline{U}$ and $\overline{V}$ have opposite signs, the scalar and vector interactions add in the spin-orbit force.

In the hydrogen atom, however, there is a factor of 3 in the spin-orbit force, all of which comes from the vector Coulomb interaction, which is not present in (6.2). The difference can be traced back to the dropping of the $\alpha_1\cdot\alpha_2$ term in eq. (4.19) in going to (6.1). If one keeps this term, reduces to large components in the two-body interactions and then sums one of the interacting particles over the nucleus, one obtains the following answer:

$$V_{SO} = \frac{\hbar^2}{8m^2 r}\left[\frac{d}{dr}(\overline{U}\rho_s) - 3\frac{d}{dr}(\overline{V}\rho)\right] \vec{L}\cdot\vec{\sigma} \qquad (6.3)$$

for the two-nucleon center-of-mass system.

Dirac phenomenology changes m to $\tilde{m}$ in the denominator. This involves introducing negative-energy states into the interaction as shown in fig. 8

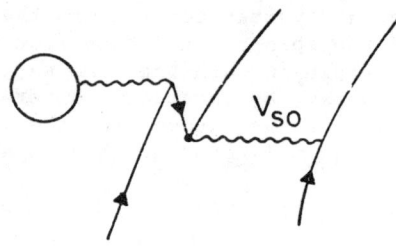

Fig. 8. Corrections from negative-energy states entering into the spin-orbit interaction

and is easily justified following our arguments of § 3.

In treating nuclear structure, one should now proceed differently from treating high-energy reactions. In the former case, screening should be introduced (See fig. 6). The 2/3 of the 3 d/dr $(\bar{V}\rho)$ which arises from the $\alpha_1 \cdot \alpha_2$ piece of vector exchange acquires a factor $(1 - \frac{1}{3}F_1)^{-1}$ from summing the screening bubbles; this is approximately $m/\tilde{m}$ -- see eq. (4.16). In the approximation where $\rho_s = \rho$, the piece of the spin-orbit interaction (6.3) acquires the factor $(1+F_o)^{-1}$; where F_o is the usual Fermi liquid interaction. Consequently, the final screened spin-orbit interaction is

$$V_{SO} = \frac{\hbar^2}{8m^2 r} \left(\frac{1}{1+F_o}\right)\left[\frac{d}{dr}(\bar{U}\rho_s) - \frac{d}{dr}(\bar{V}\rho)\right]\vec{L}\cdot\vec{\sigma}$$

$$- 2\frac{\hbar^2}{8m\tilde{m}r}\frac{d}{dr}(V\rho)\vec{L}\cdot\vec{\sigma}$$

(6.4)

The F_o here, as well as m, is highly density dependent, varying roughly as the nucleon density. At nuclear matter density, Brown and Osnes find $F_o = -0.73$, although relatively few people would agree that the compression modulus of nuclear matter is as low as they find. None the less, the enhancement by $(1 + F_o)^{-1}$ is likely to be of the same general size as that by $m/\tilde{m}$.

Dirac phenomenology deals with high-energy scattering in which the polarization bubbles will be "shaken off". Thus, the form (6.3) with m replaced by $\tilde{m}$ is directly applicable. The replacement of m by $\tilde{m}$ leads to substantial enhancement; the mass of the nucleon travelling through nuclei does really change locally in response to the scalar coupling to negative-energy states. The factor of 3 in eq. (6.3) is converted to unity in transforming from center-of-mass to lab systems, so it does not enter directly into Dirac phenomenology.

The factor of 3 noted above has been known since the original Dirac theory of the hydrogen atom. It is included in reductions of the two-body amplitude in the nucleon-nucleon interaction (See, e.g., D.Y. Wong[29]).

7. Relativistic Two Body Calculations

There is no doubt that relativistic kinematics must be included in the two-body equation as one goes to high energies. There are many equations in which this is done. Differences between results from these equations tend to cancel out, provided the calculation is carried to a sufficiently high order[30].

These equations presumably result from effective Lagrangians which remain after integrating out the quark and gluon degrees of freedom in QCD. Since they should come from QCD -- any many investigators are presently working to derive them -- they should possess the underlying invariances of QCD. In addition to Lorentz invariance, etc., they should possess chiral invariance. Many of the problems which appear in Dirac phenomenology and in relativistic two-body equations arise because chiral invariance is not enforced. We have seen how important are the constraints imposed by chiral invariance in the discussion of matrix elements of the axial operator J_A^o in §5.

My second "principle" is to not carry both spinors and interactions off shell at the same time. There is no use in compounding ignorance! From the work of ref. 30, one finds out that differences in the way the interaction is carried off shell in second order -- e.g., differences in the way retardation is handled -- tend to be evened out when the calculation is carried to fourth order.

In a relativistic two-body calculation, at least one of the nucleons must be off shell. For the two-nucleon system, like the deuteron, I prefer the Blankenbecler-Sugar equation[31] in which the nucleons are symmetrically off shell. A detailed discussion of these matters is given in ref. 32.

Calculations of elastic electron-deuteron scattering in a relativistic formalism by Arnold, Carlson and Gross[33] both violate chiral invariance and carry the interaction and the spinor of one particle off shell at the same time. The other particle is left on shell. Whereas this asymmetrical treatment of the two particles may be appropriate for the hydrogen atom, where the proton is much heavier than the electron, I find it completely unreasonable for the deuteron. Discrepancies between this theory and experiment[34] are significantly larger than in nonrelativistic calculations.

The deuteron electromagnetic form factor $A(q^2)$ has been calculated using the Bethe-Salpeter equation by Zuilhof and Tjon[35]. This equation has the virtue that it treats the particles symmetrically. However, it is well known that the Bethe-Salpeter equation in the ladder approximation does not reduce to the Schrödinger equation $m \to \infty$. Processes of the type fig. 9 are left in the

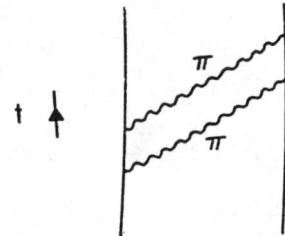

Fig. 9. Processes where two pions are in the air at the same time.

Bethe-Salpeter equation in this limit, whereas they are not included in the ladder approximation in the Schrödinger equation. Woloshyn and Jackson[30] find that one has to go to fourth order, including the crossed pion exchange, before one can make sense out of the Bethe-Salpeter equation. One might expect this, because there are important cancellations between box and crossed pion exchange in fourth order.

The treatment of Zuilhof and Tjon has the advantage over that of Arnold et al[33] that the nucleons are treated symmetrically. Zuilhof and Tjon find that contributions to $A(q^2)$ from negative-energy states are small, provided axial-vector πN coupling is used for the pion. This essential enforces chiral invariance. Almost all of the relativistic correction is kinematical in origin, coming from the boost on the one-particle propagator. These are the sorts of relativistic corrections one might expect. The corrections in the Arnold, Carlson, Gross paper probably come more from using a meson theory which is not chiral invariant and an asymmetrical treatment of nucleons than from genuine relativistic sources.

The Zuilhof and Tjon calculation gives results substantially closer to the data[34] than that of Arnold et al. (although the nonrelativistic calculation is even closer). It is clear that in any of these theories the isoscalar exchange current involving $\rho\pi\gamma$ coupling must be added, and this removes most of the discrepancy between theory and experiment. There is no low-energy theorem governing the isoscalar exchange currents, so there is necessarily some ambiguity in making this correction.

My general conclusion is that one has to work awfully hard in a relativistic calculation of a two-body problems, such as the deuteron, to do as well or better than the nonrelativistic calculation. Corrections to the latter are likely to be chiefly kinematical in origin. In any case, I would prefer, also in the two-body case, to introduce the small relativistic corrections by perturbation theory, rather than going over to a fully relativistic equation, where one doesn't always know what one's doing.

8. Dirac Phenomenology

For the sake of completeness, I wish to touch base on Dirac phenomenology. It is obvious that relativistic kinematics should be used in scattering at energies of several hundred MeV. It is equally obvious that one should make contact with the boson exchange results for the nucleon-nucleon force, which have been calculated in relativistic formalisms. Considerable progress has been made in giving Dirac phenomenology a basis by the development of the relativistic impulse approximation[36] and by the connection of this with the invariants of the boson-exchange model by Tjon and Wallace.

Major problems result from employing γ_5-coupling of pions, which mostly seem to disappear when pseudovector coupling is used. It appears to me that these are problems for which we have long known the solution in terms of enforcing chiral invariance[38]. Very different scalar mesons should be employed if one uses pseudoscalar or pseudovector pion coupling, as is known from the Weinberg transformation[39]. The much larger scalar exchange to be used with pseudoscalar coupling of pions effectively wipes out the main coupling to negative-energy states. This has been known for some years in low-energy theorems. It looks as if at least the main ambiguities will disappear, once chiral invariance is enforced.

9. Conclusions

It is clearly necessary to work through the relativistic field theory on the way from effective meson Lagrangians, having a basis in QCD, to the highly successful nonrelativistic many-body theory which describes nuclear matter and nuclei.

We find two important contributions from going through this procedure:

1) There is a relativistic correction to the energy per particle of

$$(\delta\bar{\epsilon})_{rel} = 2.4 \left(\frac{\rho}{\rho_o}\right)^{8/3} \text{ MeV}.$$

As noted in §4, this can equivalently be viewed as a medium-dependent correction to the mass of the scalar σ-particle, which gives most of the attraction in the nucleon-nucleon interaction.

2) The spin-orbit interaction is enhanced by factors involving $m/\tilde{m}$ by coupling to negative-energy states via scalar exchange. We found a factor of 3 in the spin-orbit force connected with ω-exchange, but this disappears in the center-of-mass to lab system transformation.

We point out that intricate cancellations dictated by Ward identities eliminate many of the apparent relativistic effects found by many workers.

Use of a relativistic formalism significantly complicates calculations. To date, many-body effects have been handled only crudely, essentially at the G-matrix level, if that, in relativistic calculations. Therefore, we would prefer to treat the relativistic effects as perturbations in a basically nonrelativistic theory.

As noted earlier, most of the considerations reported here were carried out together with Wolfram Weise, and we plan to publish an article together, reviewing the situation of relativistic effects. We have benefited greatly from suggestions and helpful criticism from Andy Jackson and Mannque Rho. I would like to thank Carl Shakin for much patient tuition. I am grateful to Jim Shephard for several conversations and help on the center-of-mass to lab transformation. Questions by Gerry Garvey led me to take up the matters discussed here.

Appendix

In this appendix I wish to relate the relativistic corrections, fig. 2, to the more general loop expansion of Jackson et al.[8] Two types of loops, the nucleon loop and the σ-meson loop appear in the work of these authors.

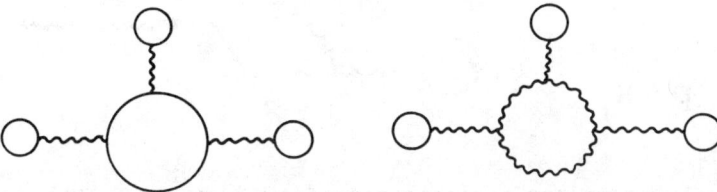

Fig. 10. The wavy lines refer to σ-mesons; the solid lines to nucleons.

The correction to the energy, fig. 2, comes from two tadpoles on the nucleon loop, fig. 11.

Fig. 11. Two tadpoles on the nucleon loop.

Fig. 2 results because the particle, in the middle bubble, is blocked from going to an occupied state. Renormalization has, of course, been carried out to remove the k_f-independent (infinite) self energy from vacuum polarization.

In ref. 8 it is shown that the nucleon loop very accurately cancels off the meson loop, leaving only a small remainder. This means that, in this procedure, our relativistic correction, fig. 2, has been swallowed up by the meson loop. A $\rho^{8/3}$ term, however, appears out of the meson loop, when nucleon loops are inserted in the meson propagators, see fig. 12. After renormalization, this nucleon

Fig. 12. Lowest-order correction to the meson loop by insertion of nucleon bubbles.

loop just represents the medium-induced change in σ mass, discussed in figs. 3 and 4. It can be represented as a loop correction to the process, fig. 2; see fig. 13. Since only the σ-field is

involved

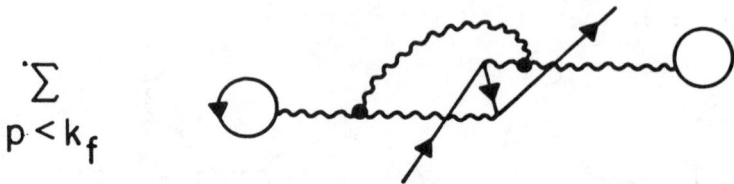

Fig. 13. Meson-loop correction to fig. 2, part of ΔU_σ in ref. 8.

in this loop correction, the density dependence is just that of the process, fig. 2. So here our $\rho^{8/3}$ reappears, but the coefficient need not be that we calculated for the other process, fig. 2.

One can see that the whole business is rather intricate. Even more intricate is the fact that one cannot calculate, at higher densities, the fluctuating scalar field by perturbation theory, but must minimize the energy with respect to it. It is vitally important to keep the ΔU_σ, the lowest order term of which is shown in fig. 13, in this minimization procedure. The growth in the effective σ mass in the medium, is important in the loop calculations. This was already noted by Nyman and Rho[11].

Jackson et al[8] have estimated the coefficients in a term

$$\frac{\Delta m_\sigma}{m_\sigma} = -\alpha\rho + \beta\rho^{5/3} \qquad A(1)$$

which, when added to Day's two-body nuclear matter calculations, would give saturation at the correct density. When translated into an energy per particle, it turns out that their β exactly reproduces our 2.4 $(\rho/\rho_\circ)^{8/3}$ MeV/particle. In view of the above discussion, this seems to be a coincidence, but it indicates that the crude approach used in the text gives about the correct many-body force for saturation.

It should be clear that in going to dense matter at several times nuclear matter density, it is necessary to retain the entire loop calculation, as in ref. 8, and not brutally truncate it as in the text.

References

1. B.D. Serot and J.D. Walecka, <u>The Relativistic Nuclear Many-Body Problem</u>, Advances in Nuclear Physics;, <u>16</u> (1985) eds. J.W. Negele and E. Vogt, (Plenum Press).
2. L.S. Celenza and C.M. Shakin, <u>Relativistic Nuclear Physics</u>: Theories of Structure and Scattering, World Scientific Publishers, Singapore, to be published.

3. B.C. Clark, S. Hama and R.L. Mercer, in *The Interaction Between Medium Energy Nucleons in Nuclei - 1982*, edited by H.O. Meyer (A.I.P., New York, 1983); L.G. Arnold, B.C. Clark and R.L. Mercer, Phys. Rev. C19, 917 (1979); L.G. Arnold, B.C. Clark, R.L. Mercer and P. Schwandt, Phys. Rev. C23, 1949 (1981).
4. John Negele, *Comments on Nuclear and Particle Physics*, to be published.
5. G.E. Brown and D.G. Ravenhall, Proc. Roy. Soc. London A208, 552 (1951).
6. J. Sucher, Phys. Rev. A22, 348 (1980); Phys. Rev. Letts. 55, 1033 (1985).
7. G.E. Brown, A.D. Jackson and T.T.S. Kuo, Nucl. Phys. A133, 481 (1969).
8. A.D. Jackson, M. Rho and E. Krotscheck, Nucl. Phys. A407, 495 (1983); A.D. Jackson, Annual Reviews of Nuclear and Particle Science 16, 105 (1983).
9. G.E. Brown and E. Osnes, Phys. Lett., to be published.
10. B.D. Day and R.B. Wiringa, 1985 Argonne Natl. Lab. Preprint.
11. T.D. Lee and G.C. Wick, Phys. Rev. D9, 2291 (1974); See also E.M. Nyman and M. Rho, Nucl. Phys. A268, 408 (1976).
12. S. Barshay and G.E. Brown, Phys. Rev. Lett. 34, 1106 (1975).
13. E. Baron, G.E. Brown, J. Cooperstein, and M. Prakash, to be published.
14. R.W. Finlay, J.R.M. Annard, J.S. Petler and F.S. Dietrich, Phys. Lett. 155B, 313 (1985).
15. C. Mahaux and R. Sartor, Symposium on Nuclear Structure 1985 (Copenhagen, May 20-24, 1985) North Holland Publ. Co.
16. C. Mahaux, P.F. Bortignon, R.A. Broglia and C.H. Dasso, Phys. Repts. 120, 1 (1985).
17. G. Baym and S.A. Chin, Nucl. Phys. A262, 527 (1976).
18. T. Matsui, Nucl. Phys. A370, 365 (1981).
19. H. Kurasawa and T. Suzuki, Yukawa Hall preprint RIFP-613, 1985.
20. G.E. Brown, S.O. Bäckman and J. Niskanen, Phys. Repts. 124, 1 (1985); S. Babu and G.E. Brown, Ann. Phys. (N.Y.) 78, 1 (1973).
21. W. Bentz, A. Arima, H. Hyuga, K. Shimizu and K. Yazaki, Nucl. Phys. A436, 593 (1985).
22. G.E. Brown and M. Rho, Nucl. Phys. A338, 269 (1980).
23. H. Esbensen and G.F. Bertsch, Phys. Rev. Lett. 52, 2257 (1984).
24. J.I. Fujita and H. Hirata, Phys. Letts. 37B, 237 (1971).
25. R. Nolte, A. Baumann, K.W. Rose and M. Schumacher, 1985 University of Gottingen preprint.
26. T. Yamazaki in *Mesons in Nuclei*, ed. by Mannque Rho and D. Wilkinson, North-Holland Publ. Co. (1979) Ch. 16.
27. J.A. McNeil and J.R. Shepard, Phys. Rev. C31, 686 (1985).
28. M.V. Hynes, A. Picklesheimer, P.C. Tandy and R.M. Thaler, Phys. Rev. Lett. 52, 978 (1984).

29. D.Y. Wong, Nucl. Phys. $\underline{55}$, 212 (1964).
30. R.M. Woloshyn and A.D. Jackson, Nucl. Phys. $\underline{B64}$, 269 (1973).
31. R. Blankenbecler and R. Sugar, Phys. Rev. $\underline{142}$, 1051 (1966).
32. G.E. Brown and A.D. Jackson, The Nucleon-Nucleon Interactions, North Holland Publ. Co., Amsterdam (1976).
33. R.G. Arnold, C.E. Carlson and F. Gross, Phys. Rev. $\underline{C21}$, 1426 (1980).
34. S. Auffret, J.M. Cavedon, J.C. Clemens, B. Frois, D. Goutte, M. Huet, Ph. Leconte, J. Martino, Y. Mizuno, X.-H. Phan, S. Platchkov and I. Sick, Phys. Rev. Letts. $\underline{54}$, 649 (1985).
35. M.J. Zuilhof and J.A. Tjon, Phys. Rev. $\underline{22}$, 2369 (1980).
36. J.A. McNeil, J.R. Shepard and S.J. Wallace, Phys. Rev. Lett. $\underline{50}$, 1439 (1983); see S.J. Wallace, Proc. Los Alamos Natl. Lab. Workshop on Dirac Approaches to Nuclear Physics, Los Alamos, N.M. Jan. 31-Feb. 2, 1985.
37. J.A. Tjon and S.J. Wallace, Univ. of Maryland preprint URO-5126-224; submitted for publication.
38. G.E. Brown, Chap. 8 "Chiral Symmetry and the Nucleon-Nucleon Interaction", p. 331, Mesons in Nuclei, eds. M. Rho and D.H. Wilkinson, North-Holland Publ. Co., 1979.
39. S. Weinberg, Phys. Rev. Lett. $\underline{17}$, 616 (1966); See, also, C.-K. Au and G. Baym, Nucl. Phys. $\underline{A236}$, 500 (1974).
40. B.D. Day, Phys. Rev. $\underline{C24}$, 1203 (1981).

COMPARISON OF ELECTROMAGNETIC AND HADRONIC PROBES OF NUCLEAR STRUCTURE

James J. Kelly
Department of Physics and Astronomy
University of Maryland, College Park, MD 20742

ABSTRACT

We have developed a versatile method for modeling direct reactions that employs a linear expansion of the transition amplitude "$t\rho$". Knowledge of either structure ρ or interaction t permits evaluation of the unknown factor. We apply this method to study three issues related to the extraction of a radial neutron transition density from hadron scattering data. First, we use pseudodata to study the intrinsic radial sensitivity of nucleon scattering. Next, by extending the methods used to analyze electron scattering data, we formulate a high-q bias that minimizes the model dependence of the analysis. Unfortunately, this analysis also demonstrates that geometric moments cannot be extracted in a model-independent manner. Finally, we study the implications of residual errors in the reaction model for the accuracy of fitted structure results.

I. INTRODUCTION

One important goal of inelastic scattering experiments is to determine the radial structure of various states of the target nucleus. To achieve this goal three issues must be addressed. First, we should study the intrinsic radial sensitivity of each available probe, assuming that its interaction with the target is understood. Second, we must understand the dependence of our results upon the choice of model used to interpret the data. Finally, we must ascertain the accuracy of our results by investigating their sensitivity to residual uncertainties in the model used to describe the reaction. The purpose of this contribution is to begin a discussion of these issues as they pertain to the extraction of radial information from data for the inelastic scattering of hadronic projectiles, specifically protons.

The extraction of radial structure is most advanced for electron scattering. The electroexcitation of discrete nuclear transitions is driven by a local one-body electromagnetic interaction that is known accurately.[1] Furthermore, the target is transparent with respect to electromagnetic probes. Therefore, the issues of accuracy and radial sensitivity are considerably simpler for electron scattering than they are for the scattering of strongly interacting projectiles. From a technical point of view, the primary issue involved in unfolding the nuclear structure content of electron scattering data is the model dependence of the analysis.[2-5]

The interpretation of electron scattering data generally relies upon a linear expansion of the radial dependence using a judicious subset of a complete basis of radial functions,[6] thereby avoiding the overt model dependence of simple, and arbitrary, analytic functions. However, experiments are performed at discrete points which span a finite range of momentum transfer. Therefore, unless the expansion is severely truncated, several <u>a priori</u> conditions must be placed upon the analysis. For example, a unique density can be determined only if the behavior of the fitted form factor beyond the measured range of momentum transfer is required to fall below an upper limit suggested by physical considerations.[2] The flexibility of the fitted density consistent with this constraint yields an estimate of the "incompleteness" or "model" error.[4]

During the last decade, many high-precision elastic and inelastic scattering experiments spanning a large range of momentum transfer have been analyzed in this manner. The precision with which charge and current densities have been measured is truly impressive -- the quoted uncertainties are often only a few percent or less where the densities are large. The availability of such precise measurements has had considerable impact upon the theory of nuclear structure.[6,7]

Yet, despite this precision, the versatility of electron scattering is somewhat limited. Its most conspicuous limitation is its insensitivity to the bulk neutron distribution, whose contribution to elastic and inelastic electron scattering is usually quite small. Hadronic projectiles, on the other hand, are about equally sensitive to both neutrons and protons. The variety of one-body densities sampled by these probes is also considerably larger.[8,9] However, the interpretability of hadronic reactions is hindered by uncertainties in both the reaction mechanism and the effective interaction. These complications are primarily responsible for the limitation of traditional analyses of hadronic reactions to models involving restrictive analytic functions of only a few parameters. Although the physical basis of most of these models is sound, the relationship between the fitted parameters and microscopic quantities is tenuous at best. For example, it is impossible to ascertain the relationship of a Woods-Saxon optical potential to an underlying effective interaction which depends upon density. Similarly, it is difficult to interpret a fitted proportionality factor between neutron and proton transition densities when a microscopic structure theory predicts substantially different radial shapes.

We have developed a versatile method for the linear expansion analysis (LEA) of direct reactions that is capable of fitting a variety of reaction or structure models to hadron scattering data while maintaining close contact with microscopic quantities of theoretical interest.[10-12] The philosophy and techniques of this approach are discussed in the next section. Several applications have appeared in Refs. 10-13. The present paper focuses upon the sensitivity of proton inelastic scattering to the radial form of the neutron transition density. The intrinsic radial sensitivity of proton scattering is analyzed in Sec. III. The model dependence of the fitted density is discussed in Sec. IV, wherein the methods used in electron scattering to constrain the asymptotic decay of fitted form factors and to estimate the incompleteness error are extended to hadron scattering. The implications of residual uncertainties in the reaction model for the accuracy of fitted densities is analyzed in Sec. V. The model dependence of geometric moments is analyzed in Sec. VI. A summary of our conclusions is finally presented in Sec. VII.

II. LINEAR EXPANSION ANALYSIS

Suppose we can represent the scattering amplitude T for the binary reaction $A(\vec{a},\vec{b})B$ as a linear expansion

$$T = \sum_n a_n H^n(\theta) \tag{1}$$

of coefficients a_n times basis amplitudes $H^n(\theta)$ where the four spin projections (m_A, m_a, m_b, m_B) associated with each H^n have been suppressed for clarity. The quadratic forms

$$X^{nn'}_{\alpha\beta}(\theta) = \text{Trace}\left[H^n(\theta)\, \sigma_\alpha\, H^{n'}(\theta)^*\, \sigma_\beta\right] \tag{2}$$

can then be constructed as traces, over spin projections, of products involving scattering amplitudes $H^n, H^{n'}$ and polarization vectors σ_α and σ_β. For spin-1/2 projectiles, σ_α and σ_β are Pauli matrices describing the incident and exit polarizations, respectively. The observables are simply contractions of these quadratic forms:

$$\sigma_0(\theta) = \frac{\mu_a \mu_b}{(2\pi)^2} \frac{k_b}{k_a} I_0(\theta) \tag{3a}$$

$$I_0(\theta) = \frac{1}{2} \sum_{nn'} a_n X^{nn'}_{00}(\theta) a_{n'}^* \tag{3b}$$

$$I_0 D_{\alpha\beta} = \frac{1}{2} \sum_{nn'} a_n X^{nn'}_{\alpha\beta}(\theta) a_{n'}^* \tag{3c}$$

where $\mu_a(\mu_b)$ is the reduced mass and $k_a(k_b)$ is the wavenumber in the incident (exit) channel. The analyzing power $A_y = D_{y0}$ and the induced polarization $P = D_{0y}$ are special cases of the polarization transfer matrix $D_{\alpha\beta}$ defined by Ohlsen.[14]

A very simple search algorithm is capable of minimizing the composite chi-square (χ^2) for an arbitrary set of observables with respect to the expansion coefficients a_n. The method is efficient in the sense that the basis amplitudes need only be calculated once and then stored — it is not necessary to recalculate distorted waves or overlap integrals during the optimization of parameters. The method is versatile in that it is applicable to any structure or interaction model that can be represented by a linear expansion.

Although our applications have been restricted to inelastic

scattering within a nonrelativistic distorted wave approximation that uses local operators, the method itself is more general and is equally applicable to nonlocal interactions or to relativistic theories. All that is really required is that the reaction be direct and that the quantity of interest can be linearized. It just happens that the nonrelativistic theory is, at present, more highly developed and more convenient to use. Furthermore, these methods can be easily extended to elastic scattering and to iterative cycles in which the distortion is varied self-consistently.

Reaction models of the folding type are most amenable to a linear expansion analysis. The transition amplitude produced by any folding model can be described by the classic "tρ" form which represents the folding of an effective interaction t with a nuclear transition density ρ. Control of either factor, interaction or structure, permits the systematic evaluation of the other factor. When the structure is known, corrections to the effective interaction can be expanded as a linear series and fitted to a global data set that can, and should, include many target transitions simultaneously. Once a suitable empirical effective interaction has been obtained, the transition densities for states of unknown structure can be fitted to all data pertaining to the same transition.

Phenomenological analyses of this type maintain direct contact with the microscopic quantities of theoretical interest. For example, the theory of nuclear matter predicts strong density dependent modifications of the two-nucleon effective interaction.[15-17] These effects can be accurately represented in a simple linear form.[10,11] However, the nuclear matter theory is subject to severe truncation and several approximations of unknown accuracy.[18] Moreover, the application of this theory to finite nuclei is fraught with ambiguity. Therefore, we do not expect quantitative accuracy. Nevertheless, the parametric form can be used to guide an empirical interaction fitted to data. If an empirical effective interaction satisfying several general requirements emerges from a systematic analysis of a broad database, then the parameters of these medium modifications may be considered measured quantities for direct comparison with theory. Such an analysis is now in progress.[10,11]

In the past, macroscopic analyses of hadron scattering data have sometimes exploited linear expansions in an effort to reduce the model dependence or bias of the results. Such analyses include optical potentials for alpha[19,20] and pion[21,22] elastic scattering and collective models of alpha inelastic scattering. Folding models with empirical interactions have also been used for alpha scattering.[20] However, the simple Gaussian forms used for the interaction between composite systems have little microscopic content.

Ground-state neutron densities have been studied using similar "model-independent" analyses of high-energy proton scattering.[23,24] The accuracy of these results has been studied using the dependence of the fitted densities upon projectile energy.[25] Unfortunately, the degree of consistency obtained in these studies is usually not sufficient to confidently interpret the relatively small differences in radial shape between ground-state neutron and proton densities.

In the present paper, we apply our linear expansion analysis (LEA) to study the radial sensitivity of proton inelastic scattering. Except in the lowest collective states, the radial shapes of neutron and proton densities will usually be considerably different. Even in the collective states, the ratio between neutron and proton matrix elements is often considerably different from the simple N/Z prediction of the hydrodynamic model.[26] Therefore, the study of neutron transition densities with inelastic scattering offers a larger signal than does elastic scattering. We may then expect to learn significant new nuclear structure information even while using an imperfect reaction theory.

III. RADIAL SENSITIVITY

When analyzing experimental data, it is important to distinguish between the intrinsic radial sensitivity of the probe and the accuracy of the fitted quantity. The intrinsic sensitivity is a measure of the range of fitted densities that is consistent with the experimental data and the physical constraints, assuming that the reaction model is perfectly accurate. However, the uncertainties in any model of hadron scattering presently available are substantial. These uncertainties limit the accuracy of the fitted density, even when the intrinsic sensitivity is excellent.

The intrinsic sensitivity of a particular probe to the radial structure of the target nucleus can be studied in a manner that is largely independent of residual uncertainties in the effective interaction. Our method relies upon the analysis of pseudodata constructed in accordance with the best reaction models available. First, we select a representative nuclear transition and construct realistic transition densities. Next, we calculate the scattering amplitudes for each radial basis function and construct the quadratic forms. Then, we produce pseudodata using a realistic set of expansion coefficients. The number and angular range of these pseudodata should reflect experimental practice. Random fluctuations must be applied according to a normal distribution whose width represents customary experimental precision. Finally, starting from different initial guesses, we refit the pseudodata and examine the uncertainties in the fitted quantities. These uncertainties are estimated using the error matrix for the fit (see Sec. IV). The fitted quantities reproduce, on average, the input used to generate the pseudodata. The scatter in the fitted results reflects the estimated uncertainty in these quantities. These uncertainties include contributions from the precision and range of the data, from the penetrability, distortion, and wavelength of the probe, and from the relative sensitivity of the probe to the quantity of interest.

This procedure has been used to study the intrinsic radial sensitivity of nucleons and pions to the neutron transition densities of collective transitions.[10-12] For our purposes, it suffices to consider a single example. The sensitivity of nucleons to the nuclear interior can be clearly illustrated using an inelastic monopole transition in a medium-mass nucleus. In contrast with transitions of higher multipolarity, monopole transition densities need not vanish at the origin. The requirement that higher multipole densities vanish at the origin inhibits the growth of interior error envelopes in a manner more characteristic of the parametrization than of the data. Therefore, we have chosen to investigate the sensitivity of nucleons to the interior using the lowest monopole transition of ^{40}Ca.

Pseudodata spanning momentum transfers between 0.4 and 2.6 fm^{-1} in steps of 0.075 fm^{-1} were produced with ±10% random fluctuations. The proton density was fixed to the electron scattering results of Harihar et al.,[27] which display a strong lobe at the origin. We assumed that the neutron and proton densities were equal and analyzed the sensitivity of incident protons to the neutron transition density. Calculations were performed for many incident proton energies between 60 and 800 MeV. For E_p ≤ 400 MeV, the density-dependent effective interaction based upon the Paris potential was used in the local density approximation.[17] For E_p ≥ 500 MeV, the Love-Franey parametrization of the free t-matrix was used.[28] A zero-range approximation was used to constrain the exchange contribution.[29] Microscopic optical potentials were employed throughout.

The energy dependence of the radial sensitivity of proton scattering is illustrated in Fig. 1. Although absorption increases steadily throughout this range of energy, the interior sensitivity improves rapidly between 60 and 200 MeV. This interior sensitivity improves because the density dependence decreases as the energy increases. At low energies, strong Pauli blocking suppresses the contribution of the interior. As this suppression declines with increasing energy, the contribution of the interior increases relative to that of the surface, thereby enhancing the sensitivity to the interior. The sensitivity is greatest between 200 and 400 MeV. In this regime, declining density dependence and increasing absorption balance. For higher energies, the growing absorption degrades the penetrability and the interior sensitivity. Nevertheless, the sensitivity at 800 MeV is still sufficient to provide a useful probe of neutron transition densities. The simplifications in reaction theory enjoyed at high energies may prove to be a decisive advantage if the medium modifications of the effective interaction cannot be predicted accurately at intermediate energies.

Similar results have been obtained by Ray and collaborators for the elastic scattering of 800 MeV protons.[23] Although the sensitivities of elastic and inelastic scattering appear to be comparable, the elastic scattering results are often more difficult to interpret. The differences between ground-state proton and neutron densities are usually small and thus are easily obscured by the scattering from the average density. The interpretation of small differences can be ambiguous with imperfect reaction models.[25] For inelastic scattering, on the other hand, the differences between neutron and proton transition densities can be quite

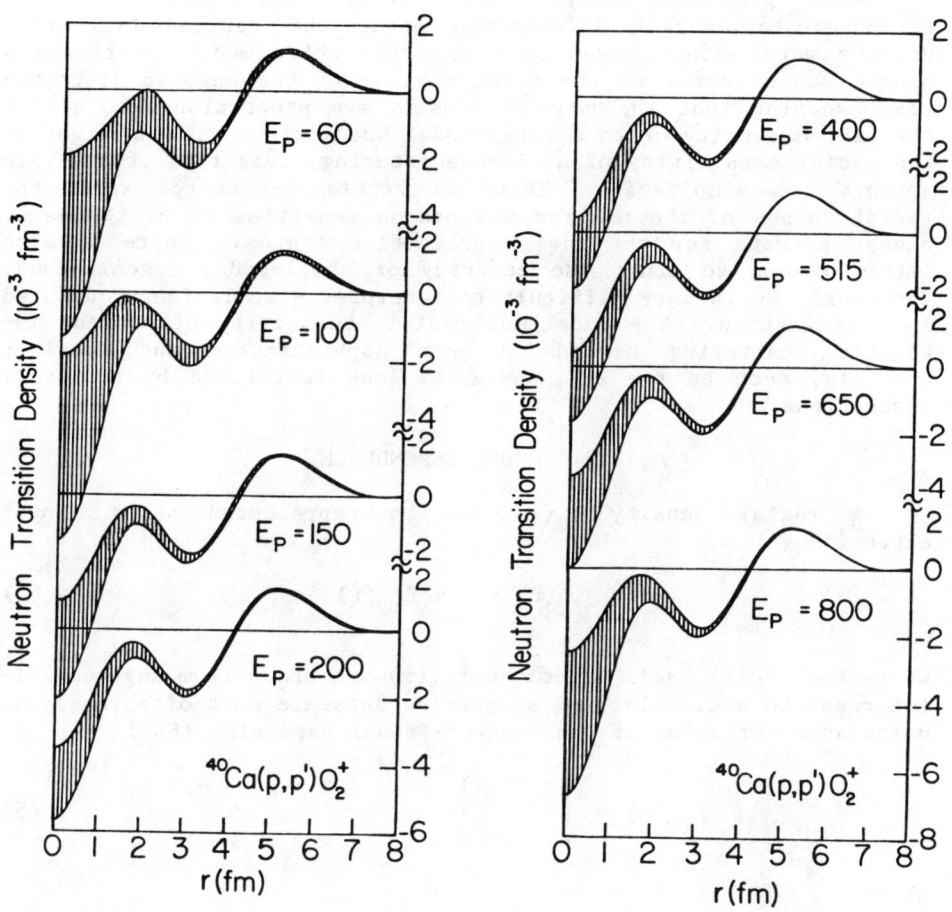

FIG. 1. Energy dependence of error envelopes for the 0_2^+ state of ^{40}Ca fitted to pseudodata for incident proton energies between 60 and 800 MeV.

large. Therefore, we expect that meaningful comparisons between fitted transition densities and theoretical models of nuclear structure will soon be possible.

Most previous analyses of hadron inelastic data which attempted to extract information about the neutron transition density were either based upon the collective model or fitted a single scale factor to the first maximum of the angular distribution assuming that the neutron density was proportional to either the proton density or to a shell-model prediction. In the light of the radial sensitivity of nucleon scattering, this type of analysis appears too simplistic. There is little reason to expect the radial shapes of the neutron and proton densities to be the same, except perhaps for the lowest collective states. There is also little reason to trust the accuracy of shell-model calculations. Therefore, it is very difficult to interpret a scale factor deduced in this fashion. The excellent radial sensitivity of nucleon inelastic scattering demands a more sophisticated and flexible analysis, such as the LEA, which is less restricted by arbitrary assumptions.

IV. MODEL DEPENDENCE

A radial density $\rho_\ell(r)$ can be represented as a linear expansion

$$\rho_\ell(r) = \sum_\nu a_\nu f_{\nu\ell}(r) \tag{4}$$

where the radial basis functions $f_{\nu\ell}(r)$ are drawn from any convenient complete set. Electron scattering data are most often analyzed using some variation of the Fourier-Bessel expansion (FBE)[6]

$$f_{\nu\ell}(r) = \begin{cases} j_\ell(q_\nu r) & r < R \\ 0 & r > R \end{cases} \tag{5}$$

where

$$j_\ell(q_\nu R) = 0. \tag{6}$$

The principal advantage of this expansion is that the dominant contribution each term of the expansion makes to the form factor is localized in momentum transfer near $q=q_\nu$. Thus, the correlation between coefficients is small.[4]

The form factor

$$F_\ell(q) = \frac{\sqrt{4\pi}}{Z} \frac{\hat{j}_f}{\hat{j}_i} \tilde{\rho}(q) h(q) \tag{7}$$

is proportional to the Fourier transform of the density

$$\tilde{\rho}_\ell(q) = \int dr \, r^2 \, j_\ell(qr) \rho_\ell(r), \tag{8}$$

where Z is the atomic number, $j_f(j_i)$ is the final (initial) spin, and $\hat{x} \equiv \sqrt{2x+1}$. The nucleon form factor is $h(q)$. Only one term of the FBE contributes at $q=q_\nu$:

$$\rho_\ell(q_\nu) = a_\nu \frac{R^3}{2} j_{\ell+1}^2(q_\nu R). \qquad (9)$$

While other expansions are useful, none has a relationship between its expansion coefficients and the form factor that is so transparent. If the plane-wave expansion were accurate, the expansion coefficients could be determined from measurements performed at $q=q_\nu$ for each $\nu=1,\ldots,N$.

However, the maximum momentum transfer q_m accessible to measurement is always limited. Therefore, only

$$N \lesssim \frac{q_m R}{\pi} \qquad (10)$$

terms can be determined from the data. In the absence of a physical constraint, the amplitude of basis functions with $q_\nu > q_m$ are undetermined. The uncertainty in the fitted density is then infinite. This problem is not unique to electron scattering, but is characteristic of q_m rather than of the probe. Therefore, unless care is taken to bias the fitted coefficients or to limit the number of terms appropriately, unphysical short wavelength behavior can creep into the results.

A second major defect of the FBE is that each basis function is oscillatory at large radius. Thus, unless sufficient data exist at small momentum transfer or unless the asymptotic radial behavior is suitably biased, undesirable oscillations will persist at unreasonably large radii. The polynomial-Gaussian expansion (PGE)

$$f_{\nu\ell}(r) = [\alpha^3 y^\ell e^{-y^2}] y^{2\nu}, \qquad (11)$$

where $y = \alpha r$, provides a useful alternative with a complementary set of advantages and disadvantages. The PGE declines exponentially for both large radius and large momentum transfer. If α is chosen according to the harmonic oscillator model, a natural radial scale is introduced which minimizes the number of coefficients required to represent any plausible radial variation. However, the form factor for each basis function is oscillatory. Moreover, the radial basis functions overlap strongly. Therefore, the coefficients tend to be highly correlated.

A third alternative can be found in the Laguerre-Gaussian expansion (LGE)[30]

$$f_{\nu\ell}(r) = x^\ell e^{-x^2} L_\nu^{\ell+1/2}(2x^2) \qquad (12)$$

where $x = r/b$ and L_ν^a is a generalized Laguerre polynomial. This expansion also enjoys exponential decay at both large radius and large momentum transfer. However, as an orthogonal series, the correlation between neighboring coefficients is considerably less than that which plagues the PGE.

In the absence of physical constraints upon the coefficients, meaningful fits to data can only be obtained by limiting the number of terms in the expansion. The results then become model dependent. Although each of these expansions is mathematically complete, and therefore equivalent, truncation to a small number of terms destroys this equivalence and demands consideration of their relative merits.

The solution to the incompleteness problem for electron scattering was suggested by Borysowicz and Hetherington[2] and then developed more fully by Dreher et al.[4] If we assume that the single-particle wave functions for the constituent nucleons can be adequately represented by the solution to a (nonrelativistic) Schrödinger equation, including a smooth well-behaved potential, the Fourier transform of the transition density $\tilde{\rho}_\ell(q)$ must fall at least as fast as q^{-4} for sufficiently large q. Moreover, because nuclei have a well-defined Fermi momentum, we expect that the probability of momentum transfers larger than twice the Fermi momentum will fall precipitously. Therefore, the onset of the q^{-4} asymptotic tail probably begins soon after $2k_F \approx 2.7$ fm^{-1}. Experiments at high-energy electron accelerators usually reach this momentum transfer. In fact, using some high-q data, Heisenberg[6] has shown that $\tilde{\rho}(q)$ often falls considerably faster than the q^{-4} limit -- thus his analyses usually assume an exponential decay beyond $2k_F$. In our work, we feel that it is better to employ the less restrictive q^{-4} decay, thereby allowing the fitted density more flexibility and a larger estimate of the uncertainty due to the incomplete span of accessible momentum transfer.

The range of $\rho(r)$ that is consistent with the postulated behavior of $\tilde{\rho}(q)$ beyond the maximum measured momentum transfer q_m can be ascertained by performing fits of the data supplemented by many random sets of permissible pseudodata beyond q_m. Dreher et al.[4] have demonstrated that a more economical prescription produces the same results. Although our approach is basically the same as that of Dreher et al., it is useful to review in some detail the application of the high-q bias and the subsequent error analysis.

The upper limit envelope can be attached at q_m using

$$\rho^{lim}(q) = \tilde{\rho}_\ell(q_m)(q_m/q)^4. \tag{13}$$

The analysis is formulated most clearly using the FBE. A uniform distribution of permissible $\tilde{\rho}(q)$ points can be represented by a supplemental set of pseudodata selected at the Bessel roots q_ν beyond q_m such that

$$\tilde{\rho}_\ell(q_\nu) = 0 \quad \text{for} \quad q_\nu > q_m \tag{14}$$

with variances

$$[\delta\tilde{\rho}_\ell(q_\nu)]^2 = \tfrac{1}{3}[\rho^{lim}(q_\nu)]^2. \tag{15}$$

The factor 1/3 occurs because a uniform, rather than normal, distribution is assumed.

The uncertainty $\delta\rho(r)$ in the fitted density can be decomposed into a statistical contribution $(\delta\rho)_{stat}$ due to the uncertainties δy_i in the measured data $\{y_i, i=1, N_d\}$ and a model contribution $(\delta\rho)_{model}$ due to the uncertainties in the high-q pseudodata $\{y_i, i=N_d+1, N\}$ by analyzing the propagation of errors formula

$$[\delta\rho(r)]^2 = \sum_{i=1}^{N} (\delta y_i)^2 \left[\frac{\partial\rho(r)}{\partial y_i}\right]^2. \tag{16}$$

If we represent the density $\rho(r)$ using a set of fitting parameters $\{a_\nu\}$ we find that

$$\frac{\partial\rho}{\partial y_i} = \sum_\nu \frac{\partial\rho}{\partial a_\nu} \frac{\partial a_\nu}{\partial y_i}. \tag{17}$$

If we denote the fit to the ith data point as $\bar{y}_i$, we find

$$\frac{\partial a_\nu}{\partial y_i} = \sum_\mu \varepsilon_{\nu\mu} \frac{\partial \bar{y}_i}{\partial a_\mu} \frac{1}{(\delta y_i)^2} \tag{18}$$

where $\varepsilon = \alpha^{-1}$ is the error matrix.

The curvature matrix $\alpha_{\mu\nu}$ can be decomposed into two parts:

$$\alpha_{\mu\nu}^{stat} = \sum_{i=1}^{N_d} \frac{1}{(\delta y_i)^2} \frac{\partial \bar{y}_i}{\partial a_\mu} \frac{\partial \bar{y}_i}{\partial a_\nu} \tag{19}$$

$$\alpha_{\mu\nu}^{model} = \sum_{i=N_d+1}^{N} \frac{1}{(\delta y_i)^2} \frac{\partial \bar{y}_i}{\partial a_\mu} \frac{\partial \bar{y}_i}{\partial a_\nu} \tag{20}$$

by assigning the measured data to α^{stat} and the pseudodata to α^{model}. We can then decompose the error matrix

$$\varepsilon_{\mu\nu} = \varepsilon_{\mu\nu}^{stat} + \varepsilon_{\mu\nu}^{model} \tag{21}$$

using

$$\varepsilon_{\mu\nu}^{stat} = \sum_{\mu'\nu'} \varepsilon_{\mu\mu'} \alpha_{\mu'\nu'}^{stat} \varepsilon_{\nu'\nu} \tag{22}$$

$$\varepsilon_{\mu\nu}^{model} = \sum_{\mu'\nu'} \varepsilon_{\mu\mu'} \alpha_{\mu'\nu'}^{model} \varepsilon_{\nu'\nu}. \tag{23}$$

Note that although ε is the inverse of α, ε^{stat} is not the inverse of α^{stat} when pseudodata have been included. Substituting all this, we find

$$[\delta\rho]^2_{stat} = \sum_{\mu\nu} \frac{\partial\rho}{\partial a_\mu} \epsilon^{stat}_{\mu\nu} \frac{\partial\rho}{\partial a_\nu} \qquad (24)$$

$$[\delta\rho]^2_{model} = \sum_{\mu\nu} \frac{\partial\rho}{\partial a_\mu} \epsilon^{model}_{\mu\nu} \frac{\partial\rho}{\partial a_\nu}. \qquad (25)$$

It is important to recognize that this analysis is not limited to the density, but applies to any function of the fitting parameters $\{a_\nu\}$.

We illustrate this high-q regularization with a couple of examples based upon the first 2^+ state of ^{18}O. The proton density was deduced from electron scattering data[31] that extended to $q_m = 2.7$ fm^{-1}. Differential cross section measurements for the excitation of this state by 135 MeV protons extend to considerably larger momentum transfer. However, an analysis of the proton scattering data should be restricted to the same range of momentum transfer for which the charge form factor is known from electron scattering data. A truncated PGE expansion was used in Ref. 13 to fit the neutron transition density to the data. Rather than cloud the present analysis with peripheral issues concerning the accuracy of this result, we shall use this density to produce pseudodata, which are free of such concerns. These pseudodata were produced for $0.4 \leqslant q \leqslant 2.7$ fm^{-1} in steps of 0.1 fm^{-1} with random fluctuations corresponding to ±5% uncertainties, using the reaction model described in the preceding section. To fit the pseudodata, we employ the Fourier-Bessel expansion with a cut-off radius R = 8.0 fm.

To illustrate the effects of a finite span of momentum transfer, fitted transition densities are shown in Fig. 2 that correspond to 6, 7, and 8 terms of the FBE. All fits were done using the same set of pseudodata and used $\rho_n=0$ as the initial guess. No high-q pseudodata were included in this group of fits. It appears that six terms of the FBE are not sufficient to obtain an adequate representation of the true density, shown as the solid line, which was parametrized in the PGE. This particular fit of the cross section pseudodata is unsatisfactory. The addition of one more term to the expansion is sufficient to obtain a good fit and to reproduce the true density. However, although the last term has a significant effect upon the quality of the fit, it is strongly correlated with the preceding term and is only weakly determined by the data. As a result, the width of the error band grows substantially. Finally, the addition of yet one more term produces a catastrophic growth in the error envelope, with little improvement in the fit to the data. The oscillatory structure of the error envelope has eight nodes and clearly reflects the final Bessel function. An unconstrained fit cannot determine the amplitude of a frequency whose primary contribution to the cross section occurs well beyond the measured range of data.

This procedure can be regularized by applying the q^{-4} bias as

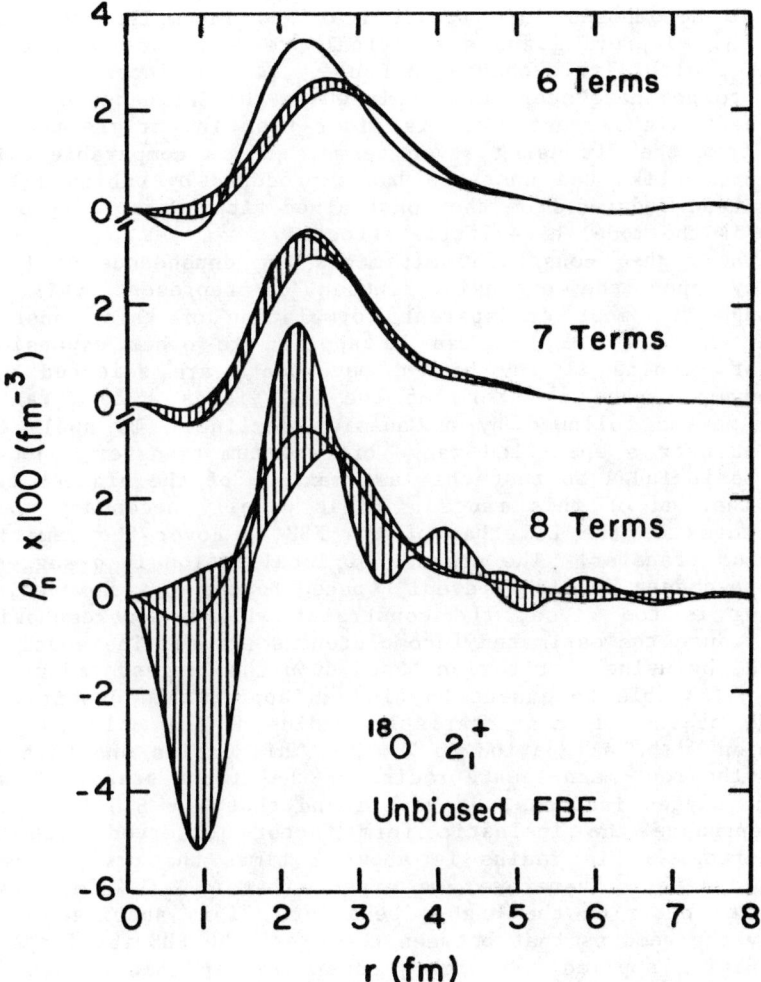

FIG. 2. Influence of number of terms of FBE in absence of high-q constraint.

described above. A biased fit using 12 terms of the FBE is shown in Fig. 3a, and provides a good fit of the data. The error envelope encompasses the true density over most of the radial range and is thus a realistic estimate of the uncertainty. The error band is decomposed into statistical and incompleteness errors in Fig. 3b. Improving the statistical precision of the data reduces $(\delta\rho)_{stat}$ with little change in $(\delta\rho)_{model}$. The incompleteness error tends to dominate near the origin where the influence of the high-q components is largest. It is wider near the origin than was the band from the fit using seven terms, and is comparable elsewhere. However, unlike the unstable bands produced by unbiased fits, the error band deduced from the constrained fit is stable -- additional terms in the model have little effect.

The high-q constraint minimizes the dependence of the fitted density upon the expansion chosen to represent this density. Although the most transparent formulation of this constraint is based upon the FBE, it can be applied to other expansions with similar results if the high-q pseudodata are selected properly. For example, the nth term of the LGE yields a form factor with (n+1) maxima followed by a Gaussian decline. To apply the constraint over a specified range of momentum transfer, enough terms must be included so that the last maximum of the highest term lies near the end of this range. It is usually necessary to include more terms of the LGE than of the FBE to cover the same range of momentum transfer. The absence of localization in q suggests that the pseudodata should be evenly spaced beyond q_m. However, if this spacing is too close, the constraint will be overemphasized and will reduce the estimated incompleteness error. The spacing can be decided by using a criterion based upon the largest radius at which it is plausible to expect to find an appreciable density. An extended object of characteristic radius R can only support frequencies with half-period $\Delta q \gtrsim \pi/R$. This radius should be several times the root-mean-square radius of the ground state. In the case of the oxygen isotopes, we have found that R = 8.0 fm is adequate to reproduce the inelastic form factors observed with electron scattering. This radius is about 3 times the rms radius of the ground state. Therefore, we expect that $\Delta q \gtrsim 0.4$ fm^{-1} is sufficient to describe the high-q behavior. This spacing is approximately the same as that between the q_ν of the FBE for large ν.

Having applied the high-q constraint in this manner, we have demonstrated that the LGE produces the same fit and similar estimates of the incompleteness and statistical uncertainties as emerge from the FBE. Therefore, this procedure minimizes the model dependence of the fitted density, while retaining sufficient flexibility to describe any physically plausible radial variation.

V. ACCURACY

Unlike electron scattering, the reaction models used to describe hadron scattering are fraught with uncertainty, ambiguity, and unverifiable approximations. A list of such problems must include:

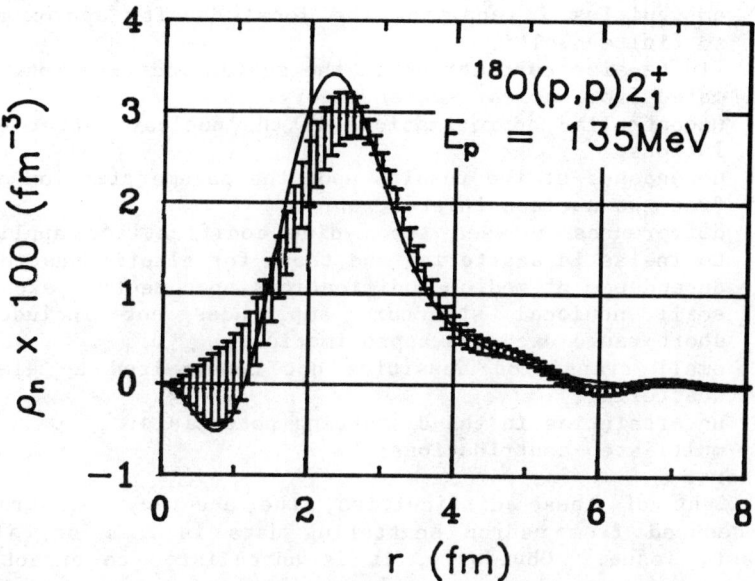

FIG. 3a. Density (band) fitted to pseudodata for 2_1^+ state of ^{18}O, using 12 terms of FBE with high-q bias, is compared with true density (line).

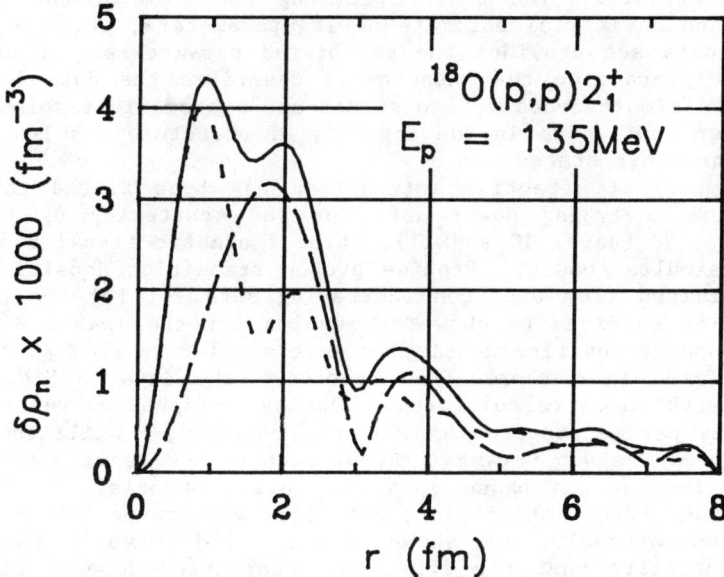

FIG. 3b. Statistical (long dashes) and model (short dashes) contributions to total uncertainty (solid line).

a) ambiguities in applying the local density approximation to finite nuclei;
b) finite-size corrections to the medium modifications estimated from nuclear matter theory;
c) uncontrolled approximations in the nuclear matter calculations;
d) dependence of the results upon the parametrization of the free two-nucleon interaction;
e) differences between the medium modifications applicable to inelastic scattering and those for elastic scattering;
f) dependence of medium modifications upon neutron excess;
g) small nonlocal structure amplitudes not included in short-range exchange approximations;
h) small transition densities not constrained by electron scattering;
i) uncertainties in the distorting potentials;
j) multi-step contributions;

among others.

In light of these difficulties, the accuracy of structure results deduced from hadron scattering data is a major, if not predominant, issue. Obviously, it is unrealistic to expect that any effective interaction presently available from theory will be sufficiently reliable to enable accurate quantitative structure to be extracted from data. Therefore, we must be prepared to calibrate an empirical effective interaction to the data for states of known structure. The nature of such an empirical effective interaction should, of course, be guided by the best theoretical interactions available. The theory provides the form of the fitting function and an initial estimate of its parameters, while a fit to a global data set provides the calibrated parameters. A suitable effective interaction that adequately describes the data for many transitions simultaneously in a smooth and systematic fashion would bolster our confidence in the accuracy of structure results fitted to data for other states.

An empirical effective interaction has been fitted to cross section and analyzing power data for the scattering of 135 MeV protons by ^{16}O (Refs. 10 and 11). Nine inelastic transitions were included simultaneously. Precise proton transition densities have been determined from electron scattering data.[32] For the present purposes, it suffices to show the results for the lowest 2^+ state of ^{16}O, whose transition density is most similar to that of the ^{18}O state analyzed in the preceding section. The data in Fig. 4 are compared with three calculations. The short-dashed curves use the Love-Franey parametrization of the free t-matrix,[28] while the long-dashed curves use von Gerambs' calculation of the density dependent effective interaction based upon the Paris potential.[17] The results of the fitted interaction, optimized with respect to all nine states simultaneously, are shown as the solid curves. These results demonstrate that the effective interaction depends strongly upon density. The Paris effective interaction explains the trend of these effects, but underestimates their magnitude. The fitted interaction has a considerably stronger dependence upon density and

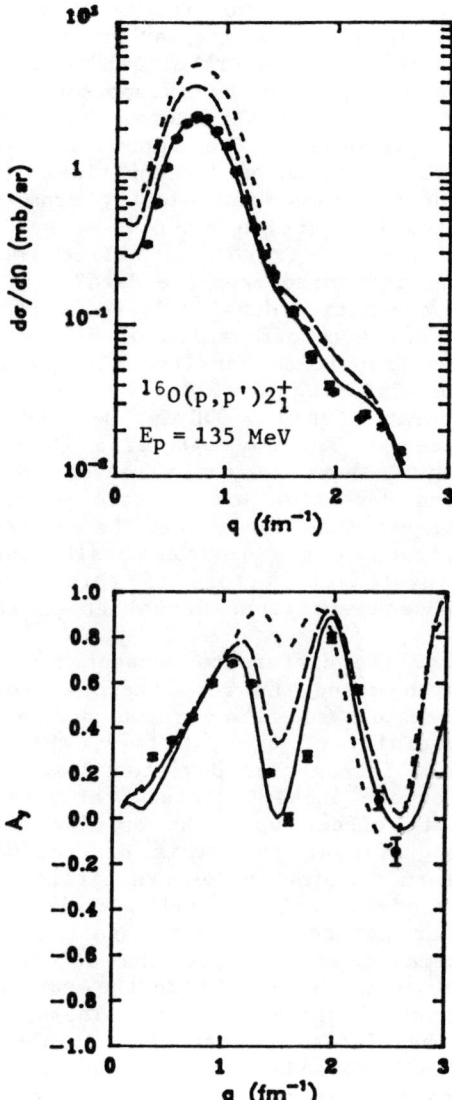

FIG. 4. Data for the excitation of the 2_1^+ state of ^{16}O by 135 MeV protons are compared with calculations using a free interaction (short dashes), a theoretical effective interaction (long dashes), and an empirical effective interaction (solid line).

provides a good description of the entire data set.

The quality of this fit is remarkable for a global optimization. The fitted interaction accurately describes the experimental cross section over most of the angular range, only deviating in the vicinity of $q = 2.0$ fm^{-1}. Nevertheless, residual discrepancies in the effective interaction can have important implications for the accuracy with which nuclear structure can be determined from the data for nearby isotopes. The importance of these systematic errors can be assessed by using the empirical effective interaction to extract a neutron transition density from the ^{16}O data. The empirical effective interaction was deduced subject to the explicit assumption that $\rho_n(r) = \rho_p(r)$ for a self-conjugate nucleus. Can this assumption be recovered from the data?

The neutron transition density fitted to the ^{16}O data using 12 terms of the FBE and a cutoff radius of 8.0 fm is compared in Fig. 5 with the proton transition density. The same results were also obtained with the LGE using $b = 2.0$ fm. The fitted density appears to be quite accurate for $r > 3.0$ fm, but there is a noticeable discrepancy inside of 3.0 fm. We find this level of agreement quite encouraging. There is ample reason to hope that reliable neutron transition densities with little model bias will soon be available for transitions to which the neutron contribution is large. However, because the neutron density must compete with the proton density, systematic errors of this size will probably be amplified for those transitions for which ρ_n is generally smaller than ρ_p.

The origin of the difference between $\rho_n(r)$ and $\rho_p(r)$ is most clearly illustrated using the form factors shown in Fig. 6. The largest difference between the proton and neutron form factors occurs in the vicinity of $q = 2.0$ fm^{-1}, which is the same place that the empirical interaction deviates from the data. The close correspondence between these positions reflects the fact that distortion has little effect upon the angular distribution. It is probably also significant that this discrepancy occurs near the minimum of the form factor. Structure effects not included in the present model can easily fill in such a minimum.

Several other methods are also available to assess the influence of systematic errors upon the accuracy of fitted densities. Any direct reaction connecting the same set of states shares a common set of structure variables. Therefore, several consistency tests are available. First, for a given probe, the fitted density should be independent of projectile energy. Second, an accurate and complete description must also describe polarization-transfer observables consistently, but an inaccurate model that fits the cross section data may fail to describe analyzing power data. Third, structure variables must be independent of the probe, such as pions versus nucleons.

VI. GEOMETRIC MOMENTS

The results of structure analyses are often presented in the form of geometric moments such as the matrix element

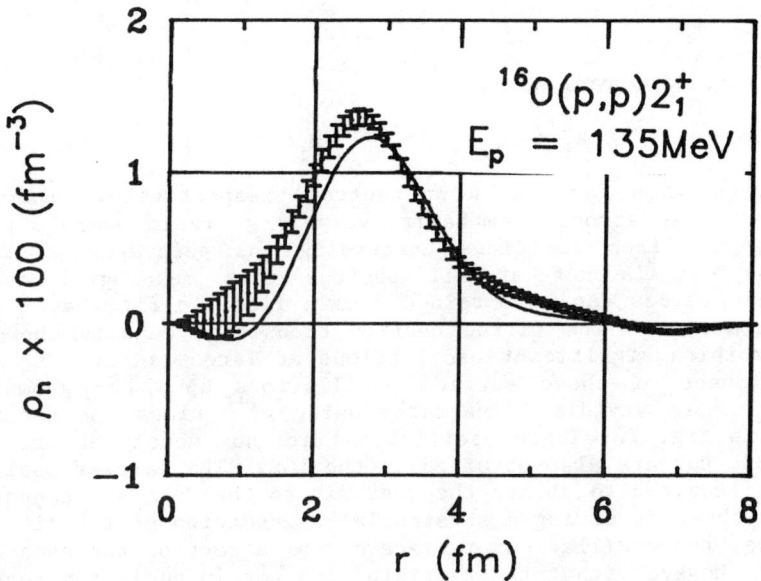

FIG. 5. Fitted neutron transition density (band) for the 2_1^+ state of ^{16}O is compared with proton transition density (line).

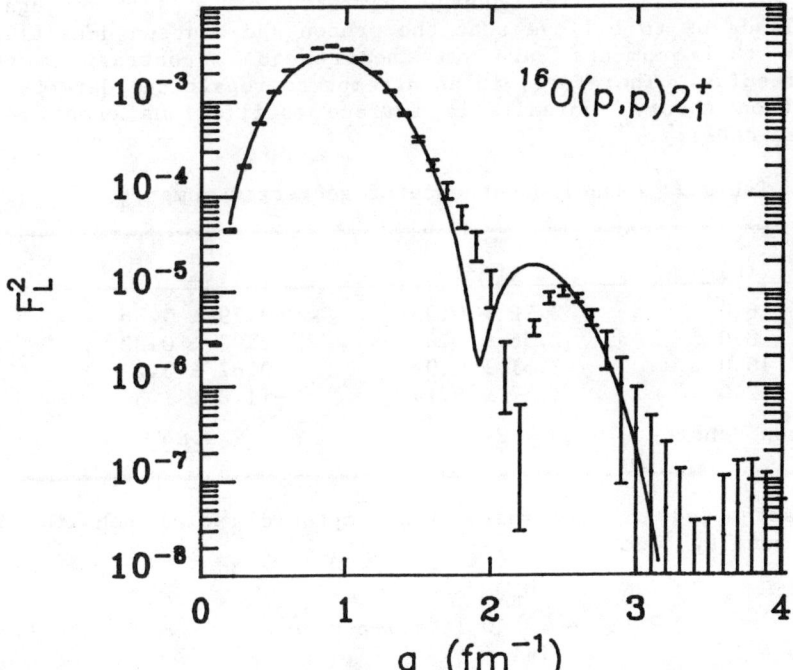

FIG. 6. Neutron (error bars) and proton (line) form factors for 2_1^+ state of ^{16}O.

$$M_i = \int dr \, r^{\ell+2} \, \rho_i(r) \tag{26}$$

and the transition radius

$$R_i^2 = M_i^{-1} \int dr \, r^{\ell+4} \, \rho_i(r) \tag{27}$$

where i = p or n for proton or neutron, respectively. However, these quantities strongly emphasize very large radii where $\rho(r)$ is small. Apart from questions concerning the relevance of these quantities, it is not at all obvious that meaningful model-independent values can be obtained from a flexible fit that uses a linear expansion. The fitted neutron transition density shown in Fig. 5 exhibits significant oscillations at large radii. The disturbing impact of these surface oscillations upon the geometric moments is more vividly illustrated using $r^{\ell+4}$ times the density, as shown in Fig. 7. These oscillations are not merely an artifact of the FBE, but are also obtained in the LGE. The large-r oscillations are required to obtain the best fit to the data and cannot be removed without incurring a substantial degradation of the fit. In this sense, the oscillations reflect a real aspect of the data. It is likely, however, that their origin lies not in nuclear structure but in inaccuracies of the reaction theory.

The geometric moments of neutron density obtained using a cutoff radius R = 8.0 fm are compared with those of the proton density in Table I. The charge symmetry of a self-conjugate nucleus leads us to believe that the proton and neutron densities, and hence their moments, are very nearly equal, contrary to the present results. Therefore, in an attempt to repair the defects of the reaction theory, unrealistic surface oscillations crept into our fitted density.

Table I: Model dependence of geometric moments

R (fm)	M_n (fm^2)	R_n (fm)
6.0	2.92 ± 0.03	3.79 ± 0.06
8.0	2.56 ± 0.05	2.30 ± 0.32
10.0	2.51 ± 0.09	0.67 ± 3.2
12.0	1.37 ± 0.19	-11.8 ± 1.3
proton density	2.79	3.84

We attempted to constrain the asymptotic radial behavior by adding a penalty function

$$\chi_t^2 = \sum_{i=1}^{N_t} w_i \left(t(r_i) - \rho(r_i) \right)^2 \tag{28}$$

to chi-square that inhibits deviations of the fitted density from a

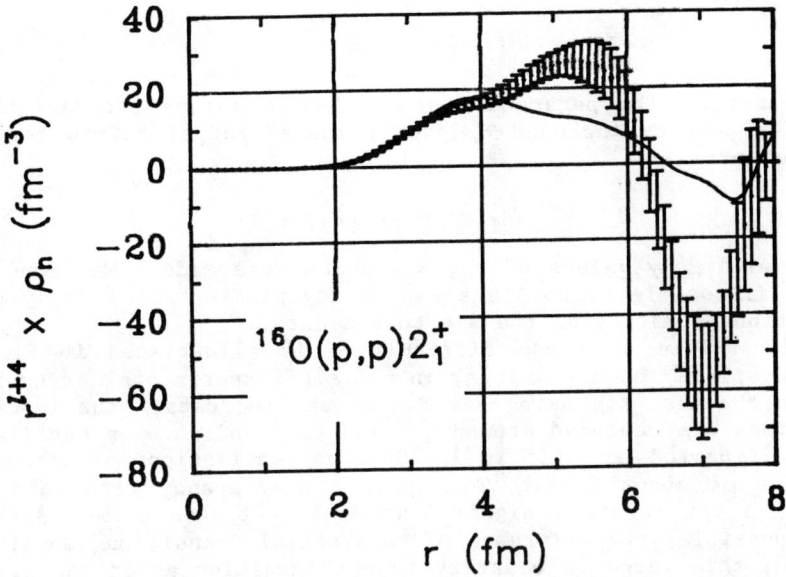

FIG. 7. Neutron (band) and proton (line) densities for 2_1^+ state of ^{16}O multiplied by $r^{\ell+4}$.

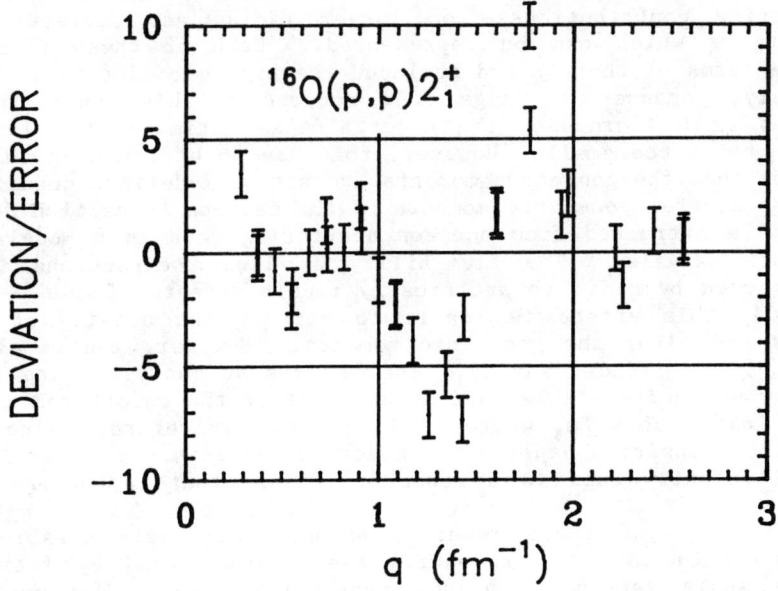

FIG. 8. Residuals $[(y_i - \bar{y}_i)/\delta y_i]$ of the fit to cross section data for 2_1^+ state of ^{16}O.

radial tail

$$t(r) = \frac{s\, e^{-dr}}{r^2} \qquad (29)$$

beyond $r=r_m$. The parameters s and d were matched to the fit at r_m. The penalty included N_t evenly spaced radial points for $r>r_i$ with weights

$$w_i = \left(w\, t(r_i)\right)^{-2}. \qquad (30)$$

Trials with many values of r_m, N_t, and w were made. We found that it was impossible to obtain a smooth asymptotic radial decay without seriously affecting the fit to the data.

The origin of these difficulties is illustrated in Fig. 8, which displays the deviations, in units of experimental error bars, between the best fit using R = 8.0 fm and the data. The important deviations are centered around $q = 1.5$ fm^{-1} and show an oscillation with half-period $\Delta q \sim 0.5$ fm^{-1}. Such an oscillation corresponds to a radius of about 6 fm. To repair a discrepancy with this frequency, a fit requires significant density beyond 6 fm. Although not impossible, the existence of substantial transition density for a radius this large is contrary to our intuition about the size of ^{16}O. Furthermore, the density fitted to the electron scattering data fell more smoothly at large radii without the intervention of a tail constraint.

As shown in Table I, the χ^2 of the fit can be improved substantially by increasing the cutoff radius R of the FBE, thereby permitting contributions from larger radius and decreasing the minimum Δq which can be represented. Each of these fits used twelve terms of the FBE and included the high-q constraint. Unfortunately, because the range of q covered by this constraint decreases as R increases, the uncertainties estimated for larger R are somewhat too small. However, this has no effect upon the conclusion that the geometric moments are not well-defined quantities.

Meaningful geometric moments of the neutron transition density cannot be extracted from nucleon scattering data in a model independent analysis. The flexibility of linear expansions is too easily used by a fit to artificially repair defects of the reaction theory. This effect is liable to corrupt the density at large radius and alter the geometric moments. However, the density is small at these radii anyway. In the present analysis, the fitted density was quite stable, and independent of the cutoff radius, for radii inside of 4 fm, where it is large. Therefore, we conclude that the transition density for moderate radii is a much more meaningful quantity than is a geometric moment that emphasizes large radii.

In light of these results, we must seriously question the interpretation of neutron matrix elements obtained by fitting a single scale factor to hadron scattering data. The excellent radial sensitivity of intermediate energy nucleon scattering mandates a more flexible analysis. However, the geometric moments are of dubious relevance to such an analysis.

VII. SUMMARY AND CONCLUSIONS

The radial structure of neutron contributions to inelastic transitions is, as yet, largely unexplored. We have become accustomed, in the last decade or so, to the precision and accuracy with which electromagnetic probes view the spatial distributions of charge and current in the nucleus. With the advent of accurate high-resolution experiments using intermediate-energy protons and pions, considerable attention has been paid to nuclear structure applications which exploit the complementarity between hadronic and electromagnetic reactions. Of particular interest is the possibility of unfolding the neutron contribution to a nuclear transition whose proton density has been measured using electron scattering. However, most analyses of hadron inelastic scattering have been content to extract from the data only a scale factor which characterizes, in an qualitative manner, the relative contributions of neutrons compared with protons.

We have developed a method for modeling direct reactions using a linear expansion analysis (LEA) of the transition amplitude. The folding model, which, schematically, represents this amplitude by the "$t\rho$" convolution of an effective interaction t with a nuclear transition density ρ, is most amenable to such an approach. States of known structure can be used to calibrate an empirical effective interaction and to assess the consistency of a reaction theory. Once the interaction is known, the structure of other transitions can be investigated.

Using a pseudodata technique that is largely independent of residual uncertainties in the reaction model, we have demonstrated that the intrinsic sensitivity of intermediate energy protons to the neutron transition density is surprisingly good, even in the interior of a large nucleus. The intrinsic sensitivity is optimal between about 200 and 400 MeV, and remains quite adequate even at 800 MeV. This sensitivity, in fact, is quite comparable to that of electron scattering. Therefore, an analysis of nucleon scattering data which presupposes a proportionality between neutron and proton densities or simply scales a shell-model wave function is thoroughly unsatisfactory and can be quite misleading. A fit using a restrictive analytic function of only a couple of parameters is little better. If we are to minimize the inherent bias of our analysis, we should employ a linear expansion in terms of radial basis functions drawn from a complete set.

Using experimental data of inevitably limited quantity and quality, it is not possible to determine a unique density without several <u>a priori</u> conditions, or physical biases, upon the result. The most important such condition describes the permissible behavior of $\tilde{\rho}(q)$ beyond the largest momentum transfer that has been measured. Learning from the experience acquired analyzing electron scattering data, we also require that $\tilde{\rho}_n(q)$ falls faster than q^{-4} beyond twice the Fermi momentum. When biased in this manner, the fitted density and its error envelope become almost independent of the number of terms and of the form of the expansion used to represent the density. The flexibility of the fitted density consistent with the high-q constraint provides an estimate of the uncertainty in the density that is due to the truncation of the span of momentum transfer accessible experimentally.

A second condition that must be placed upon a fitted density describes its behavior for large radii. However, we have found that it is not always possible, in practice, to impose a strong constraint upon the density at large radius. We attribute this difficulty to inadequacies in the reaction theory which introduce spurious structure that varies rapidly with momentum transfer. A variation with small Δq requires strength at large radius. Therefore, fits performed with an imperfect reaction model can yield a density which fails to fall at large radii as rapidly as we expect it should.

Although it is customary to quote neutron matrix elements, the commonly used geometric moments of a density emphasize the density at very large radii in a potentially misleading fashion. The density in this region is very small and is not well determined by a fit. The influence of a radial cutoff in the Fourier-Bessel expansion or the choice of an oscillator parameter in the Laguerre-Gaussian expansion also introduces a dangerous degree of model dependence. We have found, in fact, that the common geometric moments cannot be extracted from a fit to nucleon scattering data in a model-independent analysis. Systematic errors in the reaction theory affect these quantities quite strongly.

Unlike electron scattering, theories of hadron scattering are subject to considerable uncertainty. These theoretical uncertainties constitute unknown systematic errors with important implications for the accuracy with which structure results can be deduced from data. Therefore, we must calibrate an empirical effective interaction using states of known structure. We studied the propagation of the residual systematic errors by fitting a neutron transition density to data for the first 2^+ state of ^{16}O, a self-conjugate nucleus for which we expect the neutron and proton densities to be equal. Although the error band did not encompass the correct density, the fitted density did display the correct features and was moderately accurate. This result suggests that we can extract useful radial densities whenever the neutron contribution is not generally smaller than the proton contribution. However, because it is likely that the systematic errors will be amplified for states with small neutron contributions, such results should be viewed with caution.

We conclude that the good radial sensitivity of intermediate energy proton scattering merits the development of sophisticated analysis procedures that extract detailed radial information from the data with minimum bias. The quality of the data now, and soon to be, available demands such a development. The analysis of electron scattering data provides a valuable prototype and sets an inspirational standard. However, the influence of inaccuracies in the reaction theory poses a very thorny problem which requires further investigation.

REFERENCES

1. T. DeForest and J. D. Walecka, Adv. Phys. 15, 1 (1966).
2. J. Borysowicz and J. H. Hetherington, Phys. Rev. C 7, 2293 (1973).
3. I. Sick, Nucl. Phys. A218, 509 (1974).
4. B. Dreher et al., Nucl. Phys. A235, 219 (1974).
5. J. L. Friar and J. W. Negele, Nucl. Phys. A240, 301 (1975).
6. J. Heisenberg, Adv. Nucl. Phys. 12, 61 (1981).
7. J. Heisenberg and H. P. Blok, Ann. Rev. Nucl. Part. Sci. 33, 569 (1983).
8. F. Petrovich, R. J. Philpott, A. W. Carpenter, and J. A. Carr, Nucl. Phys. A425, 609 (1984).
9. F. Petrovich and W. G. Love, Nucl. Phys. A354, 499c (1981).
10. J. J. Kelly, in Advanced Methods in the Evaluation of Nuclear Scattering Data, Lecture Notes in Physics, Vol. 236, edited by H. J. Krappe (Springer-Verlag, Berlin, 1985).
11. J. J. Kelly, in Current Problems in Nuclear Physics, (to be published, 1985).
12. J. A. Carr, F. Petrovich, and J. J. Kelly, in Neutron-Nucleus Collisions - A Probe of Nuclear Structure, AIP Conference Proceedings No. 124, edited by J. Rapaport et al. (AIP, New York, 1985), p. 230.
13. J. J. Kelly et al., submitted to Phys. Lett.
14. G. G. Ohlsen, Rep. Prog. Phys. 35, 717 (1972). Our $D_{\alpha\beta}$ is the same as K_{α}^{β} defined by Ohlsen.
15. J. P. Jeukenne, A. Lejeune, and C. Mahaux, Phys. Rev. C 10, 1391 (1974); 15, 10 (1977); 16, 80 (1977).
16. F. A. Brieva and J. R. Rook, Nucl. Phys. A291, 299 (1977); A291, 317 (1977); A297, 206 (1978); A307, 493 (1978).
17. H. V. von Geramb, in The Interaction Between Medium Energy Nucleons in Nuclei, AIP Conference Proceedings No. 97, edited by H. O. Meyer (AIP, New York, 1983), p. 44; L. Rikus, K. Nakano, and H. V. von Geramb, Nucl. Phys. A414, 413 (1984).
18. C. Mahaux, in The Interaction Between Medium Energy Nucleons in Nuclei, AIP Conference Proceedings No. 97, edited by H. O. Meyer (AIP, New York, 1983), p. 20.
19. H. J. Gils et al., Phys. Rev. C 21, 1239 (1980); E. Friedman and C. J. Batty, Phys. Rev. C 17, 34 (1978).
20. H. J. Gils, H. Rebel, and E. Friedman, Phys. Rev. C 29, 1295 (1984); E. Friedman, H. J. Gils, and H. Rebel, Phys. Rev. C 25, 1551 (1982); H. J. Gils et al., Phys. Rev. C 21, 1245 (1980).
21. W. Gyles et al., Nucl. Phys. A439, 598 (1985).
22. E. Friedman, Phys. Rev. C 28, 1264 (1983).
23. L. Ray, W. Rory Coker, and G. W. Hoffman, Phys. Rev. C 18, 2641 (1978); L. Ray, Phys. Rev. C 19, 1855 (1979); G. W. Hoffman et al., Phys. Rev. C 21, 1488 (1980); L. Ray et al., Phys. Rev. C 23, 828 (1981).
24. I. Brissaud et al., Nucl. Phys. A191, 145 (1972); I. Brissaud and M. K. Brussel, Phys. Rev. C 15, 452 (1977); I. Brissaud and X. Campi, Phys. Lett. 86B, 141 (1979).

25. L. Ray and G. W. Hoffman, Phys. Rev. C $\underline{31}$, 538 (1985).
26. A. M. Bernstein, V. R. Brown, and V. A. Madsen, Phys. Rev. Lett. $\underline{42}$, 425 (1979); Phys. Lett. $\underline{103B}$, 255 (1981); A. M. Bernstein, Adv. Nucl. Phys. $\underline{3}$, 325 (1969).
27. P. Harihar et al., Phys. Rev. Lett. $\underline{53}$, 152 (1984).
28. W. G. Love and M. A. Franey, Phys. Rev. C $\underline{24}$, 1073 (1981).
29. F. Petrovich, H. McManus, V. Madsen, and J. Atkinson, Phys. Rev. Lett. $\underline{22}$, 895 (1969); W. G. Love, Nucl. Phys. $\underline{A312}$, 160 (1978).
30. H. G. Andresen, in Advanced Methods in the Evaluation of Nuclear Scattering Data, Lecture Notes in Physics, Vol. 236, edited by H. J. Krappe (Springer-Verlag, Berlin, 1985).
31. B. E. Norum et al., Phys. Rev. C $\underline{25}$, 1778 (1982).
32. T. N. Buti et al., to be published in Phys. Rev. C.

SESSION B

COLLECTIVE MODES OF MOTION AT HIGH ENERGY

ISOSPIN STRUCTURE OF GIANT RESONANCES FROM PION INELASTIC SCATTERING

Susan J. Seestrom-Morris
University of Minnesota, Minneapolis, MN 55455

ABSTRACT

A large amount of data has been accumulated for the excitation of the giant quadrupole resonance by pion inelastic scattering. The π^+ and π^- cross sections measured for the self-conjugate nucleus ^{40}Ca are equal, but for heavier nuclei with a neutron excess the π^- cross section is always larger than the π^+ cross section. Furthermore, the ratio $\sigma(\pi^-)/\sigma(\pi^+)$ increases more rapidly than does N/Z for nuclei from ^{90}Zr to ^{238}U. The analysis of the existing data to extract neutron and proton matrix elements is discussed. Recent energy-dependence data for the GQR in ^{208}Pb are used to draw some conclusions about the model dependence of the extracted strengths and the relative neutron and proton strengths.

INTRODUCTION

The comparison of π^+ and π^- scattering should be useful in the identification of the isospin structure of nuclear transitions. This utility results from the pion-nucleon [3,3] resonance, which dominates the pion-nucleus interaction at energies around T_π = 180 MeV. The isospin (T = 3/2) of this resonance results in a strong enhancement of π^+-p and π^--n scattering, i.e.

$$\frac{\sigma(\pi^+ p)}{\sigma(\pi^- p)} = \frac{\sigma(\pi^- n)}{\sigma(\pi^+ n)} = 9 \qquad (1)$$

In inelastic scattering, this enhancement in the elementary cross sections should result in π^+ preferentially exciting proton transitions and π^- preferentially exciting neutron transitions.
There is a growing body of data indicating that the elementary cross section ratios do indeed carry over to inelastic transitions. In the p-shell, transitions to stretched states have been extremely well-studied by (π,π').[1-5] For these transitions, the shell model often predicts very dramatic isospin structure, e.g. both pure neutron and pure proton excitations are predicted, sometimes in the same nucleus. In most cases these predictions have been verified by π^+/π^- comparisons.[2-4] In contrast to the stretched states, in which very large asymmetries were observed, for collective transitions in self-conjugate nuclei the π^+ and π^- cross sections are nearly equal.[6] This indicates isoscalar transitions as expected for such states. Only recently have π^+/π^- comparisons been performed for collective transitions in heavier nuclei with a neutron excess. One example of such a comparison is for the first 2^+ state in the deformed nucleus ^{152}Sm for which the ratio of π^- to π^+ cross sections is 1.4 (Ref. 7).

GIANT RESONANCES

Pion scattering studies in heavy nuclei have basically been limited by the energy resolution currently attainable, which in many cases is insufficient to resolve states of interest in heavy nuclei. Energy resolution does not present such a limitation in the study of giant resonance states which typically have widths of 2-3 MeV. A considerable amount of effort has recently been directed at obtaining high quality giant resonance data by pion inelastic scattering.

The majority of the (π,π') data are for the giant quadrupole resonance (GQR), and these are the only data that will be discussed here. The earliest giant resonance experiments with pion inelastic scattering were those of Arvieux et al.,[8] at SIN in which π^+ scattering was measured to the GQR in ^{40}Ca and ^{89}Y. Somewhat later, Ullmann, et al. measured both π^+ and π^- scattering at a pion energy of 130 MeV for targets ^{40}Ca and ^{118}Sn.[9] The cross sections for π^+ and π^- excitation of the GQR in ^{40}Ca were found to be equal, indicating a purely isoscalar transition as expected in a self-conjugate nucleus. In contrast, the π^- cross sections measured for the GQR in ^{118}Sn were larger than the π^+ cross sections, with a ratio $R = \sigma(\pi^-)/\sigma(\pi^+) = 1.9$. Spectra for π^+ and π^- scattering from ^{118}Sn are shown in Figure 1. More recently, the same group has measured π^+ and π^- scattering from ^{90}Zr and ^{118}Sn at 164 MeV.[10] Figure 2 displays spectra for π^+ and π^- scattering from ^{90}Zr from this experiment. For ^{118}Sn the π^-/π^+ ratio was measured to be $R = 2.2$, whereas for ^{90}Zr it was found to be $R = 1.7$.

The largest π^-/π^+ asymmetries observed for giant resonance states have been for the GQR in ^{208}Pb and ^{238}U measured at an incident pion energy of 162 MeV. Figure 3 displays π^+ and π^- spectra for scattering from ^{208}Pb (Ref. 11). The GQR is very strongly excited in π^- scattering with a signal-to-noise ratio comparable to that obtained in alpha-particle scattering.[12] The GQR is much more weakly excited in π^+ scattering, however, and the ratio between the π^- and π^+ cross sections measured near the peaks in the experimental angular distributions is $R = 2.7$. In Figure 4 the angular distributions extracted for the GQR are plotted for both π^+ and π^-.

Recently, single π^+ and π^- spectra were measured for scattering to the giant resonance region of ^{238}U at $T_\pi = 162$ MeV. These spectra are plotted in Figure 5, along with preliminary fits that included both the GQR and the giant monopole resonance. From these preliminary fits, the ratio of π^- to π^+ cross sections is $R = 3.0$.

All the (π,π') experiments done so far show that when there is a neutron excess the GQR is more strongly excited with π^- than π^+. Table I summarizes the ratios that have been observed. In the most naive interpretation of these data the ratio of the neutron matrix element M_n to the proton matrix element M_p is given approximately by the ratio of the π^- to the π^+ cross section, $M_n/M_p \simeq \sigma(\pi^-)/\sigma(\pi^+)$. Thus, the ratios measured for the GQR indicate that $M_n > M_p$ and therefore that the transitions are not isoscalar.

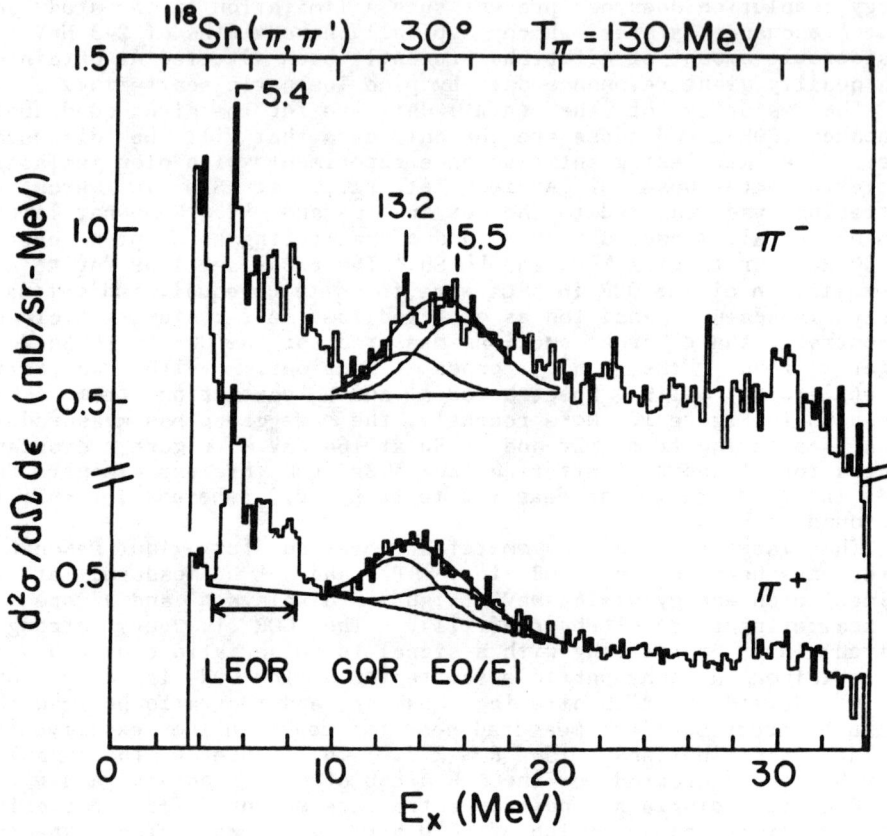

Fig. 1. Spectra for π^+ and π^- scattering from ^{118}Sn at T_π = 130 MeV from Ref. 9. These spectra are measured at angles near the peaks in the experimental angular distributions.

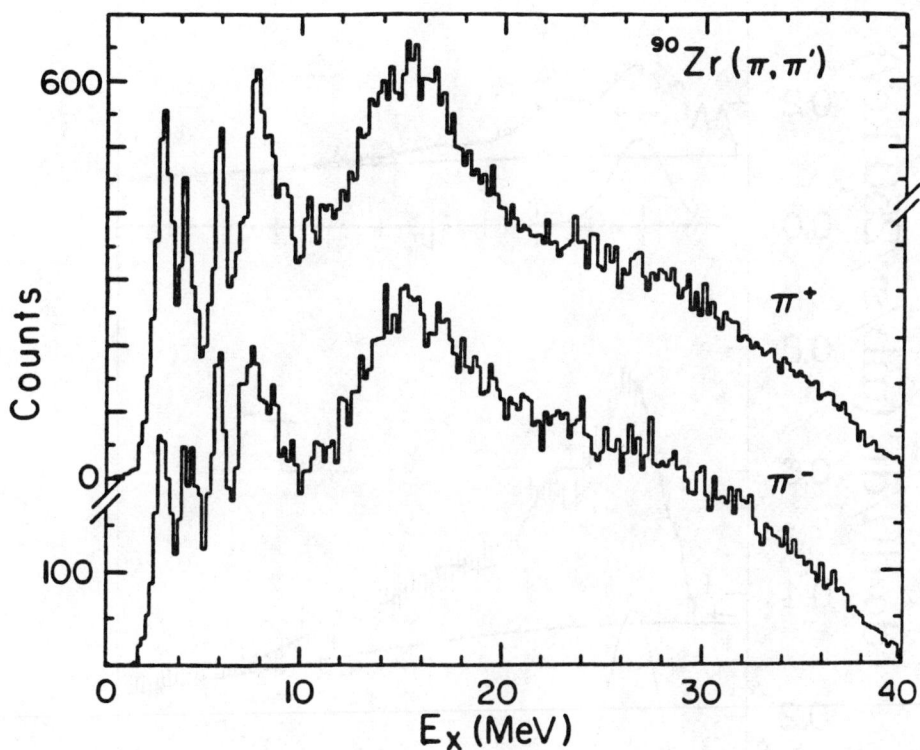

Fig. 2. Spectra for π^+ and π^- scattering from ^{90}Zr at T_π = 164 MeV from Ref. 10. These spectra are measured at angles near the peaks in the experimental angular distributions.

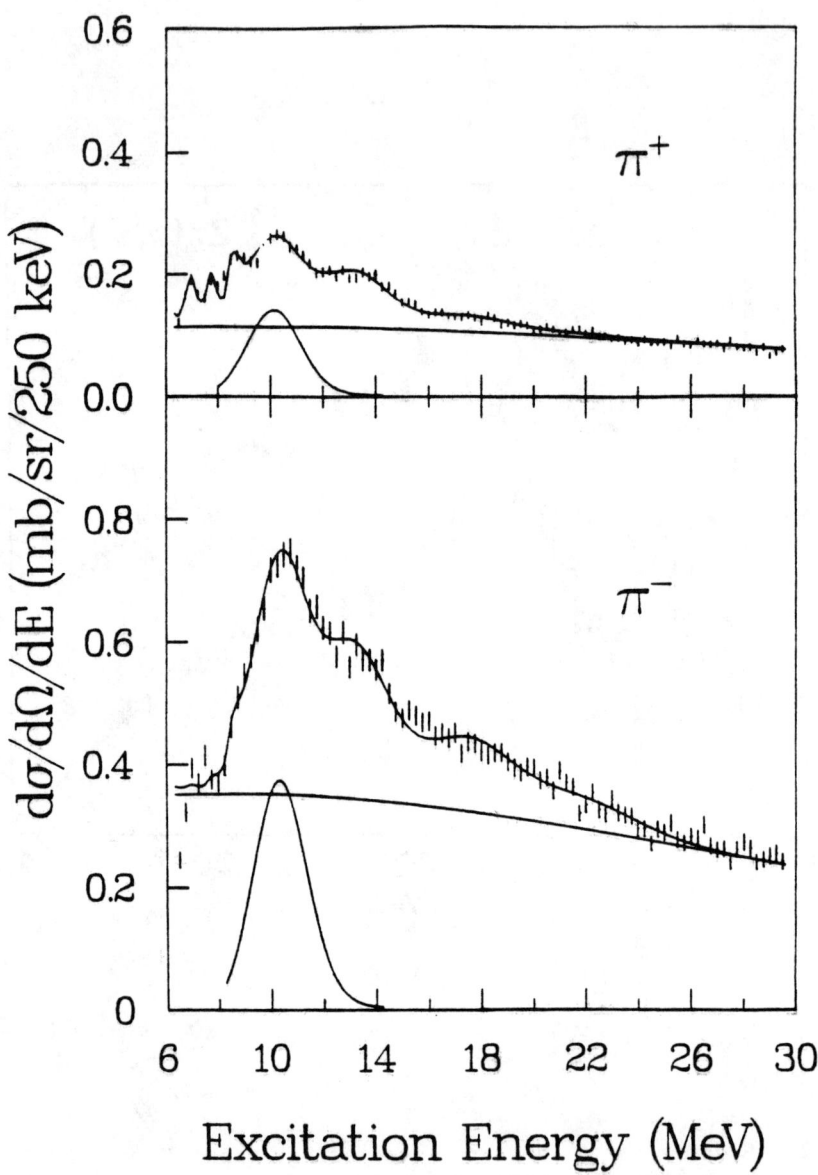

Fig. 3. Spectra for π^+ and π^- scattering from ^{208}Pb at T_π = 162 MeV from Ref. 11. These spectra are measured at angles near the peaks in the experimental angular distributions.

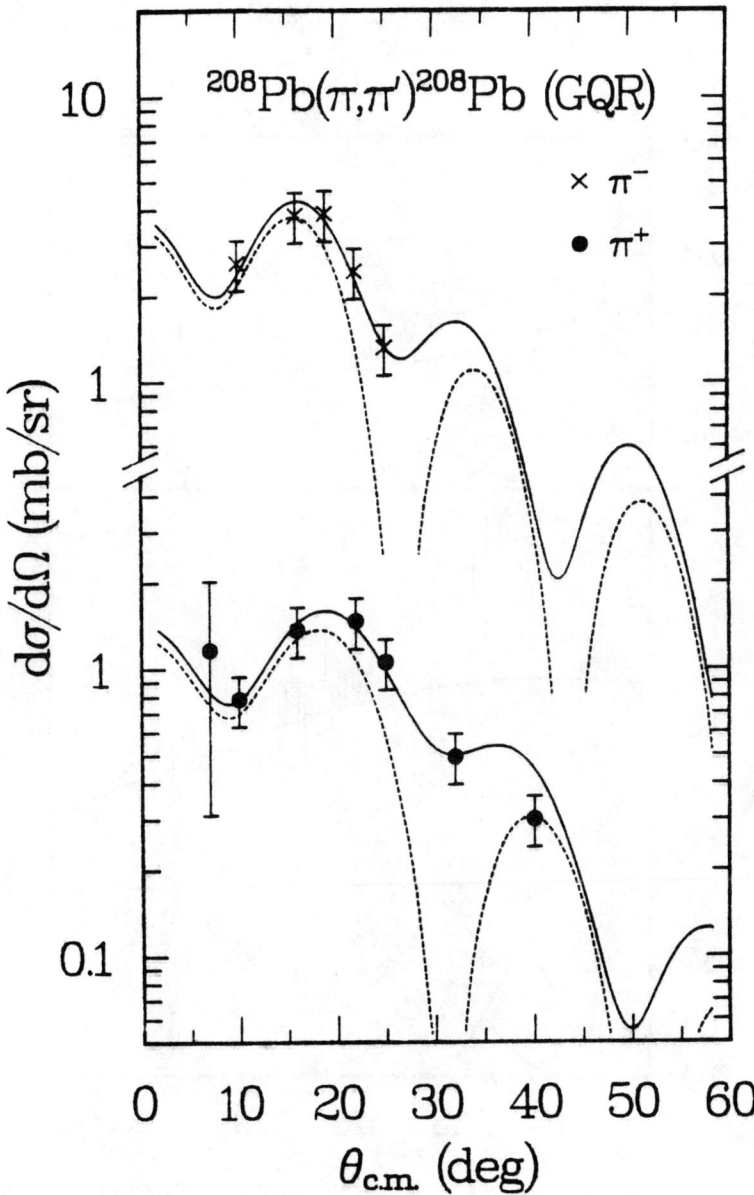

Fig. 4. Angular distributions measured at T_π = 162 MeV for the GQR in ^{208}Pb (Ref. 11). Dashed curves are DWIA calculations for pure E2 transition density and solid curves are the sum of E2 and E4 necessary to fit the data.

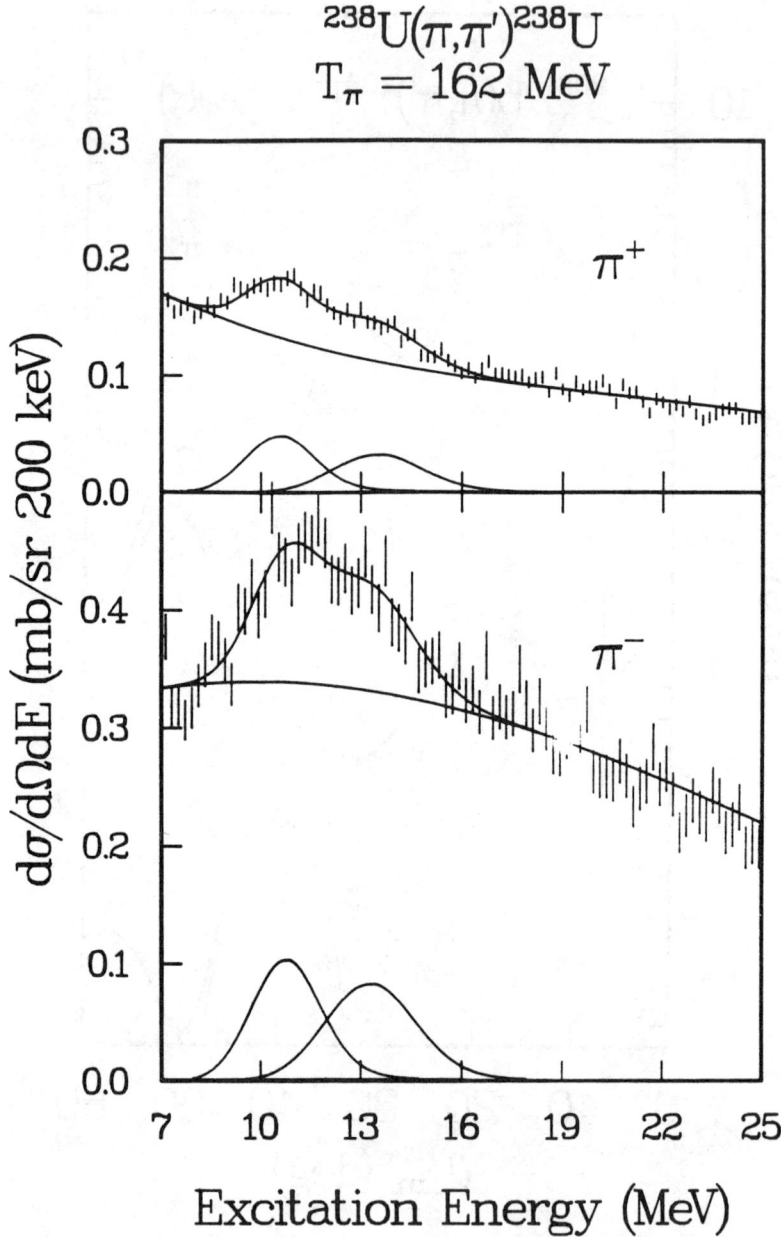

Fig. 5. Spectra for π^+ and π^- scattering from ^{238}U at T_π = 162 MeV. These spectra are measured at angles that are expected to be near the peaks in the experimental angular distributions.

Table I M_n/M_p Values for GQR extracted from analysis of (π,π')

Nucleus	N/Z	$\sigma(\pi^-)/\sigma(\pi^+)$	M_n/M_p	Ref.
^{40}Ca	1.0	1.0		9
^{90}Zr	1.25	1.65	1.6	10
^{118}Sn	1.36	2.2	2.4	10
^{208}Pb	1.54	2.7	3.4	11
^{238}U	1.59	3.0		16

In a more sophisticated analysis effects of distortions and coulomb potential are included by performing distorted wave impulse approximation (DWIA) calculations and comparing to the data to extract the neutron and proton matrix elements. The input includes parameters for the ground state densities and a model for the transition density. For all calculations presented here the transition density was calculated from either a collective model or a Tassie model, with separate neutron and proton deformation parameters β_n and β_p. These deformation parameters are adjusted to fit the data and then used to calculate the matrix elements

$$M_{n(p)} = \beta_{n(p)} \int \rho_{tr}^{n(p)}(r) r^{\lambda+2} dr \qquad (2)$$

where ρ_{tr} is the neutron or proton radial transition density and λ is the multipolarity of the transition. We have performed such an analysis for the 164-MeV ^{90}Zr and ^{118}Sn data of Ullmann, et al. and our 162-MeV ^{208}Pb data. The curves plotted in Fig. 4 are the result of this analysis. The resulting values of M_n/M_p are summarized in Table I, along with the value of N/Z for these nuclei.

The value of N/Z is included because it is the value that the hydrodynamical model[13] predicts for M_n/M_p. More sophisticated RPA calculations[14] also predict M_n/M_p to be near N/Z. From Table I it can be seen that the value of M_n/M_p from the (π,π') analysis increases more rapidly than does N/Z as we go from ^{40}Ca to ^{208}Pb. The very large ratio extracted for ^{208}Pb, 3.8, has thus far not been reproduced by any theoretical models.

Further understanding of these large values of M_n/M_p may come from looking at the fraction of the energy weighted sum rule (EWSR) that is exhausted. For ^{208}Pb we have found[11] 48% of the isoscalar EWSR and 16% of the isovector EWSR. This corresponds to 50% of the neutron EWSR but only 5% of the proton EWSR. Similarly, our analysis of the 164-MeV ^{118}Sn data indicates 55% of the neutron EWSR and 12% of the proton EWSR. Thus, it seems that the large values of M_n/M_p are due to a depletion of the observed proton strength rather than an enhancement in the neutron strength. Because of the coupling between the low-lying 2^+ states and the GQR in the RPA it is impossible to explain such a small M_p for the GQR without a large impact on the M_p

for the 2_1^+ state[15] which is well determined from electromagnetic measurements.

We are in the process of trying to understand these unexpectedly large values of M_n/M_p extracted from the (π,π') analysis and in the process to determine the uncertainties and limitations of the present analysis. One source of uncertainty arises from the fact that the pion is a strongly absorbed probe. This may result in the pion seeing only a small part of the transition density near the surface; therefore the extracted M_n/M_p could reflect only differences between the neutron and proton transition densities in this region, not overall differences in strength. Although an analysis of pion scattering from ^{18}O has indicated[16] that the pion penetrates to near the half-density point, the same will not necessarily be true in a heavy nucleus such as ^{208}Pb. If the interaction is limited to a small region in the nuclear surface, then our conclusions about M_n and M_p are dependent on the model used to describe the shape of the transition densities. Furthermore, to the extent that the model allows different shapes for the neutron and proton transition densities, the extracted ratio M_n/M_p will also be model dependent.

In an attempt to identify such effects in (π,π'), we have recently measured π^+ and π^- scattering to the GQR in ^{208}Pb as a function of incident pion energy from T_π = 120 MeV to 300 MeV. The pion mean free path in the nucleus is about the same at 120 MeV as it is at 300 MeV and is longer than at the resonance energy of 180 MeV. Spectra obtained for π^+ and π^- scattering at 120 and 300 MeV are shown in Figures 6 and 7. The GQR is strongly excited at both energies, although at 300 MeV the instrumental backgrounds are larger than at 120 MeV or 162 MeV. The large π^- enhancement observed at 120 MeV is about the same as previously measured at 162 MeV. At 300 MeV, however, the π^- cross section is only about a factor of 2 larger than the π^+. The π^-/π^+ ratios measured at 200 and 250 MeV are also near 2.

We have performed a DWIA analysis on the preliminary data at these energies, using Tassie-model transition densities, to extract values for M_n and M_p. The results are plotted in Figure 8. Although the error bars are large, there seems to be an energy dependence to the extracted values of M_n and M_p. M_n decreases somewhat with increasing energy whereas M_p increases. This results in M_n/M_p decreasing with energy. Of course, the physical matrix elements are constant and therefore this energy dependence must be an artifact of our analysis. If one considers the ratio calculated from the average values of M_n and M_p for all five energies, then $\langle M_n \rangle / \langle M_p \rangle$ = 2.8, which is still considerably larger than N/Z for ^{208}Pb. It should be noted that if isoscalar and isovector matrix elements (M_{IS} and M_{IV}) are constructed from M_n and M_p then M_{IS} is more constant than either M_n or M_p but M_{IV} has energy dependence. This points to a possible problem in the energy-dependence of the isovector part of the pion-nucleus interaction.

We have made a preliminary investigation of the effect of shape differences between the neutron and proton transition densities on the values extracted for M_n and M_p. Since the neutron ground state radius in ^{208}Pb should be about .2 fm larger than that of the protons, we have performed calculations using a neutron ground state distribution (the derivative of which is used to calculate the transition density) with the rms radius increased by .2 fm. This did not significantly change the extracted M_n or M_p at any energy. We

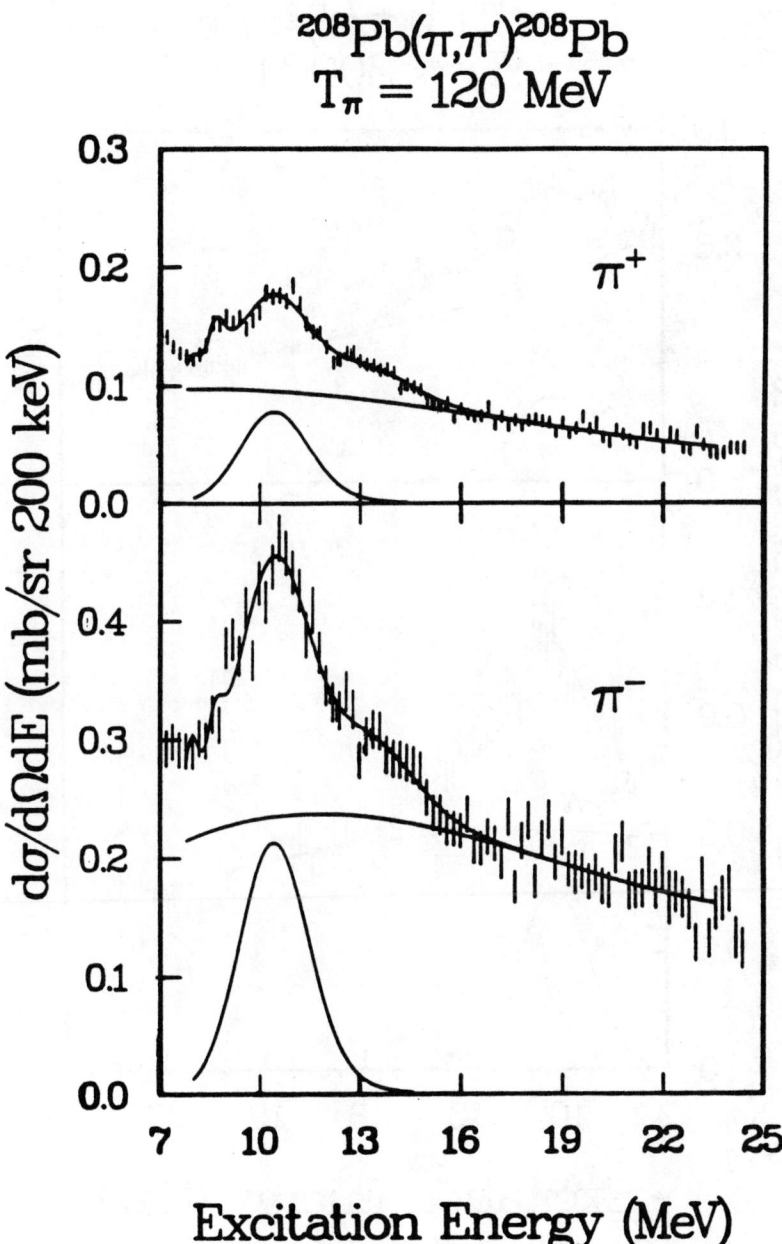

Fig. 6. Spectra for π^+ and π^- scattering from ^{208}Pb at T_π = 120 MeV. These spectra are measured at angles near the peaks in the experimental angular distributions.

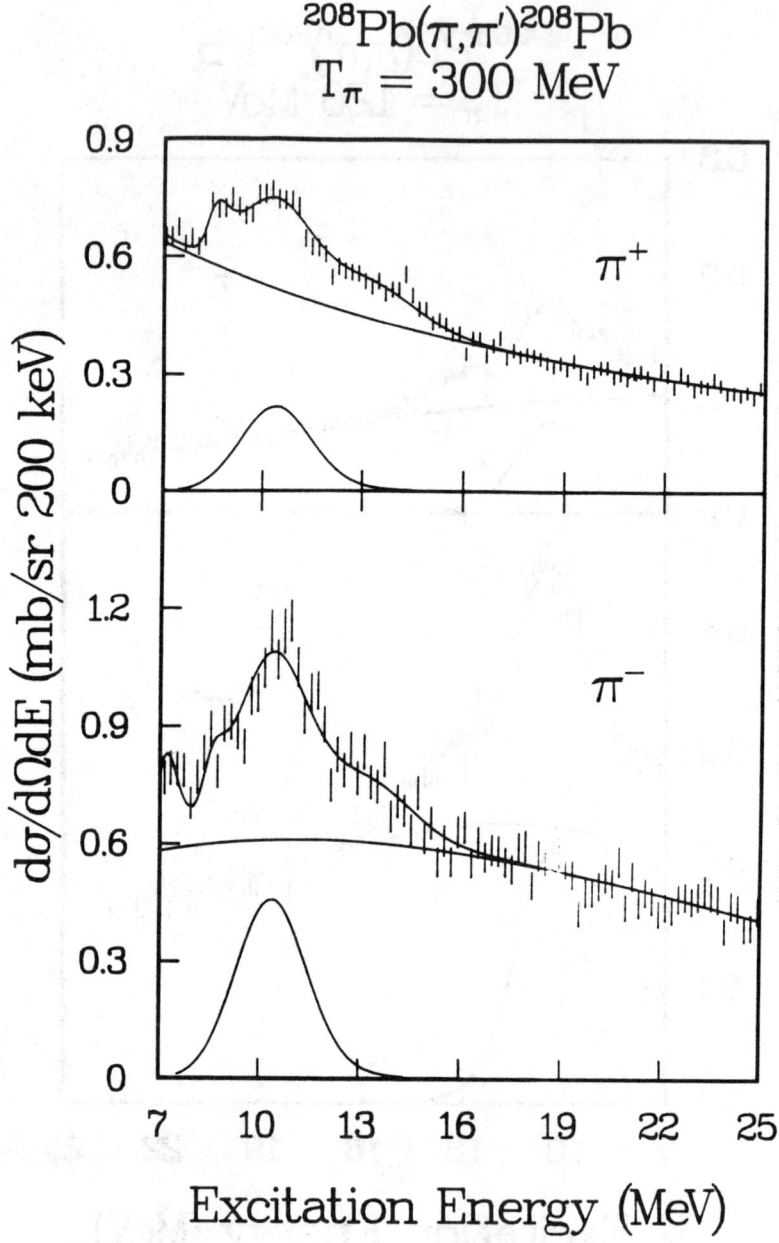

Fig. 7. Spectra for π^+ and π^- scattering from ^{208}Pb at T_π = 300 MeV. These spectra are measured at angles near the peaks in the experimental angular distributions.

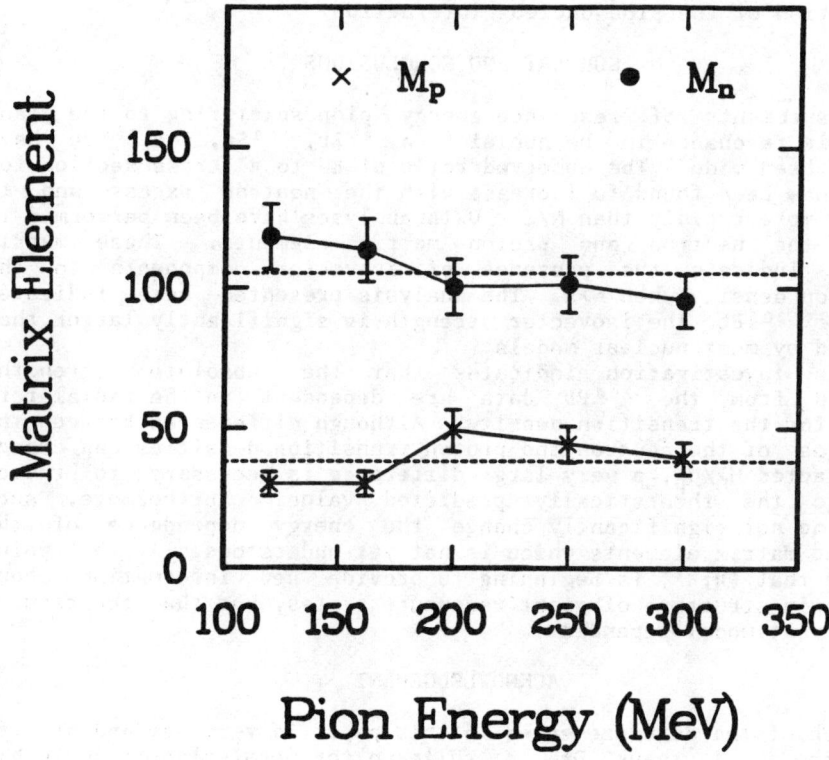

Fig. 8. Neutron and proton matrix elements M_n and M_p extracted as a function of incident pion energy for the GQR in ^{208}Pb using a Tassie-model transition density. The data for energies other than 162 MeV are still preliminary.

then introduced differences between only the neutron and proton transition densities, not the ground state densities. This was done by changing the radius parameter of the Fermi function whose derivative is used to calculate the Tassie model transition density. In order to decrease $\langle M_n \rangle / \langle M_p \rangle$ from 2.8 to 1.54 (N/Z), it is necessary to introduce a 1.0 fm difference between the neutron and proton radius parameters. This corresponds approximately to a 1 fm shift between the peaks of the neutron and proton transition densities but with both having about the same shape. While this is a very large difference between neutrons and protons, it does indicate that the extracted ratio M_n/M_p is sensitive to differences between the shapes of the neutron and proton transition densities. This large difference did not, however, alter the energy dependence of the extracted M_n or M_p values. It therefore seems likely that an explanation of this energy dependence requires an energy-dependent modification of the pion-nucleus interaction.

SUMMARY AND CONCLUSIONS

Measurements of resonance-energy pion scattering to the giant quadrupole resonance in the nuclei ^{40}Ca, ^{90}Zr, ^{118}Sn, and ^{208}Pb have recently been made. The observed ratio of π^- to π^+ cross section for the GQR has been found to increase with the neutron excess and to increase more rapidly than N/Z. DWIA analyses have been performed to extract the neutron and proton matrix elements. These matrix elements indicate the presence of isovector components to the transition density when N>Z. The analysis presented here indicates that for ^{208}Pb the isovector strength is significantly larger than predicted by most nuclear models.

This investigation indicates that the absolute strengths extracted from the ^{208}Pb data are dependent on the radial form assumed for the transition density. Although differences between the geometries of the neutron and proton transition densities can change the extracted M_n/M_p, a very large difference is necessary to reduce M_n/M_p to the theoretically predicted value. Furthermore, such changes do not significantly change the energy dependence of the extracted matrix elements which is not yet understood. At this point it seems that (π,π') is beginning to provide new information about the isospin structure of giant resonance states, but that the results are somewhat model-dependent.

ACKNOWLEDGEMENT

Much of the data presented in this paper is very new and as yet unpublished. I thank Dr. J. Ullmann for permission to quote his unpublished data and, especially, my collaborators from the Universities of Colorado, Minnesota, Paris, Pennsylvania, Texas and the Los Alamos National Laboratory. This work has been supported in part by the U. S. Department of Energy and The Robert A. Welch Foundation.

REFERENCES

1. C. L. Morris, et al., Phys. Lett. 86B, 31 (1979).
2. D. Dehnhard, et al., Phys. Rev. Lett. 43, 1091 (1979) and S. J. Seestrom-Morris, et al., Phys. Rev. C 26, 954 (1982).
3. D. B. Holtkamp, et al., Phys. Rev. Lett. 47, 216 (1981) and D. B. Holtkamp, et al., Phys. Rev. C31, 957 (1985).
4. S. J. Seestrom-Morris, et al., Phys. Rev. C31, 923 (1985).
5. D. B. Holtkamp, et al., Phys. Rev. Lett. 45, 420 (1980).
6. C. L. Morris, et al., Phys. Rev. C24, 231 (1981).
7. C. L. Morris, et al., Phys. Rev. C28, 2165 (1983).
8. J. Arvieux, et al., Phys. Rev. Lett. 42, 753 (1979) and J. Arvieux. et al., Phys. Lett. 90B, 371 (1980).
9. J. L. Ullmann, et al., Phys. Rev. Lett. 51, 1038 (1983) and Phys. Rev. C31, 177 (1985).
10. J. L. Ullmann, private communication.
11. S. J. Seestrom-Morris, et al., submitted to Physical Review C (1985).
12. H. P. Morsch, et al., Phys. Rev. C28, 1947 (1983).
13. A. Bohr and B. R. Mottelson, Nuclear Structure, Vol. 2, ed. W. A. Benjamin, Inc., Reading, Mass., P.404 (1975).
14. N. Auerbach, A. Klein, and E. R. Siciliano, Phys. Rev. C31, 682 (1985).
15. V. Brown, private communication and to be published.
16. S. J. Seestrom-Morris, et al., to be published.

ELECTROEXCITATION OF THE NUCLEAR CONTINUUM

K.T. Knöpfle
Max-Planck-Institut für Kernphysik, D-6900 Heidelberg, Germany

ABSTRACT

The power of coincidence electron scattering to probe the nuclear response functions in the continuum region is demonstrated by the example of the ^{28}Si(e,e'c) reaction in which the charged particle decay products c = p,α are measured in coincidence with the electron which has caused the inelastic excitation. A model-independent analysis of the ^{28}Si(e,e'c) coincidence data, taken at three momentum transfers, yields E1 and E2/E0 form factors and the respective strength distributions in the giant resonance region of ^{28}Si (E_x = 14-22 MeV). The analysis of the (e,e'α_o) angular correlation functions separates E0, E1 and E2 excitations. Evidence for the presence of large isovector E2/E0 contributions and sizeable E0 excitations is presented.

INTRODUCTION

The advantages in using the electromagnetic probe for studying nuclear structure are well known[1,2]. Due to the validity of QED and the weakness of the interaction, the reaction process is calculable to much better precision than in any other scattering experiment, and thus the hadronic structure of the target can be directly examined. The general response of a nucleus to an external electromagnetic field can be studied as a function of the momentum transfer q and the energy transfer ω due to the exchange of either a real photon or a virtual photon. The superiority of electroexcitation over photoabsorption consists in the decoupling of momentum transfer q from the energy transfer ω. This allows electron scattering to map out the spacelike sector, $Q^2 = \omega^2 - q^2 < 0$, of the (q,ω) plane, while photoexcitation is restricted to explore the one cross section of this surface whose trace is given by the straight line q = ω (Figure 1). In varying ω, one can select different regimes of nuclear excitation as the bound state region, the particle unbound giant resonances, quasifree scattering which roughly follows the kinematic line for a free nucleon with effective mass m*, the dip region where meson exchange currents show up, the Δ resonance, and so on. Keeping ω fixed and varying q, electron scattering allows one to map out the Fourier transforms of the transition charge and current densities.

In this paper only a small part of the response surface will be addressed, i.e the giant resonance (GR) region; ω will extend from

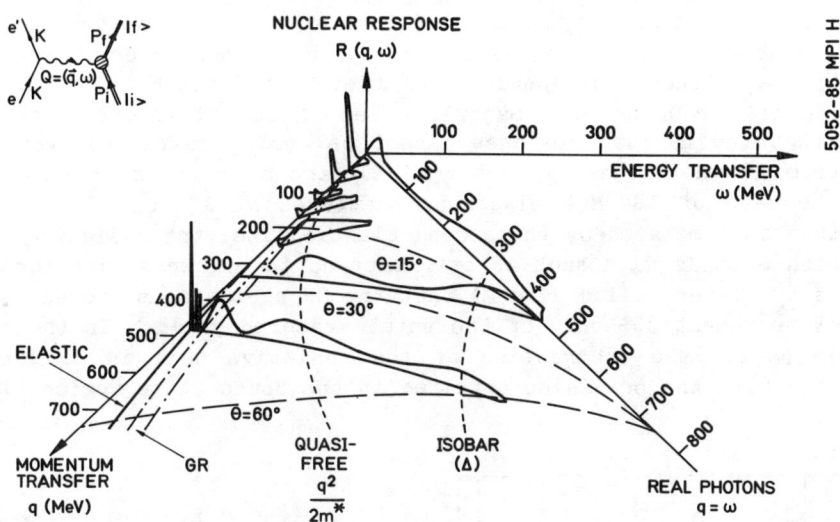

Fig. 1. The schematic view of the response function of a nucleus in an external electromagnetic field (following the example given in ref. 2).

Table I Investigations of the giant resonance region in ^{28}Si

Reaction	Beam Energy (MeV)	Resolution (keV)	References
^{28}Si(γ,Tot)[a]	10 - 50	100 - 500	7)
(γ,n)[a]	13 - 33	180 - 400	8,9)
(γ,c)[a]	15 - 23	150 - 200	10)
^{27}Al(p,γ_o)	4 - 13	15	11)
^{24}Mg(α,γ_o)	4 - 14	150 - 340	12)
^{28}Si(e,e')	92	400	13)
(e,e')	38 - 61	30	14)
(p,p')	61, 115	90 , 250	15,16)
(α,α')	120 -155	90 - 180	17-19)
(α,α')[b]	129	100 - 300	20)
^{28}Si(α,α'c)	155	180	21)
(α,α'c)[b]	129	300	22)
(e,e'c)	183	120	6)

[a] measurements with monochromatic photons;
[b] measurements at 0° scattering angle.

10 to 22 MeV while q is restricted to values of less than 1 fm^{-1}. By varying the momentum transfer, different multipole resonances can be made to dominate the inelastic electron scattering cross section due to the different q-dependence of their form factors.

So far, however, the major problem of such investigations has been that the nuclear response cannot be easily extracted from the spectrum of inelastically scattered electrons as shown in Figure 2 for the case of 180 MeV electrons scattered at 30° on ^{28}Si. This spectrum is dominated by the strong elastic line, the radiative tail of which extends with such an intensity up to higher excitation energies E_x (later called also ω) that the GR excitations around 18 to 22 MeV represent 20% only of the total measured yield. In the continuum region the subtraction of the radiative tail is no longer feasible with the precision obtained in the bound state region. This

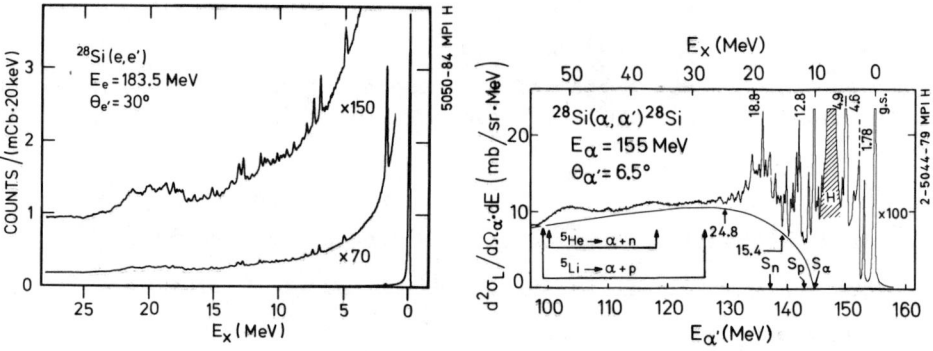

Fig. 2. The comparison of single-arm spectra from inelastic electron (left) and inelastic α scattering (right) on ^{28}Si (refs. 17,34)

situation is seemingly similar to that in inelastic hadron scattering where the GRs are also riding on a large background (see right part of Figure 2), the subtraction of which plagues the analysis of single-arm experiments. The physical origin of the backgrounds encountered in inelastic electron and hadron scattering is, however, completely different. The large physical backgrounds arising in inelastic hadron scattering from multistep processes are excluded in electron scattering by the dominance of one-photon exchange. Since there is, in particular, no nuclear excitation involved with the elastic radiative tail background, it will be effectively removed (as well as the radiative tails of all other bound states) in an experiment where the inelastically scattered electron is detected in coincidence with a nuclear particle decay. Of course, the coincidence cross section will also provide relevant information on the microscopic structure of the GRs by revealing their coupling to the different final states of the target.

(e,e'x) coincidence experiments have become feasible only recently with the advent of high duty factor electron accelerators, and the experimental data are still scarce[3,4,5]. Rather than to present a review on previous work, I will restrict myself here to the discussion of a specific ^{28}Si(e,e'c) coincidence experiment[6] in order to demonstrate by example the progress which can be achieved in the exploration of the nuclear continuum by coincidence electron scattering.

The GR region of ^{28}Si has been studied recently in great detail by numerous experiments using both hadronic and electromagnetic probes (see Table I). High precision data obtained from photoabsorption and photodisintegration experiments for the giant dipole resonance (GDR) will allow scrutiny of the new approach presented here for the model-independent extraction of multipole strength distributions and form factors in the nuclear continuum. On the other hand, the comparison with the results from (α,α') single-arm and $(\alpha,\alpha'c)$ coincidence experiments, which selectively excite isoscalar modes only, will provide new information on the isospin structure of the giant E0/E2 resonances.

THE (e,e'c) COINCIDENCE CROSS SECTION

The inclusive cross section for electroexcitation has the form[1,2]

$$d^2\sigma/d\omega d\Omega_e = (1/M_i)\cdot\sigma_M\cdot\{ v_L\widetilde{W}_L + v_T\widetilde{W}_T \}. \qquad (1)$$

σ_M denotes the Mott cross section and the electron kinematical factors are given by

$$v_L = (Q^2/q^2)^2 \text{ and } v_T = -(Q^2/2q^2)+\tan^2(\theta/2). \qquad (2)$$

The two independent response functions $\widetilde{W}_L$ and $\widetilde{W}_T$ represent the contributions which are longitudinal (L) or transverse (T) with respect to the direction of 4-momentum Q. W_L and W_T are generated by the charge and current operators, respectively, and may be separated experimentally by the familiar Rosenbluth procedure, i.e., by fixing q and ω (i.e., Q^2, $Q\cdot P_i = \omega\cdot M_i$, P_i being the 4-momentum of the initial target state) and varying energy and scattering angle θ of the electron.

The cross section for the exclusive process (e,e'x), where a particle x of 4-momentum $P_x = P$ is present in the final state and detected in coincidence with the scattered electron (Figure 3), provides additional information about the transverse-transverse (TT) and transverse-longitudinal (TL) interferences[23-26]

$$d^3\sigma/d\omega d\Omega_e d\Omega_x = (1/M_i)\cdot\sigma_M\cdot\{v_L W_L + v_T W_T + v_{TT} W_{TT} \cos 2\phi_x + v_{TL} W_{TL} \cos\phi_x\}. \quad (3)$$

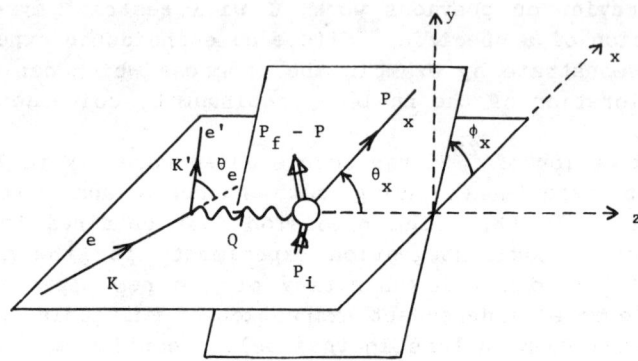

Fig. 3. Kinematics for the (e,e'x) reaction[26]. The polar coordinates (θ_x, ϕ_x) of the emitted particle x are measured in the centre-of-momentum system of the decaying nucleus. The electron scattering defines the xz-plane with the momentum transfer lying along the z-axis.

The electron kinematical factors v_L and v_T are as defined above, while the two additional factors are given by

$$v_{TT} = (Q^2/2q^2) \text{ and } v_{TL} = (Q^2/q^2)\{-(Q^2/2q^2) + (1/2)\tan^2(\theta/2)\}^{1/2}. \quad (4)$$

The four nuclear response functions W_L, W_T, W_{TT}, and W_{TL} depend now on the four Lorentz-invariant scalars $\{Q^2, Q\cdot P_i, P_x\cdot P_i, Q\cdot P_x\}$ or on the laboratory quantities (q, ω, p_x, θ_x). In particular, these response functions do not contain any dependence on the azimuthal angle ϕ_x, which appears only in the explicit factors $\cos 2\phi_x$ and $\cos\phi_x$. Taking thus advantage of both this explicit azimuthal angle dependence and the Rosenbluth method, it is possible in principle to separate all four response functions from one another.

EXPERIMENT

The 183.5 MeV cw electron beam of the MAinz MIcrotron MAMI A[27] at 10 to 25 μA current was used to bombard a natural Si target (92% ^{28}Si) of 2.89 mg/cm² thickness. A schematic view of the experimental setup is shown in Figure 4. Inelastically scattered electrons were measured at 25°, 30° and 45° by the Mainz 180° magnetic spectrometer[28] with a resolution of ~120 keV. The corresponding momen-

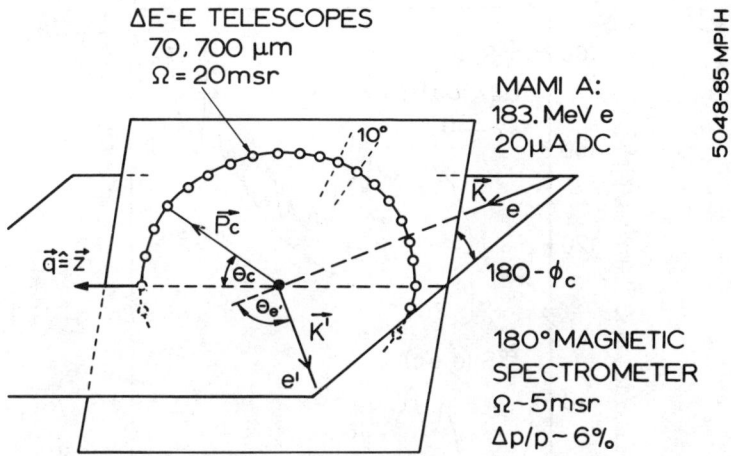

Fig. 4. Schematic view of the experimental arrangement.

tum transfers of $q = 0.39$, 0.47 and 0.68 fm^{-1} cover the maximum of the E1 and the increasing slope of the E0/E2 form factors. The 6% momentum acceptance of the spectrometer allowed study of the excitation energy range from 12 to 22 MeV with one magnetic field setting. In this energy interval one finds a major part of the GDR and of the isoscalar giant quadrupole resonance (GQR) which have centroid energies of 18 and 19 MeV and widths of 4 and 5 MeV, respectively.

Secondary charged particle decay products $c = p,\alpha$ were detected in coincidence by 8 ΔE-E surface barrier detector telescopes subtending a solid angle of 20 msr each. CoSm magnets in front of the detectors provided a 2 kG magnetic sweeping field in order to reduce the large background from low-energy Møller electrons. The thickness of 70µ and 700µ for ΔE and E detectors allowed a unique distinction between α particles and protons for energies larger than 2.5 MeV. The telescopes were mounted in a plane rotated 45° around the q axis from the scattering plane (see Figure 4). The measured angular correlation functions (ACFs) cover the complete angular range from -10° to 215° relative to the direction of momentum transfer q in steps of typically 10°.

THE ELIMINATION OF THE ELASTIC RADIATIVE TAIL

Figure 5 shows an enlarged part of the singles ^{28}Si(e,e') spectrum (top) measured at 30°. The range of excitation energies extends from the charged particle thresholds at ~11 MeV up to the region of the GDR and GQR from 16 and 22 MeV. At this momentum transfer contributions from E1 and E2/E0 GRs are expected to exhibit similar cross sections. Below, there are displayed the 4π integrated

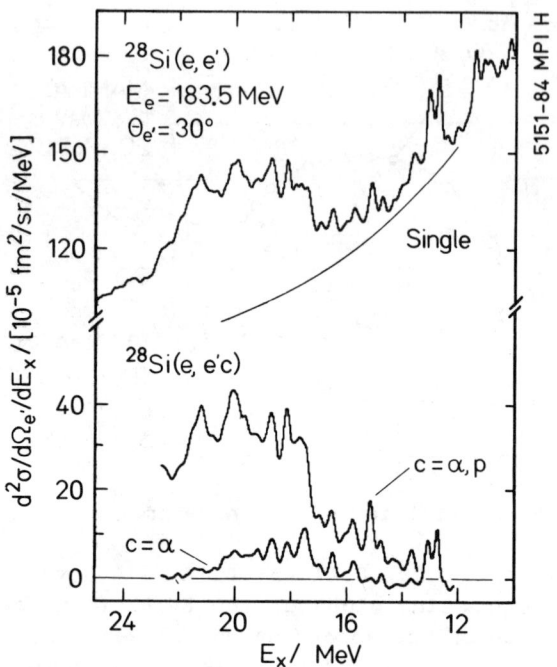

Fig. 5. Singles (e,e') spectrum in the GR region of ^{28}Si (top) and the 4π integrated ^{28}Si(e,e'c) coincidence cross sections (bottom) in all c = p,α channels as well as in the α channel. The radiative tail (smooth curve below the singles spectrum) is deduced by subtracting the ^{28}Si(e,e'{c=p,α}) cross section from the singles yield (ref. 34)

(e,e'c) coincidence cross sections in the total c=p,α channel as well as in the α channel. Obviously, the coincidence requirement has effectively removed the large elastic radiative tail, and the transition from isolated resonances to the GR region with its characteristic fine structure becomes clearly visible.

The conversion of the triple coincidence cross section to a doubly differential (e,e') cross section was done by numerical integration over θ_c from 0° to 180° and integration over ϕ_c assuming no ϕ_c dependence. This procedure is strictly correct only if the W_{TL} structure function can be neglected. W_{TT} does not contribute because $\cos 2\phi_c = 0$ for $\phi_c = 135°$. There is both theoretical and experimental evidence that this assumption is not unreasonable in the present kinematical situation. Macroscopic models for GR vibrations predict[25] the relation between the transverse electric matrix element (TJ) and the longitudinal Coulomb matrix element (CJ) as (J>0)

$$\langle J \| T^{el}(q) \| 0 \rangle = -([J+1]/J)^{1/2} (\omega/q) \langle J \| M^{el}(q) \| 0 \rangle. \tag{5}$$

For $\omega = 18$ MeV and $q = 0.7$ fm^{-1} the transverse matrix element amounts accordingly to less than 14% of the Coulomb component. An independent experimental test is provided by subtracting the 4π integrated $c = p, \alpha$ coincidence cross sections from the corresponding singles spectra. Assuming correct 4π integration, the resulting difference spectra should exhibit the smooth and monotonic decrease of the elastic radiative tail. As seen in Figure 6, the difference spectra obtained at $\theta_{e'} = 25°$ and $30°$ show indeed the postulated smooth flow with a slight increase above the neutron threshold at 17.2 MeV indicating the undetected contribution from the neutron channel. Below 13.5 MeV of excitation energy, the difference spectra reflect the incomplete detection of low-energy α particles which are in part stopped within the target. At $\theta_{e'} = 45°$, however, the difference spectrum shows much more structure than at $25°$ or $30°$ which is also clearly correlated with peaks in the singles spectrum. Since the measured angular correlation functions are covered at $\theta_{e'} = 45°$ by less data points than at $25°$ and $30°$, it remains open whether these correlations indicate a poorer integration over θ_c or an increase of transverse contributions. In any case, the fluctuations of the difference spectra from a smooth line always amount to less than 10% which conforms to the theoretical estimate presented above.

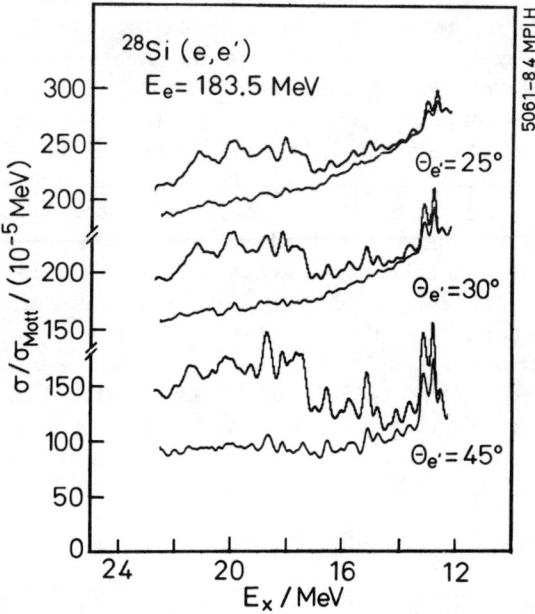

Fig. 6. Singles (e,e') spectra in the GR region of ^{28}Si at three indicated scattering angles. The respective lower curves represent the difference spectra between singles yields and the 4π integrated ^{28}Si(e,e'c={p,α}) coincidence cross sections (ref. 34).

MODEL-INDEPENDENT EXTRACTION OF Eλ FORM FACTORS
AND STRENGTH DISTRIBUTIONS

The clean preparation of the nuclear response by the (e,e'c) reaction provides the basis for the model-independent multipole analysis outlined below which does yield not only the decomposition of E1 and E2/E0 strength distributions in all measured decay channels but also the determination of the respective form factors.

The traditional multipole expansion method[29,30] assumes that the form factor in each energy bin ω_i can be expressed as the linear combinations of the longitudinal Eλ excitations:

$$|F(q_k,\omega_i)|^2 = \Sigma_\lambda \, a_{E\lambda}(\omega_i) \cdot |F_{E\lambda}(q_k,\omega_i)|^2 \quad . \tag{6}$$

The multipole coefficients $a_{E\lambda}(\omega_i)$, $\{\lambda = 1,\Lambda\}$, are determined by a least squares fit procedure to the experimental form factors $|F(q_k,\omega_i)|^2$ measured at K different momentum transfers. Obviously, the number of data points has to be larger than the number of multipoles ($\Lambda < K$). This method of multipole expansion requires an assumed q dependence for the form factor $|F_{E\lambda}(q_k,\omega_i)|^2$ of each multipole which is usually calculated from the transition charge densities provided by a chosen nuclear model. The model dependence of the resulting form factors is known to be strong in particular for the case of E1 excitations where the Goldhaber-Teller (GT), Steinwedel-Jensen (SJ) and Myers-Swiatecki models[31-33] predict a largely different q-dependence. At the highest momentum transfer, q = 0.7 fm^{-1}, of the present ^{28}Si(e,e'c) experiments, GT and SJ form factors deviate already by nearly a factor of 2 (see Figure 7).

The present approach[34] to a multipole decomposition is exclusively based on measured quantities, i.e., the form factors which are given by the 4π integrated (e,e'c) coincidence cross sections discussed above. For each arbitrary decay channel c, the experimental input thus consists of 750 data points $\sigma^c_{exp}(q_k,\omega_i)$ measured at 3 momentum transfers q_k, (k = 1,3), between ω = 12 to 22 MeV in 250 energy bins ω_i, (i = 1,250), of 40 keV width. In analogy to the traditional method, the measured form factors are expressed in each energy bin ω_i as the sum of the different contributing multipole components [equations (7)]:

$$\begin{aligned}
\sigma^c_{exp}(q_1,\omega_i)/\sigma_M &= \quad\quad\quad \left\{ |F_{E1}(q_1)|^2 \right. \quad\quad\quad \left\{ |F_{E2/E0}(q_1)|^2 \right. \\
\sigma^c_{exp}(q_2,\omega_i)/\sigma_M &= a^c_{E1}(\omega_i) \cdot \left\{ |F_{E1}(q_2)|^2 \right. + a^c_{E2/E0}(\omega_i) \cdot \left\{ |F_{E2/E0}(q_2)|^2 \right. \\
\sigma^c_{exp}(q_3,\omega_i)/\sigma_M &= \quad\quad\quad \left\{ |F_{E1}(q_3)|^2 \right. \quad\quad\quad \left\{ |F_{E2/E0}(q_3)|^2 \right. .
\end{aligned}$$

This ansatz assumes (i) that only E0, E1 and E2 excitations contribute and (ii) that E2 and E0 form factors are identical. In order to determine both the multipole coefficients $a_{E\lambda}^c(\omega_i)$ and the respective form factors $|F_{E\lambda}(q_k)|^2$ a third assumption is introduced, namely (iii) that between $12 < \omega < 22$ MeV states of identical multipolarity exhibit identical form factors, i.e.,

$$|F_{E\lambda}(q_k, \omega_i)|^2 = |F_{E\lambda}(q_k)|^2, \quad \lambda = 0, 1, 2, \quad \text{and}$$

$$|F_{E2}(q_k)|^2 = |F_{E0}(q_k)|^2.$$
(8)

Assumption (iii) leads so from $I = 250$ independent systems of three linear equations with six unknown quantities to a single system of $3 \cdot I = 750$ nonlinear equations with $2I + 4 = 504$ unknown quantities

$$a_{E1}(\omega_i), \quad a_{E2/E0}(\omega_i) \quad i=1,250$$

$$|F_{E1}(q_k)|^2, \quad |F_{E2/E0}(q_k)|^2 \quad k=2,3,$$
(9)

which is overdetermined for $I > 4$ and can thus be solved by a least squares fit procedure; (the products $a_{E\lambda}(\omega_i) \cdot |F_{E\lambda}(q_1)|^2$, $\lambda = 1$ and $\{0,2\}$, can arbitrarily be normalized to 1).

This approach has only become feasible because of the availability of experimental data, the precision of which is unaffected by any background subtraction. Any possible failure of assumptions (i) through (iii) in any energy bin ω_i will be revealed with extreme sensitivity by a corresponding failure of the fit to reproduce the measured data points of the respective energy bin ω_i.

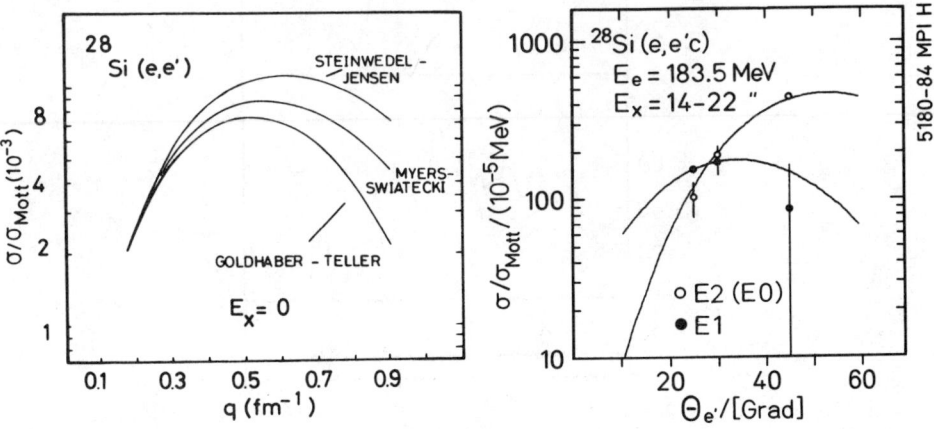

Fig. 7. The predictions of different models[31-33] for the dependence of the E1 form factor on the momentum transfer q (ref.13).

Fig. 8. Deduced E1 and E2/E0 form factors (dots) compared with the respective predictions (curves) of the Goldhaber-Teller and Tassie models (ref. 34).

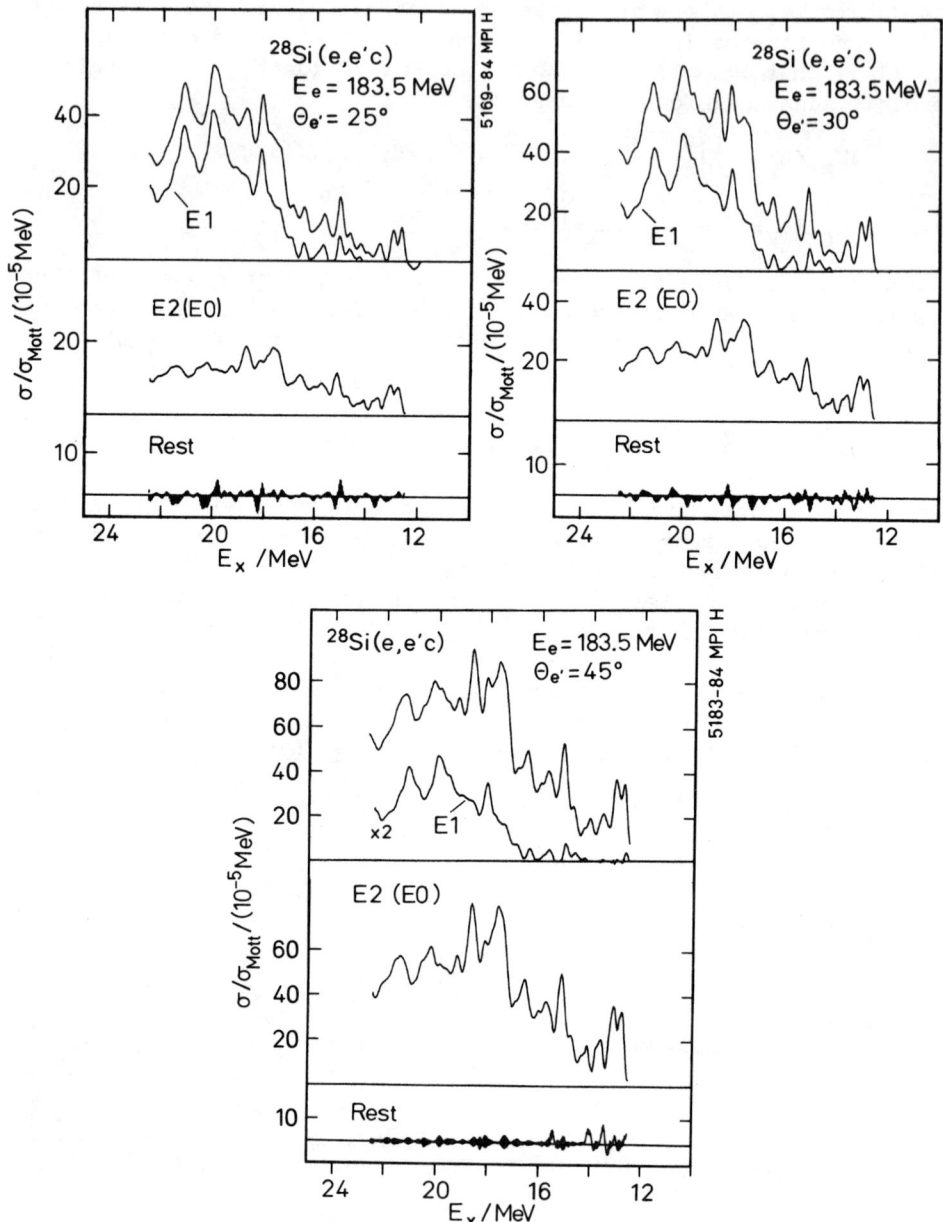

Fig. 9. The 4π integrated ^{28}Si(e,e'c) coincidence cross sections at the three measured momentum transfers in all c=p,α decay channels and their decomposition into the respective E1 and E2/E0 components. "Rest" denotes the difference spectra between measured yields (top curves) and the sum of deduced E1 and E2/E0 cross sections (ref.34).

The deduced integral E1 and E2/E0 form factors are shown in Figure 8. Within their corresponding error bars they are compatible with the respective predictions of the GT and Tassie[35] models calculated[36] in distorted wave Born approximation (DWBA). The SJ prediction is not adequate to describe the form factor of the GDR in ^{28}Si. Using the GT prediction for the extrapolation to the photon point, the E1 strength is found to exhaust $(37\pm2)\%$ of the energy weighted sum rule (EWSR) between $14<\omega<22$ MeV. This value is in good agreement with the results from photodisintegration experiments where the difference $[\sigma(\gamma,abs) - \sigma(\gamma,n)]$ amounts to $[43\%-8\%] = 35\%$ of the EWSR. The E2/E0 strength exhausts $(42\pm2)\%$ of the electromagnetic E2 EWSR, that is, $(84\pm4)\%$ of the isoscalar E2 EWSR, and I will discuss the implications of this result later.

Figure 9 shows the 4π integrated (e,e'c) cross sections of the total $c = p,\alpha$ charged particle channel at the three measured momentum transfers together with their decompositions into the respective E1 and E2/E0 components as determined in the fit procedure. The "error" or "rest" spectra at the bottom represent the difference between the measured cross sections and the sum of the extracted E1 and E2/E0 contributions. These error bands--being small anyway--exhibit for a given energy bin in general no correlated deviations at the three different momentum transfers. This establishes the validity of our assumption (iii) that the E1 and E2/E0 form factors are largely independent of excitation energy. Under the condition that E2 and E0 strengths are uncorrelated, the good agreement between data and fit also establishes that E2 and E0 form factors are identical within the error bars shown in Figure 8. In fact, it remains an open question whether some subtle deviations are due to a difference in form factors or rather due to the presence of higher multipolarities $E\lambda, \lambda > 2$--a problem which could be solved by extending the measurements to higher momentum transfer.

Another independent check of the present multipole decomposition method is provided by comparison of the extracted E1 strength distributions in various decay channels c with the corresponding results from a ^{28}Si(γ,c) photodisintegration experiment[10] using tagged photons. For this comparison, the (γ,c) results were converted to a corresponding (e,e') cross section by a normalization factor which was common for all decay channels c. Using again the GT form factor for the extrapolation to the photon point, the absolute E1 strengths determined in both experiments are found to agree within 20%. Figure 10 reveals the nice detailed agreement between the (e,e'c) results and the (γ,c) data, which is remarkable in view of the strongly structured p_0 and $p_{1,2}$ strength distributions. Reasonable agreement is also observed for the tiny amount of E1 strength in the α_0 channel, even if discrepancies in the reproduction of several fine structure peaks are present.

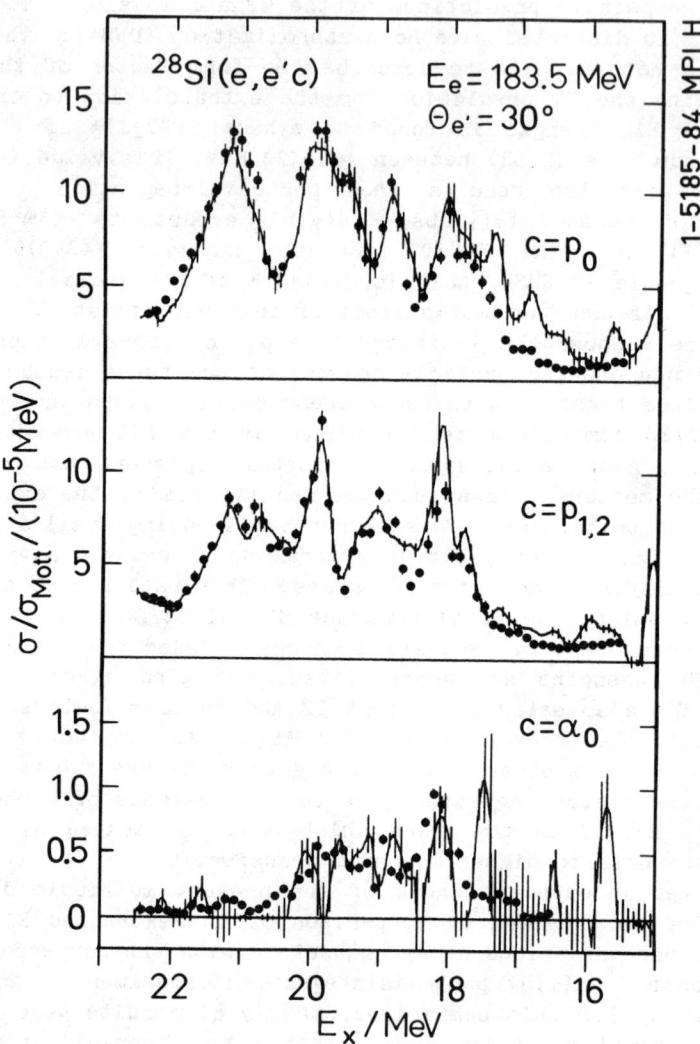

Fig. 10. Deduced E1 strength distributions in various decay channels c (ref. 34). Dots represent for comparison the E1 strength distributions deduced from a $^{28}Si(\gamma,c)$ experiment[10] with tagged photons.

EVIDENCE FOR ISOVECTOR E2/E0 STRENGTH

Between $14 < \omega < 22$ MeV the total extracted E2/E0 strength exhausts (84 ± 4)% of the isoscalar E2 EWSR, which is more than twice the value of (34 ± 6)% deduced from inelastic α scattering[17]. In view of the excellent agreement between the E1 strength distributions from the present multipole analysis and recent photodistintegration results, this observation assumes indisputable significance. Three relevant physical origins come immediately into mind for the explanation of this rather interesting result: (i) extracted transition rates from (α,α') scattering to the GR region are incorrect because of an inadequate subtraction of the hadronic background and/or problems with the DWBA analysis, (ii) a much stronger E0 vs E2 excitation in the (e,e') scattering than in the (α,α') scattering, and (iii) the existence of considerable isovector E2/E0 strength in the region of the isoscalar GQR which would indeed be excited in (e,e') scattering but inaccessible by (α,α') scattering due to the isoscalar nature of this hadronic probe.

Based on a recent ^{28}Si$(\alpha,\alpha'c)$ coincidence study[21] in which the E2 strength in the $c = \alpha_0$ channel has been found to agree reasonably well with the E2 strength deduced from the ^{24}Mg(α,γ_0) capture reaction[12], proposition (i) can be discarded. As to conjecture (ii), recent analysis[4,29] of (e,e') scattering data indicates for low momentum transfer that a given (e,e') cross section corresponds to an E0 EWSR fraction which is roughly twice the value one would obtain if the excitation is assumed to be of E2 instead of E0 multipolarity. A very similar situation is encountered[37] in case of inelastic α scattering at scattering angles not too close to 0°, which seems to rule out issue (ii), too. This judgement should, however, be considered with reservation since it is based only on model assumptions of the relevant transition densities which never have been thoroughly tested by experiment. In fact, the E2/E0 strength distributions of Figure 9 clearly show a prominent peak at the same excitation energy of ~17.6 MeV where in 0° inelastic α scattering an outstanding giant monopole resonance (GMR) structure is observed which at larger scattering angles is no longer prominent (see Figure 2 of ref. 20 and Figure 11, respectively).

In general, however, the E2/E0 strength distributions extracted from the (e,e'c) reaction do **not** resemble the resonance structures observed in the (α,α') reaction at either $\theta_{\alpha'} = 6.5°$ (Figure 11) where the E2 strength shows up strongly, nor at $\theta_{\alpha'} = 0°$ where monopole excitations dominate[20]. While both the GQR and GMR structures observed in (α,α') scattering exhibit a distinct falloff at around 21 MeV of excitation energy, the E2/E0 strength distributions deduced from (e,e') scattering keep rather a flat and high level in

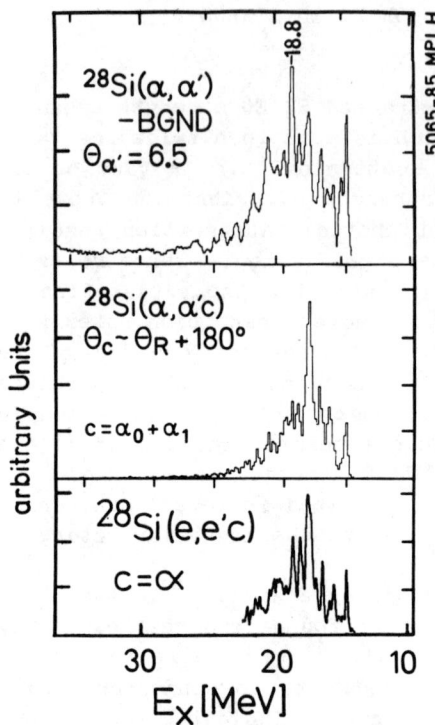

Figure 11. The comparison of the (α,α') singles spectrum and the $(\alpha,\alpha'x)$ and $(e,e'x)$ coincidence spectra in the GQR region of ^{28}Si. In the singles spectrum the background indicated in Fig. 2 is subtracted.

Table II Comparison of the E2/E0 strengths S_2 in various decay channels c as deduced from the present $(e,e'c)$ study[6,34] and a previous $(\alpha,\alpha'c)$[21] experiment in the giant resonance region of ^{28}Si. Quoted values of S_2 are given in fractions of the isoscalar E2 energy weighted sum rule.

c	$S_2(e,e'c)$ (%)	$S_2(\alpha,\alpha'c)$ (%)	Appr. Ratio
α_0	3.5 ± 0.2	3.5 ± 0.9	1.0
α_1	16.9 ± 0.9	8.9 ± 1.9	1.9
α_{rest}	3.1 ± 0.2	3.8 ± 0.9	0.8
p_0	17.8 ± 0.9	7.0 ± 1.6	2.5
$p_{1,2}$	15.6 ± 0.8	6.4 ± 15	2.4
$p_{3,4}$	8.4 ± 0.4	2.9 ± 0.8	2.9
p_5	4.8 ± 0.2	1.6 ± 0.7	3.0
p_{rest}	13.6 ± 0.7	–	–
α_{sum}	24 ± 1	16 ± 3	1.5
p_{sum}	61 ± 3	18 ± 3	3.4
Total	84 ± 4	34 ± 6	2.5
Γ_p/Γ_α	2.5	1.1	2.3

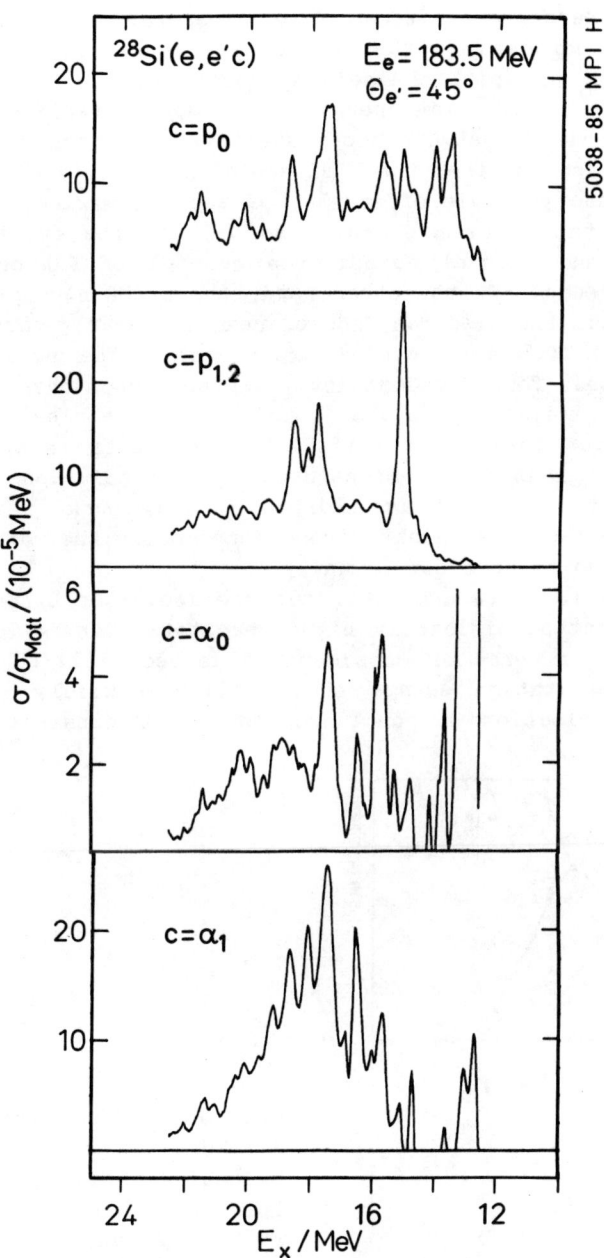

Fig. 12. Deduced E2/E0 strength distributions in various decay channels c (ref. 34).

this energy region (Figure 9). A similar behaviour is observed for the E2/E0 strengths in the c = p_o and $p_{1,2}$ channels (Figure 12). Only the α_1 channel exhibits an obvious resonant shape. Again, this is in marked contrast to the ($\alpha,\alpha'c$) results where both the α_1 as well as the p_o and $p_{1,2}$ channels exhibit a distinct resonant shape below 21 MeV of excitation energy. Even more revealing is the comparison in Table II between the respective strengths in various decay channels deduced from the ^{28}Si(e,e'c) and ^{28}Si($\alpha,\alpha'c$) reactions. Considering the possible effects[38] of isospin mixing, both the absolute strength in the α channels as well as the overall shape of the (e,e'α) and ($\alpha,\alpha'\alpha$) coincidence spectra of Figure 11 exhibit little difference. On the other hand, the E2/E0 strengths in the p channels are in the electron-induced reaction nearly three times the value deduced from the ($\alpha,\alpha'p$) experiment. The excess of E2/E0 strength is thus found predominantly in the proton channels [Γ_p/Γ_α ~ 2.5 and ~1, in (e,e') v. ($\alpha,\alpha'c$)], which provides clear evidence for the presence of large isovector E2/E0 components in the region of the isoscalar giant E2/E0 resonances. This conclusion is confirmed by the results of still unpublished ($\alpha,\alpha'c$) work[22] at $\theta_{\alpha'} = 0°$ where for the isoscalar giant monopole resonance the ratio Γ_p/Γ_α has been determined to be close to unity.

This result is remarkable since the isovector E2/E0 resonances are expected at significantly higher excitation energy of typically 40 MeV. The presence of considerable isovector E2/E0 strength at much lower excitation energy thus implies a widely spread E2/E0 strength distribution--a conclusion which is consistent with the

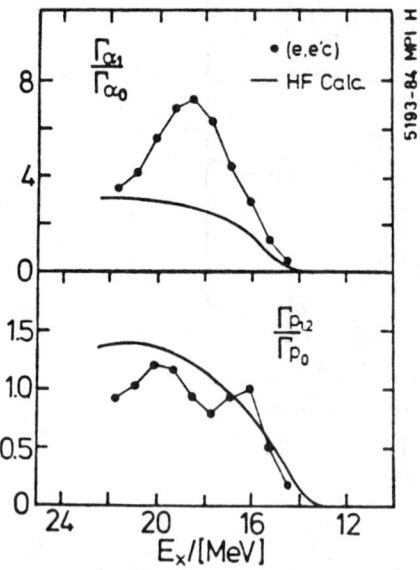

Fig. 13. The comparison (ref. 34) of the measured relative branching ratios $\Gamma\alpha_1/\Gamma\alpha_o$ and $\Gamma p_{1,2}/\Gamma p_o$ (dots) of the giant E2/E0 resonance decays in ^{28}Si with the respective predictions by Hauser-Feshbach calculations (curves).

failure of previous scattering experiments to identify compact isovector E2 strength (see, e.g., the discussion in ref. 39). A large spreading width of the isovector GQR would imply a predominant statistical decay of this mode. Figure 13 shows that the $\Gamma_{p1,2}/\Gamma_{po}$ ratio indeed follows roughly the Hauser-Feshbach prediction. A corresponding comparison for the isoscalar modes is provided by the ratio $\Gamma_{\alpha 1}/\Gamma_{\alpha o}$ due to the isospin selectivity of the α channel. In agreement with the conclusions of previous $^{28}Si(\alpha,\alpha'c)$ studies[21,22], the large deviation of this ratio from the Hauser-Feshbach calculation provides evidence for strong nonstatistical decay modes of the isoscalar GMR and/or GQR.

E0, E1 AND E2 STRENGTH DISTRIBUTIONS FROM THE $(e,e\alpha_o)$ ANGULAR CORRELATION FUNCTIONS

A most model-independent decomposition of the various multipole excitations, including the E0 component, has to resort to the analysis of the measured angular correlation functions (ACFs) since the close similarity of E0 and E2 form factors prevents an application of the multipole expansion method outlined above. ACFs of **isolated** resonances are indeed known to show characteristically different patterns for different spin values of the resonance, but the present study of the nuclear continuum region is naturally confronted with a regime of many overlapping and coherently interfering resonances. Depending thus on the respective decay channel, the general theoretical expression for the ACF may easily contain a prohibitively large number of free parameters which have to be determined from a relatively small number (~25 in the present case) of data points. Due to the particularly simple and unique coupling of angular momenta, I therefore focus here on the analysis of ACFs in the α_o channel only, i.e., the decay of an intermediate electric resonance excitation of spin $J = \lambda$ via emission of a 0^+ α particle to the 0^+ ground state of ^{24}Mg.

In case of an **isolated** resonance of **pure** Coulomb character, the α_o ACF is given in the centre-of-mass (cm) system of the decaying nucleus by

$$W_\lambda(\theta,q,\omega) = |\tilde{\lambda} \cdot C\lambda(q,\omega) \cdot P_\lambda(\cos\theta)|^2 \quad \text{with} \quad \tilde{\lambda} = (2\lambda+1)^{1/2} \quad (10)$$

exhibiting no ϕ dependence. $P_\lambda(x)$, $x = \cos\theta$, is the Legendre polynomial of order λ. If the width of the energy bin ω is sufficiently large with respect to the width of the resonance, the coefficient $|C\lambda(q,\omega)|$ represents—apart from known kinematical factors—the absolute value of the reduced Coulomb matrix element. The **general** expression for the α_o ACF includes the contributions from the transverse electric component with the angle dependence of the squared

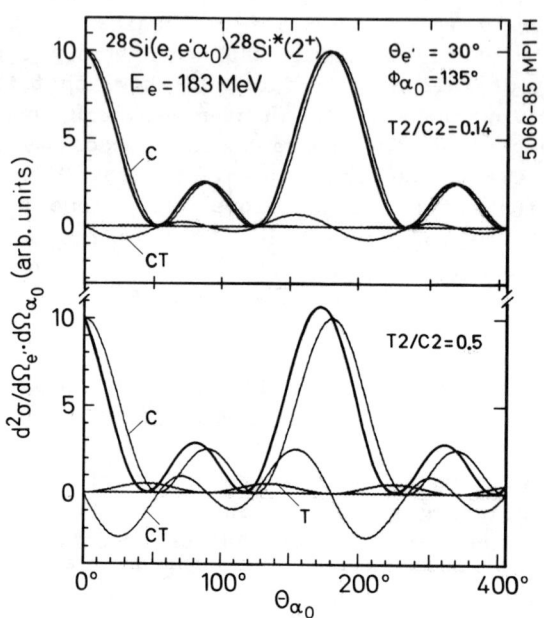

Fig. 14. The theoretical e'α₀ angular correlation functions (dark curves) of a 2^+ resonance excitation with different ratios of the transverse (T2) and longitudinal (C2) matrix elements. Various contributing interference terms are indicated.

first associated Legendre polynomial of order λ, $P_\lambda^1(x)$, as well as the transverse-transverse $[\propto \{P_\lambda^1(x) \cdot \cos 2\phi\}^2]$ and longitudinal-transverse $[\propto \{P_\lambda(x) \cdot P_\lambda^1(x) \cdot \cos \phi\}]$ interference terms (see equation 4.1 of ref. 25 which uses, however, a different (θ,ϕ) coordinate system). A numerical evaluation of this general expression for the present kinematical situation is shown in Figure 14 assuming different ratios of longitudinal and transverse matrix elements. The inclusion of the transverse component results essentially in an angular shift of the ACF pattern obtained for the pure Coulomb case. For a reasonable ratio of transverse and longitudinal matrix elements (see eq. 5 and Figure 14) this shift is less than 1°, i.e., beyond the angular resolution of the present experiment. Larger shifts of a few degrees were indeed not observed and, therefore, the transverse terms were not included in the analysis of the measured ACFs. So far, these considerations are strictly correct for isolated resonances only. In case of several **overlapping** resonances, their decay amplitudes add up coherently, leading to an ACF of the form

$$W(\theta,q,\omega) = | \Sigma_\lambda \ \tilde{\lambda} \cdot C\lambda(q,\omega) \cdot P_\lambda(\cos\theta) |^2. \tag{11}$$

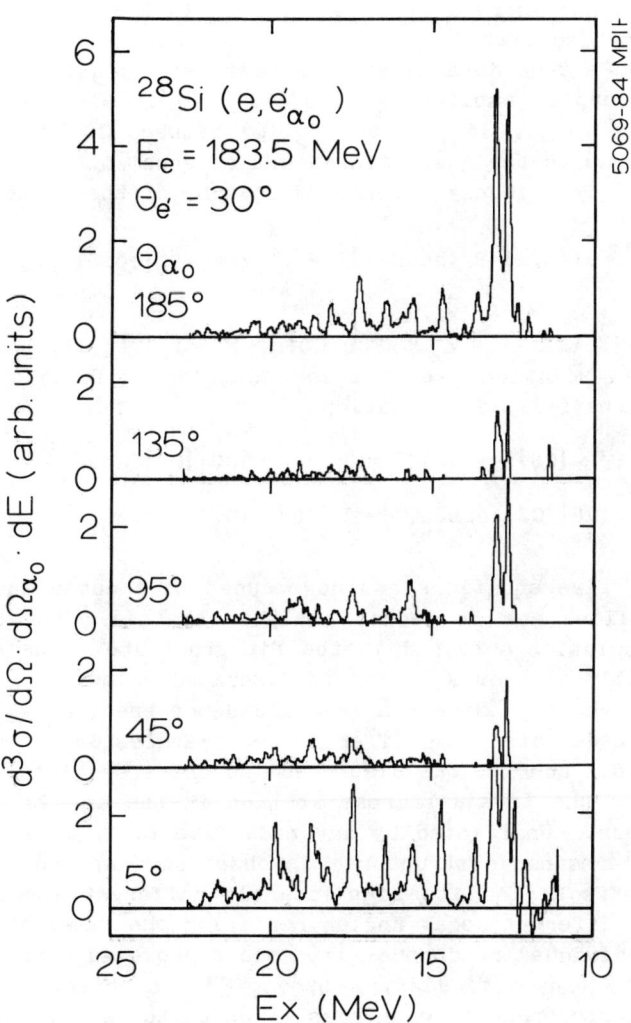

Fig. 15. Spectra of inelastic electron scattering in the GR region of ^{28}Si measured at 30° in coincidence with α_o decay detected at the different indicated angles $\theta\alpha_o$. $\theta\alpha_o$ is measured with respect to the direction of momentum transfer corresponding to $\theta\alpha_o = 0°$ (see figs. 3 and 4).

For interfering resonances of opposite parity, this expression will lead in accordance with Bohr's theorem[40] to a large forward [W(θ = 0°)]-backward [W(θ = 180°)] asymmetry (FBA) of the ACF, even if one resonance amplitude is relatively small[38]. In the present analysis the sum is extended over λ values of 0, 1, and 2. The fit to the experimental ACFs thus determines 5 parameters: the absolute values of the three complex amplitudes $C0(q,\omega)$, $C1(q,\omega)$, $C2(q,\omega)$, and two relative phases, e.g., $\delta_{01}(\omega)$ and $\delta_{02}(\omega)$ between C0 and C1 or C2, respectively. Due to the quadratic nature of equation (11), only the term $|C2(q,\omega)|^2$ is uniquely determined because of the identity

$$\left| \sum_{\lambda=0}^{2} \tilde{\lambda} \cdot C\lambda(q,\omega) \cdot P_\lambda(\cos\theta) \right|^2 = \sum_{\lambda=0}^{4} a_i \cdot P_i(\cos\theta) \quad (12)$$

with 5 $|C(q,\omega)|^2$ (18/7) = a_4. For $C0(q,\omega)$ and $C1(q,\omega)$, however, in general several solutions exist which identically fulfill equation (11) and obey the following relations[41]

$$|C2|^2 = k_1 \; ; \quad |C0|^2 + |C1|^2 = k_2 \; ; \quad |C0||C1|\cos\delta_{01} = k_3 \; ;$$

$$|C0|^2 - 5^{1/2}|C0||C2|\cos\delta_{02} = k_4 \; ; \quad |C1|\cos(\delta_{02} - \delta_{01}) = k_5 \; ;$$

This system of five equations can be reduced to a cubic equation in $|C0|^2$ which allows one to deduce analytically all other solutions from a first solution obtained in the fit procedure. The selection of the physically correct solution is discussed below.

Figure 15 shows ^{28}Si(e,e'α_o) coincidence spectra with the α_o particle being detected at different decay angles $\theta\alpha_o$. Due to the small ^{28}Si recoil energy, the displayed lab cross sections are nearly identical to the (e,e'α_o) cross section in the cm system of the decaying nucleus. Unaffected by any radiative tail, a large number of overlapping resonance structures is observed. Apparently strong interference effects lead to significantly different spectral distributions at different decay angles rendering the immediate identification of individual resonances impossible above an excitation energy of 14 MeV. In particular, a strong FBA is observed, indicating[40] the interference of resonances of opposite parity. A similar FBA has been observed in a recent ^{28}Si(α,α'α_o) experiment[21,42] at q = 0.7 fm^{-1}; other than in this hadron-induced reaction, however, α knockout contributions providing a trivial FBA appear in (e,e'α_o) to be much weaker. This is obvious from the apparent lack of appreciable continuous strength in the spectra close to the q direction and understandable, at least in part, from the lower momentum transfer in the present (e,e') experiment.

Figure 16 reveals the rich variety of ACF patterns observed in different regions of excitation energy. The solid curves represent

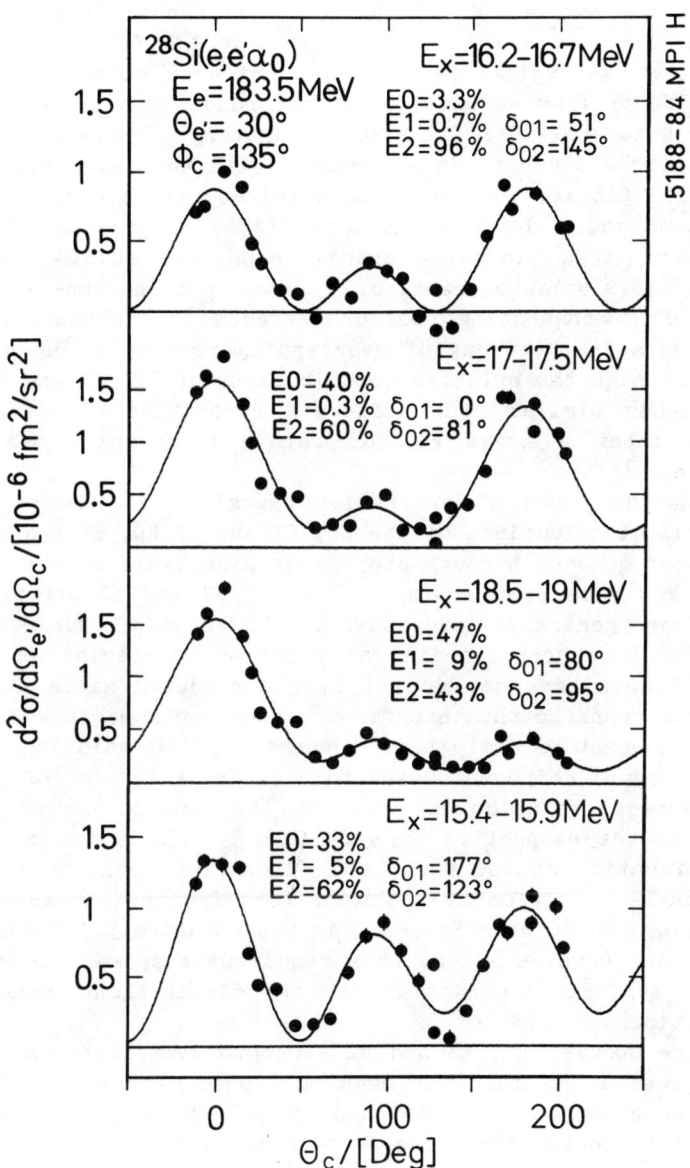

Fig. 16. Typical (e,e'α_o) angular correlation functions measured in different regions of excitation energy E_x and the comparison with the best fits (curves) to the theoretical expression for interfering E0, E1 and E2 resonances. The relative E0, E1 and E2 contributions to the 4π integrated cross section are indicated as well as the relative phases δ_{01}, δ_{02} between E0 and E1 or E2 amplitudes (ref. 34).

fits to the expression of equation (11) assuming interfering resonances of 0^+, 1^+ and 2^+ spin. Between $16.2 < E_x < 16.7$ MeV, the measured ACF shows the typical $[P_2(\cos\theta)]^2$ pattern which is characteristic for α_0 decay from an isolated 2^+ resonance. A similar pattern is observed between $17.0 < E_x < 17.5$ MeV indicating the presence of appreciable monopole strength in accordance with the naive expectation; an identical fit is, however, also obtained with a much smaller (2%) E0 component and a largely increased (38%) E1 contribution. The two remaining ACFs in the energy bins of $18.5 < E_x < 19.0$ MeV and $15.4 < E_x < 15.9$ MeV show distinctly asymmetric patterns with respect to $\theta\alpha_0 = 0°$ demonstrating possible effects of non-vanishing interference terms in the decay of overlapping resonances of opposite parity; although the relative contributions of E0, E1 and E2 amplitudes are rather similar, the different respective phases are causing, e.g., either a large FBA or rather a strongly oscillatory ACF pattern.

In the range of excitation energies from 14 to 22 MeV, the spectral distributions of the E0, E1 and E2 E0, E1 and E2 multipoles have been deduced through step-by-step analysis of the $(e, e'\alpha_0)$ ACFs in 80 keV wide energy bins. The E0, E1 and E2 strength distributions were generated independently for the two lower measured momentum transfers corresponding to electron scattering at 25° and 30°; the 45° data have so far not been considered since at this higher momentum transfer the interference contributions from E3 excitations can no longer be neglegted. Generally, two solutions for each of the E0 and E1 components were found. The selection of the physically "correct" solution is based on the deduced **correlated** q dependence of the respective form factors. This allows the unambiguous classification of "correct" and "incorrect" solutions according to the single criterion that the E1 form factor increases more slowly than does the E0 form factor. At the relatively low momentum transfers under consideration, this requirement appears indeed to represent a sufficient condition for the distinction between E0 and E1 excitations.

The deduced E0, E1 and E2 strength distributions are displayed in Figures 17 to 19. For each multipole $E\lambda$ the average of the results obtained at $\theta_{e'} = 30°$ and 25° after integral normalization of the latter to the 30° cross section is given:

$$\bar{\sigma}_{E\lambda}(\theta_{e'}=30°, \omega) = (1/2)[\sigma_{E\lambda}(30°, \omega) + f_{E\lambda} \cdot \sigma_{E\lambda}(25°, \omega)] \,. \quad (14)$$

$f_{E\lambda}$ represents essentially an **integral** form factor of the considered $E\lambda$ excitation. The displayed error bars are the deviation of the respective distributions:

$$\Delta\bar{\sigma}_{E\lambda}(\theta_{e'}=30°, \omega) = (1/2)[\sigma_{E\lambda}(30°, \omega) - f_{E\lambda} \cdot \sigma_{E\lambda}(25°, \omega)] \,. \quad (15)$$

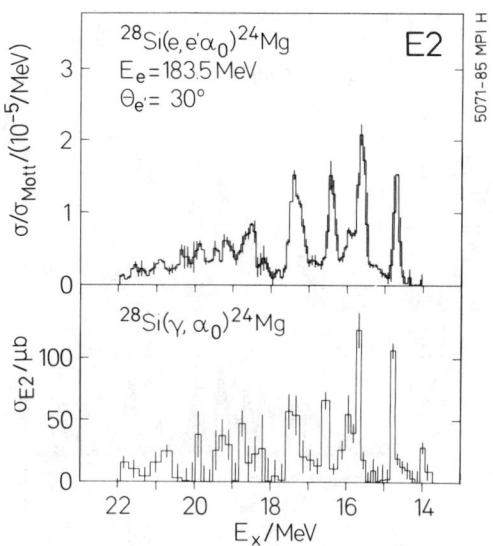

Fig. 17. The E2 strength distributions in ^{28}Si from the analysis of e'α_0 ACFs (ref. 41) in comparison with the results deduced from the ^{24}Mg(α,γ_0)^{28}Si capture reaction[12].

Larger deviations occur in the slopes of the resonance structures only, indicating a remarkable overall state independence of the E0, E1 and E2 form factors. The integral E1 and E2 form factors are found to be in excellent agreement with predictions of the GT and Tassie models (see Figure 8). The integral E0 form factor exhibits a slightly smaller slope than does the E2 form factor. The deduced E2 strength distribution is in excellent agreement with the corresponding results obtained by the ^{24}Mg(α,γ_0)^{28}Si capture reaction[12] (Figure 17). The E1 strength distribution extracted from the e'α_0 ACFs confirms the results from the multipole decomposition of the 4π integrated (e,e'α_0) coincidence cross sections (Figure 18). For E_x > 18 MeV good agreement is also observed with the E1 strength distribution obtained from the ^{28}Si(γ,α_0) study[10]; the origin of distinct deviations at lower excitation energies has not yet been found. Figure 19 provides clear evidence for the existence of rather compact E0 strength throughout the region of the GQR in ^{28}Si. The present analysis confirms the existence of a 0^+ resonance state observed in inelastic α scattering[19] at E_x = 15.9 MeV. In terms of the respective EWSRs, the E0 and E2 giant resonances exhibit in ^{28}Si a coupling to the α_0 channel of comparable strength. Both being $2\hbar\omega$ excitations, this observation may indicate a very similar microscopic structure of GMR and GQR in ^{28}Si.

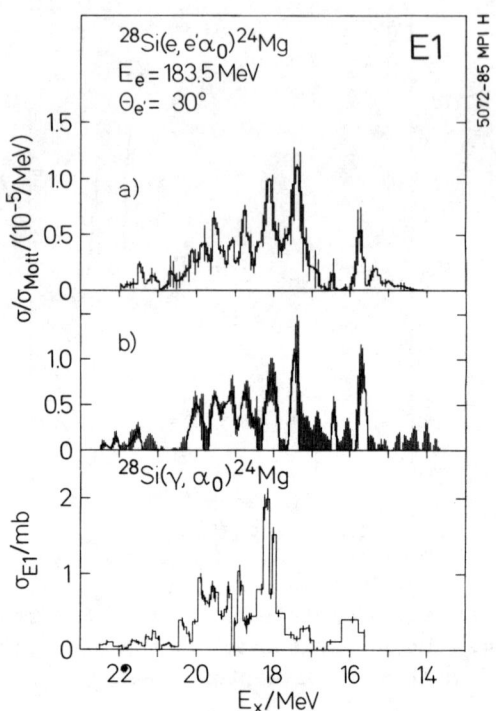

Fig. 18. The E1 strength distributions in ^{28}Si from the a) analysis of e'α_o ACFs (ref. 41) in comparison with b) the results[34] from a multipole decomposition of the 4π integrated (e,e'α_o) coincidence cross sections as well as the results from a ^{28}Si(γ,α_o) study[10].

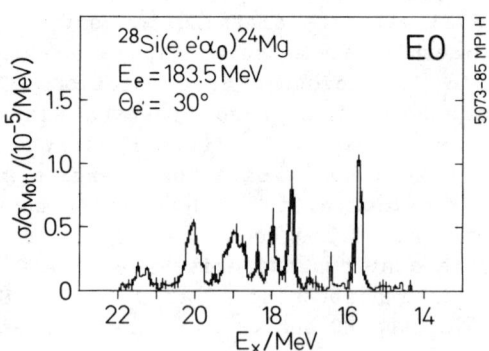

Fig. 19. The E0 strength distributions in ^{28}Si deduced from the analysis of e'α_o ACFs (ref. 41).

ACKNOWLEDGEMENTS

It is a pleasure to acknowledge all my colleagues who have contributed to this work. The experiments were carried out at the Mainz microtron MAMI A in collaboration with H.J. Emrich, G. Fricke, R. Neuhausen, R.K.M. Schneider (Institut für Kernphysik der Universität Mainz), J.R. Calarco (University of New Hampshire), Th. Kihm, H. Riedesel, and P. Voruganti (Max-Planck-Institut für Kernphysik, Heidelberg). I wish to extend my thanks to my other present associates at Heidelberg, Sh. Khan, M. Spahn and H.J. Schulz, for the always pleasant and highly motivated collaboration in this exciting new field of coincidence electron scattering. The professional typing of the manuscript by J. Long is gratefully acknowledged.

REFERENCES

1. T.W. Donnelly and J.D. Walecka, Ann. Rev. Sci. **25**, 9 (1975).
2. H. Überall, "Electron Scattering from Complex Nuclei," Academic Press, New York, (1979) Volumes A and B.
3. D.H. Dowell, L.S. Cardman, P. Axel, G. Bolme, and S.E. Williamson, Phys. Rev. Lett. **49**, 113 (1982).
4. K.A. Griffioen, P.J. Countryman, K.T. Knöpfle, K. Van Bibber, M.R. Yearian, J.G. Woodworth, D. Rowley, and J.R. Calarco, Phys. Rev. Lett. **53**, 2382 (1984).
5. J.R. Calarco, J. Arruda-Neto, K.A. Griffioen, S.S. Hanna, D.H.H. Hoffmann, B. Neyer, R.E. Rand, K. Wienhard, and M.R. Yearian, Phys. Lett. **146B**, 179 (1984).
6. Th. Kihm, K.T. Knöpfle, H. Riedesel, P. Voruganti, H.J. Emrich, G. Fricke, R. Neuhausen, K.R.M. Schneider, and J.R. Calarco, Proc. XIth Europhys. Div. Conf. on Nucl. Phys. with Electromagnetic Probes, Paris 1985, p.240; and to be published.
7. J. Ahrens, H. Borchert, K.H. Czock, H.B. Eppler, H. Gimm, H. Gundrum, M. Kröning, P. Riehn, G. Sita Ram, A. Zieger, and B. Ziegler, Nucl. Phys. **A251**, 479 (1975).
8. A. Veyssiere, H. Beil, R. Bergere, P. Carlos, and A. de Miniac, Nucl. Phys. **A227**, 513 (1974).
9. R.E. Pywell, B.L. Berman, J.W. Jury, J.G. Woodworth, K.G. McNeill, and M.N. Thompson, Phys. Rev. **C27**, 960 (1983).
10. R.L. Gulbranson, L.S. Cardman, A. Doron, A. Erell, K.R. Lindgren, and A.I. Yavin, Phys. Rev. **C27**, 470 (1983).
11. P.P. Singh, R.E. Segel, L. Meyer-Schützmeister, S.S. Hanna, and R.G. Allas, Nucl. Phys. **65**, 577 (1965).
12. E. Kuhlmann, K.A. Snover, G. Feldmann and M. Hindi, Phys. Rev. **C27** 948 (1983).
13. R. Pitthan, F.R. Buskirk, J.N. Dyer, E.E. Hunter and G. Pozinsky, Phys. Rev. **C19**, 299 (1979).
14. A. Friebel, doctoral thesis, Technische Hochschule Darmstadt, (1981), unpublished.
15. F.E. Bertrand, Ann. Rev. Nucl. Sci. **26**, 457 (1976).

16. S. Kailas, P.P. Singh, A.D. Bacher, C.C. Foster, D.L. Friesel, P. Schwandt, and J. Wiggins, Phys. Rev. **C25**, 1263 (1982).
17. K.T. Knöpfle, G.J. Wagner, A. Kiss, M. Rogge, C. Mayer-Böricke, and Th. Bauer, Phys. Lett. **64B**, 263 (1976).
18. D.H. Youngblood, C.M. Rosza, J.M. Moss, D.R. Brown, and J.D. Bronson, Phys. Rev. **C15**, 1644 (1977).
19. K. van der Borg, M.N. Harakeh, S.Y. van der Werf, A. van der Woude, and F.E. Bertrand, Phys. Lett. **67B**, 405 (1977).
20. Y.-W. Lui, J.D. Bronson, D.H. Youngblood, Y. Toba, and U. Garg, Phys. Rev. **C31**, 1643 (1985).
21. K.T. Knöpfle, H. Riedesel, K. Schindler, G.J. Wagner, C. Mayer-Böricke, W. Oelert, M. Rogge, and P. Turek, Phys. Rev. Lett. **46**, 1372 (1981).
22. Y. Toba, U. Garg, Y.-W. Lui, D.H. Youngblood, P. Grabmayr, K.T. Knöpfle, H. Riedesel, and G.J. Wagner, to be published.
23. T. de Forest, Ann.Phys. **45**, 365 (1967).
24. D. Drechsel and H. Überall, Phys. Rev. **181**, 1383 (1969).
25. W.E. Kleppinger and J.D. Walecka, Ann.Phys. **146**, 349 (1983).
26. T.W. Donnelly, in Symmetries in Nuclear Structure, edited by K. Abrahams, K. Allaart, and A.E.L. Dieperink (Plenum, N.Y., 1983) p. 1.
27. H. Herminghaus, A.Feder, K.H.Kaiser, W. Manz, and H.v.d.Schmitt Nucl. Instr. Meth. **138**, 1 (1976).
28. H. Ehrenberg, H. Averdung, B. Dreher, G. Fricke, H. Herminghaus, R. Herr, H. Hulzsch, G. Lürs, K. Merle, R. Neuhausen, G. Nöldecke, H.M. Stolz, V. Walther, and H.D. Wohlfahrt, Nucl. Instr. Meth. **105**, 253 (1972).
29. S. Fukuda and Y. Torizuka, Phys. Lett. **62B**, 146 (1976).
30. M. Sasao and Y. Torizuka, Phys. Rev. **C15**, 217 (1977).
31. M. Goldhaber and E. Teller, Phys. Rev. **74**, 1046 (1948).
32. H. Steinwedel and H. Jensen, Z. Naturforschung, **5a**, 413 (1950).
33. W.D. Myers, W.J. Swiatecki, T. Kodama, L.-J. El-Jaick, and E.R. Hilf, Phys. Rev. **C15**, 2032 (1977).
34. Th. Kihm, doctoral thesis, Heidelberg (1985), unpublished.
35. L.J. Tassie, Austr. J. Phys. **9**, 407 (1956).
36. H.J. Emrich, private communication.
37. M.N. Harakeh, K. van der Borg, T. Ishimatsu, H.P. Morsch, A.van der Woude, and F.E. Bertrand, Phys. Rev. Lett. **38**, 676 (1977).
38. K.T. Knöpfle, Lecture Notes in Physics **108**, 311 (1979).
39. A. Erell, J. Alster, J. Lichtenstadt, M.A. Moinester, J.D. Bowman, M.D. Cooper, F. Irom, H.S. Matis, E. Piasetzky, U. Sennhauser, and Q. Ingram, Phys. Rev. Lett. **52**, 2134 (1984).
40. A. Bohr, Nucl. Phys. **10**, 486 (1959).
41. M. Spahn, Diplomarbeit, Heidelberg, (1986), unpublished.
42. K.T. Knöpfle and G.J. Wagner, Phys. Rev. **C27**, 2422 (1983).

DECAY OF THE ISOSCALAR GIANT MONOPOLE RESONANCE IN ^{208}Pb

A. VAN DER WOUDE
Kernfysisch Versneller Instituut, 9747 AA Groningen
The Netherlands

1. INTRODUCTION

The existence of an isoscalar giant monopole resonance (GMR) in heavy nuclei (A ≳ 100) is now well established[1,2]. Not much is known though about the GMR decay properties[3,4] nor about the decay of the other resonances[5-10]. Yet this is of great importance for a good understanding of the microscopic structure of these fundamental excitations, as has been emphasized in several theoretical papers[11-14].

As a first approximation one can write the total width Γ of a resonance as the sum of three different components:

$$\Gamma = \Delta\Gamma + \Gamma^{\uparrow} + \Gamma^{\downarrow}$$

Here $\Delta\Gamma$, the Landau damping, is due to the fact that in a realistic RPA calculation the intrinsic (1p-1h) collective state, being the doorway state for the giant resonance, may be already fragmented. This is especially true for deformed nuclei. The component $\Gamma^{\uparrow}$ is due to the coupling of the (1p-1h) doorway state to the continuum while $\Gamma^{\downarrow}$ results from coupling to (2p-2h) states. These (2p-2h) configurations can decay by nucleon emission ($\Gamma^{\downarrow\uparrow}$) or act as doorway states to even more complicated (3p-3h) ... (np-nh) configurations ($\Gamma^{\downarrow\downarrow}$). In the last case the final result is an equilibrated system which decays statistically. Thus $\Gamma^{\downarrow} = \Gamma^{\downarrow\uparrow} + \Gamma^{\downarrow\downarrow}$, where $\Gamma^{\downarrow\uparrow}$ is the pre-equilibrium component.

In heavy nuclei like ^{208}Pb for which $\Delta\Gamma$ is small, various calculations predict that $\Gamma^{\downarrow} \gg \Gamma^{\uparrow}$, so that $\Gamma \simeq \Gamma^{\downarrow}$ is mainly determined by coupling to the appropriate (2p-2h) states. Of special interest is the excitation of low-lying surface vibrations by the particle (hole) of the (1p-1h) state constituing the giant resonance[7,11]. This process is also important in the damping of single particle excitations like the deep-hole states[11]. The coupling between the high-lying giant resonances and the low-lying surface vibrations is more complicated though than for these hole-states because of the possible interference between the (particle-vibration) coupling with the hole as a spectator and the (hole-vibration) one with the particle as spectator. Detailed calculations indicate[11] that for ^{208}Pb the coupling occurs mainly to the 3^- state at E_x=2.65 MeV and that especially for the very collective GMR the interference effect cancels to a large extent the particle-phonon and hole-phonon coupling contributions.

In nucleon decay the various components can be experimentally distinguished, at least in principle, by measuring the residual (A-1) spectrum. Semi-direct decay will populate (neutron) hole states which can be easily located by means of a neutron pick-up reaction on the nucleus A. Statistical decay can be recognised as such by comparing the measured spectrum with the one calculated by means of a Hauser-Feshbach calculation. As has recently been pointed out[15,16], in order to calculate the population of low lying states in the (A-1) nucleus, one should in such calculations use the actual level scheme and not rely on an extrapolated level-density formula as has often been done[4,7,8]. The result of pre-equilibrium decay (the $\Gamma^{\downarrow\uparrow}$ component) is difficult to predict except for the part due to the coupling of such vibrations which in nucleon decay reveals itself by enhanced population of (phonon-hole) coupling states in the (A-1) nucleus.

A special case is the fission decay of giant resonances in actinide nuclei. Fission is assumed to be the result of a complicated process from an equilibrated system and as such fission decay can be identified with the statistical decay component $\Gamma^{\downarrow\downarrow}$.

This paper is concerned with the GMR neutron decay in ^{208}Pb[17]. Both from the experimental and theoretical point of view ^{208}Pb is an attractive nucleus to study. For decay studies it has the advantage that there are several well-known giant resonances[1] for which the Landau damping is small so that $\Gamma = \Gamma^{\uparrow} + \Gamma^{\downarrow}$. It also has a well-known vibrational state, the $J^{\pi}= 3^-$ state at E_x = 2.65 MeV,

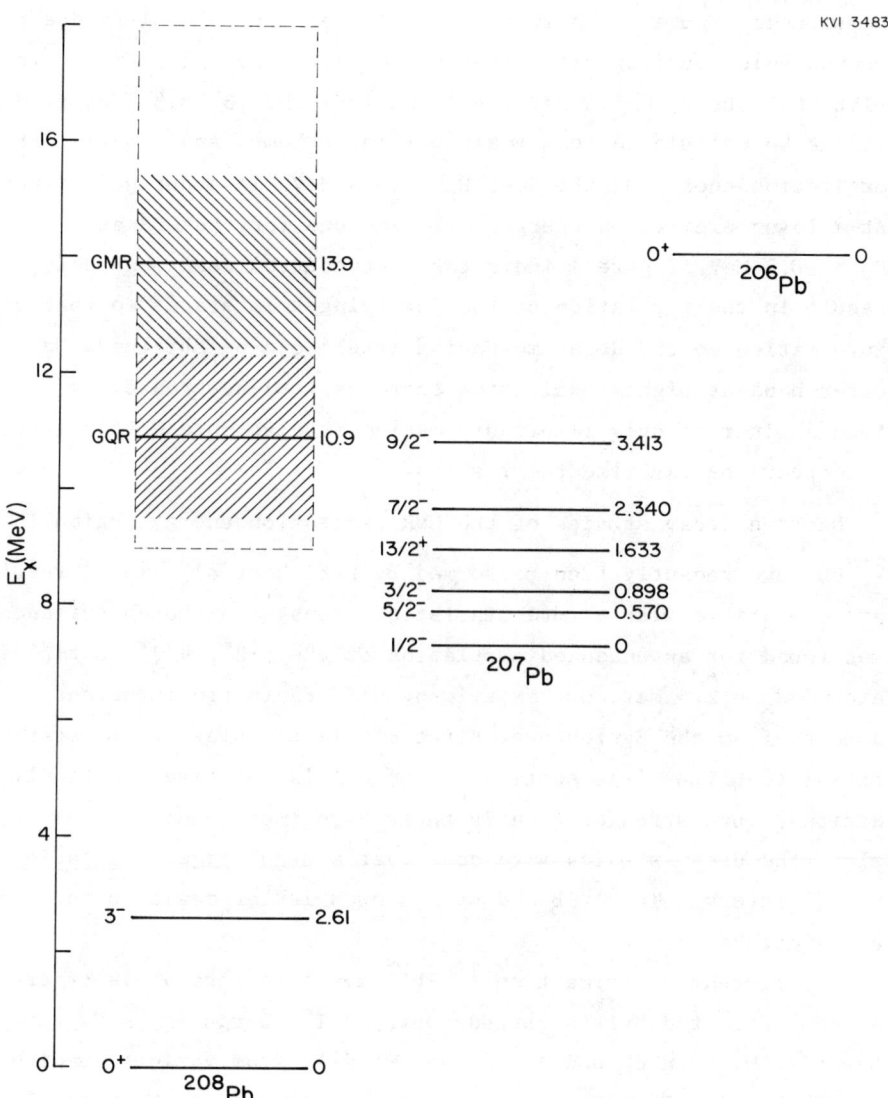

Fig. 1. The relevant level scheme of the Pb isotopes involved in this experiment.

which makes the ^{208}Pb nucleus an attractive one to study the role of the (quasi-particle phonon) coupling in the damping. The level scheme of the residual nucleus ^{207}Pb with its hole state is well known (see figure 1). Also the location of the multiplets due to phonon-hole coupling are either known, like the $5/2^+$, $7/2^+$ multiplet at about E_x = 2.5 MeV resulting from the $[3^- \times \frac{1}{2}^-]$ coupling, or can be calculated in a weak-coupling scheme. And finally, the excitation energy of the GMR, $E_x \sim 13.5$ MeV, is about just right. At a lower excitation energy, like the one for the GQR at $E_x \sim 10.9$ MeV, figure 1 indicates that neutron decay can only result in the population of the low-lying hole states so that no information on the decay mechanism itself can be obtained. On the other hand at higher excitation energies, two-neutron decay becomes increasingly important, making it also more difficult to interpret the experimental results.

Neutron decay studies of the GMR excitation energy region in ^{208}Pb have recently been performed by Eyrich et al[4]. Their results are compatible with a pure statistical decay[15] although evidence was found for an enhanced population of the $5/2^+$, $7/2^+$ multiplet around E_x = 2.6 MeV. Our experiment differs in two important aspects from the Eyrich one. First of all by using the 0° scattering technique (see section 2) the GMR is relatively strongly excited, much stronger than in their experiment. And in the second place the decay studies were done over a much larger excitation energy interval in ^{208}Pb and over a much larger region of neutron energies.

The present experiment on ^{208}Pb[17] was a collaborative effort of a group from the KVI (S. Brandenburg, W.T.A Borgholt, A.G. Drentje, M. Harakeh and A. van der Woude), from various Swedish institutes (T. Ekström, A. Hakanson, L. Nilsson and N. Olsson) and Italian universities (R. de Leo and M. Pignanelli).

2. EXPERIMENTAL METHOD

It is well-known that the GMR is selectively and strongly excited by inelastic alpha-scattering at 0°. Inelastic scattering at very small angles requires a magnetic spectrograph to separate the scattered particles from the beam. This technique has been pioneered by Youngblood et al[18] and has also been extensively used by the Grenoble group in their (^{3}He,^{3}He') experiments[2]. It requires a very careful set-up of (cyclotron + beamline) system in order to avoid beam halo, which would give rise to a high instrumental background. At the KVI we have developed a set-up for 120 MeV α-scattering at 0° which gives a small if not negligible instrumental background. Recently we have done the same for 150 MeV α-beam.

Our system allows us to measure the energy-bite $9 < E_x < 18$ MeV over the angular range $0° < \theta_{\alpha'} < 3°$, while $\theta_{\alpha'}$ can be reconstructed on an event by event base[3,17] with an accuracy of about 0.6°. Thus we are able to separate the data in two parts, a "core" part and a "lateral" part corresponding to $0 < \theta_{\alpha'} < 1.5°$ and $1.5° < \theta_{\alpha'} < 3°$ respectively. The difference between the "core" and "lateral" spectra is mainly due to GMR excitation[3,17]. This difference-of-spectra technique to enhance the GMR has been applied in both cases, ^{208}Pb and ^{238}U.

The Pb (α,α'n) experiment was performed[17] by measuring the inelastically scattered α-particles over $0° < \theta_{\alpha'} < 3°$ in coincidence with neutrons detected with several liquid scintillator detectors with a thickness of 5 cm and a diameter of 30 cm at a distance of 150 cm from the target. The neutron detectors were at 45°, 90, 107, 121, 135, 150 and 160° with respect to the beam direction. Pulse shape discrimination was used to separate neutrons from γ's. The neutron time-of-flight was measured with respect to the cyclotron RF. The overall time resolution obtained was better than 2 ns which corresponds for 7 MeV neutrons to an energy resolution of better than 500 keV. The ^{208}Pb target had a thickness of 5.5 mg/cm^2 and was enriched to 97.8%. In order to keep a good ratio of true to accidental coincidences the beam current was limited to 10 nA.

3. THE ^{208}Pb (α,α'n) EXPERIMENT

3.1. The experimental results

A useful way to present the three-dimensional α-neutron

coincidence data is by constructing on an event by event basis Q-value spectra according to:

$$E_x = \Delta E_\alpha - B_n - E_n$$

Here $\Delta E_\alpha = E_{exc}$ is the excitation energy in ^{208}Pb and B_n, E_n are the binding and the kinetic energy of the neutron respectively. These spectra show the final state population in ^{207}Pb and can be constructed as a function of E_{exc}, the α-scattering angle θ_α and the neutron detection angle θ_n.

One of the difficulties in determining the characteristics of particle decay of resonances is the possibility of interference with other processes like the knock-out process[8]. In light nuclei clear evidence for interference in charged particle decay has been found[16,19,20]. The knock-out process is especially important at forward angles as demonstrated in figure 2. Here the Q-value spectra are shown for the full opening angle and over the whole excitation energy range of 9 to 18 MeV for θ_n =45° and 135° respectively, while figure 2^c shows the difference between the two spectra. Clearly at forward angles the knock-out process strongly populates the known hole states. Above E_x~3.5 MeV, the two spectra are very similar, indicating that for these excitation energies no concentrated hole strength is present. In order to minimise knock-out effects only the data for $\theta_n \geq 90°$ were used in the resonance-decay studies.

Figure 3 shows Q-value spectra for $0° \leq \theta_\alpha \leq 3°$, summed over all neutron detectors with $\theta_n \geq 90°$ and for the ^{208}Pb excitation energy regions $10 \leq E_x \leq 12.5$ MeV (GQR), $12.5 \leq E_x \leq 15.5$ MeV (GMR) and $15.5 \leq E_x \leq 18$ MeV respectively. Decay from the GQR region only populates the low-lying hole states which can be understood from figure 1. Remarkably is the relative strong population of the $13/2^+$ state which is an indication of the presence of $L \geq 4$ strength in this excitation energy range[4,5].

For the highest excitation energy range, $15.5 \leq E_x \leq 18$ MeV the spectrum shows a typical strong evaporation-like component but in addition there is a small surplus of counts populating the region of the low-lying hole states ($E_x < 3.5$ MeV) and also the region

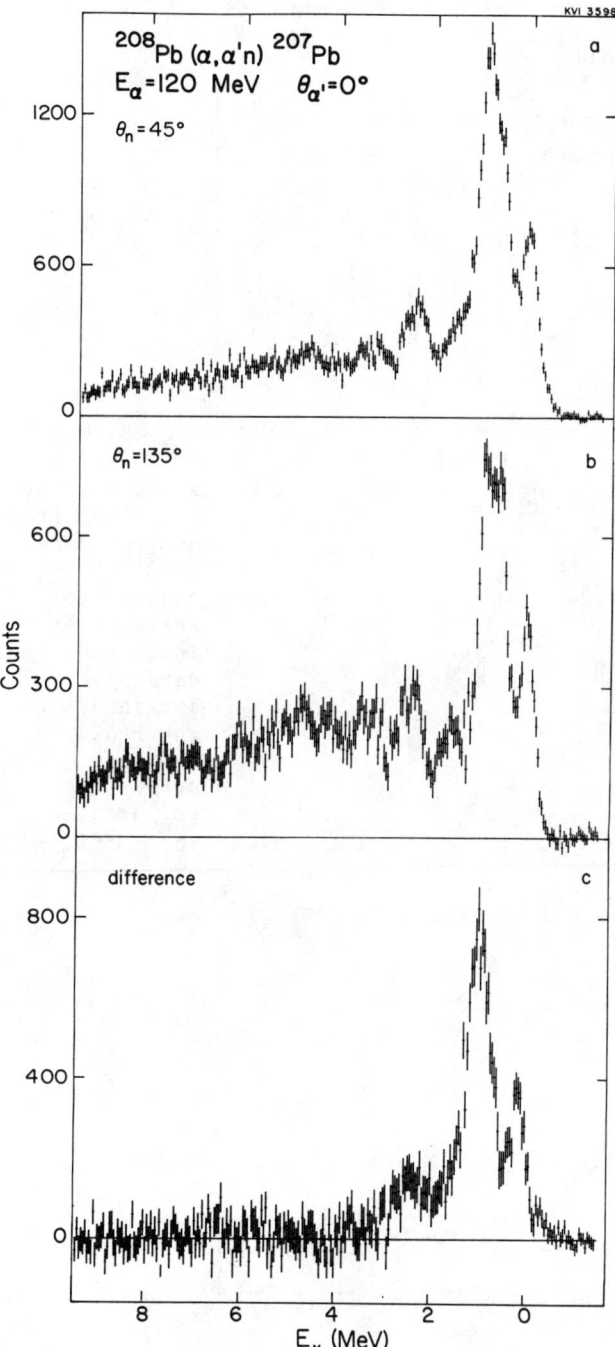

Fig. 2. The ^{207}Pb final state spectrum resulting from neutron decay of the 15.5 ≤ E_x ≤ 18 MeV excitation energy range in ^{208}Pb and for −3° < θ_α < +3°. The difference spectrum is presumably due to knock-out processes.

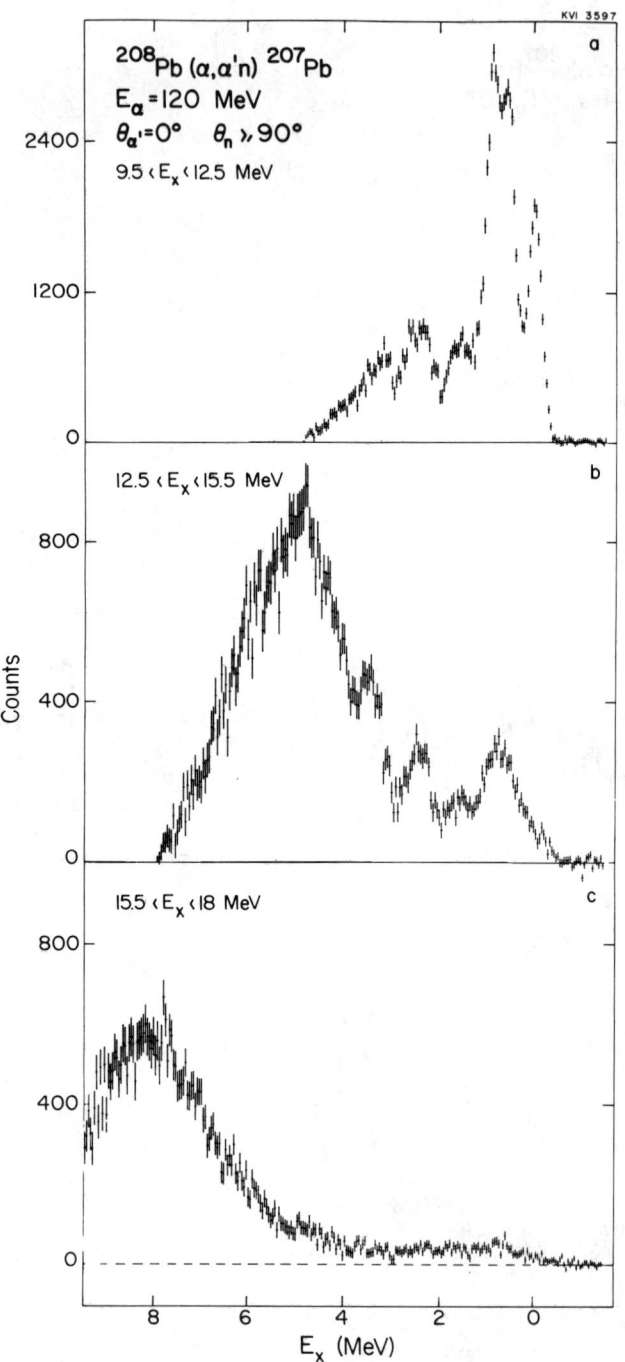

Fig. 3. The ^{207}Pb final state spectrum resulting from neutron decay of ^{208}Pb excited to energies as indicated. The data are obtained with the backward neutron detectors ($\theta_n > 90°$) and for $-3° < \theta_\alpha < +3°$.

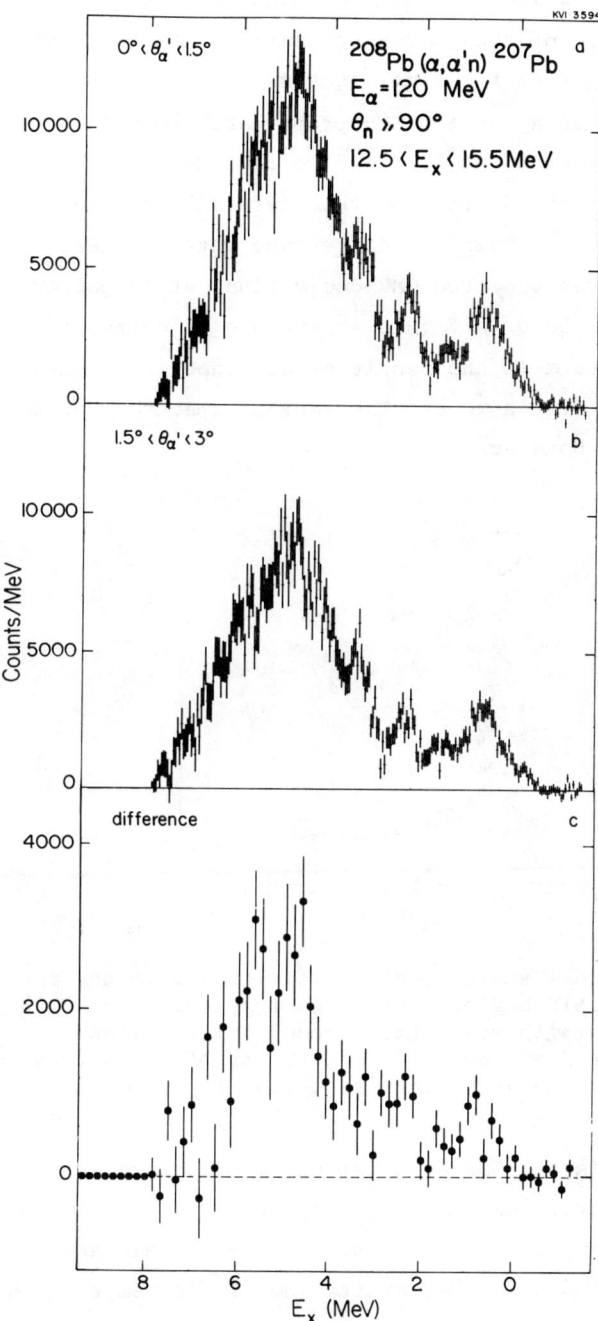

Fig. 4. The ^{207}Pb from neutron decay of the GMR region in ^{208}Pb for a) the 'core' part, b) the 'lateral' part and c) the difference between the 'core' and 'lateral' spectra, corresponding to the decay of the GMR.

between 4 and 5 MeV excitation in ^{207}Pb. The data for the GMR-region show a mixture of the two: a statistical component and a component populating specific final states.

The data for the decay of the GMR proper are shown in figure 4 which are Q-value spectra for $12.5 \leq E_x \leq 15.5$ MeV in ^{208}Pb. Figures 4^{a,b} are for the "core" and the "lateral" part respectively, while figure 4^c shows the difference between these two spectra and thus gives the true GMR-decay final state pattern. The relative large error bars in figure 4^c are due of course to the fact that this spectrum is the result of a channel by channel subtraction of two large numbers. The various spectra show the same overall trend. However,

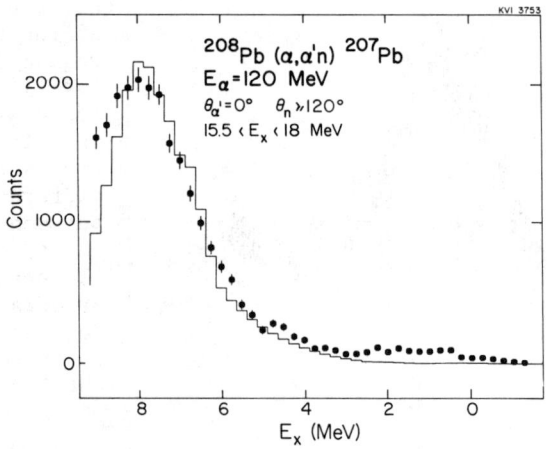

Fig. 5. The ^{207}Pb final state spectrum resulting from neutron decay of the $15.5 \leq E_x \leq 18$ MeV excitation energy region in ^{208}Pb (dots) in comparison with statistical model calculations (histogram) assuming $J^\pi = 2^+$ decay. The data are obtained from the neutron detectors in the very backward region ($\theta_n > 120°$).

there are features in the GMR-decay spectrum which are not observed in the lateral one and which may be significant; namely the rather sudden jump in the spectrum at $E_x \sim 4.5$ MeV and a hint of structure between 4 and 6 MeV excitation. It is tempting to connect possible structure around $E_x \sim 4.5$ MeV with the hole-phonon coupling states predicted around that energy (see figure

1). Another factor causing differences between the spectra of figure 4b and 4c may be the presence of large multipole strength in the "lateral" part decaying to the high-spin states in ^{207}Pb.

3.2 Comparison with statistical calculations

For a more quantitative analysis we calculated the Q-value spectra assuming a pure statistical decay mode. In these calculations known levels up to 3.5 MeV were used. For the higher excitation energy range the level densities were calculated using the parameters of Dilg et al[21]. A different prescription was used by Dias and Wolynec[15] but the final results for the region above 3.5 MeV are very similar. There is also some ambiguity in the choice of the optical model parameters for calculating the transmission coefficients but it turns out that this has only a slight effect. The spectra were calculated by folding with the energy resolution of the neutron detectors and the shape of the inelastic alpha spectrum. Except for the GMR proper, the calculations were performed assuming $J^{\pi} = 2^+$ decay.

Figure 5 shows the results for the region $15.5 < E_x < 18$ MeV in ^{208}Pb. The statistical calculations reproduce the data reasonably well for the high excitation energy range in ^{207}Pb but fail to reproduce the data for the low-lying hole states. Figure 6[a,b] shows the results for the "lateral" part containing no 0^+-strength and for the GMR proper respectively. For both, the data are reasonably well in agreement with the statistical model calculations, for the lateral part perhaps somewhat better than for the GMR. Part of the discrepancies observed in figure 6[a] between data and calculations can be explained by the presence of $J^{\pi} > 2^+$ strength in this energy region populating high spin states in ^{207}Pb.

The data of Eyrich et al. for the GMR region[4] are, if properly analysed[15], also in agreement with statistical decay. However their data refer to the decay of the GMR proper plus the continuum underlying the GMR and are thus equivalent to our combined data of the GMR proper (fig. 6[b]) and the "lateral" part (fig. 6[a]). Our

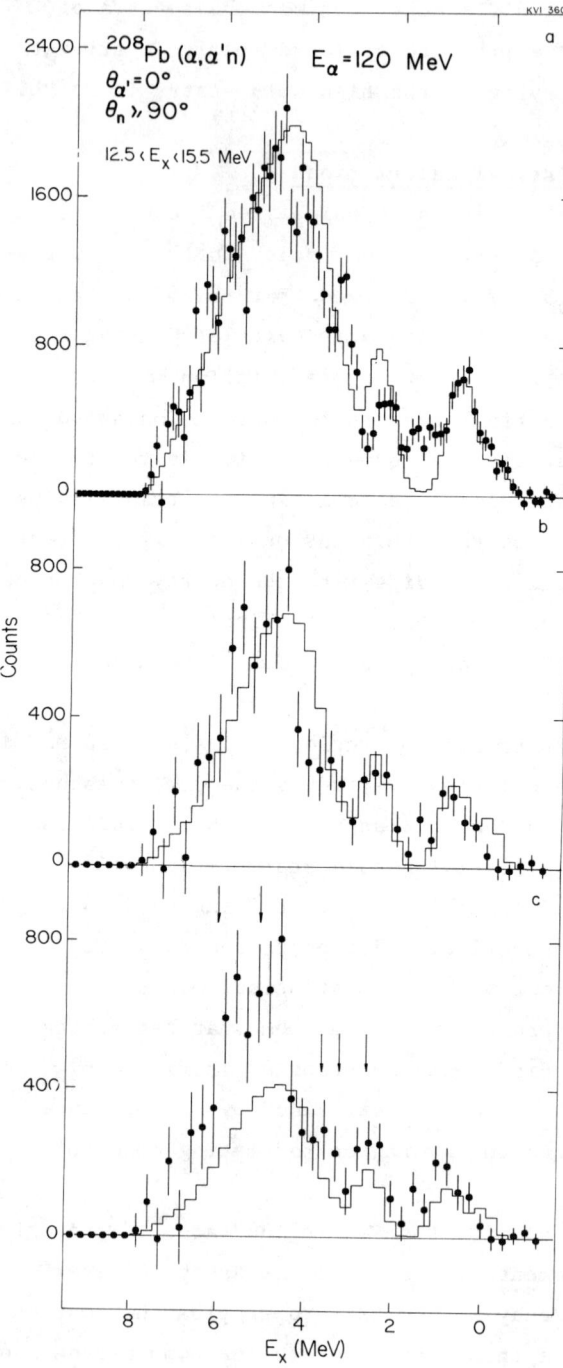

Fig. 6. Comparison of the measured (dots) and, on basis of a statistical model, calculated (histogram) final state spectrum of ^{207}Pb resulting from neutron decay of the GMR region in ^{208}Pb for a) the 'lateral' part with residual 0^+ strength subtracted and assuming $J^\pi = 2^+$ decay, b) for the GMR proper assuming pure statistical decay and c) for the GMR proper assuming 60% statistical decay.

data show that not only for the (GMR + continuum) but also for the GMR proper a pure statistical decay gives a reasonable account of the observations.

3.3 Is there any evidence for pre-equilibrium decay?

An interesting question is to what extent our data on the GMR contains any information on pre-equilibrium decay, like structures which cannot be explained by statistical calculations. Especially the sudden jump in level population around $E_x \sim 4.3$ MeV, which is not repreoduced by the data, and the possible structure between 4 and 6 MeV might indicate that mechanisms other than a pure statistical one are present.

Taking these structures for granted, one can try to decompose the observed spectrum in various components; a statistical component chosen in such a way that it fits the data between 3 and 4 MeV, a direct component which is then the difference between the observed population of the hole states and the adjusted statistical one and a pre-equilibrium component, where the last one may well be due to the particle(hole)-phonon damping mechanism. Such a decomposition, which in its details is of course quite uncertain, is shown in figure 6^c. For the case shown the statistical component is about 60%, the direct one about 10% and the pre-equilibrium one about 30% of the total spectrum.

It is clear that in order to make a more definite statement on possible deviations from statistical decay one would need data with smaller uncertainties. With the present set-up this would require a very long running time estimated at about 2500 hrs.

REFERENCES

1) J. Speth and A. van der Woude, Rep. Progr. Phys. 44 (1981) 719.
2) M. Buenerd, Journ. de Physique 45 (1984) C4-115.
3) S. Brandenburg et al. Phys. Rev. Lett. 49 (1982) 1687; R. de Leo et al. to be published in Nucl. Phys. A (1985).
4) W. Eyrich et al. Phys. Rev. C24 (1984) 418.
5) W. Eyrich et al. Phys. Rev. Lett. 43 (1979) 1369.
6) H. Steuer et al. Phys. Rev. Lett. 47 (1981) 1702.

7) L.S. Cardman, Nucl. Phys. A 354 (1981) 173c.
8) H. Egiri, Journ. de Physique, 45 (1984) C4-135 and references therein.
9) M.N. Harakeh, Journ. de Physique, 45 (1984) C4-155; A. van der Woude, Lecture Notes in Physics Vol. 168 (1982) 351.
10) F.E. Bertrand, Journ. de Physique, 45 (1984) C4-99.
11) P.F. Bortignon, R.A. Broglia and G.F. Bertsch, Phys. Lett. B148 (1984) 20; P.F. Bortignon, KR.A. Broglia and Xia Ke-ding, Journal de Physique, 45 (1984) C4-209; G.F. Bertsch, P.F. Bortignon and R.A. Broglia, Rev. Mod. Phys. 55 (1983) 287; P.F. Bortignon and R.A. Broglia, Nucl. Phys. A371 (1981) 405; R.A. Broglia and P.F. Bortignon, Phys. Lett. B101 (1984) 135; G.F. Bertsch et al. Phys. Lett. B80 (1979) 161.
12) R. de Haro, S. Krewald and J. Speth, Nucl. Phys. A388 (1982) 265.
13) N. Van Giai and H. Sagawa, Nucl. Phys. A371 (1981) 1.
14) J. Wambach and B. Schwesinger, Journal de Physique 45 (1984) C4-281;
B. Schwesinger and J. Wambach, Nucl. Phys. A426 (1984) 253.
15) H. Dias and E. Wolynec, Phys. Res. C30 (1984) 1164.
16) F. Zwarts, A.G. Drentje, M.N. Harakeh and A. van der Woude, to be published in Nucl. Phys. A (1985); F. Zwarts, thesis, Groningen, 1983.
17) S. Brandenburg et al. to be published; S. Brandenburg, thesis, Groningen, 1985.
18) D.H. Youngblood et al. Phys. Rev. Lett. 39 (1977) 1188; D.J. Youngblood, in: Giant Multipole Resonances, ed. F.E. Bertrand (Harwood, New York, 1980).
19) F. Zwarts et al. Phys. Rev. C25 (1982) 2139.
20) K.T. Knöpfle and G.J. Wagner, Phys. Rev. C27 (1983) 2422.
21) W. Dilg et al. Nucl. Phys. A217 (1973) 269.
22) J.T. Caldwell et al. Phys. Rev. C21 (1980) 1215.

SESSION C

EXCITATION OF STRETCHED CONFIGURATION HIGH-SPIN STATES WITH VARIOUS PROBES AND REACTIONS

EVIDENCE OF PARTIAL OCCUPANCY OF ORBITS FROM ELECTRON SCATTERING

C. N. Papanicolas
Nuclear Physics Laboratory and Department of Physics
University of Illinois, Champaign, IL 61820

Abstract

Partial occupancy of shell orbits is expected to lead to the quenching of single particle transitions at high momentum transfers. A procedure for the extraction of quenching factors from electron scattering data that can be interpreted in terms of occupation numbers is suggested. Examination of the quenching factors of a wide range of transitions in the lead region finds them to be in accord with recent theoretical suggestions that the occupation probabilities of single particle states just below and above the fermi energy are ~0.7 and ~0.1 respectively.

INTRODUCTION

Experimental evidence accumulated over the last decade has brought the realization that nuclear mean field theory (MFT) cannot adequately describe a wide range of nuclear phenomena. It has gradually become evident that the deficiency is systematic throughout the periodic table and more serious than the early successes of Hartree-Fock calculations have led us to hope. Theoretical attempts to resolve the discrepancy have largely focused on estimating higher order processes (such as core polarization, 2p-2h corrections, meson exchange currents, etc.). They have shown such effects to be non-negligible. However, these attempts have been sporadic and concerned with specific reaction channels.

It has recently been proposed by Pandharipande, Papanicolas and Wambach[1] and by Jaminon, Mahaux and Ngô[2] that the systematic suppression (as compared to MFT calculations) of single particle phenomena can be explained to a large degree by invoking partial occupancy of shell orbits.

I shall review in this paper evidence offered by single arm electron scattering experiments in support of this thesis. I will restrict myself to the examination of elastic and inelastic scattering from bound states. For reasons that will become apparent later, the data from bound excitations are the least susceptible to model error.

THEORETICAL CONCEPTS

Quenching implies a comparison, a comparison between a theoretical result (in our case the result of a mean field calculation) and data. We need to critically evaluate both legs of this comparison to insure that our conclusions are not masked by either huge model error or extraneous experimental signal.

The first concern is about the theoretical result. Can a comparison be performed with mean field theoretical results that is free of model error? And if not, how much is the particular choice of model space going to influence our results?

The second concern, about the experimental data, is even more difficult to address. How do we insure that the experimental observable is the appropriate one for a comparison? How successful can we be in selecting data which have little influence from processes irrelevant to the question of interest?

A successful resolution of both concerns is, I believe to a large degree, possible.

The Liege-Orsay group[2] provides substantial insight into the bias that is introduced by the selection of a particular model basis by examining the properties of the one-body density of nuclei:

Given a model choice for the single particle wavefunction basis $[\phi_\alpha]$ which has the appropriate completeness and orthonormality properties, it is always possible to write a given one-body density distribution as:

$$\rho(r) = \frac{1}{4\pi r^2} \sum_{\alpha,\alpha'} (2j+1)\, \rho_{\alpha\alpha'}\, \phi_\alpha(r)\, \phi^*_{\alpha'}(r) \qquad (1)$$

where α is an all encompassing quantum number and $\rho_{\alpha\alpha}$ is the appropriate one-body density matrix for that basis. Under most circumstances a unitary transformation U can be found that will render ρ diagonal. This transformation defines[3,4] then the "natural orbitals" ς_α:

$$\varsigma_\alpha \equiv U_{\alpha\beta}\, \phi_\beta$$

In the unique basis of natural orbitals the density matrix is by definition diagonal and the one-body density distribution takes its simplest form:

$$\rho(r) = \frac{1}{4\pi r^2} \sum_\alpha (2j+1)\, n_\alpha\, \varsigma^*_\alpha(r)\, \varsigma_\alpha(r) \qquad (2)$$

where the diagonal elements $\rho_{\alpha\alpha'}$ have been expressed more economically as "occupation probabilities" n_α.

The basis of natural orbitals provides an excellent theoretical concept and very useful language for the discussion of occupation probabilities.[3] In practice, to the best of my knowledge, it is not known how to calculate natural orbitals exactly, even in the fields where the concept was originally developed like atomic or chemical physics. A prescription of how to approximate natural orbitals in the nuclear many body problem can be found in the recent paper of Jaminon et al.[2]

Several approximations to the one-body density are also of considerable value as they are more easily realizable in model calculations. One can define a Fermi gas density $\rho^{FG}(r)$:

$$\rho^{FG}(r) = \frac{1}{4\pi r^2} \sum_\alpha (2j+1)\, \phi_\alpha^*(r)\, \phi_\alpha(r) \quad (3)$$

which is obviously defined within the given model space. The suppressed occupation probability has the known Fermi-Dirac distribution at zero temperature:

$$n = \begin{cases} 1 & \varepsilon < \varepsilon_f \\ 0 & \varepsilon > \varepsilon_f \end{cases} \quad (4)$$

A subset of this class of one-body densities are the various types of Hartree-Fock distributions, $\rho^{HF}(r)$.

When the truncation implied by the Fermi gas distribution is imposed on the true (or physical) one-body density expressed in the basis of natural orbitals one obtains the truncated physical one-body density, $\rho^0(r)$:

$$\rho^0(r) = \frac{1}{4\pi r^2} \sum_{\varepsilon_\alpha < \varepsilon_F} (2j+1)\, n_\alpha\, \varsigma_\alpha(r)\, \varsigma_\alpha^*(r) \quad (5)$$

The fraction of density to be found below the Fermi surface is a characteristic parameter of the system. It indicates how successful an independent particle model can be in describing this many body system. Obviously:

$$\eta = \frac{\int \rho^0(r)\, d^3r}{\int \rho(r)\, d^3r} \quad (6)$$

A different way of expressing the same statement is to say that η is a measure of the influence of correlations in the system.

Partial occupancy of shell orbits is required by our very basic notions about the nuclear force and medium. Its manifestations are numerous, perhaps the most irrefutable being the existance of the imaginary part of the nuclear optical potential. This realization is explicit from the very beginning of nuclear mean field theory, as it is thoroughly dealt with, for instance, in the work of Brueckner and coworkers.

The necessity of partial occupancy in the particle hole space leads to two distinct theoretical paths. In the first, one deals with the particle hole distribution and proceeds to calculate matrix elements using bare operators and occupation numbers. In the second, that of the Landau theory[5,6], the Fermi distribution is restored by a quasiparticle transformation which allows the calculation of matrix elements using a Fermi gas quasiparticle distribution and effective operators.

As the title of this talk reveals, we will stay within the framework of particle distribution and matrix elements calculated with bare operators.

MEASURING OCCUPATION PROBABILITIES

The remarks of the previous section would lead us to believe that the occupancy of shell orbits must have been mapped by now with high precision and the limits of validity of mean field theory be precisely known. The fact that we are nowhere close to such a state of affairs can be attributed to two reasons: i) most of the available data concern phenomena dominated by the nuclear surface where it is particularly difficult to test the concept of the mean field. ii) of the measurements that possess this sensitivity only a small fraction can be interpreted reliably because of large uncertainties in the reaction mechanism. Knowledge of the imaginary part can, in principle, yield occupation probabilities through the use of dispersion relations.[7,8] However, as is invariably the case with dispersion theorems, it is difficult to realize the delicate mathematical relations with the imperfect and restricted data of the real world. C. Mahaux and H. Ngô[8], taking advantage of the improved knowledge of the imaginary part of the optical potential resulting from recent elastic neutron scattering, has performed such an evaluation. This estimate shows that we should expect for valence orbitals in the lead region occupation probabilities of the order 0.8.

It was recently claimed[1] that a direct estimate of occupation probabilities can be obtained from high precision elastic and inelastic measurements in the lead region. Before we proceed to examine the evidence, we need to briefly review the arguments that lead to this claim.

In electron scattering one measures form factors, $F(q)$, which can be mapped as a function of the momentum transfer q. The form factor that characterizes a transition between initial state $|i\rangle$ and final state $|f\rangle$ can be separated into two parts

$$F(q) = Q\, F_{sp}(q) + F_{bg}(q) \qquad (7)$$

where $F_{sp}(q)$ is the "single particle" form factor and will have the identical shape as that predicted by an independent particle model based on natural orbitals. Its amplitude will be different from that of the MFT since the occupation probabilities will reduce the probability of finding the particle in the particle orbit and will block the transition by the partial occupation of the hole. The momentum tranfer independent amplitude Q is solely a function of the occupation numbers. In heavy nuclei (in the limit of $A \to \infty$ and where $|f\rangle \simeq |i\rangle$) the quenching factor Q is given by:

$$Q \simeq n(p)[1 - n(h)] \qquad (8)$$

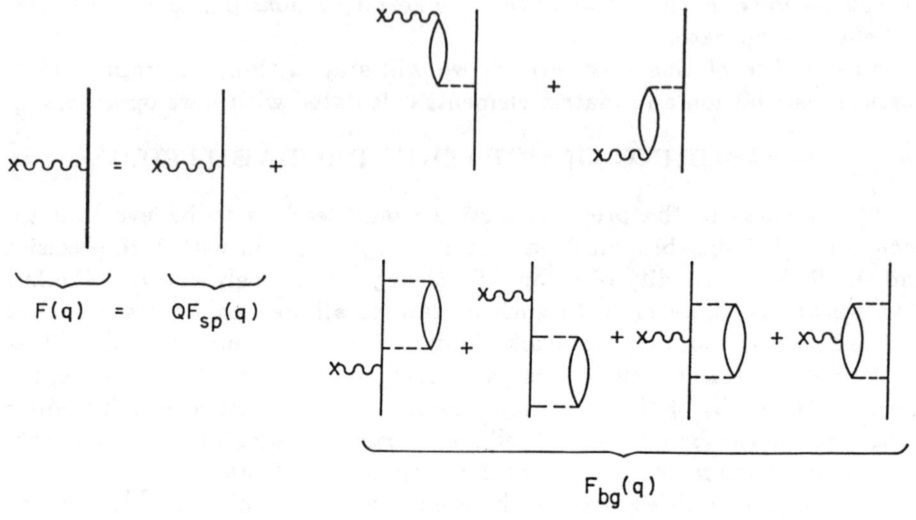

Figure 1: The separation of the observed form factor $F(q)$ into a quenched single particle piece $QF_{sp}(q)$, and a bakground piece $F_{bg}(q)$ can be easily understood in a diagramatic expansion. Terms up to second order in the interaction are shown.

In the limit where MFT accurately describes reality then $n(p) = 1$, $n(h) = 0$ and $Q=1$, as expected.

The presence of the background form factor $F_{bg}(q)$ is a necessary consequence of correlations. Correlations create complicated multiparticle-multihole configurations which cause partial occupancy and quenching ($Q < 1$). Scattering from the complicated multiparticle-multihole configurations leads to the background response, $F_{bg}(q)$, which exhibits an entirely different behavior than that of $F_{sp}(q)$, the "single particle" response. The situation here is quite analogous to that encountered in the familiar case of scattering from a state with configuration mixing. The observed form factor again can be separated into the piece associated with the single particle state and the term arising from configuration mixing. In Figure 1 a diagramatic expansion up to 2p-2h contributions is shown. The identification of the terms that bring about the quenching and contribute to the "background" form factor $F_{bg}(q)$ is indicated.

Because the multiparticle-multihole state is quite complicated no characteristic signature in momentum space should be observed. In heavy nuclei one expects the background form factor to be featureless and peaked at low momentum transfers. We should expect then to measure at high momentum transfer $QF_{sp}(q)$ and at low momentum tranfers $F(q)$. This underlying assumption is crucial and its validity permeates the analysis of the experimental data that follows.

The presence of the background term is also necessary for the restoration of a number of conservation laws at the long wavelength limit, implied by the appropriate Ward identities. For instance at low momentum transfers we must recover the Mott limit as conservation of charge requires. On the other hand its high momentum behavior is not guaranteed by any rigorous theorems, it can only be revealed through a microscopic calculation.

The above considerations then clearly define the establishment of the following experimental procedure for the measurement of Q (quenching factor) in a heavy nucleus:

1. Measure the form factor of a bound single particle transition in a heavy nucleus at high q.

2. Assure that extraneous contributions (such as MEC) are not significant or that they have been rendered so through subtraction.

3. Calculate within the framework of a successful mean field theory the expected single particle form factor.

4. If the shape of the calculated form factor closely resembles the measured one, a simple scaling of the theoretical F(q) will yield the quenching factor Q.

High momentum transfers are taken to mean $q \simeq 2k_f$, the Fermi momentum, typically the region up to which form factors of a single particle state stay flat before taking an exponential dive. At these high momentum transfers the background term is assumed to be insignificant.

"Heavy nucleus" in this paper will always be taken to be ^{208}Pb, its isotopes and isotones. To facilitate the discussion of the various transitions involved we show in Figure 2 the single particle spectrum of ^{208}Pb. The requirement of a bound state provides some degree of assurance that the model wavefunctions of a "successful" mean field theory (such as those of Hartree-Fock) do not differ drastically from the natural orbitals for bound states.

The above procedure is best demonstrated through an example. As such we choose the M14 and M12 excitation of ^{208}Pb.[9] Lead, being the heaviest doubly-closed shell system available, constitutes the best testing ground for the principles outlined above. The $1\hbar\omega$ stretched states are well isolated and therefore relatively immune to mixing. As such the $J^\pi = 14^-, 12^-$ states satisfy the criterion of a single particle transition being widely recognized as resulting from the $\nu(i_{13/2}^{-1}, j_{15/2})$ configuration. The measured form factor at high q (where it naturally peaks) exhibits the same shape as the one calculated using Hartree-Fock (DME)[10] wavefunctions. As shown in Figure 3 when the F_{sp} form factor is scaled by $Q = 0.71 \pm 0.5$ reasonable agreement with the data is achieved.

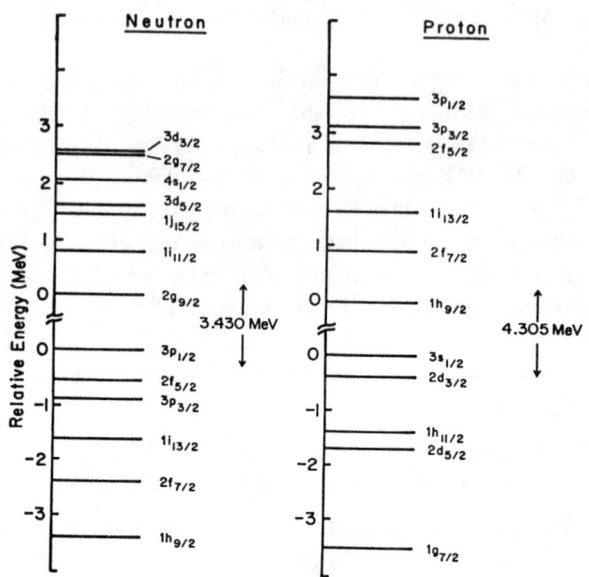

Figure 2: The empirical single particle (hole) energy spectrum for ^{208}Pb

In the following sections we proceed to examine the available elastic (magnetic and charge) data in the lead region and then certain types of inelastic transitions that conform to our criteria. We will restrict ourselves to this mass region for the reasons we have already stated. However, we will try to demonstrate that the types of transitions discussed exhibit similar behavior throughout the periodic table and that we are not just faced with some peculiarity of the lead region.

Admittedly the suggested procedure may involve substantial model error which we have no way of estimating at this time. However, it appears reasonable, and its reliability must eventually be judged by the consistency of the results it yields and their compatibility with related quantities deduced by alternative methods.

ELASTIC SCATTERING

A. Elastic Magnetic Scattering

Unpaired nucleons in odd-even nuclei give rise to elastic magnetic scattering. In the case of systems neighboring doubly closed shells elastic magnetic form factors can be taken as an excellent example of single particle excitations.

Elastic magnetic scattering has been systematically studied during the last decade and it was recently reviewed by Donnelly and Sick.[11] Mean field

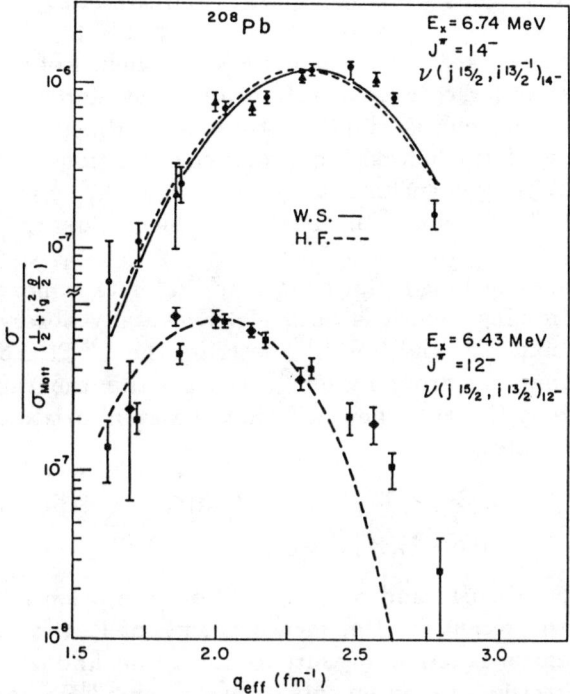

Figure 3: Extracting quenching factors for states of single particle character. The high spin states of ^{208}Pb satisfy all criteria (see text). The choice of Hartree-Fock or Woods-Saxon basis has little effect on the deduced quenching factor (see Table II).

theory is found to predict the shape of the observed form factors rather accurately throughout the periodic table. In most cases the calculated form factors overestimate the measured values. This is especially true for nuclei neighboring magic shells.

For the reasons we have extensively discussed in the previous section, the lead region is of particular interest. The elastic magnetic form factors of ^{209}Bi ($\pi h_{9/2} \otimes |0^+\rangle_{208}$) and ^{207}Pb ($\nu 3p_{1/2}^{-1} \otimes |0^+\rangle_{208}$) have been measured recently.[12,13] ^{209}Pb ($\nu 2g_{9/2} \otimes |0^+\rangle_{208}$) and ^{207}Tl ($\pi 3s_{1/2}^{-1} \otimes |0^+\rangle_{208}$) are unstable and therefore inaccessible to experimental investigation. Due to the unavailability of ^{207}Tl, ^{205}Tl ($3s_{1/2}^{-1} \otimes |0^+\rangle_{206}$) is of special interest and it has been the subject of many investigations recently. Configuration mixing is known to play a significant role in ^{205}Tl and must be explicitly accounted for. The measured elastic magnetic form factors and the corresponding MFT predictions, obtained using H.F. wavefunctions[10,14] are shown in Figures 4, 5 and 6. As we have already anticipated, MFT overestimates the experimental results.

T. Suzuki and coworkers[15] have explored the influence of meson exchange currents and first order core polarization and they find them significant. This may indicate that our assumption of vanishing $F_{BG}(q)$ at high q is not entirely true in these form factors. Ignoring such corrections and applying our procedure for obtaining quenching factors at $q \simeq 2.0$ fm^{-1} we deduce[1] $Q = 0.7 \pm 0.1$ for both ^{207}Pb and ^{209}Bi. The error reflects only the experimental uncertainty.

Extracting the quenching factor from the ^{205}Tl data is more involved due to configuration mixing, which is both experimentally observed and theoretically understood. Co and Speth[16] describe the ^{205}Tl ground state by perturbatively expanding around the ^{208}Pb core within the framework of the finite Fermi systems theory. They find that the appropriate expression for the ^{205}Tl ground state is:

$$|1/2\rangle^+_{205} = \alpha(|3s_{1/2}\rangle \otimes |0^+\rangle_{206}) + \beta(|2d_{3/2}\rangle \otimes |2^+\rangle_{206}) + \gamma(|2d_{5/2}\rangle \otimes |2^+\rangle_{206}) \qquad (9)$$

with $\alpha = 0.86$, $\beta = -0.41$ and $\gamma = 0.3$. This description is in reasonable agreement with the recent (d,^{3}He) measurements of P. Grabmayer et al.[17] (and previous transfer reaction measurements[18]) that find $\alpha^2 = 0.89 \pm 0.08$. The very high precision measurements of (e,e'p) on ^{208}Pb and ^{206}Pb from NIKHEF[19] are also consistent with such admixtures. It is important to note that the above depletion of the $3s_{1/2}$ orbital is due to configuration mixing and has little to do with the partial occupancy in ^{208}Pb we have been discussing. This is manifestly clear since Co and Speth start with a Woods Saxon description of ^{208}Pb, in which the particle states are fully occupied.

The more complicated configurations (second and third terms in the expansion given above) will be insignificant at high momentum transfers because the ^{206}Pb quadrupole phonons are known[20] to have very little high momentum content. Scaling then the MFT form factor, to match the first maximum of the form factor will yield $\tilde{Q}$; $\tilde{Q}$ is found to be 0.55 ± 0.1. $\tilde{Q}$ accounts for the quenching arising from both the influence of configuration mixing and partial occupancy. If we take the experimental value [17] of $\alpha^2 (= 0.89 \pm 0.08)$, we can then derive a value of $Q \simeq 0.6 \pm 0.15$ consistent with the ^{207}Pb and ^{209}Bi results. The quenching factors derived from the three magnetic scattering experiments discussed above are summarized in Table I.

B. Elastic Charge Scattering

Elastic charge scattering has been pursued at various electron scattering facilities for over thirty years now. As a result very extensive data on most nuclear systems exist and they provide very demanding tests on nuclear structure theories. The precision that can be achieved is nicely exhibited in

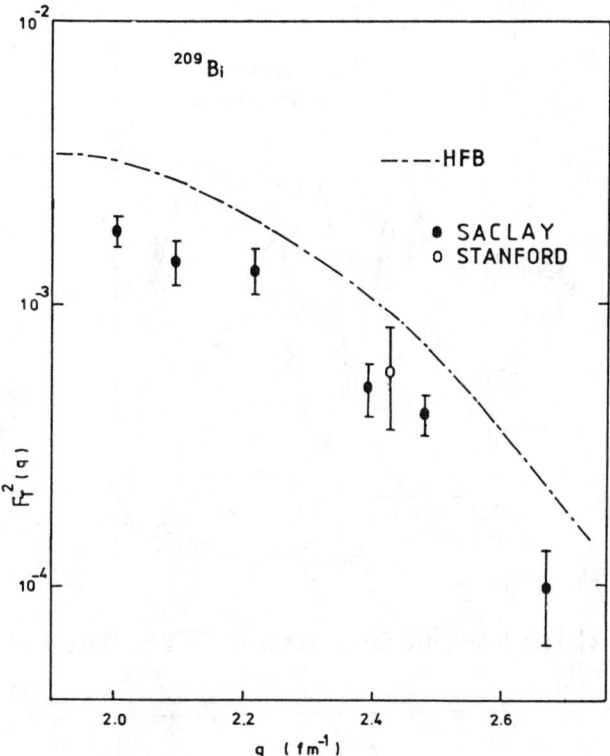

Figure 4: The elastic magnetic form factor of ^{209}Bi. Data and MFT result.

A	h	Q	Ref.
^{205}Tl	$\pi 3s_{1/2}$	$0.6 \pm .15$	13
^{209}Bi	$\pi 1h_{1/2}$	$0.7 \pm .1$	12
^{207}Pb	$\pi 3p_{1/2}$	$0.7 \pm .1$	13

Table I: Quenching factors Q derived from elastic magnetic scattering. The character of the unpaired nucleon and the nucleus in which it was studied are also given. The stated uncertainty reflects only the experimental error.

Figure 7 where the reconstructed ground charge densities of nuclei with closed shells are shown.[21] The uncertainty characterizing the experimental result is of the order of a few percent throughout the nuclear volume. It is indicated in the figure by the width of the line depicting the empirical density. In the same figure the charge densities derived by mean field theory (Hartree-Fock calculations of Decharge and Gogny[14]) are also shown. It is a well known fact that such calculations nicely reproduce the surface behavior but overestimate the density of the interior. This is most evident in the case of ^{208}Pb

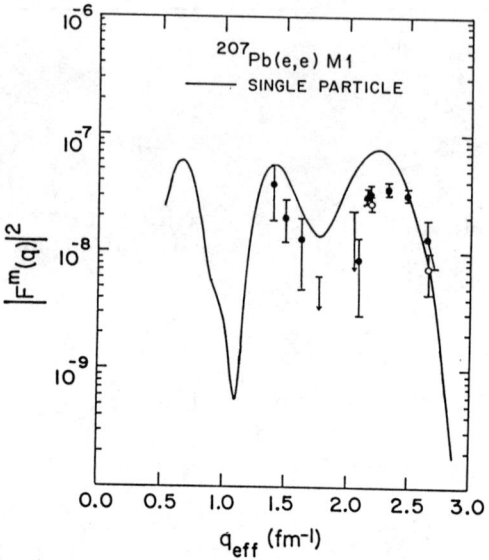

Figure 5: The elastic magnetic form factor of ^{207}Pb. Data and MFT result.

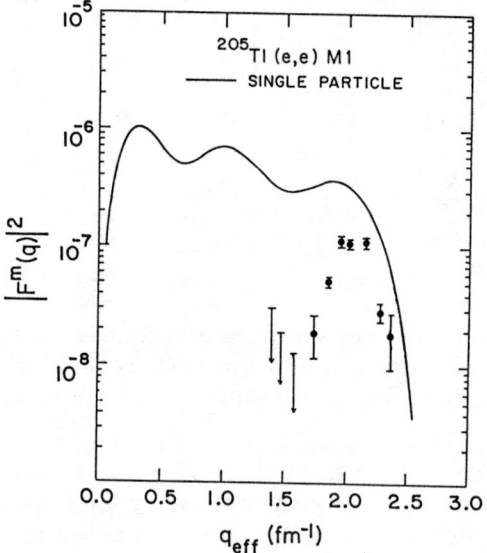

Figure 6: The elastic magnetic form factor of ^{205}Tl. Data and MFT result. No configuration mixing is included in the mean field estimate.

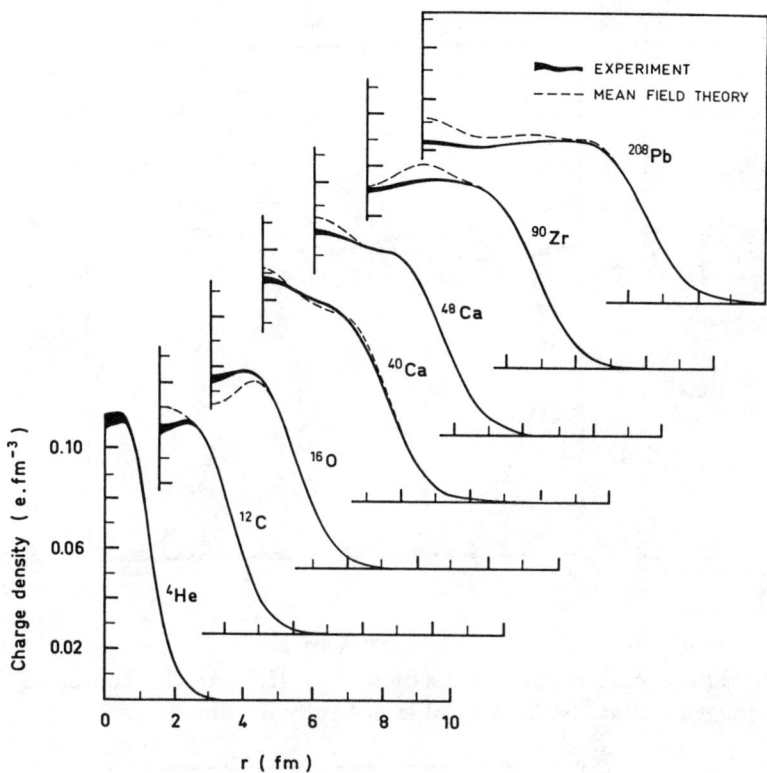

Figure 7: Very high precision characterizes the empirical charge distribution. Mean field calculations (H.F.)[14] systematically overestimate the charge density at the nuclear interior.

where the interior region is quite extensive. Elastic charge scattering is the prime example of a collective excitation. The procedure established earlier for dealing with transitions of single particle character are obviously not appropriate for dealing with charge scattering. However, the change in the charge density brought about by the addition or removal of a single nucleon is related to a single particle density and one might hope to establish contact with the phenomenon we are concerned with through it. The observation that the suppression of the empirical charge density of ^{208}Pb at the origin can be brought about only by the partial occupation of the s orbitals and in particular of the $3s_{1/2}$ orbital, see Figure 8, lead to the systematic study of the isotopic and isotonic charge differences in the lead isotopes. The measured charge difference between ^{206}Pb and ^{205}Tl has the qualitative features of a $3s_{1/2}$ density, (see Figure 9) thus providing the most direct evidence for the validity of the concept of the independent particle at the nuclear interior. However, as one can observe in the same figure the prediction of

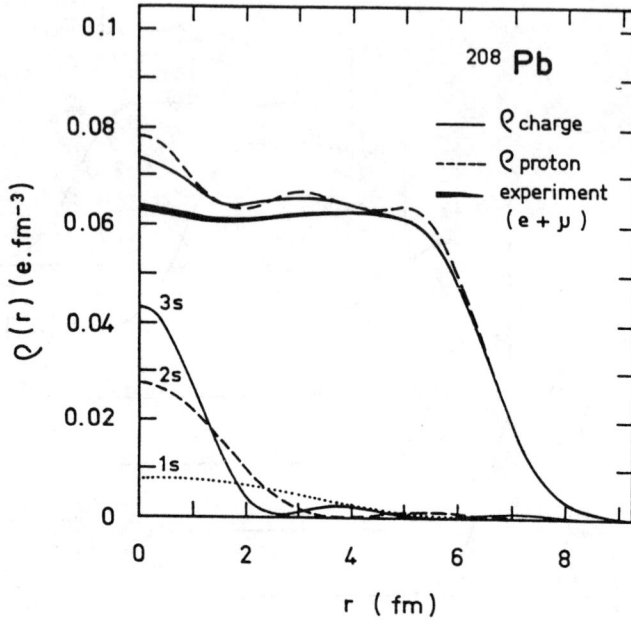

Figure 8: The excess of charge predicted by H.F. at the center of ^{208}Pb strongly suggests that the 3s orbital is not fully occupied.

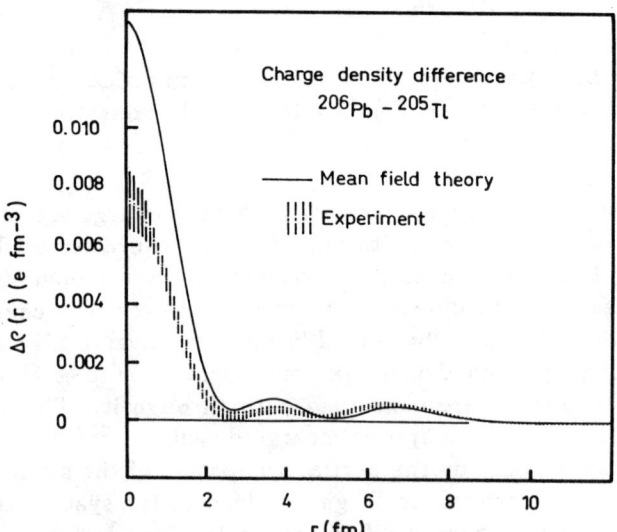

Figure 9: The reconstructed isotopic charge difference between ^{206}Pb and ^{205}Tl, direct evidence for the presence of the $F_{sp}(q)$ term. The Hartree-Fock estimate of the same quantity overestimates the data.

mean field theory[22] overestimates the concentration of charge at the origin by ~80%. This effect is undoubtedly partly due to configuration mixing (see discussion in the previous section). It is only partially accounted for by the Hartree-Fock result. While the evidence for partial occupancy in the charge scattering from ^{208}Pb, ^{206}Pb and ^{205}Pb is quite strong, quantitative information on the occupation number for the $3s_{1/2}$ orbital cannot be deduced until the influence of mixing has been precisely accounted for.

INELASTIC SCATTERING

Inelastic scattering to states of known single particle character has provided a sensitive test of MFT predictions and naturally leads to the extraction of quenching factors. Complementary data for occupation probabilities to those derived from elastic scattering can be obtained from such transitions. Inelastic excitations actually constitute a richer source of information because many more orbitals can be tested besides the valence orbitals to which elastic magnetic scattering and isotonic charge differences are necessarily limited. We shall examine here form factors derived from inelastic scattering to the two most obvious classes of single particle (hole) transitions: High spin transitions in ^{208}Pb and neutron hole transitions in ^{207}Pb.

Excitation of stretched configurations involves $1\hbar\omega$ transitions that yield maximum realignment of the total angular momentum, e.g. $(i_{13/2} \to j_{15/2})_{14^-}$ in ^{208}Pb. It is simply impossible to construct excitations of such high angular momentum in any other way at nearby energies. This implies absence of mixing and thus insures the relative purity of the single particle character of the transitions. These states have been extensively studied throughout the periodic table with a variety of probes. A compilation of the resulting strengths and form factors can be found in the recent review paper by R. A. Lindgren.[24]

It is generally the case that mean field theory predicts the excitation energy and the shape of the form factor correctly but invariably it overestimates the strength. Our prescription for the extraction of occupation probabilities is directly applicable. Actually we have already used the case of the M14 and M12 transitions of ^{208}Pb to demonstrate this prescription (see Fig. 3). In Figure 10 the extracted quenching factors (squared) for a variety of nuclear systems is shown. It is evident that quenching factors vary widely throughout the periodic table, but the found strength never comes close to the mean field result. These transitions are found at high excitation energies where the density of states is quite high. If fractionation is at work, i.e., if the strength is concentrated in a cluster of states instead of a single peak the quenching of the dominant observed peaks (shown in Figure 10) cannot possibly be related to an occupation number. Substantial evidence exists indicating that the lack of found strength in all but the doubly closed shell nuclei is partly due to fractionation. This is quite evident in the spectra of ^{208}Pb, ^{207}Pb,

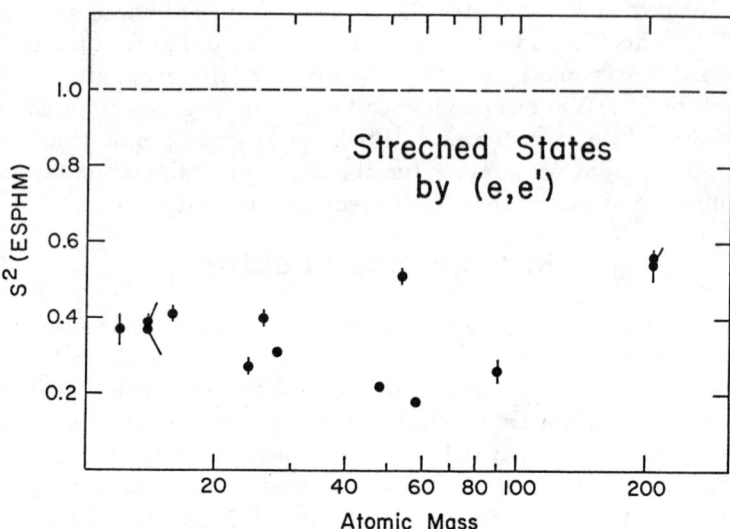

Figure 10: The observed strength of transitions to stretched configurations is invariably quenched throughout the periodic table.[24] ^{208}Pb offers the most favorable case.

^{206}Pb and ^{204}Pb shown in Figure 11. The successive removal of neutrons from the ^{208}Pb core results in fractionation of the high spin states. This has prevented us from recovering in ^{206}Pb and ^{204}Pb[25] the strength observed in ^{208}Pb. In the case of ^{140}Ce we have not been able to identify any high spin states, presumably due to fractionation, despite the fact that this nucleus has a closed shell for neutrons (N = 82). However, we do believe that no fractionation exists in ^{208}Pb, beyond that which is expected from the same mechanisms that generate partial occupancy in the first place (such as 2p-2h excitation, etc.).

The high resolution inelastic electron scattering at ^{208}Pb has mapped not only the form factor of the 14^- but also of other high spin states such as 12^+, 12^-, 10^+, etc. They are as immune to mixing as the 14^- and the shape of their form factors are nicely predicted by Hartree-Fock wavefunctions. In Table II we summarize the quenching factors obtained from all the known high spin states in ^{208}Pb.

Single particle (hole) transitions in odd-even nuclei neighboring doubly closed shells share certain features with the stretched states examined above. They have a clearly identified single particle (hole) character and in general they are well isolated which prevents any mixing. They are found at very low excitation energies where resolution problems do not arise. Once again these states have been thoroughly studied with inelastic electron scattering and they are found to be rather accurately described in shape but overestimated in amplitude by mean field theory. Typical is the case of the 6.33 MeV state

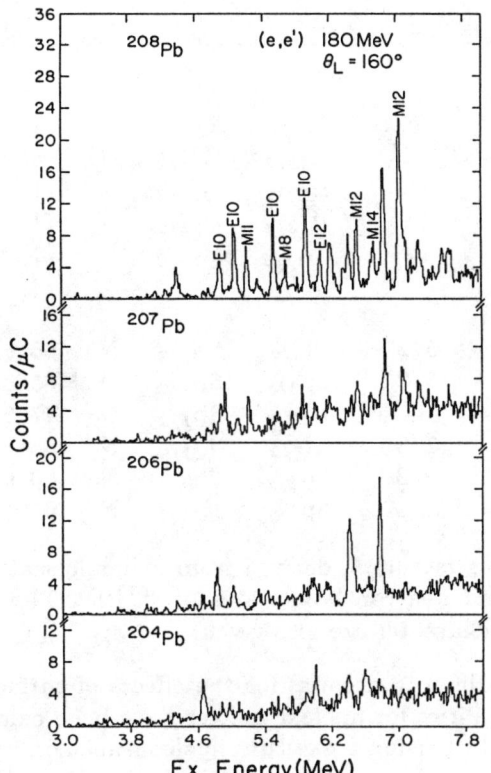

Figure 11: Fractionation of high spin states can experimentally be confused with quenching. A comparison of spectra of ^{208}Pb, ^{207}Pb, ^{206}Pb, and ^{204}Pb strongly suggests that fractionation already occurs in ^{206}Pb and ^{204}Pb.

of ^{15}N identified as due to the $(p_{1/2}^- \to p_{3/2}^-)$ proton hole transition. Its measured form factor is shown in Figure 12.

In ^{207}Pb these states have been studied but no corresponding measurements exist for ^{209}Bi. The low energy excitation spectrum of ^{207}Pb is shown in Figure 13. It is dominated by single neutron hole states. It can be seen in Figure 14 that the form factors of these states can be adequately described by Hartree-Fock transition densities after the amplitudes have been adjusted.[28] The quenching factors extracted for these states are catalogued in Table II, along with those of the stretched states of ^{208}Pb.

A THEORETICAL ESTIMATE

A theoretical estimate of the occupation probabilities characterizing the shell orbits of ^{208}Pb is offered in reference 1. It is assumed that they can be reasonably approximated by the occupation numbers of nuclear matter,

E (MeV)	J^π	h	p	t	Q	Ref.
^{208}Pb (e,e')						
4.04	7^-	$2f_{5/2}$	$2g_{9/2}$	N	.51 ± 0.5	26
6.10	12^+	$1i_{13/2}$	$1i_{11/2}$	N	.65 ± .04	9
6.43	12^-	$1i_{13/2}$	$1j_{15/2}$	N	.71 ± .05	9
6.74	14^-	$1i_{13/2}$	$1j_{15/2}$	N	.71 ± .05	9
7.06	12^-	$1h_{11/2}$	$1i_{13/2}$	P	.71 ± .05	9
^{207}Pb (e,e')						
0.57	$5/2^-$	$2f_{5/2}$	$3p_{1/2}$	N	.65 ± .05	28
0.90	$3/2^-$	$3p_{3/2}$	$3p_{1/2}$	N	.65 ± .05	28
1.63	$13/2^+$	$1i_{13/2}$	$3p_{1/2}$	N	.47 ± .05	28
2.34	$7/2^-$	$2f_{7/2}$	$3p_{1/2}$	N	.55 ± .05	28
2.73	$9/2^+$	$3p_{1/2}$	$2g_{9/2}$	N	.50 ± .05	28
3.51	$11/2^+$	$3p_{1/2}$	$1i_{11/2}$	N	.65 ± .05	28

Table II: Quenching factors Q derived from inelastic scattering to high spin states of ^{208}Pb and neutron hole states of ^{207}Pb. The spin-parity of the transition and its character are also given.

n_{NM}, suitably modified to account for the effects of surface vibrations. The occupation probabilities for nuclear matter have been calculated by Fantoni and Pandharipande[29] using a realistic nucleon-nucleon interaction. In this calculation $n_{NM}(K)$ is found to have little density dependence. The strong energy dependence is assumed to also apply in the case of ^{208}Pb where the following prescription is adopted

$$n_\alpha(\varepsilon_\alpha - \varepsilon_F) = n_{NM}(\varepsilon_\alpha - \varepsilon_F) + \delta n_\alpha^{RPA} \quad (10)$$

where n_α is the occupation number characterizing the shell orbital ϕ_α of ^{208}Pb whose single particle energy is ε_α. The nuclear matter $n_{NM}(\varepsilon - \varepsilon_F)$ occupation number is further reduced (enhanced) for the particle (hole) orbits of ^{208}Pb by δn_α^{RPA} due to the new degrees of freedom associated with the nuclear surface. δn_α^{RPA} has been taken from the calculation of Decharge, Gogny and Sips.[30] Most of the nuclear matter effect is found to be either due to short range correlations or due to interactions induced by the tensor force. The RPA result is derived using a central density dependent interaction (no tensor part). It is thus assumed that double counting effects are small.

The resulting occupation numbers are shown in Figure 15. This distribution indicates that substantial amount of correlations are present in the ground state of ^{208}Pb; it yields an $\eta = 0.8$ (see Eq. 6). Most of the effect, as can be seen in the figure, is due to mechanisms already at work in nuclear matter. It is true however that the surface vibrations contribute substantially to the depletion of the valence orbitals.

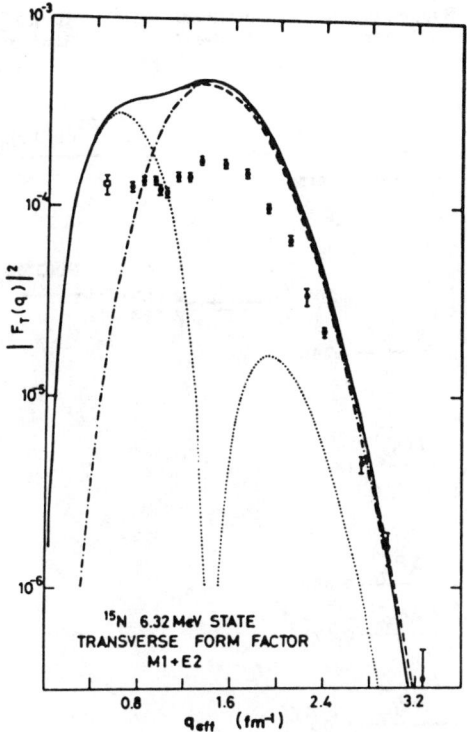

Figure 12: The transverse inelastic scattering from states of known particle (hole) character in nuclei neighboring doubly closed shells is invariably quenched. Example: $p_{1/2} \to p_{3/2}$ proton hole transition[27] in ^{15}N.

In this estimate the surface orbitals are characterized by an occupation of ∼0.65 for the particles and of ∼.08 for hole states lying immediately below the Fermi sea. This according to Eq. 8 will yield a quenching of the single particle transitions associated with them (such as the data we have examined in the previous section) of 0.6 ± 0.1 in good agreement with the data.

CONCLUSIONS

Our basic understanding of nuclear physics requires partial occupancy of the particle shell orbits. Until recently, theoretical uncertainties and the lack of data that can directly yield absolute spectroscopic factors has prevented us from obtaining even a rough empirical estimate of occupation numbers.

The very precise and extensive investigation of the lead region through electron scattering and the inability of mean field theories to account for an almost universal suppression of single particle strength has lead to recent at-

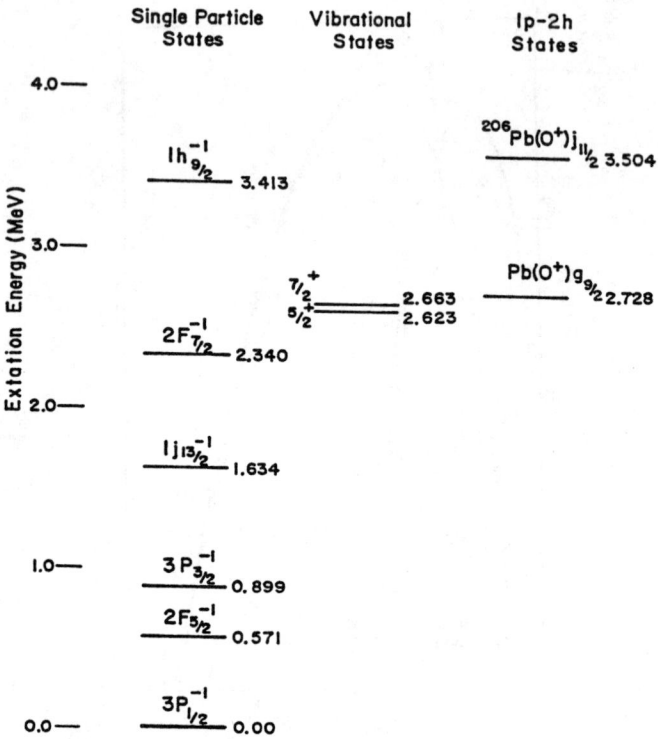

Figure 13: The low energy excitation spectrum of ^{207}Pb. A number of neutron hole states can be studied and quenching factors obtained

tempts to investigate the influence of partial occupancy on our understanding of these data.

By reviewing the available data we find that a procedure can be established for the extraction of quenching factors from inclusive electron scattering data concerning excitations of single particle character. These quenching factors, measured at high momentum transfers, can be related directly to occupation probabilities of shell orbits. However, this procedure has substantial theoretical bias and it is therefore subject to considerable model error. Its power and validity will have to be established from the consistency of its results and agreement with complementary probes.

The examination of a rather wide array of transitions in the lead region, elastic and inelastic, magnetic or electric in character, and ranging in multipolarity from $\lambda = 1$ to $\lambda = 14$ yields, consistently, quenching factors of 0.60 ± 1.5 (see Tables I and II). This result gives credence to recent theoretical estimates[1] that find occupation probabilities of single particle states just below (above) the Fermi energy of ~ 0.7 (~ 0.1).

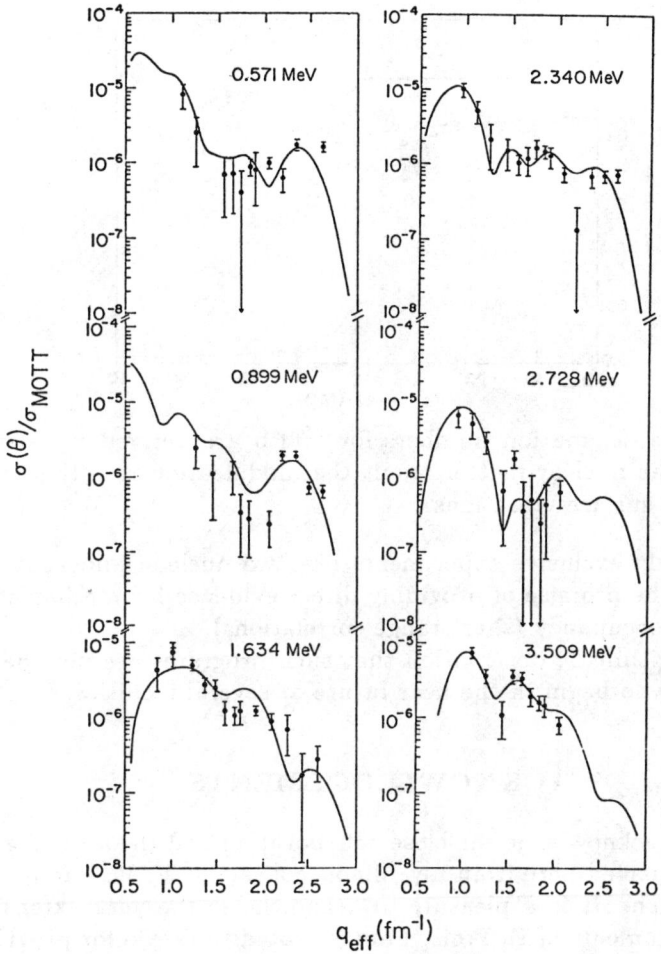

Figure 14: The neutron hole states of ^{207}Pb can be adequately described by the mean field calculation after rescaling.[28] The resulting quenching factors are given in Table II

More precise and more exclusive experiments are needed to get a more accurate and less model dependent handle on occupation numbers. The $(e,e'p)$ program[19] promises to provide very important and complementary information to the data discussed here. The short range correlations which are largely responsible for partial occupancy are also modifying the momentum content of the nucleus, enriching it with high momentum components. This implication can be directly tested by $(e,e'\gamma)$ experiments. The $(e,e'\gamma)$ probe promises to provide the ideal tool for the exploration of the particle (hole) transitions in odd-even systems neighboring doubly closed shells. Fi-

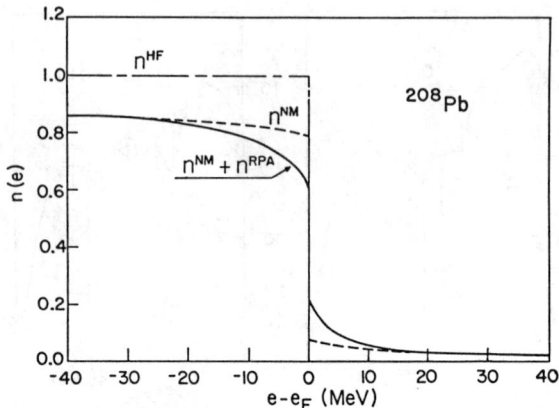

Figure 15: The occupation numbers for ^{208}Pb are derived in reference 1 by adding to the nuclear matter result the modification resulting from the coupling to the surface vibrations.

nally very highly exclusive experiments like two nucleon knockout process (e, e'2N) hold the promise of providing direct evidence for mechanisms that lead to partial occupancy (short range correlations).

With the optimistic observation that such programs are now beginning or are expected to begin in the near future at several facilities, I would like to conclude.

ACKNOWLEDGEMENTS

I would like to acknowledge the close collaboration and theoretical guidance in the work reported here from my Illinois colleagues V. R. Pandharipande and J. Wambach. It is a pleasure to acknowledge the very extensive and penetrating comments of B. Frois. I am indebted to G. Co for providing me with the early results of his calculations and useful discussions. The critical comments and extensive help of S. Williamson during the preparation of this manuscript is gratefully acknowledged. This work has been supported in part by the National Science Foundation under grant NSF PHY 83-11717.

REFERENCES

1. V. R. Pandharipande, C. N. Papanicolas and J. Wambach, Phys. Rev. Lett. **53**, 1133 (1984).
2. M. Jaminon, C. Mahaux and H. Ngô, Nucl. Phys **A440**, 228 (1985).
3. P. O. Lowdin, Phys. Rev. **97**, 1474 (1955).
4. L. Schafer and H. A. Weidemuller, Nucl. Phys. **A174**, 1 (1971).
5. A. B. Migdal, Theory of Finite Fermi Systems and Applications to Atomic Nuclei, (John Wiley & Sons, NY 1967).
6. J. Speth, E. Werner, and W. Wild, Phys. Rep. **33C**, 127 (1977).
7. C. Mahaux and H. Ngô, Nucl. Phys. **A378**, 205 (1982) and C. Mahaux and H. Ngô, Nucl. Phys. **A410**, 271 (1983).
8. C. Mahaux and H. Ngô, Nucl. Phys. **A431**, 486 (1984).
9. J. Lichtenstadt, J. Heisenberg, C. N. Papanicolas, C. P. Sargent, A. N. Courtemanche, and J. S. McCarthy, Phys. Rev. **C20**, 497 (1979).
10. J. W. Negele and P. Vautherin, Phys. Rev. **C5**, 1472 (1972).
11. T. W. Donnelly and I. Sick, Rev. Mod. Phys. **56**, 461 (1984).
12. S. K. Platchkov, J. B. Bellicard, J. M. Cavedon, B. Frois, D. Goutte, M. Huet, P. Leconte, Phan Xuan Ho, P. K. A. deWitt Huberts, L. Lapikas, and I. Sick, Phys. Rev. **C25**, 2318 (1982); P. K. A. deWitt Huberts, in Proceedings of the Conference on Modern Trends in Elastic Electron Scattering, Amsterdam (1978), edited by C. de Vries (Instituut voor Kernphysisch Onderzoek, Amsterdam, The Netherlands, 1978).
13. C. N. Papanicolas, L. S. Cardman, J. Heisenberg, O. Schwentker, T. Milliman, B. Hersman, R. Hicks, G. Peterson, J. S. McCarthy, J. Wise, and B. Frois, to be published.
14. J. Decharge and D. Gogny, Phys. Rev. **C21**, 1568 (1980).
15. T. Suzuki and H. Hyuga, Nucl. Phys. **A402**, 491 (1983).
16. G. Co and J. Speth, private communication and to be published; also see L. Zamick, V. Klemt, and J. Speth, Nucl. Phys. **A245**, 365 (1975).
17. P. Grabmayer, S. Klein, H. Clement, K. Reiner, W. Renter, G. J. Wagner, and G. Seegert, Phys. Lett. **164B** 15 (1985); G. J. Wagner, this volume.
18. E. R. Flynn, R. A. Hardekopt, J. P. Sherman, J. W. Sunier, and J. P. Coffin, Nucl. Phys. **A279**, 394 (1977).
19. P. K. A. deWitt Huberts, this volume and to be published.
20. C. N. Papanicolas, J. Heisenberg, J. Lichtenstadt, D. Goutte, B. Frois, J. M. Cavedon, P. Leconte, M. Huet, Phan Xuan Ho, S. Platchkov, J. S. McCarthy, and I. Sick, Phys. Rev. Lett. **52**, 247 (1984).

21. J. M. Cavedon, Ph.D. thesis, University of Paris (1982) unpublished and B. Frois in Proceedings of the Niels Bohr Centennial Conference, Copenhagen (1985), R. Broglia, E. Hagemann and B. Herskind, eds. N. Holland, Amsterdam, 1985.
22. J. M. Cavedon, B. Frois, D. Goutee, M. Huet, Ph. Leconte, C. N. Papanicolas, X. H. Phan, S. K. Platchkov, and S. Williamson, Phys. Rev. Lett. **49**, 978; B. Frois et al. Nucl. Phys. **A396**, 409c (1983).
23. X. Campi and D. W. L. Sprung, Nucl. Phys. **A194**, 401 (1972).
24. R. A. Lindgren, Journal de Physique **45**, C4-433 (1984).
25. C. N. Papanicolas, J. Heisenberg, J. Lichtenstadt, and J. S. McCarthy, Phys. Lett. **B99**, 96 (1981).
26. J. Heisenberg, J. Lichtenstadt, C. N. Papanicolas, and J. S. McCarthy, Phys. Rev. **C25**, 2292 (1982).
27. R. P. Singhal, J. Dubach, R. S. Hicks, R. A. Lindgren, B. Parker, and G. A. Peterson, Phys. Rev. **C28**, 513 (1983).
28. C. N. Papanicolas, J. Lichtenstadt, C. P. Sargent, J. Heisenberg, and J. S. McCarthy, Phys. Rev. Lett. **45**, 106 (1980).
29. S. Fantoni and V. R. Pandharipande, Nucl. Phys. **A427**, 473 (1984).
30. J. Decharge and L. Sips, Nucl. Phys. **A407**, 1 (1983) and J. Decharge, L. Sips, and D. Gogny, Phys. Lett. **98B**, 229 (1981).

EFFECTS OF MESON EXCHANGE CURRENTS, UNBOUND WAVE FUNCTIONS, AND
KNOCKOUT EXCHANGE AMPLITUDES ON THE EXTRACTION OF PARTICLE-HOLE
STRENGTHS FOR $1\hbar\omega$ STRETCHED STATES

R.A. Lindgren[*]
University of Virginia, Charlottesville, VA 22901

M. Leuschner
University of New Hampshire, Durham, NH 03824

B.L. Clausen and R.J. Peterson
University of Colorado, Boulder, CO 80309

M.A. Plum
Los Alamos National Laboratory, Los Alamos, NM 87545

F. Petrovich[†]
Lawrence Livermore National Laboratory, Livermore, CA 94550

ABSTRACT

It is well known that the strength for excitations of $1\hbar\omega$ high spin, stretched states observed via inelastic scattering is generally much smaller than that predicted by spherical shell model calculations. In addition, results obtained from electromagnetic and hadronic studies have discrepancies at the 20% level. To gain a better understanding of reduced magnetic strength in electron scattering and hopefully close the gap between experiment and theory, calculations of the electron scattering form factors have been performed including the effects due to meson exchange currents in the transition amplitude and the effects due to unbound wave functions for the valence nucleon. The effect of the meson exchange current contributions is to uniformly enhance the form factors near the first maximum resulting in a 16 to 20% further reduction of the stretched particle-hole strength. The effect due to the radial wave functions deduced from Woods-Saxon potentials in which the nucleon is not bound is to reduce the form factors, thereby resulting in an increase in the spectroscopic strength. In regard to the comparison of results obtained with electromagnetic and hadronic probes, the implied sensitivity to higher order current and spin-current transition densities associated with the non-locality due to the tensor knockout exchange amplitudes in nucleon-nucleus scattering is considered explicitly. It is found that the simplest correspondence between electron and nucleon-nucleus scattering is preserved for isovector excitations but not for isoscalar excitations under the usual assumptions for the tensor interaction. It is clear that precise comparisons between experiment and theory (or between probes) cannot be made unless these and related effects are consistently included.

I. INTRODUCTION

The scattering of electrons, nucleons, and pions from the nucleus has provided a large body of information on nuclear structure and effective nucleon- and pion-nucleon interactions.[1] In the simple distorted wave impulse approximation treatment of the scattering of these probes from the nucleus, the interaction is taken to be that between the probe and a nucleon in isolation, the reaction is treated as a one-step direct process between the projectile and individual target nucleons, and the important effects of the scattering of the projectile from the mean field of the target in the entrance and exit channels is taken into account through the use of distorted waves. In electron-nucleus scattering the basic coupling is provided by the weak electromagnetic interaction and distortion effects are small so one expects to obtain rather reliable nuclear structure information. The situation is less clear in the case of the hadronic probes where the coupling interaction is strong and the distortion effects are generally large. Nonetheless, a reasonably coherent qualitative picture of the electromagnetic and hadronic excitation of a variety of classes of nuclear levels has emerged in recent years.

In particular, the available experimental and theoretical information on the electromagnetic and hadronic excitation of the high spin, $1\hbar\omega$ stretched levels observed in light to heavy nuclei has been recently summarized in Refs. 2,3. These levels are characterized by unique, active particle-hole configurations in a "$1\hbar\omega$" spherical shell model basis. Electron scattering is the primary source of information on the isovector stretched strength. Typically, only about 60% of the isovector strength predicted by state of the art spherical shell model calculations is observed experimentally as may be seen in Table 1 drawn from Refs. 2,3. This strength is also more fragmented than the theoretical calculation suggests. The isoscalar stretched strength is most clearly observed with nucleons and pions and the observed isoscalar strength is typically 50% of the isovector strength. In cases where comparisons can be made the stretched strengths deduced from the electromagnetic and hadronic reactions are consistent to within 20% as may be seen in Table 2, also drawn from Refs. 2,3. The nucleon studies which provide information on the effective nucleon-nucleon tensor interaction yield strengths which are on the average 15% lower than those deduced from electron scattering while the pion work which gives information on the effective pion-nucleon spin-orbit interaction yields strengths that are somewhat higher than electron scattering.

The discrepancies between experiment and the shell model results as well as those between the electromagnetic and hadronic probes for stretched excitations noted above may well be related to deficiencies in the structure calculation and/or the oversimplified distorted wave impulse approximation treatment of the hadronic reactions. Indeed, the largest shell model calculations performed to date are still quite restricted[4] and different interaction models for nucleon scattering yield a substantial range of

Table 1. Deduced spectroscopic strength S^2 from (e,e') for stretched states with strong isovector components. HO radial wave functions are assumed.

Nucleus	E_x (MeV)	J^π, T	CONF	b	S^2 (EXP)††	S^2 (THY)††	Z^2 (ESPHM)
^{12}C	19.50	4^-, 1	$d_{5/2}p_{3/2}^{-1}$	1.50	0.37±0.04	0.60	1
^{14}C	24.30	4^-, 2	$d_{5/2}p_{3/2}^{-1}$	1.54±0.08	0.37±0.07	0.42	1/2
^{14}N	16.91	5^-, 1	$d_{5/2}p_{3/2}^{-1}$	1.54±0.03	0.39±0.02	0.41	11/27
^{16}O	18.98	4^-, 1	$d_{5/2}p_{3/2}^{-1}$	1.63±0.03†	0.41±0.02	0.71	1
^{24}Mg	15.05	6^-, 1	$f_{7/2}d_{5/2}^{-5}$	1.85±0.04†	0.27±0.02	0.47	2/3
^{26}Mg	18.1	6^-, 2	$f_{7/2}d_{5/2}^{-3}$	1.82±0.03	0.40±0.02	0.41	1/3
^{28}Si	14.36	6^-, 1	$f_{7/2}d_{5/2}^{-1}$	1.77±0.02†	0.31±0.01	0.55	1
^{48}Ca	9.28	8^-, 4	$g_{9/2}f_{7/2}^{-1}$	1.78±0.03	0.22±0.01	----	1
^{54}Fe	13.26	8^-, 2	$g_{9/2}f_{7/2}^{-3}$	1.90±0.02†	0.51±0.02	0.72	3/8
^{58}Ni	12.50	8^-, 2	$g_{9/2}f_{7/2}^{-1}$	1.93±0.03†	0.18±0.01	0.31	1/2
^{90}Zr	7.36	10^-, 5	$h_{11/2}g_{9/2}^{-1}$	2.08±0.04	0.26±0.03	----	1
^{208}Pb	6.43	12^-, 22	$j_{15/2}i_{13/2}^{-1}$	2.18±0.05	0.54±0.04	1.00	1
^{208}Pb	7.02	12^-, 22	$i_{13/2}h_{11/2}^{-1}$	2.30±0.03	0.56±0.02	1.00	1
^{208}Pb	6.74	14^-, 22	$j_{15/2}i_{13/2}^{-1}$	2.22±0.04	0.56±0.03	1.00	1

† Harmonic oscillator parameter b and normalization S^2 were determined by a least square fit of the DWBA calculated cross section to the data. The center of mass and finite size effects have been included in the form factor during the fitting procedure.

†† $S^2(\ ) = Z^2(\ )/Z^2$(ESPHM) where Z^2 is defined in Refs. 2 and 3 and ESPHM is extreme single particle-hole model.

Table 2. A comparison of S^2 deduced from hadronic probes to S^2 from (e,e') for strong isovector stretched states. HO radial wave functions are assumed.

Target	Probe (x,x')	E_x (MeV)	J^π, T	CONF	S^2 (EXP)	$b^\dagger$	$\frac{S^2(x,x')}{S^2(e,e')}^{\dagger\dagger}$
^{16}O	(p,p')	18.98	4^-, 1	$(d_{5/2}p_{3/2}^{-1})$	0.33	1.70	0.81±0.04
^{16}O	(π^+,π^+)	18.98	4^-, 1	$(d_{5/2}p_{3/2}^{-1})$	0.50	1.62	1.22±0.06
^{16}O	(π^-,π^-)	18.98	4^-, 1	$(d_{5/2}p_{3/2}^{-1})$	0.50	1.62	1.22±0.06
^{16}O	(p,n)	6.37	4^-, 1	$(d_{5/2}p_{3/2}^{-1})$	0.50	1.72	0.80±0.06
^{24}Mg	(p,p')	15.05	6^-, 1	$(f_{7/2}d_{5/2}^{-6})$	0.24	1.86	0.88±0.06
^{28}Si	(p,p')	14.36	6^-, 1	$(f_{7/2}d_{5/2}^{-1})$	0.24	1.78	0.78±0.03
^{28}Si	(π^+,π^+)	14.36	6^-, 1	$(f_{7/2}d_{5/2}^{-1})$	0.32	1.76	1.03±0.03
^{28}Si	(π^-,π^-)	14.36	6^-, 1	$(f_{7/2}d_{5/2}^{-1})$	0.32	1.76	1.03±0.03
^{28}Si	(p,n)	14.36	6^-, 1	$(f_{7/2}d_{5/2}^{-1})$	0.25	1.74	0.81±0.03
^{48}Ca	(p,p')	9.31	8^-, 4	$(g_{9/2}f_{7/2}^{-1})$	0.22	1.90	1.00±0.05
^{54}Fe	(p,p')	13.26	8^-, 2	$(g_{9/2}f_{7/2}^{-3})$	0.45	1.92	0.88±0.03
^{208}Pb	(p,p')	6.74	14^-, 22	$(j_{15/2}i_{13/2}^{-1})$	0.50	2.30	1.12±0.06
^{208}Pb	(p,p')	6.43	12^-, 22	$(j_{15/2}i_{13/2}^{-1})$	0.80	2.30	0.67±0.05
^{208}Pb	(p,p')	7.02	12^-, 22	$(i_{13/2}h_{11/2}^{-1})$	0.20	2.30	2.88±0.88

†The center of mass correction factor has been included in the final determination of S^2 and b.
††Errors on ratios only include uncertainties in S^2(e,e').

results.[5] Nonetheless, any discussion of the extracted strengths at the 20% level must examine a number of effects heretofore not considered. In this paper we consider in some detail the effects of meson exchange currents (MEC)[6-8] on the electron scattering form factors and the associated strength for stretched excitations. The discussion and results are based on the recent approximate scheme for including MEC due to Dehesa and collaborators.[9-12] Since many of the stretched excitations, particularly in light nuclei, are unbound we also consider the effects of using appropriate unbound wave functions on the extraction of the observed strengths from experiment. We employ the prescription of Vincent and Fortune[13] for treating the unbound nucleon in this section and again consider only the electro-excitation of the stretched states. Finally, we note that the usually assumed correspondence[1-3] between the electromagnetic and hadronic reaction mechanisms for exciting stretched levels is based on a "quasi-local" description of the hadronic processes. Non-localities in the latter translate into a sensitivity to higher order current and spin-current transition densities which do not enter in electron scattering. We examine this problem explicitly for the case of the tensor knockout exchange amplitudes in nucleon-nucleus scattering using the recent mixed (coordinate and momentum space) density expansion of the off diagonal density matrix discussed in Refs. 14-17.

II. MESON EXCHANGE CURRENT CORRECTIONS

Firm evidence for two-body meson exchange-current contributions in electron scattering is restricted mainly to the few-nucleon systems (A<4)[8]. In heavier nuclei our knowledge of nuclear structure is too uncertain to allow completely unambiguous conclusions to be drawn about exchange currents. Although such currents must be present at some level, the results of any calculation must be viewed with some caution.

We do know for certain[6-8] that in order to satisfy the continuity equation

$$\vec{\nabla} \cdot \vec{J}(\bar{x}) = - i \, [H, \rho(\bar{x})] \tag{1}$$

where $H = T + V$ and V contains the one-pion exchange potential, the current $J(\bar{x})$ must contain at least two types of meson exchange-current terms in addition to the normal nucleon convection current terms. One type is called the "seagull" or pair current where a virtual nucleon-antinucleon pair is created by a photon (pion) and subsequently destroyed by a pion (photon). The second is called the pionic current where the photon interacts directly with the exchanged pion. The Feynman graphs for these processes are shown in Fig. 1a and 1b. Another type of meson exchange current that may contribute is called the isobar current, which trivially satisfies the continuity equation because it is divergence free. The isobar current arises from excitation of nucleon resonances like the intermediate $\Delta(1232)$. The Feynman diagram for this process if shown in Fig. 1c.

Fig. 1a. Pair (seagull) current.

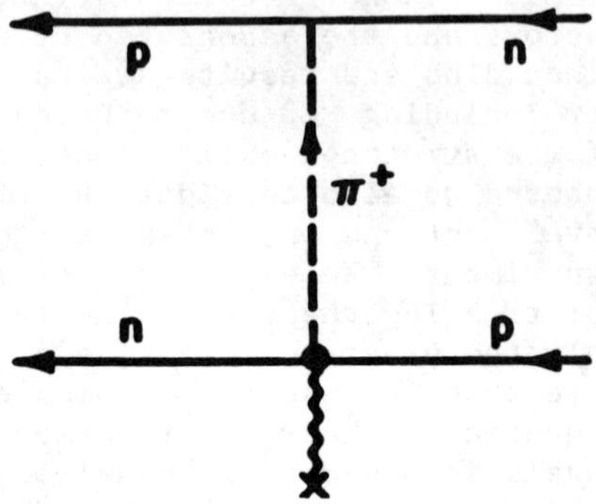

Fig. 1b. Pionic current.

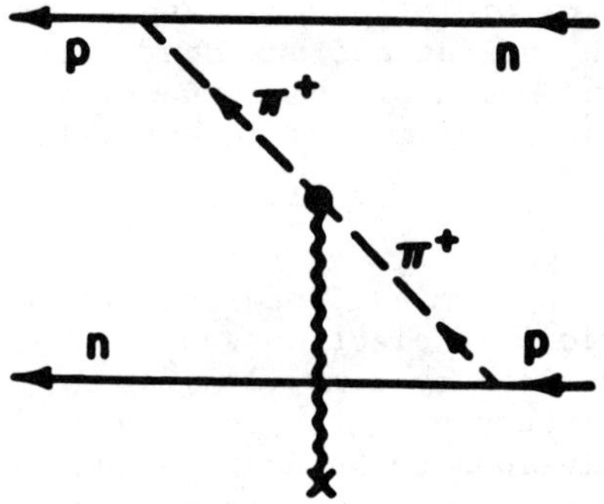

Fig. 1c. Isobar current.

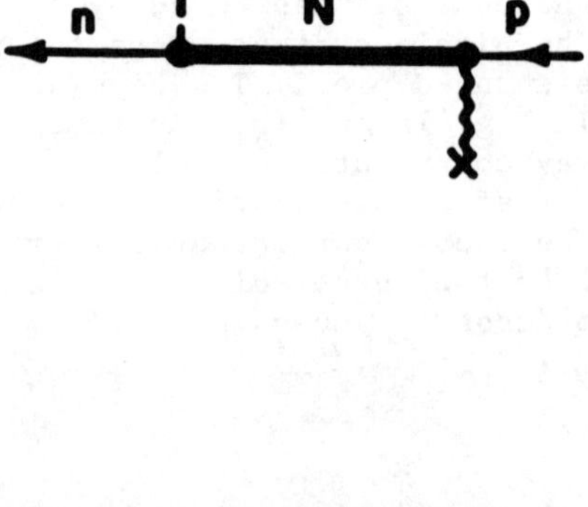

In this section we discuss the pair and pion meson-exchange current corrections to the electron scattering form factor for the high spin magnetic stretched particle-hole states. We have estimated these corrections using the effective operator approach recently devised by Dehesa and co-workers.[9-12] Specifically, from Refs. 9-12 we write the electron scattering cross section for magnetic particle-hole transitions as

$$\frac{d\sigma}{d\Omega} = Z^2 \sigma_M (\frac{1}{2} + \tan^2 \frac{\theta}{2}) F_T^2 \qquad (2)$$

where Z is the nuclear charge, σ_M is the Mott cross section, and F_T is the transverse form factor, which is given by

$$F_T = i \frac{\hat{J}}{Z} f_{cm} f_{ns} \sum_{ph} [X^*_{ph} + (-1)^{J+1} Y^*_{ph}] \hat{j}_p \hat{j}_h (-1)^{\ell_p}$$

$$\zeta(\ell_p + \ell_h + J + 1) \begin{pmatrix} j_p & j_h & J \\ -\frac{1}{2} & -\frac{1}{2} & 1 \end{pmatrix} \int_0^\infty dx\, x^2 j_J(qx) G_{ph}(x) \qquad (3)$$

with

$$G_{ph}(x) = G^{spin}_{ph}(x) + G^{orbital}_{ph}(x) + G^{pair}_{ph}(x) + G^{pionic}_{ph}(x) \qquad (4)$$

$$G^{spin}_{ph}(x) = \frac{\mu}{2M} \left\{ -\frac{d}{dx} + [1 + \frac{J(J+1)}{K_p + K_h}] \frac{1}{x} \right\} R_p(x) R_h(x) \qquad (5)$$

$$G^{orbital}_{ph}(x) = \frac{\varepsilon}{2M} [K_p + K_h + 1 - \frac{J(J+1)}{K_p + K_h}] \frac{1}{x} R_p(x) R_h(x) \qquad (6)$$

$$G^{pair}_{ph}(x) = -\frac{f_\pi^2}{m_\pi^2} \bar{v}_\pi \sum_{h'<F} \hat{j}_{h'}^2 (\delta_{ph,\upsilon} \delta_{h',\pi} - \delta_{ph,\pi} \delta_{h',\upsilon})$$

$$[R_h, R_h, I^{(2)}_{ph} + R_p R_h, I^{(2)}_{h'h}] \qquad (7)$$

$$G^{pionic}_{ph}(x) = \frac{f_\pi^2}{m_\pi^2} \bar{v}_\pi \sum_{h'<F} \hat{j}_{h'}^2 (\delta_{ph,\upsilon} \delta_{h'\pi} - \delta_{ph,\pi} \delta_{h'\upsilon})$$

$$[K_p + K_h + K_{h'} - \frac{J(J+1)}{K_p + K_h}] \frac{1}{x} I^{(2)}_{ph'} I^{(2)}_{h'h} \qquad (8)$$

where

$$I^{(2)}_{ab} = \frac{d}{dx} + [(K_a + K_b + 2)/x]R_a R_b \qquad (9)$$

$$K_a = (\ell_a - j_a)(2j_a + 1) \qquad (10)$$

and f_{cm} and f_{ns} are the center of mass and nucleon finite size correction factors. M and m_π are the nucleon and pion mass, respectively. The constants μ and ε are the magnetic moment and charge of the particle-hole excitation. The forward going and backward going RPA amplitudes are X_{ph} and Y_{ph}, the pion-nucleon coupling constant $f_\pi^2 = 0.079$, and $\bar{v}_\pi$ is a phenomenological effective pion propagator given by

$$\bar{v}_\pi = \frac{1}{(\frac{1}{40}Q)^{1/2} q + 1.2 + 0.24\ Q^2} \qquad (11)$$

where Q is identified as that momentum transfer at which the one-body form factor has its dominant maximum. The summation on h' runs over all occupied states in the nucleus and $\upsilon(\pi)$ is the conventional notation for neutron (proton). The factor $\zeta(a) = 1$ or 0 depending on whether a is even or odd.

The pair and pionic parts of the M4 form factor for the $(d_{5/2}p^{-1}_{3/2})4^-$ transition in ^{16}O calculated using this effective operator approach are shown in Figs. 2a and 2b and compared to an exact calculation. Radial wave functions were determined using a Woods-Saxon potential. The effective operator calculation for the pair term is slightly higher than the exact result at the first (<2%) and second maximum (<10%). Somewhat poorer agreement is obtained for the smaller pionic term. Similar comparisons for the M14 $(j_{15/2}i^{-1}_{13/2})14^-$ transition in ^{208}Pb are shown in Figs. 3a and 3b. Similar discrepancies are noted for the pair term with slightly larger discrepancies occurring for the weaker pionic terms. We conclude that the effective operator approach is sufficiently accurate for making pair and pionic MEC corrections to the transverse form factor for stretched states particularly in the lighter targets and near the first maximum. The results of calculations (based on the effective operator approach) of the form factors for the $(f_{7/2}d^{-1}_{5/2})$ M6 transition to the $E_x = 14.36$ MeV, T=1, $J^\pi = 6^-$ state in ^{28}Si are shown in Fig. 4 and compared to experiment. The total, spin, pair, and pionic contributions are shown separately. The pair and pionic terms come in with positive and negative signs relative to the spin term roughly in the proportion 1.13: -1.05: 1.00. The resultant effect is to increase the dominant peak of the form factor over the spin part by about 18% using radial wave functions generated from a Woods-Saxon Well (WS). A 16% increase is obtained when harmonic oscillator radial wave functions (HO) are used. Therefore, the fraction of the extreme single particle shell model strength decreases from 0.43 to 0.36 using WS and 0.31 to 0.27 using HO when MEC are included. This, of course, increases

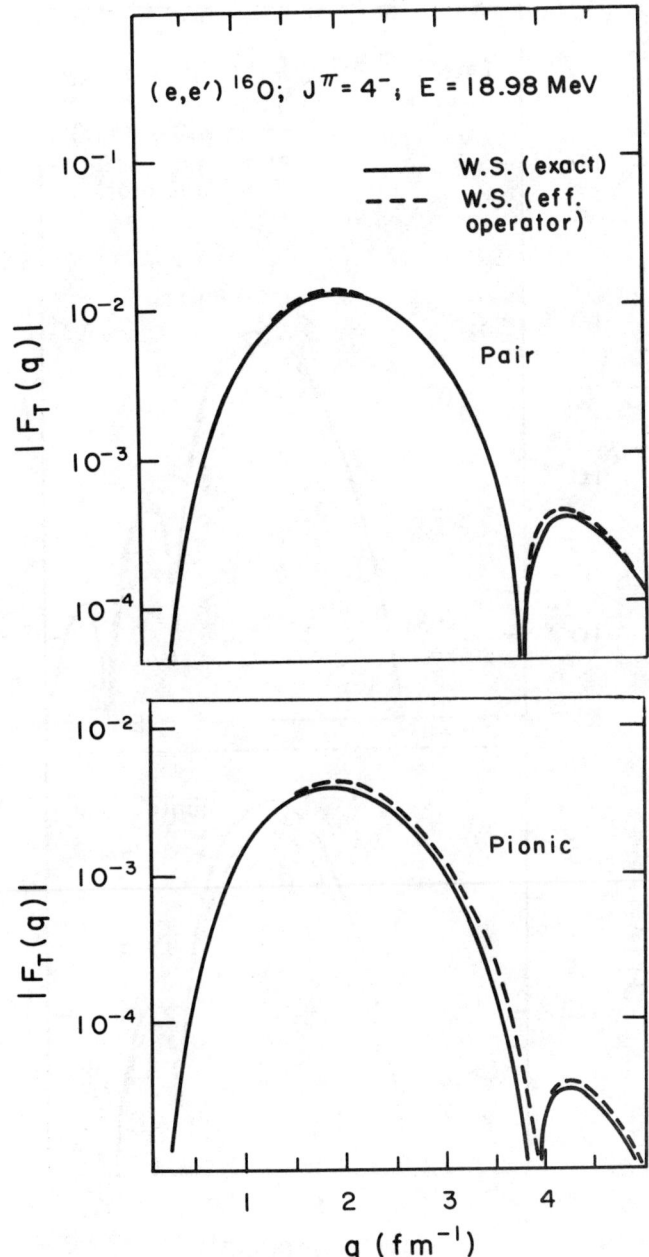

Fig. 2. A comparison of the exact evaluation of the electron scattering form factor due to the pair (2a) and pionic (2b) currents with an approximate evaluation using effective operators (Ref. 10).

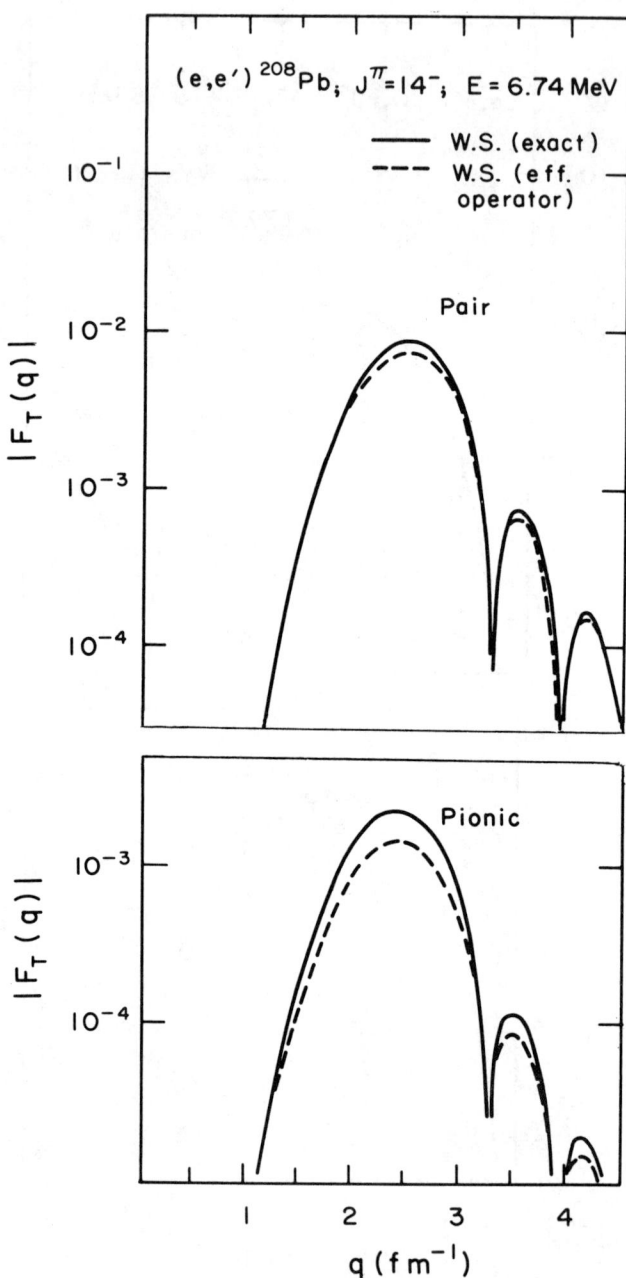

Fig. 3. A comparison of the exact evaluation of the electron scattering form factor due to the pair (3a) and pionic (3b) currents with an approximate evaluation using effective operators (Ref. 10).

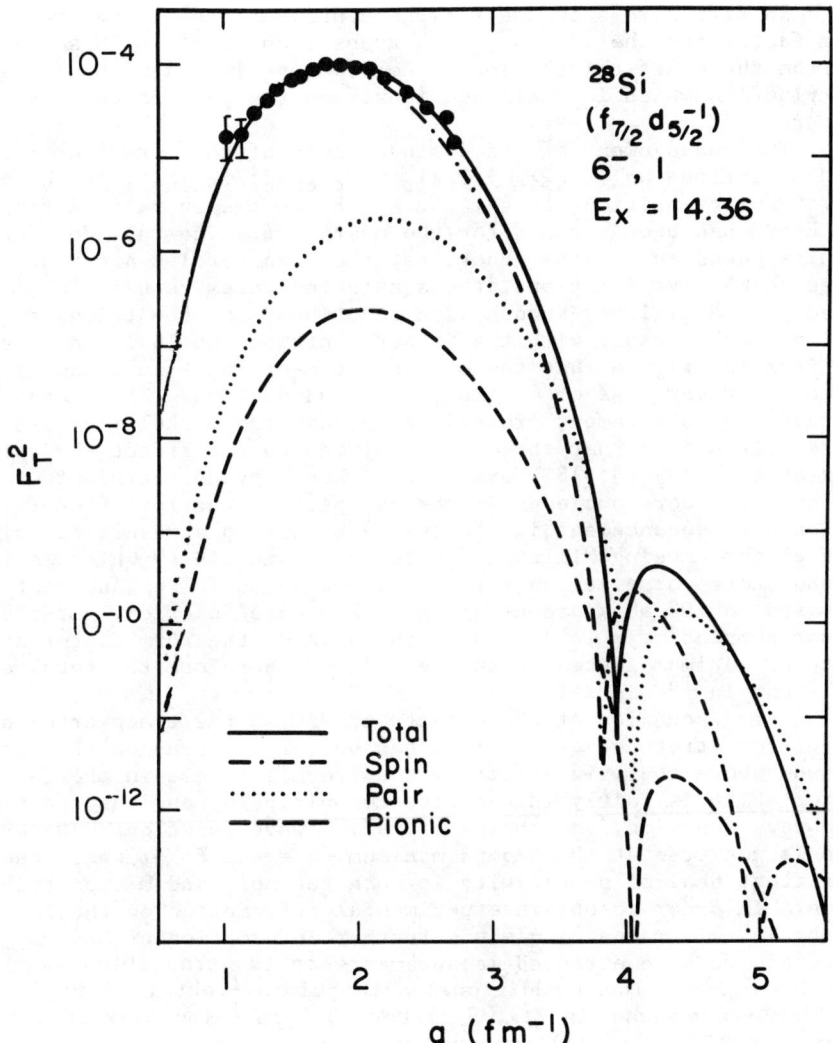

Fig. 4 A plot of the square of the electron scattering form factor due to the one-body nucleon spin current, the two-body pair and pionic currents and the total are compared for the $J^\pi = 6^-, E_x = 14.36$ MeV state in ^{28}Si. A $(f_{7/2}d_{5/2}^{-1})6^-$ particle-hole Woods-Saxon wave function is assumed for the valence pair.

the discrepancy between experiment and shell model theory. In
addition it is noted that MEC enhancement is not too sensitive to
the choice of radial wave functions.

In Fig. 5 we show the various contributions to the transverse
form factor for the $(d_{5/2}p_{3/2}^{-1})$ M4 transition in ^{14}C. This result
is from the exact calculation of Ref. 18–19 and includes the isobar
contribution which is small and increases the peak of the form
factor by 3%.

The enhancement of the dominant peak of the form factor due
to MEC, defined as $[(F_{spin} + F_{pair} + F_{pionic})/F_{Spin}]^2$, is shown in
Fig. 6 for transitions to 4^-, 6^-, and 8^- states in various targets.
The ESPHM has been assumed for the most part. The full dots are
results based on HO wave functions, the open circles are results
based on WS wave functions, the square indicates results for ^{14}C
based on the Millener-Kurath wave functions,[20] and the triangle
indicates the result with the isobar contribution included. We
see from the figure that the MEC enhancement falls in a band from
15 to 20% regardless of multipolarity and nucleus. The gradual
increase in enhancement proceeding through the nickel isotopes
reflects the fact that the added neutrons do not affect the spin
part of the $(g_{9/2}f_{7/2}^{-1})8^-$ form factor, but they do contribute to the
sum over the core nucleons in the MEC part of the form factor. The
increase in enhancement in ^{14}C over ^{12}C is also a result of this
same effect. Using Millener-Kurath wave functions[20] which include
ground state correlations for ^{14}C we obtain an 18% enhancement
compared to 20% obtained using the ESPHM wave function. If the
isobar current term is included, the peak of the form factor at the
dominant maximum increases another 3% and therefore the total enhancement in ^{14}C is 21%.

Our discussion of the effects of MEC on the transverse form
factor for stretched excitations has so far centered on the first
maximum where the main effect is a uniform increase in the form
factor which is fairly insensitive to multipole, nucleus, particle-
hole wave function, and choice of radial wave function. However,
this is not true at the second maximum, $q = 4.5$ fm^{-1}, where there
is a great deal of sensitivity in both the spin and MEC contributions. In order to obtain experimental information on the form
factor in this range of q where further information of MEC may be
expected, we have extended measurements on the transition in ^{28}Si
to this region. These additional data points[21] obtained for
$q > 3$ fm^{-1} are shown in Fig. 7. Above 3.5 fm^{-1} low peak cross
sections (<10^{-38} cm^2/sr) combined with background prevented unambiguous measurement of the yield to the 14.36 MeV state. The
arrow at the lower end of the error bars indicates this uncertainty.
The sources of the background are under study and plans are being
made to reduce it by placing more stringent requirements on the
electron trajectories that are considered bonafide electrons scattered from the target.

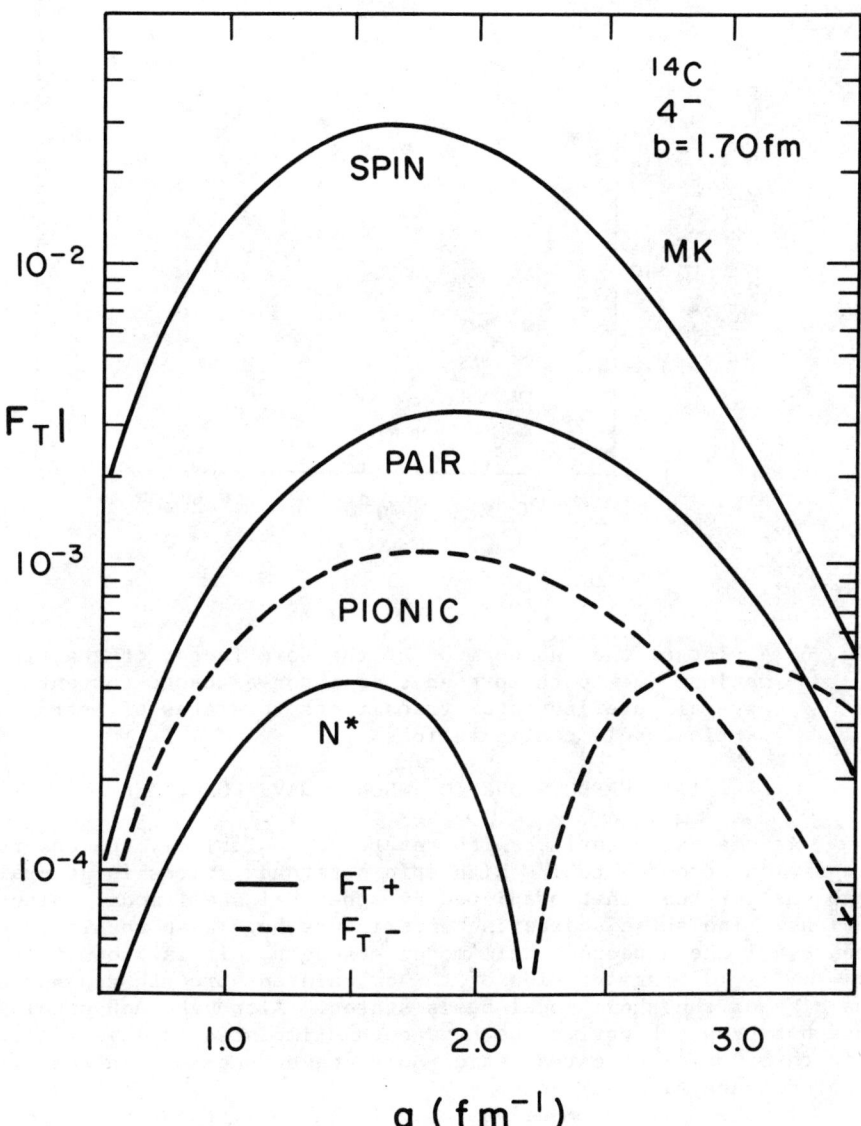

Fig. 5 A plot of the electron scattering form factor illustrating the relative importance of the spin, pair, pionic, and isobar current contributions for a pure isovector M4 transition in ^{14}C (Ref. 18-19).

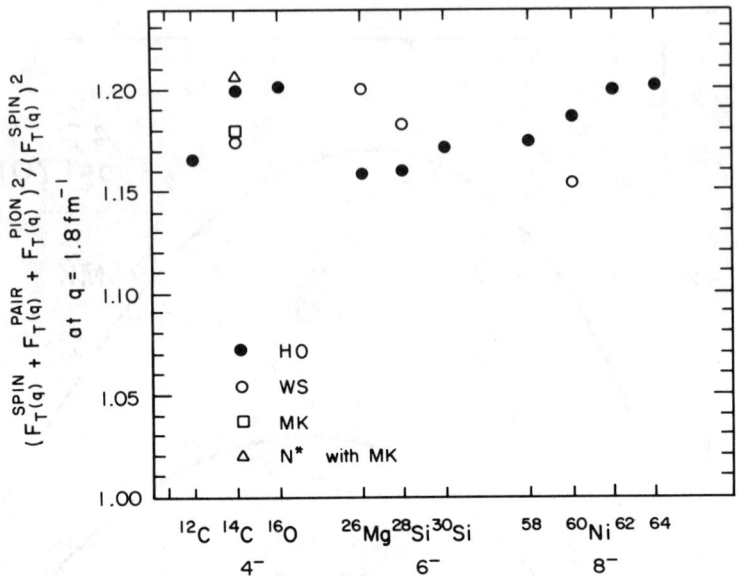

Fig. 6 A plot of the enhancement of the form factor at the first maximum due to the presence of meson-exchange currents for several pure isovector transitions to states of stretched particle-hole configurations.

III. EFFECTS DUE TO UNBOUND WAVE FUNCTIONS

As discussed earlier with reference to Table 1, the observed transition strength to $1\hbar\omega$ high-spin stretched states is generally much smaller than that predicted by spherical shell model calculations. Inelastic scattering experiments have been unable to locate all the expected shell model strength. It is known that the degree of fragmentation of the calculated strength depends on the size of the shell model basis states. Although such studies have not been exhaustive, we have been stimulated to investigate the effects of using more realistic radial wave functions on the extracted transition strength.

In the simplest model as used in Table 1, the proton and neutron radial wave functions are taken to be the same and calculated from a harmonic oscillator potential well. Obviously, this is inadequate since the nuclear mean field near the nuclear surface is not harmonic and the Coulomb potential has not been taken into account. Furthermore, many of the stretched states of interest lie above the proton or neutron separation energies and, therefore, are unbound to nucleon emission. Since the neutron is not constrained by the Coulomb barrier, its radial distribution can be quite

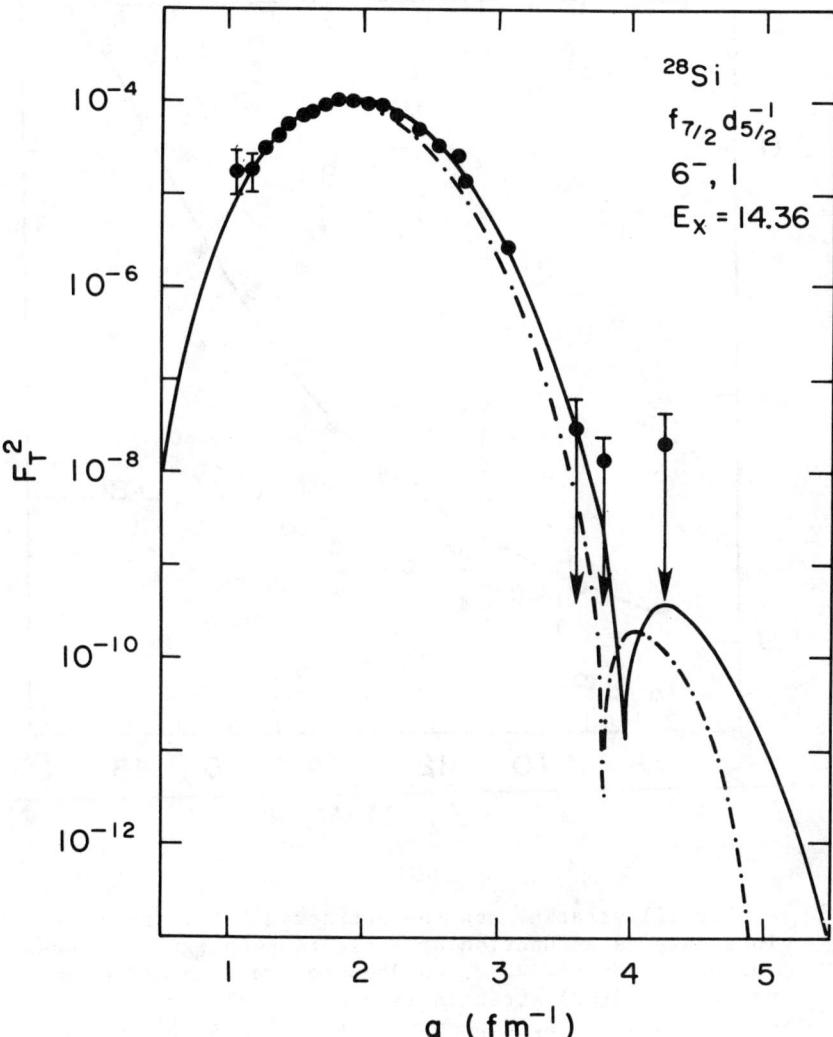

Fig. 7. A plot of the square of the electron scattering form factor calculated with the meson-exchange currents (full line) and without meson-exchange currents (dot-dash line) is compared to experiment.

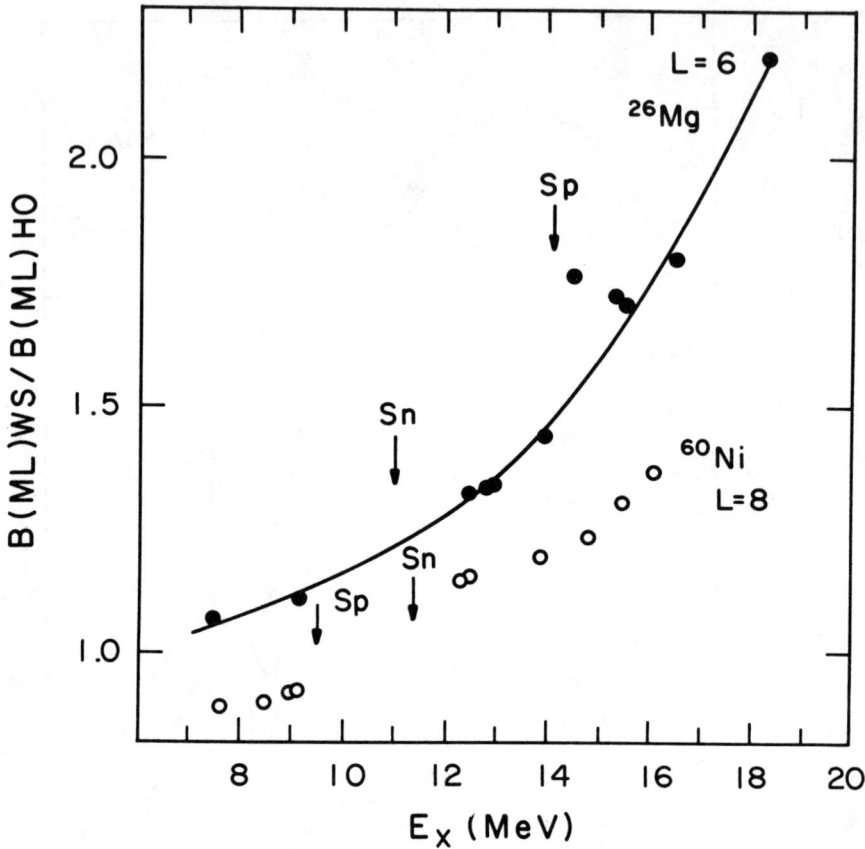

Fig. 8. A plot illustrating how the extracted B(ML) strength increases as a function of excitation energy when bound and unbound (Ref. 22) Woods-Saxon wave functions are used. The B(ML) strength is plotted relative to that extracted from the data using harmonic oscillator wave functions, B(ML)HO, which is assumed to be independent of energy. Each point corresponds to a measured transition in ^{26}Mg or ^{60}Ni. S_n and S_p indicate the neutron and proton separation energies, respectively, for ^{26}Mg and ^{60}Ni.

different from the proton and quite different from that calculated in a harmonic oscillator well. In this section of the paper the form factors are calculated for M6 transitions in ^{26}Mg and M8 transitions in ^{60}Ni using radial wave functions computed for nucleons in a harmonic oscillator potential and for nucleons in a Woods-Saxon potential. The initial wave function $d_{5/2}$ for ^{26}Mg and $f_{7/2}$ for ^{60}Ni is computed by adjusting the well depth until the proton or neutron hole is bound at -1 MeV. The unbound final wave function $f_{7/2}$ for ^{26}Mg and $g_{9/2}$ for ^{60}Ni is computed by adjusting the well depth until a resonance is produced at the excitation energy relative to the A-1 target. The algorithm established by Vincent and Fortune[13] in the computer code DWUCK4[23] was used to calculate the radial wave functions. A radial parameter r_o = 1.25 fm, diffuseness a = 0.65 fm and spin-orbit strength λ = 25 was used for ^{26}Mg and r_o = 1.173, a = 0.65, and V_{so} = -28 MeV was used for ^{60}Ni. The magnitude of the form factor was adjusted until it fit the experimental one for both harmonic oscillator and Woods-Saxon wave functions. The ratio of the extracted strength using the two models is plotted for ^{26}Mg and ^{60}Ni as a function of excitation energy in Fig. 8. Since the strength B(ML)HO is independent of excitation energy, the monotonically increasing curve shows that the B(ML)WS transition strength increases by more than a factor of two over the range E_x = 8 to E_x = 18 MeV in ^{26}Mg.

Using WS wave functions the extreme shell model M6 T=1 → T=2 strength is exhausted by the transition to the J^π = 6^-, T=2 18.1 MeV level in ^{26}Mg whereas if HO wave functions are used only 41% is exhausted as shown in Table 1. This difference mainly arises because the neutron radial wave function peaks further out on the nuclear surface resulting in lower intrinsic cross sections (or form factors) when integration over the nucleus is performed. The proton radial wave function does not change much since it is quasi-bound by the Coulomb potential. The increase in strength is not as dramatic in ^{60}Ni. In light of this sensitivity to the choice of the nuclear mean field to calculate wave functions for unbound levels, further studies are required before any conclusive statements concerning missing strength can be made, particularly for light nuclei.

IV. NON-LOCALITY IN HADRONIC SCATTERING

The electroexcitation of a stretched level proceeds through a unique diagonal or local transition density to the extent that $3\hbar\omega$ components in the wave functions for these levels are negligible.[1-3] This transition density is characterized by orbital and spin angular momentum transfer to the target nucleus L = J - 1 and S = 1 where J represents the total angular momentum transfer. The explicit nuclear matrix element corresponding to this density is given by[1-3]

$$\rho^s_{JJ-1}(r) = \langle J_f || \frac{\delta(r - r')}{r'^2} [Y_{J-1}(\hat{r}') \times \vec{\sigma}]^J || J_i \rangle \qquad (12)$$

where reference to isospin has been suppressed. The correspondence between the electromagnetic and hadronic excitation of stretched levels is based on "quasi-local" descriptions of the hadronic reactions which result in expressions for the hadronic cross section that involve ρ_{JJ-1}^s as the lone nuclear structure factor.[1-3]

Any non-locality in the transition amplitude for hadronic scattering, whether it arises from a basic non-locality in the effective interaction, channel coupling, recoil, or the exchange amplitudes associated with antisymmetrization, introduces an approximate dependence on additional local transition densities which may be classified as higher order current or spin-current densities. To see this explicitly we introduce the off-diagonal transition density which enters the hadronic transition amplitude when a source of non-locality is present,

$$\rho(\bar{r}, \bar{r}') = \sum_{\alpha\beta} C_{\alpha\beta} \phi_\beta(\bar{r}) \phi_\alpha(\bar{r}') \tag{13}$$

where the ϕ are single nucleon target wave functions and the $C_{\alpha\beta}$ are amplitudes containing the Z spectroscopic coefficients and some geometrical factors. We then write the mixed density expansion of $\rho(\bar{r},\bar{r}')$ taken from Refs. 14-17.

$$\rho(\bar{r}, \bar{r}') = \sum_{\substack{L'\\LL''\\s}} \int A_L(k)\, j_{L'}(kr'')\, \rho_{JL''}^{sL'L}(k,r)\, k^2 dk$$

$$\times \left[[Y_{L'}(\hat{r}'') \times Y_L(\hat{r})]^{L''} \times \sigma^s \right]^J \tag{14}$$

Here $\bar{r}'' = \bar{r}' - \bar{r}$ and the mixed density is given by

$$\rho_{JL''}^{sL'L}(k,r) = \sum_{\alpha\beta} C_{\alpha\beta} <\phi_\beta(\bar{r}') \left|\left| \frac{\delta(r-r')}{r'^2} \right.\right.$$

$$\times \left[[Y_L(\hat{r}') \times T_{L'}(\nabla')]^{L''} \times \sigma^s \right]^J \left|\left| \phi_\alpha(k,\hat{r}') \right> \tag{15}$$

with

$$T_{L'}(\nabla') = [\nabla' \times \nabla' \times \ldots]^{L'} \tag{16}$$

representing the stretched gradient operator and

$$\phi_\alpha(k,\bar{r}') = j_{\ell\alpha}(kr') R_\alpha(k) [Y_{\ell\alpha}(\hat{r}') \times \chi_{1/2}(\bar{\sigma})]^{j\alpha} \tag{17}$$

denoting a radially transformed single particle wave function. The mixed density satisfies the condition

$$\int \rho_{JL''}^{sL'L}(k,r)k^2 dk = \rho_{JL''}^{sL'L}(r) \tag{18}$$

with $\rho_{JL''}^{sL'L}(r)$ defined in analogy with eq. 12 and

$$\rho_{JL''}^{s0L}(r) = \delta_{LL''} \rho_{JL}^{s}(r). \tag{19}$$

To the extent that the factor $A_{L'}(k)j_{L'}(kr'')$ in eq. (14) can be expanded in powers of k or factored from the integral over dk, i.e., $A_{L'}(k)j_{L'}(kr'') \cong F_{L'}(r,r'')$, then $\rho(\bar{r}, \bar{r}')$ can be written approximately as

$$\rho(\bar{r}, \bar{r}') \cong \sum_{\substack{L' \\ LL'' \\ s}} F_{L'}(r,r'') \rho_{JL''}^{sL'L}(r)$$

$$\times [[Y_{L'}(\hat{r}'') \times Y_L(\hat{r})]^{L''} \times \sigma^s]^J \tag{20}$$

which establishes the initial contention. Whenever terms with $L' \neq 0$ in eq. (20) contribute substantially to a hadronic cross section, the reaction is sensitive to higher order current and spin-current correlations in the nucleus as represented by the local densities $\rho_{JL''}^{sL'L}$ with $L' \neq 0$.

As an explicit example of the above considerations we show in Fig. 9 the decomposition of the plane wave born approximation tensor knockout exchange cross sections for the 4^- T=0 and T=1 stretched levels in ^{16}O excited by 135 MeV protons. The results are from Ref. 14, 15, 17 and assume the free, local nucleon-nucleon t matrix interaction of Love and Franey.[24] It is clear from Fig. 9 that there are substantial contributions to the exchange cross sections from terms in eq. (14) or (20) with $L' \neq 0$. Not shown are the direct cross sections which depend only on ρ_{JJ-1}^{s}. For the T=1 transition this peaks at 35° with a magnitude of 1.51 mb (about 37 times greater than the peak exchange cross section) and for the T=0 transition this peaks at 55° and has a magnitude of .074 mb (about 12 times less than the peak exchange cross section). From these results we conclude that the (p,p') reaction is providing a reasonable measure of ρ_{JJ-1}^{s} in the case of T=1 stretched excitations and that T=0 stretched excitations are sensitive to details of nuclear structure not contained in ρ_{JJ-1}^{s}. These conclusions depend on the assumed properties of the tensor force and do not address questions associated with other possible sources of non-locality. This example has been presented mainly to illustrate a useful method for discussing the relationship between non-locality and conventional spectroscopy. We note in passing that Ref. 12 contains expressions for the MEC in terms of nuclear densities associated with higher order derivative operators which has some parallels with the discussion of the exchange non-locality presented here. In addition, there have been some related discussions of non-

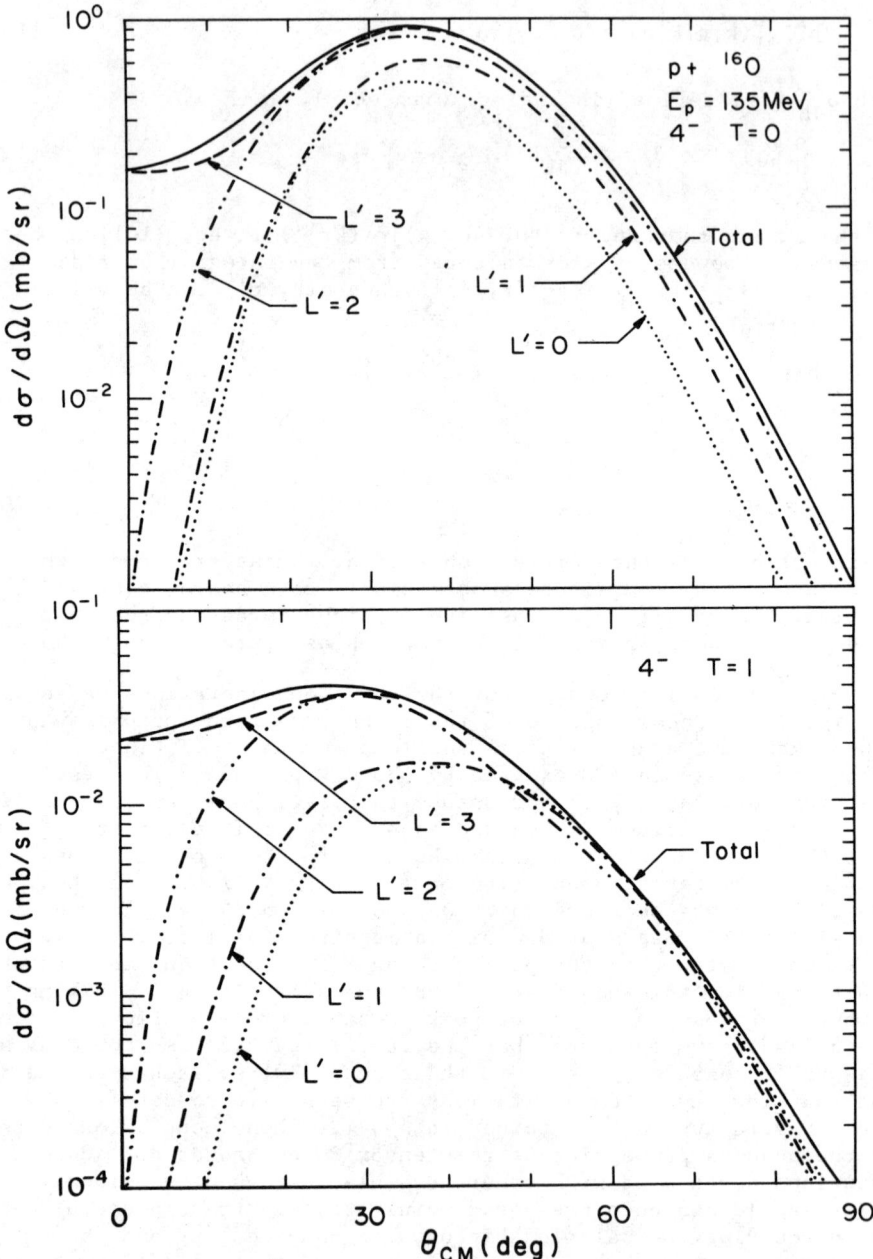

Fig. 9. The plane wave Born approximation tensor knockout exchange cross section for the excitation of the 4^- T=0 and T=1 stretched level in ^{16}O by 135 MeV protons. Curves labeled $L' = x$ contain the contributions from terms in eq. (14) with $L' = 0,1,...x$.

localities in pion-nucleus scattering in the recent literature.[25-27]

V. SUMMARY

The effects of meson-exchange currents on the form factors for stretched excitations are to increase the square of the form factor near the first maximum by about 20% and produce weaker secondary maxima at higher q. The increase in cross section results in a decrease in extracted spectroscopic strength by the same percentage (20%). This nominal figure is approximately independent of multipolarity, mass number, and the details of the shape of the radial wave function. On the other hand, for levels above the neutron separation energy, the intrinsic strength due to the one-body spin-current density decreases in almost direct proportion to increases in the excitation energy when Woods-Saxon wave functions are used to calculate the form factors. This results in an increase in extracted spectroscopic strength and strongly suggests a re-examination of the question of quenching for high-spin stretched states.

Including the effects of MEC in the analysis of electron scattering data yields spectroscopic strengths in closer agreement with those extracted from (p,p') for the same transitions. Results herein show that effects due to the exchange non-locality do not strongly affect isovector transitions to stretched states as much as they do for isoscalar transitions. Consequently, even when such effects are included, the simple relationship between (e,e') and (p,p') is mostly preserved for isovector transitions to stretched states. This is not to say, however, that there are not other sources of non-locality and even effects due to MEC that could be important in (p,p'), but not have been calculated. Similar remarks apply to pion-nucleus scattering as well. In conclusion, in order that precise comparisons can be made between experiment and theory (and between probes), the effects discussed herein such as meson exchange currents, unbound wave functions, non-localities, and related effects must be consistently treated in these reactions.

ACKNOWLEDGMENTS

The authors would like to thank J. Dubach, R.S. Hicks, R.L. Huffman, and G.A. Peterson, University of Massachusetts, and J.A. Carr, R.J. Philpott, and A.W. Carpenter, Florida State University, who have collaborated in obtaining some of the unpublished data and theoretical results that appear in this paper.

REFERENCES

*   On leave from University of Massachusetts, Amherst, MA 01003.
†   On sabbatical leave from Florida State University, Tallahassee, FL 32306.

1. See for example Spin Excitations in Nuclei, ed. F. Petrovich, G.E. Brown, G.A. Garvey, C.D. Goodman, R.A. Lindgren, and W.G. Love (Plenum, New York, 1984) and references therein.
2. R.A. Lindgren and F. Petrovich, Spin Excitations in Nuclei, ed. F. Petrovich et al. (Plenum, New York, 1984) p. 323.
3. R.A. Lindgren, Journal de Physique Colloque C4, Supplement No. 3, Tome 45, Mars 1984, p. 433.
4. A. Amusa and R.D. Lawson, Phys. Rev. Lett. 51, 103 (1983).
5. C. Olmer, The Interaction Between Medium Energy Nucleons in Nuclei, AIP Conf. Proc. No. 97, ed. H.O. Meyer (AIP, New York, 1983) p. 176.
6. J. Dubach, J.H. Koch, and T.W. Donnelly, Nucl. Phys. A271, 279 (1976).
7. J. Dubach, Nucl. Phys. A340, 271 (1980).
8. J.L. Friar, New Vistas in Electronuclear Physics, 1985 NATO Advanced Study Institute, Banff, Alberta, Canada, to be published.
9. A.M. Lallena, J.S. Dehesa, and S. Krewald, Phys. Lett. 146B, 294 (1984).
10. J.S. Dehesa, S. Krewald, A. Lallena, and T.W. Donnelly, Nucl. Phys. A436, 573 (1985).
11. J.S. Dehesa, A.M. Lallena, and S. Krewald, Anales de Fisica A81, 169 (1985).
12. A.M. Lallena, J.S. Dehesa, and S. Krewalk, Phys. Rev. C, in press.
13. C. Vincent and H.T. Fortune, Phys. Rev. C2, 782 (1970).
14. F. Petrovich, A. Carpenter, J. Philpott, and J.A. Carr, Bull. Am. Phys. Soc. 30, 1269 (1985).
15. J. Philpott, A. Carpenter, F. Petrovich, and J.A. Carr, Bull. Am. Phys. Soc. 30, 1269 (1985).
16. F. Petrovich, J.A. Carr, R.J. Philpott, A.W. Carpenter, and J. Kelly, Phys. Lett. 165B, 19 (1985).
17. F. Petrovich, A.W. Carpenter, R.J. Philpott, and J.A. Carr, to be published.
18. M.A. Plum, R.A. Lindgren, J. Dubach, R.S. Hicks, R.L. Huffman, B. Parker, G.A. Peterson, J. Alster, J. Lichtenstadt, M.A. Moinester, and H. Baer, Phys. Lett. 137B, 15 (1984).
19. J. Dubach, private communication.
20. J. Millener and D. Kurath, Nucl. Phys. A255, 315 (1975).
21. R.A. Lindgren, M. Leuchsner, R.S. Hicks, R.L. Huffman, G.A. Peterson, and R. Singhal, to be published.
22. R.J. Peterson, B.L. Clausen, and R.A. Lindgren, to be published.
23. P.D. Kunz, DWUCK Program, University of Colorado, unpublished.
24. W.G. Love and M.A. Franey, Phys. Rev. C24, 1073 (1981).
25. C. Wilkin, Nucl. Phys. A220, 621 (1974).
26. E.R. Siciliano and G.E. Walker, Phys. Rev. C23, 2661 (1981).
27. F. Lenz, Spin Excitations in Nuclei, ed. F. Petrovich et al., (Plenum, New York, 1984) p. 267.

STRETCHED STATE EXCITATIONS IN (P,N) REACTIONS

B.D. Anderson, J.W. Watson and R. Madey
Kent State University, Kent, Ohio

ABSTRACT

Studies of stretched-state excitations with the (p,n) reaction above 100 MeV both complement and significantly extend studies of stretched states performed with other reactions. Because of the isospin selectivity of the (p,n) reaction, it provides a valuable suppression of isoscalar excitations from self-conjugate nuclei. This suppression has been used successfully to identify previously unidentified $(f_{7/2}, d_{5/2}^{-1})$ 6^- fragmented, stretched-state strength in the A=32 and 40 systems. For target nuclei with a neutron excess, the (p,n) reaction can excite T-1 strength that is unavailable in simple inelastic-scattering reactions. Such excitations include not only "1 $\hbar\omega$" but also "0 $\hbar\omega$" stretched-state strength. The latter involves even parity, proton-particle, neutron-hole couplings within the same orbital. The "0 $\hbar\omega$" excitations dominate the (p,n) spectra at large momentum transfer. Usually they are concentrated in a single state (even in medium- and heavy-mass nuclei) because both the particle and the hole states are at the Fermi level. Normalization factors required for comparison with distorted wave impulse approximation calculations are greater than 0.5 and are significantly larger than those required for similar comparisons with "1 $\hbar\omega$" stretched-state excitations. We find that large basis shell-model calculations, when available, are able to account for most of the "missing" strength in these stretched-state excitations.

1. INTRODUCTION

The (p,n) reaction above about 100 MeV has been shown to be impulsive and dominated by single-step reactions in most nuclei.[1-3] This is similar to the case for the (p,p') inelastic-scattering reaction,[4] and it is appropriate to consider the (p,n) reaction as inelastic-scattering with charge exchange. However, the (p,n) reaction offers some important differences from (p,p') which can be used to advantage for studying simple excitations in nuclei. First of all, the (p,n) reaction is completely isovector, while the (p,p') reaction can be both isovector and isoscalar. Thus the (p,n) reaction can provide an isospin selectivity not posible with the (p,p') reaction. This selectivity can, and should, be used to unambiguously identify isovector excitations. We will present some examples of this use of the (p,n) reaction, including to identify fragmented isovector stretched-state strength.

Perhaps the "best" form of inelastic scattering available today is the (e,e') reaction. Inelastic electron scattering offers two important advantages over inelastic scattering with hadronic probes: (1) the basic (Coulomb) force is well known, and (2) because the basic force involved is about 100 times weaker than the strong nuclear force, the reaction is more obviously impulsive and single step. Backward-angle inelastic electron scattering is preferentially isovector and can also be used to suppress isoscalar strength; however, as we will see below, because the (p,n) reaction involves a different force than does (e,e'), the two reactions excite normal and non-normal parity transitions in differing ratios, so that it turns out that the (p,n) reaction provides the cleanest excitation of stretched states. Certainly, we do not wish to imply that the (e,e') studies are not required. In fact, we need the (e,e') results in order to check the absolute normalizations of spectroscopic factors.

We note here also that the (p,n) reaction offers the important possibility of exciting strength in the final nucleus with isospin one less than the target (T-1). Simple inelastic scattering is of course limited to isospins equal to or greater than that of the target. The significance of this difference lies in the fact that various effects, such as configuration mixing, sometimes depend strongly on isospin. For example, in stretched-state excitations, we generally see less fragmentation in the lower isospin components.

Thus the (p,n) reaction provides important differences compared with other types of inelastic-scattering reactions.

2. "1 $\hbar\omega$" STRETCHED-STATE EXCITATIONS FROM SELF-CONJUGATE NUCLEI

The normal stretched state is a "1 $\hbar\omega$" excitation formed by taking a particle from the orbital with largest J-value in one shell and promoting it to the orbital with largest J-value in the next major shell. If the excited particle couples with the resulting "hole" to form the maximum possible J-value, then the state so formed is referred to as a stretched state. (Actually it should be called the maximum stretched state between the two major shells.) Perhaps the two most widely studied examples are the 4^- state formed with the $(d_{5/2}, p_{3/2}^{-1})$ particle-hole configuration and the 6^- state formed with the $(f_{7/2}, d_{5/2}^{-1})$ configuration. The first example can be formed in inelastic scattering on the spherical nucleus ^{16}O, and the second example on the nucleus ^{28}Si. In both cases, there is no other way to couple a single particle with a single hole from the two shells considered to make such a state. Excitations to the next major shell will have the wrong parity, so that 3 $\hbar\omega$ excitations, or multparticle-multihole excitations, are required to form another state with the same spin and parity.

In Fig. 1 we show comparisons of (p,n), (p,p'), and (e,e') excitations at large momentum transfer for targets of ^{16}O and ^{28}Si.[5] All three of these reactions are seen to excite the 4^- and 6^-, "1 $\hbar\omega$" stretched states in A = 16 and 28, respectively. The important thing to note in Fig. 1 is how much alike the (p,n) and (e,e') spectra are, in contrast to the (p,p') spectra. As discussed earlier, the (p,n) and (e,e') reactions will excite only $\Delta T = 1$ strength on a self-conjugate nucleus, whereas the (p,p') reaction will excite a large background of $\Delta T = 0$ strength as well. In some nuclei, as we shall see, suppression of the T = 0 background makes it possible to identify fragmented T = 1 strength which otherwise could not be clearly seen. It is noteworthy just how similar are the (p,n) and (e,e') spectra in Fig. 1. The (e,e') reaction must necessarily proceed through the relatively weak Coulomb force, so

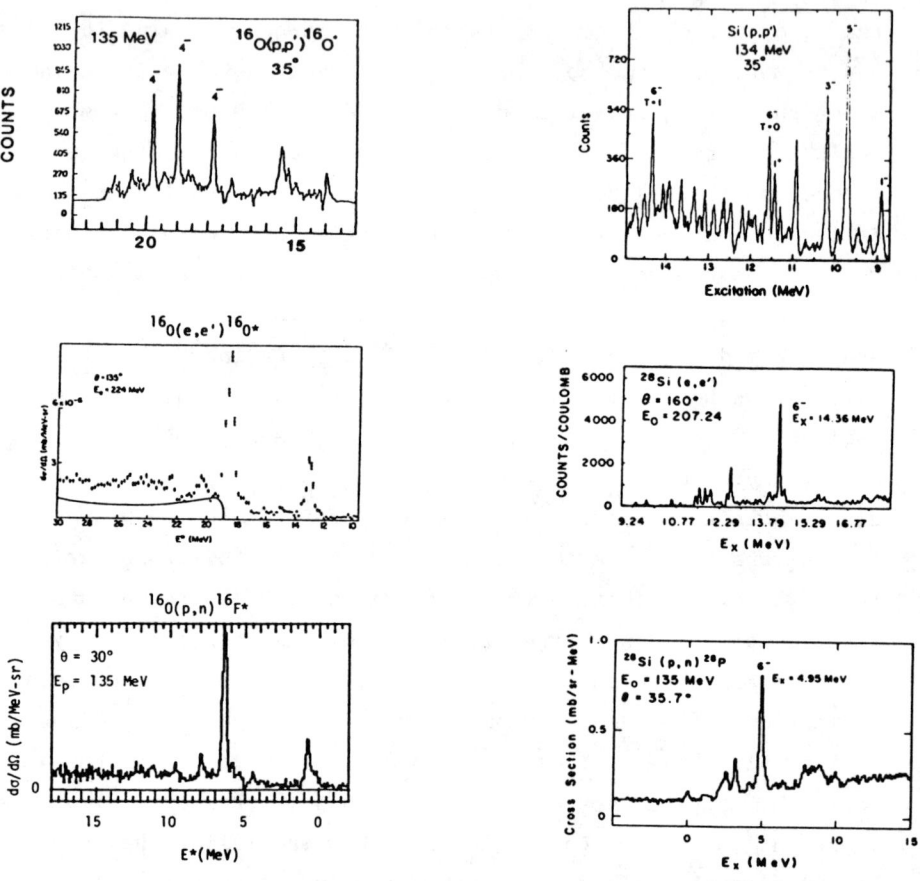

Fig. 1. Excitation-energy plots for the (p,p'), (e,e') and (p,n) reactions at large momentum transfer on ^{16}O and ^{28}Si.

that one can be confident that (e,e') is a single-step impulsive reaction. The similarity of the (p,n) and (e,e') spectra indicates that both reactions are exciting the same kind of strength.

The extracted (e,e'), (p,p'), and (p,n) angular distributions for the T = 1, 4^- and 6^- stretched state excitations are shown in Fig. 2 (from Ref. 5). All three angular distributions for each transition are seen to be described well by impulse approximation calculations. Note that the (p,p') and (p,n) normalization factors (i.e., normalization factors required to make the DWIA calculations agree with experiment) agree with each other very well, but are about 20% smaller than those obtained with the (e,e') reaction. The (p,n) and (p,p') normalization factors are expected to be the same since the transitions are to analog states excited by the same interaction, viz., primarily by the isovector tensor term of the N-N effective interaction; however, this term is poorly determined in the analysis of free N-N scattering and may be in error in the N-N t-matrix used in the DWIA. On the other hand, the interaction involved for the (e,e') reaction is well-known, so that the normalization factor extracted is probably more reliable. Thus these comparisons may indicate that the tensor strength in the N-N effective interaction[6] used in the DWIA calculations is too strong by about 10% (in amplitude).

A good example of taking advantage of the isospin selectivity of the (p,n) reaction is shown in Fig. 3 for the ^{40}Ca(p,n)^{40}Sc reaction at 36 . A broadly distributed amount of strength is seen centered near E_x = 7 MeV, which we identify as the ($\pi f_{7/2}, \nu d_{5/2}^{-1}$), 6^-; T = 1 strength for this reaction. The angular distribution extracted for this region is shown in Fig. 4. Although there is clearly some ΔL = 1 strength also in this region, the wide-angle part of the spectrum is dominated by ΔL = 5, consistent with a transition to a 6^- state as shown. Earlier studies[7] with the (p,p') reaction on this nucleus were unable to identify any T = 1, 6^- strength because of a large background of T = 0 states in this region, and for a while it was thought that such strength was missing for this reaction. The reason why this strength is

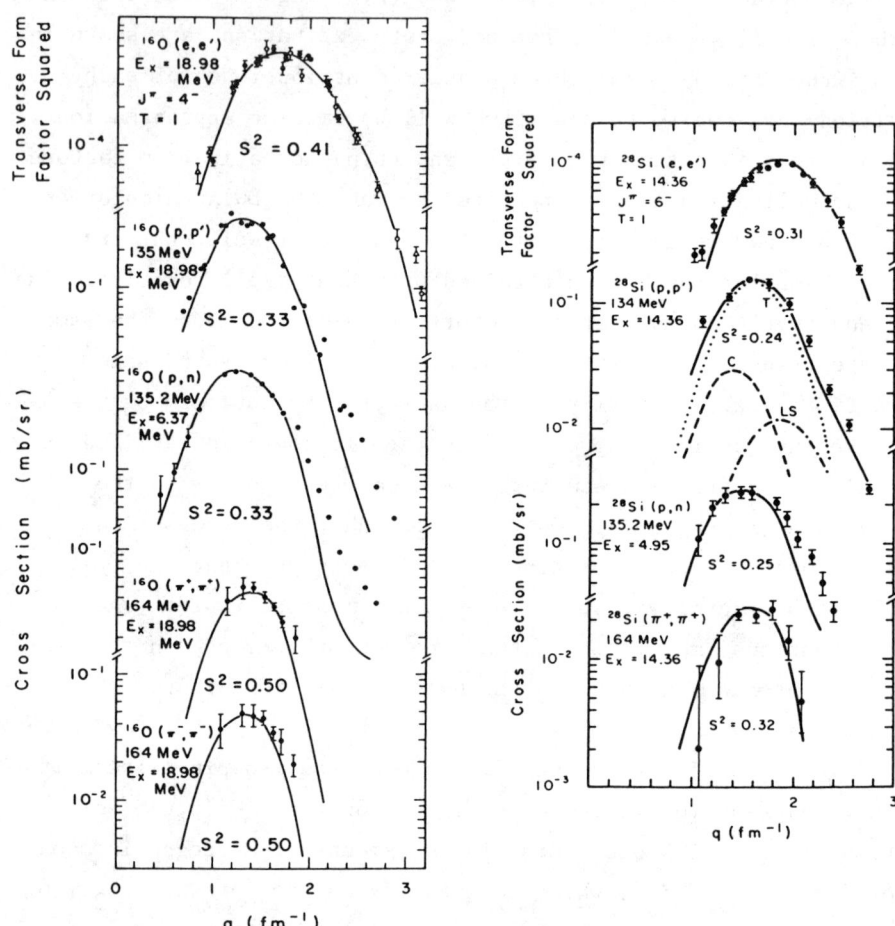

Fig. 2. Momentum-transfer distributions for the excitation of the 4^- and 6^- stretched states in A = 16 and 28 with various inelastic-scattering reactions.

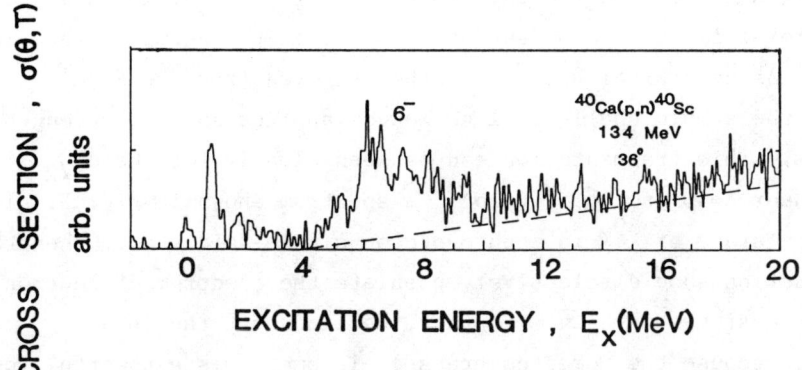

Fig. 3. Excitation-energy spectrum for the ^{40}Ca(p,n)^{40}Sc reaction at 134 MeV and 36°.

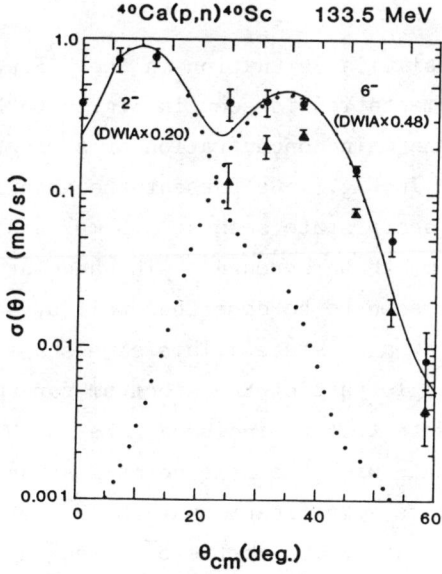

Fig. 4. Angular distribution for the excitation of the 6⁻ complex in the ^{40}Ca(p,n)^{40}Sc reaction.

fragmented in A = 40 but is apparently concentrated in A = 28 is almost certainly due to the fact that the $d_{5/2}$ orbital in mass 40 is several MeV below the Fermi surface, but is at the Fermi surface in mass 28. As an orbital becomes farther removed from the Fermi surface, the simple shell-model no longer applies and the strength fragments. This fragmentation can be seen clearly for the $d_{5/2}$ orbital near A=40 in the ^{39}K(p,n)^{39}Ca spectrum shown in Fig. 5. If the ground state of ^{39}K is a pure proton single-hole state, then the (p,n) reaction should selectively populate the spectrum of neutron single-hole states in ^{39}Ca. In fact, we find that the (p,n) reaction, because the momentum transfer is small, is a powerful tool for mapping hole-state strength distributions in odd-A nuclei. From Fig. 5 we see that while the $d_{3/2}$ orbital is concentrated in the ground state, the $d_{5/2}$ orbital is fragmented over several MeV. Considering this, and similiar evidence from pickup reactions,[8] we would be surprised if the 6^- stretched state were not fragmented also.

We observe a similiar situation in the ^{32}S(p,n)^{32}Cl (6^-;T=1) reaction. The fragmentation is seen in Fig. 6 to be spread out over several MeV, with the main concentration of strength between 3 and 8 MeV of excitation. In Fig. 7 we present the angular distributions for the relatively sharp state seen at 5.6 MeV and for all of the strength (above a smooth background). The angular distribution for the single state is seen to be described well by a DWIA calculation for the excitation of a 6^- state. This same shape is seen to describe the wide-angle part of the spectrum for the total strength quite well also. Note that we included here the DWIA prediction for strength to a 5^- state with the same normalization factor as for the 6^- calculation. It is significant that the 5^- strength is predicted to be only about 10% of the 6^- strength in this reaction. This is an important feature of the (p,n) excitation of stretched states and is due to the fact that the (p,n) reaction, when dominated by the tensor term of the N-N effective interaction, predominantly excites non-normal parity states. We will return to the experimental verification of this theoretical prediction below.

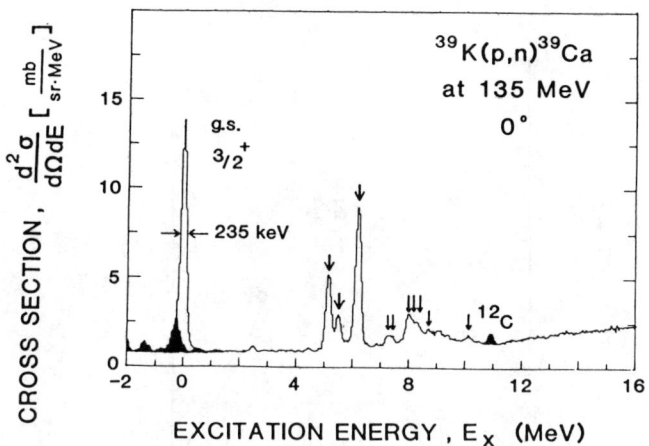

Fig. 5. Excitation-energy plot for the ^{39}K(p,n)^{39}Ca reaction at 0°. The arrows indicate peaks identified to be 5/2$^+$ strength.

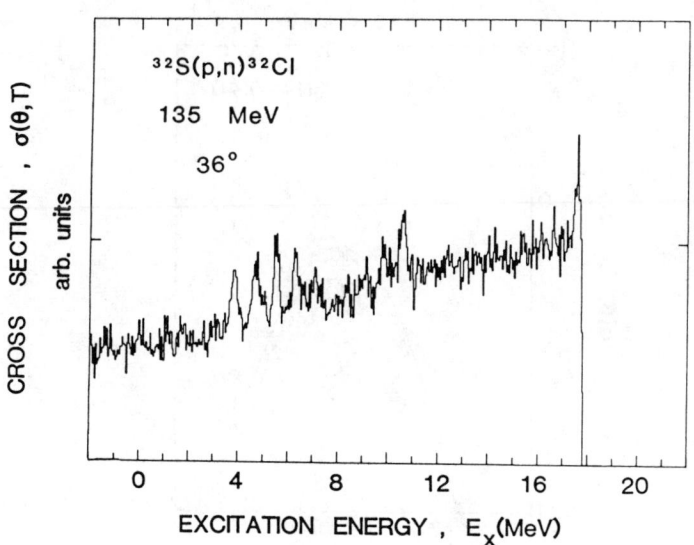

Fig. 6. Excitation-energy spectrum for the ^{32}S(p,n)^{32}Cl reaction at 135 MeV and 36°.

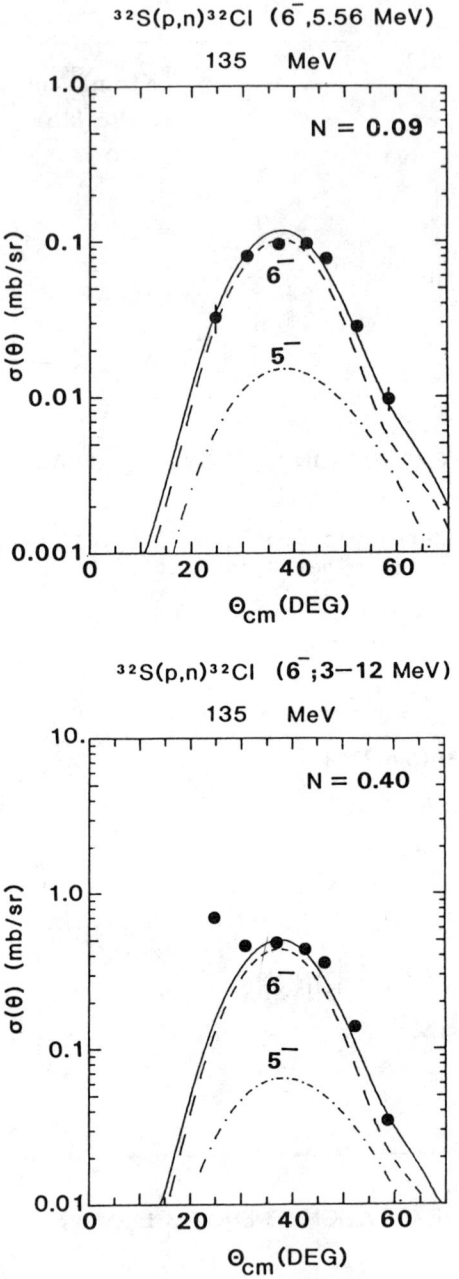

Fig. 7. Angular distributions for the sharp state seen at 5.6 MeV (top) and the entire 6⁻ complex (bottom), in the ^{32}S(p,n)^{32}Cl reaction.

It is worth noting that the total 6^- strength observed with the (p,n) reaction on ^{40}Ca and ^{32}S is somewhat greater than that observed on ^{28}Si (40% vs 25% of the simple 1p-1h strength). A plausible explanation is that the $d_{5/2}$ orbital is more nearly full in the two heavier nuclei than in ^{28}Si, where it is the valence orbital.

3. "0 $\hbar\omega$" STRETCHED-STATE EXCITATIONS

A charge-exchange reaction, such as a (p,n) reaction, offers the possibility of exciting "0 $\hbar\omega$ stretched states. These excitations involve a 1p-1h configuration with both particle and hole in the same orbital, coupled to the maximum possible J value. If this coupling occurs in the orbital with maximum J in a major shell, then this stretched state is also unique within 3 $\hbar\omega$ of excitation, similar to a 1 $\hbar\omega$ stretched state. Such excitations can be formed only by a charge-exchange reaction on a nucleus with a neutron (or proton) excess. In Fig. 8 we show the excitation of four such 0 $\hbar\omega$ stretched states with the (p,n) reaction at 135 MeV. The states are seen to be always at low excitation energy and to dominate the wide angle spectra for these nuclei. The extracted angular distributions for these transitions are shown in Fig. 9, and are seen to be described well by DWIA calculations assuming the simple 1p-1h structures for the appropriate 0 $\hbar\omega$ state. The DWIA calculations are "standard" calculations performed with the code DWBA70[9] that use the global optical model parameters of Schwandt et al.[10] and the N-N effective interaction of Love and Franey.[6] The angular distributions are all peaked at large angle (large momentum transfer) and are dominated by the strength from the tensor term in the N-N effective interaction, just as "1 $\hbar\omega$" stretched states are. The most significant fact about these excitations is that the strength is highly concentrated. This fact is seen both in the excitation-energy spectra and in the relatively large normalization factors extracted from the DWIA analyses. For 1 $\hbar\omega$ stretched-state excitations, the extracted normalization factors are always less than 0.5 (see Fig. 2). For the 0 $\hbar\omega$ stretched states shown in Fig.

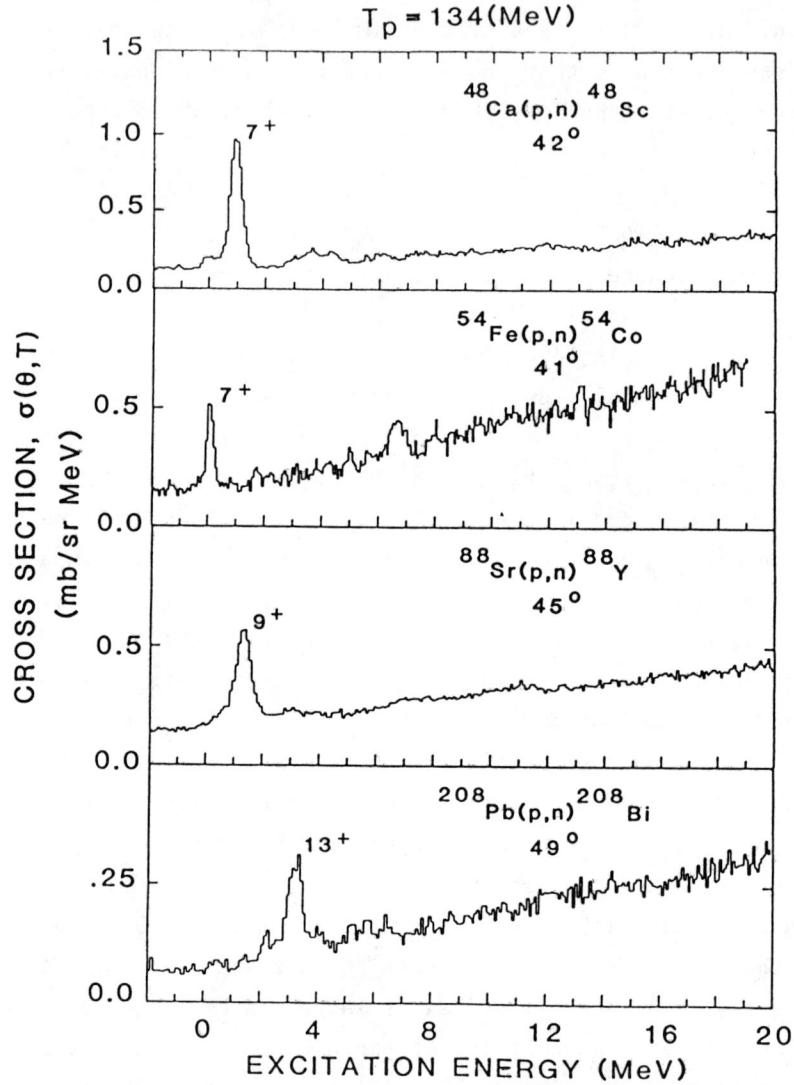

Fig. 8. Excitation-energy spectra at large momentum transfer showing "0 $\hbar\omega$" states in medium- and heavy-mass nuclei.

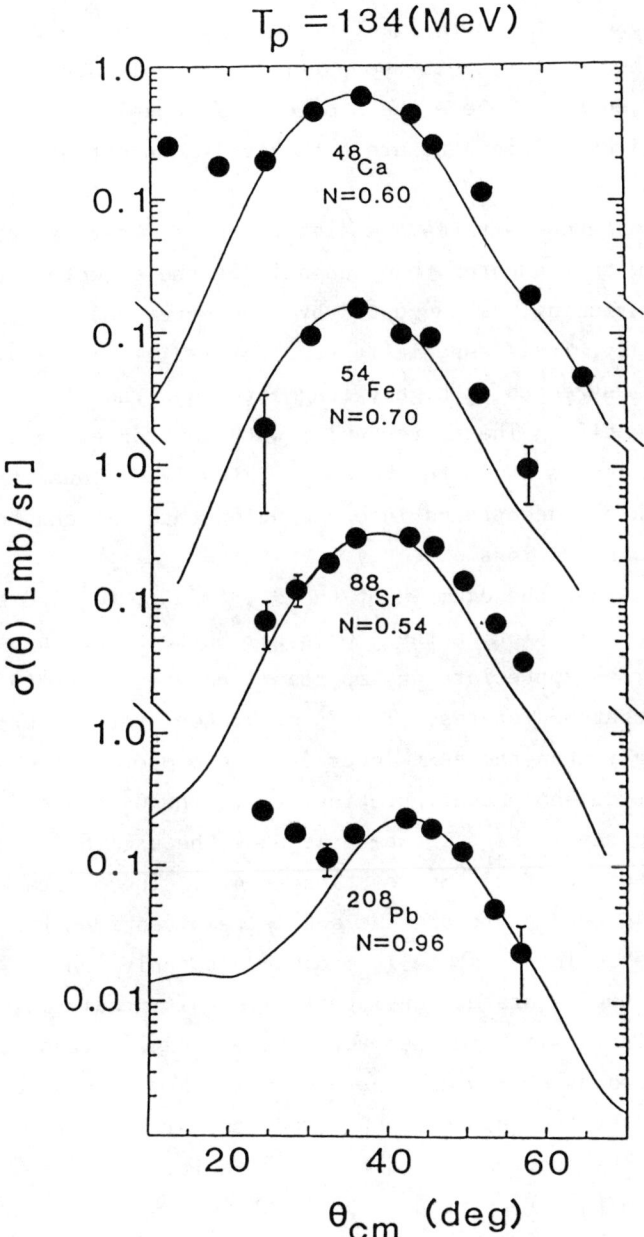

Fig. 9. Angular distibutions for the "0 $\hbar\omega$" stretched-state excitations of Fig. 8.

9, we see that the normalization factors are all greater than 0.5. We attribute these larger normalization factors to the fact that for the 0 $\hbar\omega$ stretched states, both the particle and hole are at (or near) the Fermi surface (where the simple shell model applies) and the particle and hole strengths are both highly concentrated into single states.

The 0 $\hbar\omega$ stretched-states are significant for several reasons. First, they show that greater than one-half of the expected strength is seen and not missing, as suggested by 1 $\hbar\omega$ stretched state studies. Secondly, it is especially true for medium-and heavy-mass nuclei that 1 $\hbar\omega$ strength is highly fragmented and therefore difficult to identify. These are nuclei with neutron excesses and just where 0 $\hbar\omega$ states can often be seen. Finally, because the 0 $\hbar\omega$ strength is highly concentrated into a single state, it can be used to study core polarizations directly.

Let us consider the example of the 7^+, $(\pi f_{7/2}, \nu f_{7/2}^{-1})$ 0 $\hbar$ stretched state excitation in the ^{48}Ca(p,n)^{48}Sc reaction in some detail, in order to appreciate an important feature of the (p,n) excitation of stretched states. The 7^+ stretched state is excited much more strongly than the next lower J-value member of the same $(f_{7/2}, f_{7/2}^{-1})$ particle-hole configuration, viz., the 6^+ state. This fact is seen clearly in Fig. 10 where we show the excitation-energy spectrum from the ^{48}Ca(p,n)^{48}Sc reaction at 42. The 6^+ state is known to be the ground state of ^{48}Sc and is resolved from the 7^+ state at 1.10 MeV. The 6^+ state is excited with only about 5% of the strength of the 7^+ state, consistent with DWIA predictions for these two transistions. Note that both the 6^+ and 7^+ transitions would be dominated by $\Delta L = 6$ angular momentum transfer and could not be distinguished by the shapes of the experimental angular distributions. The suppression of the 6^+ transition is due apparently to the fact that natural parity transitions dominated by the tensor term of the effective interaction will be excited only via exchange processes, which are expected to be weak.[11] This suppression is important because it means that nearly all of the strength observed with the correct L-transfer for the transition to

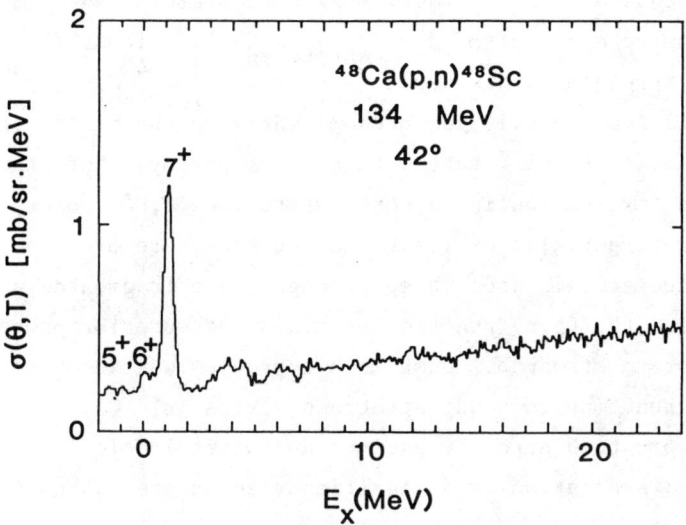

Fig. 10. Excitation-energy spectrum for the ^{48}Ca(p,n)^{48}Sc reaction at 42°.

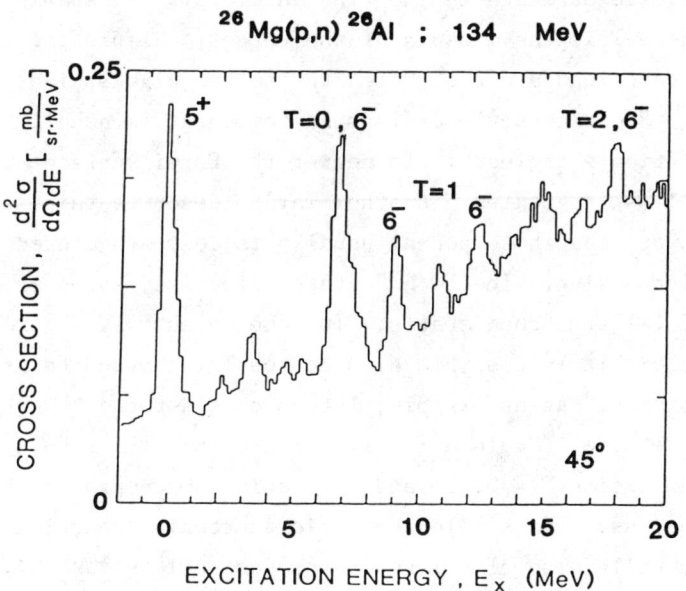

Fig. 11. Excitation-energy spectrum for the ^{26}Mg(p,n)^{26}Al reaction at 45°.

the stretched state is, in fact, stretched-state strength and not strength to the state with $J = J_{stretched} - 1$. Thus, for example, the $\Delta L = 5$ strength we discussed earlier for the $^{40}Ca(p,n)^{40}Sc$ and $^{32}S(p,n)^{32}Cl$ reactions is due predominantly to the 6^- stretched state and not to the 5^- state with natural parity. Note that this preferential excitation is in contrast to the (e,e') reaction which does not preferentially excite the stretched-state over the next lower J-value state. When these strengths are fragmented and unresolvable, the (p,n) reaction provides the superior probe of the stretched-state strength. Just such a problem was encountered using the (e,e') reaction to study stretched states in ^{40}Ca; the 5^- and 6^- strengths were both strongly excited and unresolvable.[12] This preferential excitation is a significant advantage of the (p,n) reaction.

4. EXCITATION OF STRETCHED STATES WITH $T = T_{Target} - 1$

For a target with a neutron excess, the (p,n) reaction (with $\Delta T=1$) can excite strength in the final nucleus with isospin T-1, T, and T+1, where T is the isospin of the target. Simple inelastic scattering reactions are restricted to the two highest isospin components. Because the lower isospin component is usually found at lower excitation energies, it is nearer the Fermi surface and usually is less fragmented. Another way to describe this is to say that the lower isospin component usually is less fragmented by configuration mixing. The "0 $\hbar\omega$" states discussed above are good examples of T-1 stretched states. In some cases, usually for lighter nuclei, it is possible also to see T-1 components of "1 $\hbar\omega$" stretched states. As an example, let us consider the $^{26}Mg(p,n)^{26}Al$ reaction. Because ^{26}Mg is a T = 1 nucleus, the $\Delta T = 1$ (p,n) reaction can excite T = 0, 1, and 2 isospin components in the ^{26}Al residual nucleus. The simple inelastic scattering reactions (p,p') and (e,e'), while also $\Delta T = 1$ reactions (as well as $\Delta T = 0$), can only excite states in ^{26}Mg, which has $T_z = 1$; thus these reactions excite only T = 1 and T = 2 components in the final nucleus. Both the (p,p') and (e,e') reactions[13,14] have been used to study the isovector 6⁻ stretched state with the principal configuration $(f_{7/2}, d_{5/2}^{-1})$. The T = 1 strength is seen to be highly fragmented, which is probably not surprising in this deformed nucleus. A state at 18.2 MeV was tentatively identified as a T = 2 state.[14] Recently, we have studied this same strength with the (p,n) reaction, and although we cannot achieve the energy resolution possible in the charged particle reactions, we do see the general fragmentation of the T = 1, 6⁻ strength and even the T = 2 state (at E_x = 18.1 MeV in ^{26}Al), as shown in Fig. 11; however, because we can excite T = 0 strength as well, we see the T = 0 component of the 6⁻ stretched state also. It is observed to be at lower excitation energy and significantly less fragmented than the T = 1 strength. Furthermore, we excite also the (T = 0) 0 ℏω, 5⁺ stretched state with the configuration $(d_{5/2}, d_{5/2}^{-1})$. This state is known to be the ground state of ^{26}Al and we see the 0 ℏω stretched-state strength to be highly concentrated into this single state. The angular distributions for some of these transitions are shown in Fig. 12. Characteristically, we see a significantly larger spectroscopic factor for the 5⁺, "0 ℏω" stretched state than for any other transition. The T=0, 6⁻ strength is known from transfer-reaction studies to be concentrated primarily into two states at E_x = 6.9 and 7.5 MeV. The spectroscopic strength we observe in these two states is about 30% of the expected 1p-1h strength and is in reasonable agreement with that observed recently in (α,t) transfer-reaction studies.[15] At higher excitation energies we see the analog of T=1, 6⁻ strength identified in (p,p') and (e,e') studies,[13,14] as well as in the (α,t) studies.[15] This strength is highly fragmented and has been identified to be split into at least five different states. The (p,n) reaction, with relatively poor energy resolution, is not the reaction of choice to study such highly fragmented strength. Finally, near E_x = 18.1 MeV, we see the analog of the T=2, 6⁻ state.

The important information the (p,n) reaction provides to the study of stretched-state strength in the A=26 system is the excitation of the T=0, 5⁺ and 6⁻ strengths. These strengths are seen to be more concentrated than that for the higher isospin

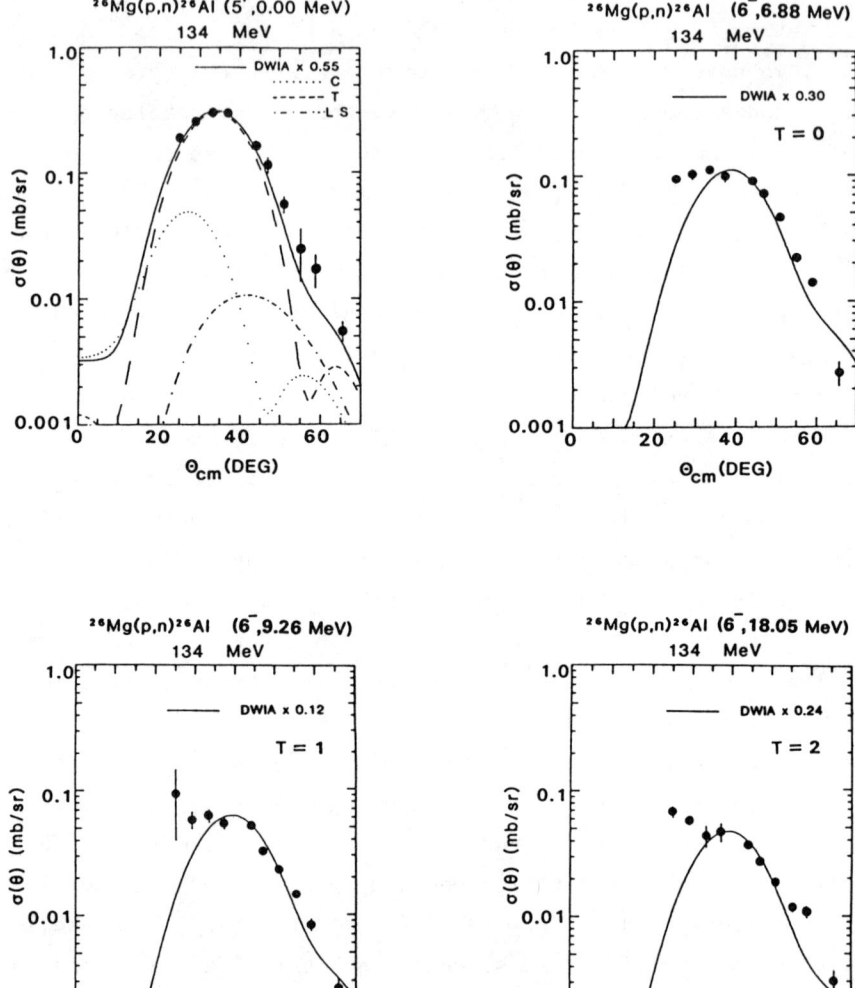

Fig. 12. Angular distributions for four stretched-state excitations in the ^{26}Mg(p,n)^{26}Al reaction.

components. We have seen that this is true in general. Thus, the
lower isospin components provide better tests of the reaction
mechanisms and assumed structure of the target nucleus. The higher
isospin components are clearly more sensitive to configuration
mixing in the final state and provide important tests of structure
calculations which try to describe such mixing.

5. DISCUSSION

We have seen that the (p,n) reaction can be used to study stretched-
state excitations in unique ways, and these studies can extend our
knowledge of such strength in various nuclei. Perhaps now it is
appropriate to ask just how reliable are the normalization factors
obtained. We should not expect to get a final answer at this time,
because we know there are important theoretical questions which are
still being explored.

First of all, I think it is important to recognize that the
(p,n) reaction above about 100 MeV is dominated by impulsive,
single-step processes. The evidence for this exists in several
places. First of all, as we saw in Fig. 1, the (p,n) reaction
apppears to excite the same kind of strength as the (e,e') reaction,
in cases where it would be expected to. Next, angular distributions
of strong transitions are described well by simple DWIA
calculations. Actually, this extends also to analyzing-power
angular distributions as well. For example, in Fig. 13 we show the
measured analyzing-power angular distributions for three (p,n)
stretched-state transitions compared with "standard" DWIA
calculations. Although the agreement is not perfect, the general
shape is indicated in each case. In fact, this is true not only for
stretched-state transitions, but for many strong transitions to
states whose structures are reasonably well known. We have measured
analyzing-power angular distributions for some 15 transitions with
targets of ^{16}O, ^{17}O, ^{18}O, ^{40}Ca, and ^{48}Ca. Every case, but one, is
described reasonably well by a standard DWIA calculation (and that
one is a weakly excited state in ^{18}F whose structure is poorly
known). This good agreement is well known also for the (p,p')

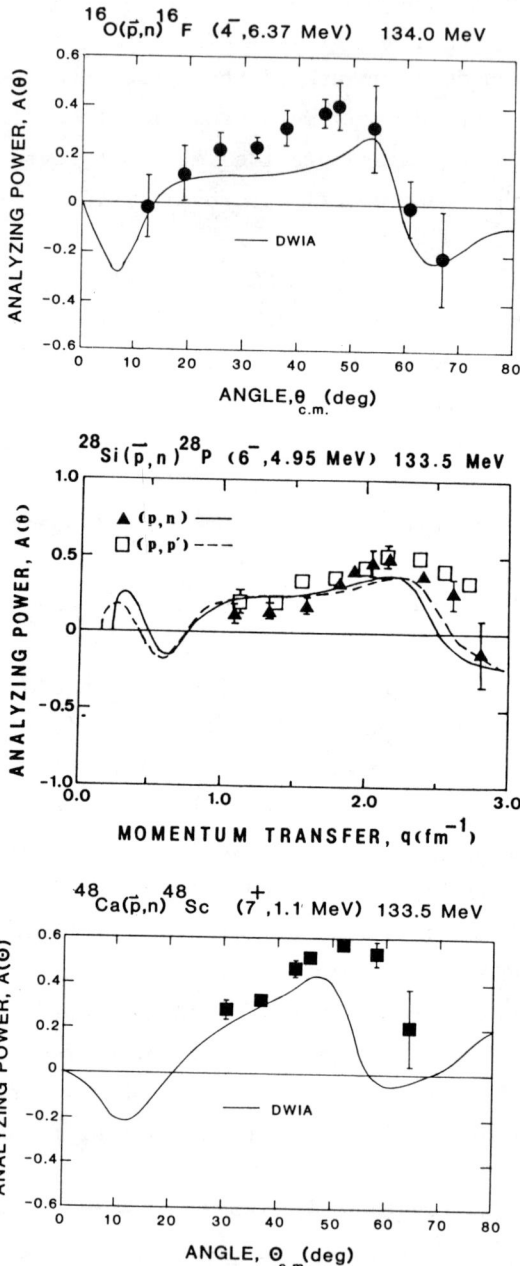

Fig. 13. Analyzing-power angular distributions for three stretched-state excitations in the (p,n) reaction.

reaction above about 100 MeV.[4] Analyzing powers are a good test of the basic reaction mechanism description and the standard DWIA passes this test quite well. Actually, we find that the calculations are sensitive to including all of the major ingredients, viz., the use of distorted waves, the correct nuclear structure, and all of the terms of the N-N effective interaction; however, we also find that the calculated analyzing powers are not very sensitive to small "reasonable" changes in any of these ingredients. The conclusion is that the standard DWIA is a good "first order" theory.

Certain corrections to the DWIA are known to be required. These include possible density-dependent effects (Pauli blocking), the need for non-standard shapes in the optical-model potentials, the consideration of certain meson-exchange currents not presently included, and the use of a true relativistic formulation. Kelly and Carr[16] investigated the effects of density dependence in isovector transitions and found these effects to be relatively small for transitions involving spin transfer. Since the stretched-state transitions involve spin transfer, it may be that density dependence is not significant. This conclusion is supported also by the work of Olmer et al.,[17] who performed DWIA calculations for the excitation of the 6^-, stretched state in the $^{28}Si(p,p')$ reaction, using a t-matrix derived from the density-dependent Paris potential.[18] They found little difference in such calculations compared with standard DWIA calculations. The need for non-standard optical potentials (e.g., "wine-bottle" shapes) is more clearly indicated at higher energies (>200 MeV). The requirement for such shapes was observed to decrease as the proton energy decreases towards 100 MeV.[19] The effects of including certain meson-exchange terms explicitly and/or adopting a relativistic approach have not yet been specifically investigated for stretched-state transitions.

Thus we are left with the conclusion that the DWIA is probably a good first-order theory which, for stretched states at least, has no known large problems. But the theory certainly is not exact. Fortunately, we have the (e,e') excitations of stretched states to help calibrate the (p,n) reaction. If we assume that the PWIA calculations for inelastic electron scattering are more reliable, then we can normalize the DWIA by requiring that we obtain the same normalization factors for 1 $\hbar\omega$ excitations. Earlier, we discussed this point briefly, when we compared the various inelastic reaction excitations of the 4^- and 6^- isovector stretched states (see Fig. 2). We saw there that both the (p,n) and (p,p') reactions yield normalization factors about 20% smaller than does the (e,e') reaction. Thus, assuming that the (e,e') results are more correct, we need to increase the inelastic hadronic normalization factors by about 20%. This discrepancy could be due to the isovector tensor strength in the N-N effective interaction being too strong, or a combination of errors in the various ingredients of the DWIA.

In Table 1 we present the results of our (p,n) studies of stretched-states for eleven different nuclei. Seven 1 $\hbar\omega$ and seven 0 $\hbar\omega$ stretched-state excitations have been observed. Note that the normalization factors presented here have been corrected for center of mass motion (see Ref. 5) and a renormalization of the strength of the tensor term in the N-N effective interaction, as discussed above. These corrections make the (p,n) normalization factors agree with the (e,e') normalization factors for A = 16 and 28. In Table 1 we see, using only the simple shell-model, that the 1 $\hbar\omega$ normalization factors are all < 0.50; whereas the 0 $\hbar\omega$ normalization factors are all > 0.55. Again, we attribute this qualitative difference to the fact that the 0 $\hbar\omega$ excitations lie at or near the Fermi surface and are more highly concentrated into a single state. This interpretation is consistent with the results obtained if a realistic shell-model calculation is used to try to account for core polarizations and/or configuration mixing. As shown in Table 1, such calculations increase the extracted normalization factors significantly, especially for the 1 $\hbar\omega$ excitations where the increase is nearly a factor of two. For the isovector 4^- excitation in A=16, Gareev et al.[20] performed a shell-model calculation which includes multiparticle-multihole configurations for both the ^{16}O target and the 4^- final state. They use a truncated PSD space and

Table 1. Normalization factors for (p,n) excitations of stretched states.

NUCLEUS	CONFIGURATION	N^*(DWIA) Simple Shell Model		N^*(DWIA) Shell-Model Calc.
		"1 $\hbar\omega$"	"0 $\hbar\omega$"	
^{16}O/^{16}F	$(d_{5/2}, p_{3/2}^{-1})4^-$	0.41		0.71[1]
^{18}O/^{18}F	$(d_{5/2}, d_{5/2}^{-1})5^+$		0.55	0.76[2]
^{26}Mg/^{26}Al	$(d_{5/2}, d_{5/2}^{-1})5^+$		0.58	0.91[3]
	$(f_{7/2}, d_{5/2}^{-1})6^-;0$	0.28		
	" ;1	>0.20		
	" ;2	0.24		
^{28}Si/^{28}P	$(f_{7/2}, d_{5/2}^{-1})6^-$	0.31		>0.6[4]
^{32}S/^{32}P	$(f_{7/2}, d_{5/2}^{-1})6^-$	0.50		
^{40}Ca/^{40}Sc	$(f_{7/2}, d_{5/2}^{-1})6^-$	0.45		
^{48}Ca/^{48}Sc	$(f_{7/2}, f_{7/2}^{-1})7^+$		0.66	0.72[5]
^{54}Fe/^{54}Co	$(f_{7/2}, f_{7/2}^{-1})7^+$		0.77	0.86[5]
^{88}Sr/^{88}Y	$(g_{9/2}, g_{9/2}^{-1})9^+$		0.61	
^{120}Sn/^{120}Sb	$(h_{11/2}, h_{11/2}^{-1})11^+$		0.60	
^{208}Pb/^{208}Bi	$(i_{13/2}, i_{13/2}^{-1})13^+$		1.12	

N^*(DWIA) = N(DWIA) x $[(A-1)/A]^{J-1}$ x 1.24

[1] Gareev et al. (truncated PSD) [SJNP40(1984)]
[2] Zuker, Buck and McGrory ($1p_{1/2}, 1d_{5/2}, 2s_{1/2}$) [PRL21(1968)39]
[3] Brown and Wildenthal (full SD) [PRC28(1983)2397]
[4] Amusa and Lawson ($2s_{1/2}, 1d_{5/2}, 1f_{7/2}$) [PRL51(1983)103]
[5] Brown (truncated 1f2p) [PRC31(1985)1147]

find that normalization factors for isovector transitions increase by about a factor of two relative to a simple 1p-1h (TDA) calculation. For the 5^+, 0 $\hbar\omega$ excitation in A=18, we performed a calculation including multiparticle-multihole configuarations using the shell-model code OXBASH[21] with the "ZBM" model space ($1p_{1/2}$, $1d_{5/2}$, $2s_{1/2}$) and interaction.[22] These wavefunctions increase the normalization factor obtained from 0.55 to 0.76. For the 5^+ stretched state in ^{26}Al, we used the full SD space interaction of Brown and Wildenthal,[23] and the normalization factor increases to 0.91. Amusa and Lawson[24] studied the effect of increasing the model space for the description of the 6^- stretched state in A=28. They found that the expected strength would decrease by at least a factor of two by including multiparticle-multihole excitations in a model space which includes the $2s_{1/2}$ orbital as well as the $1d_{5/2}$ and $1f_{7/2}$ orbitals. (It is significant that they also found that the expected strength for the 6^- isoscalar excitation would decrease by more than a factor of four.) Finally, for the 7^+ stretched state in A=48 and 54, we used a truncated 1f-2p model space due to Brown[25] which increases the extracted normalization factors to 0.72 and 0.86, respectively.

Perhaps the "best" of these extended-basis structure calculations are for A=26 and 54; these two nuclei are near the middle of major shells and the significant degrees of freedom are probably described well by the bases assumed. For these two cases, the extracted normalization factors are about 0.9. This result indicates that structure effects are important and may, in fact, account for most of the "missing" strength observed when using only simple 1p-1h wavefunctions.

6. CONCLUSIONS

The (p,n) reaction can provide unique information in comparison to the other simple inelastic reactions (p,p') and (e,e'). Because of its isospin selectivity on self-conjugate targets, the (p,n) reaction provides a superior probe of T = 1 strength compared to the (p,p') reaction which can excite T = 0 strength as well. Additionally, for the study of stretched states, the (p,n) reaction has the advantage compared to the (e,e') reaction that the natural-parity transitions will be suppressed. This property is an advantage especially when trying to separate the stretched-state strength from the strength to states with J-value one less than the stretched-state in cases where both are highly fragmented. The (p,n) reaction has the advantage over any non charge-exchanging reaction that it can access strength in the final nucleus with isospin one less than the target, including the important class of "0 $\hbar\omega$" stretched-state excitations. The spectroscopic factors observed for the 0 $\hbar\omega$ stretched states are significantly larger than for 1 $\hbar\omega$ stretched states and the strength remains highly concentrated even in medium and heavy mass nuclei. Large-basis structure calculations may be able to account for most of the "missing" strength in these excitations.

ACKNOWLEDGEMENTS

This work was supported in part by the U.S. National Science Foundation. The measurements were sometimes performed in collaboration with other workers, most notably Dr. C.C. Foster of the IUCF. The data reported here were analyzed by various graduate students at Kent State, including T. Chittrakarn, C. Lebo, and A. Fazley. It is a pleasure to thank several theorists for considerable help with calculations and interpretation; these include Dr.s B.A. Brown, R.J. McCarthy, P.C. Tandy and B.H. Wildenthal.

REFERENCES

1. B.D. Anderson et al., Phys. Rev. Lett. $\underline{45}$, 699 (1980).
2. C.D. Goodman et al., Phys. Rev. Lett. $\underline{44}$, 1755 (1980), and D.E. Bainum et al., Phys. Rev. Lett. $\underline{44}$, 1751 (1980).
3. A. Fazely et al., Phys. Rev. C$\underline{25}$, 1760 (1982).
4. A.D. Bacher, in "Polarization Phenomena in Nuclear Physics - 1980," ed. by G.G. Ohlsen et al., (AIP, New York, 1981), p.220.
5. R. Lindgren and F. Petrovich, in "Spin Excitations in Nuclei,"

ed. by F. Petrovich et al., (Plenum, New York, 1984), p. 323.
6. W.G. Love and M.A. Franey, Phys. Rev. C$\underline{25}$, 1073 (1981).
7. G.S. Adams et al., IUCF Progress Report (1977) 55.
8. J. Kallne and B. Fagerstrom, Physica Scripta $\underline{11}$, 79 (1975).
9. J. Raynal and R. Schaeffer, computer code DWBA70. The version we used supplied to us by W.G. Love.
10. P. Schwandt et al., Phys. Rev. C$\underline{26}$, 55 (1982).
11. F. Petrovich and W.G. Love, Nucl. Phys. A$\underline{354}$, 499c (1981).
12. K. Seth, Northwestern University, private communication.
13. D.F. Geesaman et al., IUCF Progress Report (1982), p.19.
14. R. Lindgren, University of Mass., private communication.
15. R.J. Peterson et al., U. of Colorado, private communication and to be published.
16. J. Kelly and J.A. Carr, in "Spin Excitations in Nuclei," ed. by F. Petrovich et al., (Plenum, New York, 1984), p.253.
17. C. Olmer et al., Phys. Rev. C$\underline{29}$, 361 (1984).
18. H.V. von Geramb and K. Nakano, in "Interaction Between Medium Energy Nucleons in Nuclei - 1982," ed. by H.O. Meyer (AIP, New York, 1983), p.44.
19. H.O. Meyer et al., Phys. Rev. C$\underline{27}$, 459 (1983).
20. F.A. Gareev et al., private communication and to be published.
21. B.A. Brown et al., computer code OXBASH, private communication.
22. Zuker, Buck and McCrory, Phys Rev. Lett. $\underline{21}$, 39 (1968).
23. B.A. Brown and B.H. Wildenthal, Phys. Rev. C$\underline{28}$, 2397 (1983).
24. A. Amusa and R.D. Lawson, Phys. Rev. Lett. $\underline{51}$, 103 (1983)
25. B.A. Brown, private communication; see also B.D. Anderson et al., Phys. Rev. C$\underline{31}$, 1147 (1985).

HIGH-SPIN STATES WITH THE (p,π) REACTION

W. W. Jacobs
Indiana University Cyclotron Facility
and Department Physics
Indiana University, Bloomington, Indiana 47405

ABSTRACT

Data from systematic studies of the $A(p,\pi^-)A+1$ reaction leading to discrete and continuous nuclear final states have revealed both expected and unexpected signatures of the dominance of a two nucleon mechanism (TNM -- NN → NNπ inside the nucleus) in near-threshold pion production. A distinctive manifestation of this dominance is the systematic and selective excitation by TNM (p,π^-) of high-spin 2p-1h states on target nuclei throughout the periodic table, representing a universal and fundamental response of the nucleus to this high momentum transfer probe. Questions as to whether TNM is also dominant in $A(p,\pi^+)A+1$, what are the corresponding implications for high-spin excitations, and to what extent is the overall observed (p,π^+) behavior related to what is known about fundamental NN → NNπ$^+$ processes are addressed through investigation of the $^{13}C(p,\pi^\pm)$ reactions populating the mirror nuclei ^{14}C and ^{14}O. In the course of these studies, comparative (p,π^+) vs. (p,π^-) data were used to sugggest a $J^\pi = 5^-$ assignment for a previously unobserved state near $E_x = 15$ MeV in the (p,π^+) spectra. Also of interest is a strong and sharp (p,π^+) transition at relatively high excitation ($E_x \cong 23.2$ MeV) in ^{14}C. While we find most observed features in the $(p,\pi^\pm)$ spectra are consistent with TNM expectations, one possible explanation for the latter state's unique behavior involves the anomalous enhancement of normally weak free NN → NNπ isospin channels when immersed in the nuclear medium.

INTRODUCTION

This report concerning recent near-threshold (p,π) results obtained at the Indiana University Cyclotron Facility (IUCF) is quite appropriate for this workshop even though the flavor of many of the sessions is representative of more standard (and much more solidly based) approaches to questions of nuclear stucture. This is because, as you probably recall, pion production is inherently a high momentum transfer process. In addition, some of the more striking and selectively excited residual states that I will be describing to you from the (p,π^-) studies are states of high spin with stretched (or nearly stretched) configurations. And, near the end of my talk I will present some interesting data, and associated conjectures, about the nature of narrow states recently observed at high excitation in several light target (p,π^+) reactions. Thus, in the course of this short presentation, I will touch briefly on all three components making up the theme of this workshop! More generally (and more seriously), I hope to be able to impart to you some appreciation as to the nature of the progress achieved in the

IUCF near-threshold (p,π) work, and some perspective as to how such studies are indeed beginning to complement other more standard probes in the investigation of nuclear structure.

Interest in exclusive pion production studies has several main components. There is "intrinsic" interest in understanding high momentum transfer processes in the nuclear environment, particularly those which in principle probe individual nuclear wave functions at momenta well in excess of the Fermi momentum. Pion production studies also complement investigations aimed at characterizing the π-nucleus interaction mechanism (πNN vertex in the nuclear medium) as is, for example, one of the goals of pion absorption studies being carried out at the meson factories. These aspects and the many results stemming from the large amount of experimental and theoretical work that has taken place over the past fifteen years or so since the first appearance of a "modern", broad range, exclusive pion production spectra,[1] are chronicled in a number review articles.[2] In these introductory remarks I shall make only the few general and historical observations which are pertinent to the subject matter of this talk and the particular bias I bring to it.

Pions can in principle be absorbed or produced by nucleons (N ↔ Nπ), but not by a free nucleon because energy and momentum conservation can't simultaneously be satisfied. Pion production can occur, however, for a nucleon inside a nucleus since the nucleus can then carry away the appropriate recoil momentum -- but the resulting momentum transfer is very large ($|k_\pi| \ll |k_p|$), typically 500-600 MeV/c or greater, even near threshold. In a plane wave model of pion production involving only one active nucleon, the cross section is proportional to $|\Phi(q)|^2$, where $\Phi(q)$ is the bound state wave function for the captured nucleon in momentum space. Clearly in this model, (p,π) would be sensing the very high momentum components of the bound state wave function, at momenta significantly larger than the Fermi momentum (≈ 200-250 MeV/c).

Early hopes in fact centered on the one-nucleon model (ONM) as depicted in Fig. 1a. However, many of the calculational ingredients: H_{int}, the residual nuclear wave function, p and π distortions (which reduce sensitivity to high momentum bound-state components), etc., were uncertain due to the large q involved, leading to extensive parameter uncertainties and little systematic success in comparison to the data. Note that in Fig. 1a, the (p,π⁻) reaction could be thought of as involving the transfer of a Δ^{++}, which at the very least appears to be "complicated" and weak, and perhaps explains why π⁻'s were so little studied initially.

A generalized two-nucleon view (TNM) of pion production, NN → NNπ in the nuclear medium, is depicted in the lower part of Fig. 1b. Clearly this sort of mechanism is more effective in sharing the large momentum transfer among several particles (and holes) in the nucleus, and is perhaps conceptually easier to understand for (p,π⁻), since π⁻ production now takes place on an equal footing with π⁺. However, TNM theory here is still quite complicated and may in practice involve many diagrams, some with questionable approximations -- but parameter sensitivities are in principle somewhat reduced.

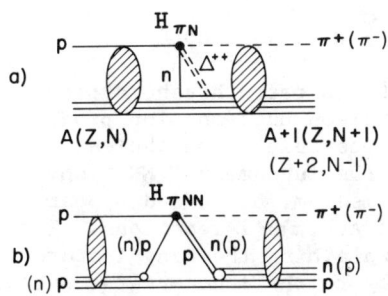

Fig. 1. Schematic diagrams of one nucleon ONM a) and two nucleon TNM b) pion production models

Up until a few years ago, although there was a fair amount of beautiful data, particularly for the (p,π^+) reaction, few systematic trends were evident. There were hints of a TNM at work in nuclear pion production, e.g., 1) the similarity[3] of analyzing power data from the reactions $A(p,\pi^+)A+1$ and $pp \to d\pi^+$ (although this similarity was not universal, see Ref. 4, and 2) the dissimilarity of the results[5] for $^{28}Si(p,\pi^+)^{29}Si$ pionic "stripping" as compared to the $^{28}Si(d,p)^{29}Si$ deuteron stripping reaction. On the other hand, during this period, π^--absorption results[6] in the region dominated by the π-N "delta" resonance seemed to indicate that 3 - 6 nucleons actively share the momentum in this process, casting some doubt as to the relevance of a TNM for pion production even though these two processes probe distinctly different regions of q-space.

Thus the nature of even the crude aspects (and certainly many of the details) of the (p,π) reaction mechanism have over the years been open to a great deal of discussion -- theoretical investigations in the past have not led the way in resolving the various uncertainties, nor have they suggested experiments that could provide answers to specific theoretical questions. (Note: realistic TNM (p,π) calculations are now beginning to be made, but have yet to reach the stage of being even qualitatively reliable in a systematic sense.[7,8])

I believe that we have made considerable recent progress experimentally in this field at IUCF. The insight for our experimental studies at each step has relied heavily on broad range ($p_{max}/p_{min} \approx 1.5$) data systematics made possible by the commissioning of a dedicated quadrupole-quadrupole-split-pole (QQSP) pion spectrometer,[9] with large solid angle (20-30 msr), short flight path (2.5 meters), relatively high resolution ($\Delta E_\pi/E_\pi \approx 0.2\%$), and excellent pion identification and background suppression.[10] The resultant experimental data base we coupled with a rather simple and heuristic view (i.e. minimal theory) of the pion production mechanism, discovering in a rather classic way, that progress in understanding features of the reaction mechanism indeed led to certain (spectroscopic) applications in our investigations of both the (p,π^-) and (p,π^+) reactions. Given the largely empirical nature of our present understanding of these processes, I will provide a somewhat broader background than usual in the discussion of this recent work, while trying to point out as I go along where the various successes, failures, (need for theoretical attention!), and ideas for future investigations lie.

SIMPLE FEATURES FOR (p,π⁻) NEAR THRESHOLD

At bombarding energies appropriate to near-threshold pion production from a nucleus, the <u>free</u> nucleon-nucleon pion production process NN → NNπ is not energetically possible. Nonetheless, a theme of these investigations is that the fundamental NN → NNπ amplitudes, "immersed" in the nuclear medium, do indeed appear to play a dominant role in near-threshold A(p,π)A+1 reactions. The success of such a simple picture in explaining the many features we observe is undoubtedly related not only to the inherent two-nucleon nature of the reaction, but would seem also to imply that the off-shell behavior of the NN → NNπ amplitudes must be qualitatively similar to that of the on-shell amplitudes.

The initial question as posed was whether, despite the complexities associated with realistic TNM calculations, there were any purely <u>experimental</u> signatures of a dominant two-nucleon mechanism for (p,π). Independent of the exact details of the TNM, we note that the possible two-nucleon (NN) channels that can contribute to A(p,π)A+1 charged pion production are:

a) $p + p \to d + \pi^+$,

b) $p + p \to p + n + \pi^+$,

c) $p + n \to n + n + \pi^+$,

d) $p + n \to p + p + \pi^-$,

where a) and b) may differ in their final state spin and/or isospin couplings -- a point I shall return to later. We see immediately, that in contrast to the situation for (p,π⁺), only a single NN process d) can contribute to (p,π⁻). Also unique to d) is that, because of a different isospin projection, the two final state nucleons can't fill the hole created by the initial NN interaction. Thus for (p,π⁻), when the configurations of initial and final 2-particle (protons), 1-hole (neutron) [with respect to the target] states are known, the shell model orbital of the "struck" target neutron is uniquely determined. Thus the reaction path for this latter process is constrained in a significant way, and we expect only the population of $|(\pi)^2(\nu)^{-1}>$ states. (Even so, at a somewhat more detailed level, i.e., in a real TNM calculation, there would still be a number of (p,π⁻) diagrams to consider.) In contrast, for (p,π⁺), even in a two-nucleon model, many different transition paths may contribute coherently, involving nucleons from a variety of orbitals. These restrictions should serve to make (p,π⁻) easier to interpret than (p,π⁺), in spite of the fact that (p,π⁺) transitions are usually stronger, not only as a result of the possible coherence just mentioned, but also because free NN processes a) and b) have the largest cross sections near threshold.[11]

These arguments concerning the restrictive nature of (p,π^-) in a TNM led to specific predictions regarding $^{12,13,14}C(p,\pi^-)^{13,14,15}O_{g.s.}$ $d\sigma/d\Omega(\Theta)$ and $A_y(\Theta)$ distributions.[12] Here the nuclear configurations force the (p,π^-) reaction to proceed in the case of ^{12}C by the interaction of the incident proton with a $1p_{3/2}$ target neutron with the resulting two final-state $1p_{1/2}$ protons coupled to spin zero in the residual ^{13}O nucleus. In contrast, for $^{13,14}C$, the interaction is with the neutrons from the $1p_{1/2}$ orbital. Since these ground state transitions involve the same neutron orbital for the ^{13}C and ^{14}C targets we expect that the resulting $d\sigma/d\Omega(\Theta)$ and $A_y(\Theta)$ would have a very similar behavior -- a correspondingly different behavior for the ^{12}C target, which involves the $1p_{3/2}$ orbital, is anticipated. This is indeed what was qualitatively observed (Fig. 2b of Ref. 12) for the $^{12,13,14}C(p,\pi^-)^{13,14,15}O$ cross section and analyzing power distributions. More quantitatively, we might expect a "scaling" of the ^{13}C versus ^{14}C cross section according to the relative $1p_{1/2}$ neutron subshell occupancy (a factor of 2.04 according to Cohen and Kurath[13]). We find that multiplication of the measured ^{13}C cross section by this factor does in fact make the $^{13,14}C$ cross sections nearly identical in magnitude (and shape) over the entire angular range of the measurement.[12]

The TNM behavior of $A_y(\Theta)$ for (p,π^-) ground-state transitions is strongly influenced by the required spin zero coupling of the two final-state protons. Angular momentum and parity conservation require the interacting proton and neutron in $p + n \rightarrow (pp)_{0+} + \pi^-$ to be in a spin triplet state.[12] Further considerations, such as the requirement that near-threshold TNM nuclear pion production requires the Fermi momentum of the struck nucleon to be predominantly towards the projectile (since this is below threshold for free $NN \rightarrow NN\pi$), and the supposition that the nucleus has a "sidedness", brought about for example by distortions, lead to a predicted[12] sign dependence of $A_y(\Theta)$ depending on $j_n = \ell \pm 1/2$ of the interacting target neutron. This sign difference for $A_y(\Theta)$ was dramatically observed for the (p,π^-) transitions involving $1p_{3/2}$ neutrons ($^{13,14}C$) versus $1p_{1/2}$ neutrons (^{12}C) as shown in Fig. 2a of Ref 12. We note here for later reference that the actual sign observed for A_y would seem to indicate that the π^-'s preferentially emerge on the side of the nucleus "opposite" from that of the incident proton, contrary to what would be expected from arguments based purely on pion absorption effects. (Some preliminary calculations in a TNM[14] seem to correctly reproduce this observed sign difference and also qualitatively describe the $A_y(\Theta)$ data, although only a few TNM diagrams are considered.) This j-dependent effect is of course superimposed upon, and is in addition to, any analyzing power associated with the "fundamental" $NN \rightarrow NN\pi$ process, although the latter is not the free analyzing power because of the final-state spin coupling constraints.

These simple predictions for the cross section and analyzing powers are based on a view of nuclear pion production in which two and only two nucleons are actively involved. The excellent agreement with the predicted behavior must be viewed as strong support for the dominant role played by fundamental $NN \rightarrow NN\pi$

amplitudes in (p,π^-). These results also support the view which attributes an inherent simplicity to the highly constrained reaction path of the (p,π^-) reaction.

HIGH SPIN SELECTIVITY OF THE (p,π^-) REACTION

In an attempt to pursue the cross section and j-dependent $A_y(\Theta)$ observations of the previous section into the higher spin orbitals found in heavier target nuclei, we soon discovered that the cross sections for these highly constrained transitions become too weak (ground states ~ .1 nb/sr) for systematic study. However, a distinctly new phenomenon emerged, which is the strong and selective concentration of (p,π^-) reaction strength in one or several low-lying ($E_x \approx 3 - 7$ MeV) discrete states as displayed for several selected targets in Fig. 2.

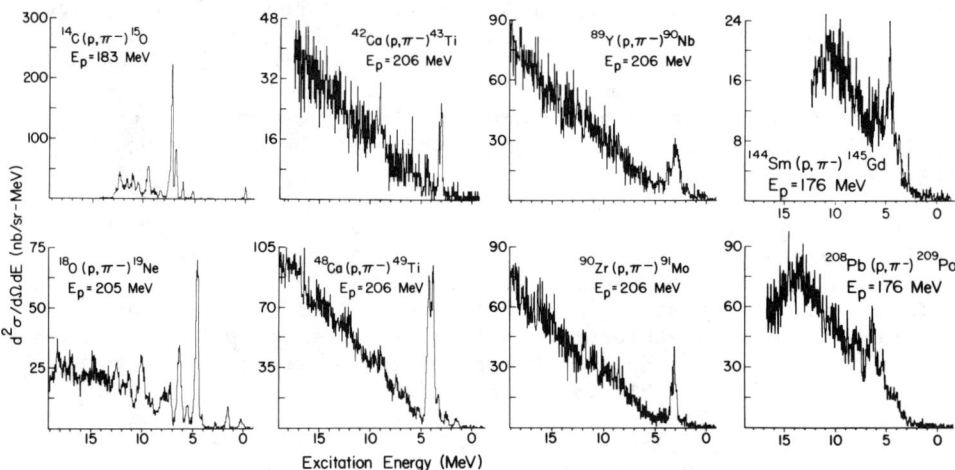

Fig. 2. Spectra for the (p,π^-) reaction at $\theta_{lab} = 30°$ (28° and 60° for ^{14}C and ^{144}Sm, respectively) for targets spanning the entire periodic table, showing selective excitation of one or a few low-lying discrete states.

The systematics of these highly selective excitations for targets ranging from the 1p shell to the Zr region have been studied and described by Vigdor et al.[15] Some common features are that, for example, 1) the discrete strength is found to be greatest for targets with $j_> = (\ell + 1/2)$ neutron subshells filled and the corresponding proton high-spin subshell empty, 2) the strength weakens rapidly — but not always monotonically — as $j_>$ neutrons are removed or $j_>$ protons are added to the target, and 3) the continuum, observed to dominate the spectrum for all but the lightest targets studied, begins a nearly linear rise close to the discrete state position, with a strength relative to these low-lying peaks that appears to grow steadily with increasing target mass. A qualitative understanding of many features of these discrete, prominent, 2p(proton)-1h(neutron) states, interpreted[14] to have a high-spin stretched, or nearly stretched character, can be found once again in a simple two-nucleon mechanism view of the (p,π) reaction near threshold.

The (p,π^-) reaction in a general TNM involves the effective charge exchange of the target neutron and capture of the incident proton, with a high degree of selectivity arising from both the constrained reaction path and the inherently large momentum mismatch. When the large q is shared among three "effective" nucleons (two particles and one hole), none need to carry more than the Fermi momentum. As opposed to transitions where the final state protons are coupled to spin zero, "optimal" accommodation of the high linear momentum transfer in the TNM view of (p,π^-) strongly favors high spin (preferably $j_>$) configurations where the two protons and neutron hole are coupled to produce a 2p-1h state of the highest spin possible in the orbital space at hand. Hence in $^{48}Ca(p,\pi^-)^{49}Ti$, the strong discrete states in this view would correspond to a configuration of the form[15] $\{(\pi f_{7/2})^2_{6+}(\nu f_{7/2})^{-1}\}_{19/2^-}$, or the same configuration coupled to slightly lower J. (Note, a $J^\pi = 21/2^-$ state is forbidden by the Pauli principle).

Confirmation of this scenario comes in part from the observed dominance of the stretched (7^+) transition in the $^{40}Ca(p,n)^{40}Sc$ charge exchange reaction at $E_p = 160$ MeV ($q \approx 200$-400 MeV/c).[16] A further element in the confirmation is the excellent consistency of observed (p,π^-) discrete strength excitation energies with those of states having known high-spin J^π assignments in several of the lighter nuclei studied (e.g., the $13/2^+$ state in ^{19}Ne at $E_x = 4.64$ MeV, with configuration $\{(\pi d_{5/2})^2_{4+}(\nu d_{5/2})^{-1}\}_{13/2^+}$, very strongly excited in the $^{18}O(p,\pi^-)$ reaction).[15,17] In addition, at the time of these investigations there were suggestions[18] from data employing the $^{48}Ca(\alpha,3n\gamma)$ reaction of an yrast $19/2^-$ state at an excitation of 4.38 MeV in ^{49}Ti, in good agreement with the location of a high-E_x strong peak observed (Fig. 3) in high resolution $^{48}Ca(p,\pi^-)^{49}Ti$.

The high-spin nature and structural configuration of these excitations in ^{49}Ti were in fact confirmed by a study[19] of the cross section and vector analyzing power distributions for states populated in the $^{51}V(d,\alpha)^{49}Ti$ reaction. As we will hear in the next talk of this session,[20] the (d,α) reaction at $E_d = 80$ MeV on $1f_{7/2}$ shell nuclei preferentially transfers a proton-neutron pair in a completely aligned $(1f_{7/2})^2_{J=7,T=0}$ configuration and can selectively populate high spin residual nuclear states. Since ^{51}V can be predominantly described as $(\pi f_{7/2})^3$ with respect to ^{48}Ca as a core, the (d,α) reaction should selectively populate high-spin two-particle one-hole (2p-1h) states in ^{49}Ti with configuration $[(\pi f_{7/2})^2(\nu f_{7/2})^{-1}]$ with respect to that core — the same states we claim to see strongly in the (p,π^-) reaction. A spectral comparison of the population of states in ^{49}Ti by (p,π^-) and (d,α) is shown in Fig. 3. Although (p,π^-) is more selective in the excitation of states with a high-spin character, peaks at $E_x = 3.29$, 3.96, and 4.38 MeV are common to both spectra. These three excitations are in fact just the strong states for which the (d,α) distributions

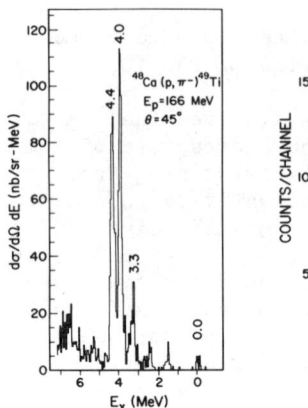
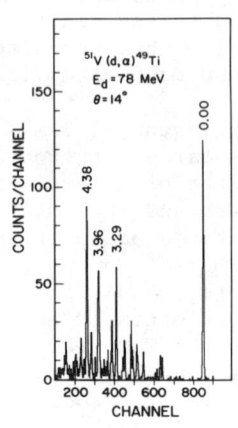

Fig. 3. Spectra for the (p,π^-) and (d,α) reactions populating states in ^{49}Ti. High-spin 2p-1h states with configuration $(\pi f_{7/2})^2 \times (\nu f_{7/2})^{-1}$ at E_x = 4.38, 3.96, and 3.29 MeV are prominent in both spectra.

indicate L = 6, J = 7 transfer, and hence population of the high-spin $|(\pi f_{7/2})^2(\nu f_{7/2})^{-1}\rangle$ configurations in ^{49}Ti as conjectured. As an additional observation, we note that this high degree of selectivity observed for (p,π^-) results not only from the large momentum mismatch, but also from the two-nucleon nature of the reaction, as evidenced by the absence in similar ^{44}Ca$(p,\pi^-)^{45}$Ti spectra[10] of states excited with a more complicated configuration (hence unreachable in a TNM) and spin greater than 19/2.

General features of systematic (p,π^-) data obtained in the calcium region have also been accounted for quite well in a comparison with calculations performed by Brown, Scholten, and Toki.[21] They have used a simplified plane-wave model which assumes that 1) the (p,π^-) reaction exclusively excites high-spin $(j_>)^3$ states, 2) the interaction is short range — so that a zero-range approximation is valid (this also implies that the final-state coupling of the two protons is spin singlet), and 3) the reaction is surface-peaked and a lower radial cutoff can be used somewhat arbitrarily to simulate this effect. Calculated "spectra" using reliable shell model wave functions (including a curve representing a Gaussian averaged resolution smearing with FWHM of 0.3 MeV) are shown in the left-hand portion of Fig. 4, with the corresponding experimentally observed (p,π^-) spectra plotted on the right-hand side. Results for the various isotopes of calcium are shown in the lower part of the figure, with the occupation number of the target neutron $1f_{7/2}$ shell increasing from ^{42}Ca to the ^{48}Ca. Thereafter, targets with increasing proton occupation of the $1f_{7/2}$ shell (isotones) are displayed.

There is very good qualitative agreement between the behavior of the calculated and measured spectra in Fig. 4, lending additional support to the basic two-nucleon view of (p,π^-) we have been discussing. We see that in the ^{49}Ti spectrum (Fig. 3), that the higher lying state at 4.4 MeV is indeed to be associated with the stretched 19/2$^-$ configuration, while the largest peak, located at 3.96 MeV, is to be associated with the second 15/2$^-$ state. Note the 17/2$^-$ (J = L-1/2) state at 3.3 MeV is suppressed (as observed experimentally)[10] in the calculations because of model-dependent jj to LS coupling coefficients.[21] It is interesting to note further

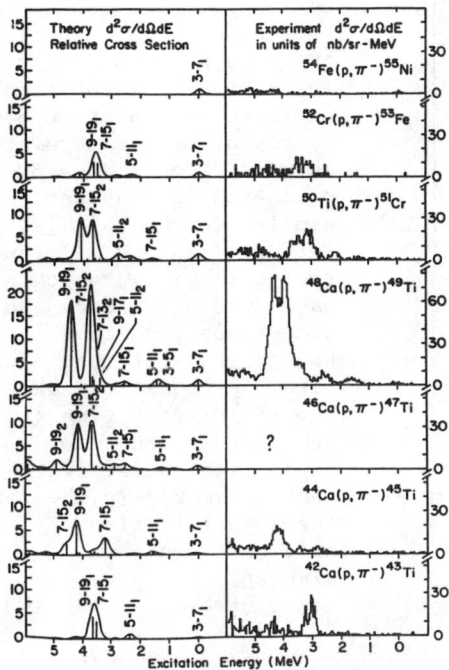

Fig. 4. Calculated (Ref. 21) and experimentally measured (p,π^-) spectra for $1f_{7/2}$ nuclei at E_p = 206 MeV and θ_{lab} = 30°. Individual states in the theoretical spectra are indicated by the lines and labeled by L-2J (see text).

(as discussed in Ref. 21), that the calculated spectra exhibit many details which have not been previously understood, and that they rather faithfully reproduce the relative amount of observed high-spin strength in this array of $1f_{7/2}$-shell targets. In addition, the qualitative description, given by such simple theory, of the cross section angular distribution behavior for the 19/2⁻ and 15/2⁻ states in ^{49}Ti is also suprisingly good as we shall soon see in the context of similar measurements to be discussed shortly.

At present, detailed experimental studies of the discrete state excitations have been carried out with sufficiently good resolution to delineate the fine structure in the oxygen,[17] calcium,[10] and most recently the zirconium regions,[22] corresponding to 2p-1h configurations $(1d_{5/2})^3$, $(1f_{7/2})^3$, and $(1g_{9/2})^3$, respectively. These measurements were made with considerably better experimental resolution than those shown in Fig. 2, allowing extraction of cross section and analyzing power angular distributions for states of different total spin within the same multiplet. A high degree of spin selectivity of the relatively few states that are excited has been observed which results from an effective "ℓ-window" imposed on the lower side by the large momentum mismatch in (p,π^-) and on the upper side by the maximum spin transfer available from the three interacting "particles" in a given orbital.[21]

Indeed, a further test of these simple concepts (and relatively simple calculations) was made by investigating the (p,π^-) high-spin state behavior near mass 90, a region of closure for the $g_{9/2}$ neutron orbital (i.e., configurations of the form $\{(\pi g_{9/2})^2_{8+}(\nu g_{9/2})^{-1}\}_{25/2+}$, and lower spin members of the same configuration are anticipated). A good (and indeed striking!) case was found for study in the ^{88}Sr$(p,\pi^-)^{89}$ reaction,[22] where two strong, yet distinct peaks, separated by ~ 370 keV were observed at

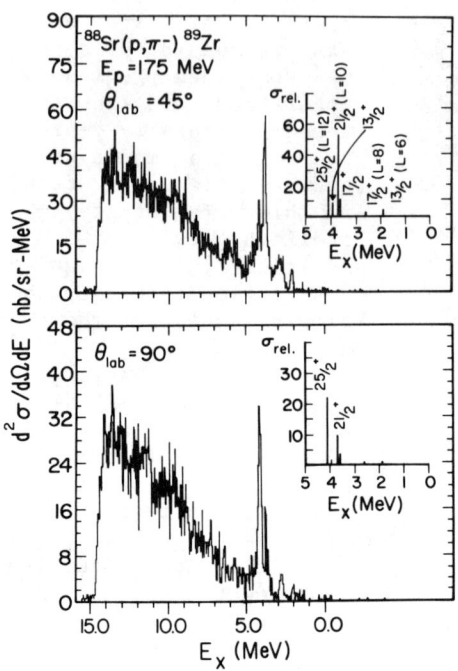

Fig. 5. Spectra for ^{88}Sr(p,π^-)^{89}Zr at two angles and E_p = 175 MeV compared to calculations (inserts) of B.A. Brown (Refs. 21,22).

the beginning of the continuum region as indicated in Fig. 5. In the insets to the figure are shown the theoretical relative $(\pi g_{9/2})^2(\nu g_{9/2})^{-1}$ multiplet member population as predicted by Alex Brown, evaluated in a manner similar to the calculations performed in the calcium region (Ref. 21) which we have just discussed. We used these predicted excitation energies, along with the cross section angular distribution behavior in order to exploit the selectivity of near threshold (p,π^-) for the identification of previously unknown high-spin $(g_{9/2})^3$ configuration states in ^{89}Zr.

The cross section distributions in these simple plane-wave calculations are proportional to the square of a Bessel function of order L (= total orbital angular momentum transfer) averaged over the surface region of the nuclear wave functions. (The lowere cutoff radius employed was extrapolated from that value found to be appropriate for the calcium region.) The higher-lying state in Fig. 5 at E_x = 4.18 MeV is identified as the stretched $25/2^+$ state (predicted to be at E_x = 4.1 MeV).[22] It clearly dominates the spectrum at the more backward angles, having a relatively flat angular distribution (see Fig. 6), as expected in the model for the transition with largest total L transfer (L = 12 for the $[g_{9/2}]^3$ configurations). The state at E_x = 3.81 MeV is identified as the (L = 10) $21/2^+$ state (predicted at 3.71 MeV) and has a cross section distribution which is forward-peaked, falling off sharply at the larger angles -- again, for reasons similar to those presented for the calcium region, the $23/2^+$ (J = L-1/2) state is subject to an angular momentum coupling suppression. Within the 70-80 keV uncertainties of both measured and calculated excitation energies, the relative E_x agreement is very good.

Hence we have compelling evidence for new spin and parity assignments in this mass region, an assignment based first of all on the <u>empirically</u> observed selectivity for stretched or nearly stretched high-spin 2p-1h states excited in the (p,π^-) reaction, secondly on the agreement in excitation energy of the observed strongly excited states with the high-spin state predictions, and finally on the cross section angular distribution behavior for these states, which is described qualitatively, and to some extent quantitatively (there is an overall normalization factor), by simple shell model calculations in a plane-wave model.[21,22] In fact the curves representing the results of these calculations in Fig. 6, give a remarkably (fortuitously?) good description of the data, and show at the very least, that future realistic TNM calculations must somehow "touch base" with this rather simplistic picture of the pion production process if they are to have any hope of reproducing the data.

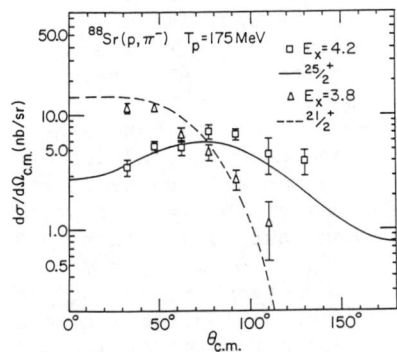

Fig. 6. Cross section angular distributions for the ^{88}Sr(p,π^-) reaction populating high-spin states in ^{89}Zr. Curves are plane wave calculations of B.A. Brown.

Before going on to the next topic I should mention that the continuum region of these spectra (by far the dominant cross section at the higher excitation energies) has also received some attention. Theoretical calculations by Scholten et al.,[23] with a contiuum shape given by a Fermi gas approximation to the 2p-1h state density at the appropriate q and E^*, do quite well in describing the overall behavior of the experimental contiuuum results in these medium mass targets (in particular, the case of ^{42}Ca(p,π^-) at E_p = 206 MeV is shown in Ref. 23). The absolute (scaling) value of these cross sections is calculated within a microscopic TNM (target emission only) using a Glauber estimate of the effective number of interacting surface neutrons. Once again, rather simple TNM calculations are seen to go a long way in describing the salient features of these data.

HIGH-SPIN (p,π^-) ANALYZING POWER SYSTEMATICS

It is interesting to note that the combined data for all the stretched configuration studies[10,17,22] for 0ħω transitions on ^{18}O ^{42,48}Ca, and now the stretched 25/2+ state in ^{89}Zr, reveal a universal $A_y(\Theta)$ pattern as indicated in Fig. 7 — a result which is all the more striking since states of slightly lower spin have a significantly different $A_y(\Theta)$ behavior. The spin system here is not as severely constrained as for the ground-state transitions

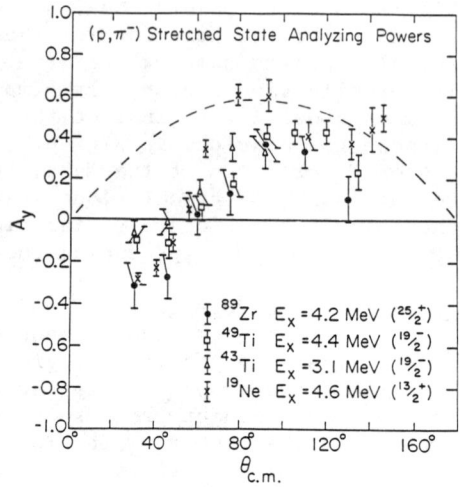

Fig. 7. Analyzing power distributions for high-spin 2p-1h stretched states showing a mass independent systematic behavior. The curve is the model dependent calculation of Toki and Kubo.[24]

described earlier (i.e., where final-state protons have $S_{pp} = 0$ from the structure, and entrance channel spin $S_{pn} = 1$ from conservation laws). However, because of the restricted reaction path generally accessible to (p,π^-) stretched transitions, many constraints, although different in nature, may apply and are perhaps responsible for the observed striking universality of the stretched state analyzing powers.

Toki and Kubo[24] have used model-dependent arguments to conclude that in fact the initial proton and neutron are in a spin triplet -- i.e., the short-range nature of the reaction implies a relative s-wave for the final state protons, requiring $S_{pp} = 0$. The pseudo-scalar nature of the interaction then requires $S_{pn} = 1$. After invoking a "sidedness" to the reaction process (in a manner similar to the ground-state $A_y(\Theta)$ arguments) resulting in this case from pion absorption effects, a calculation incorporating some TNM elements is given by the dashed curve shown in Fig. 7 which produces some of the gross features of the stretched-state analyzing power systematics. This result, however, when compared to the behavior of the ground-state transitions discussed earlier, presents a potential problem in that the "distortions" appear to cause the interaction with j> neutrons to produce a negative analyzing power in one case (ground state), but a positive analyzing power in the other (stretched state), even though the entrance channel spin S_{pn} is equal to 1 in both arguments. This apparent discrepancy might be explained if the two reactions, one well matched and the other poorly matched in angular momentum, take place in very different parts of the nucleus[25] (i.e., presumably the pion absorption has a significant effect only on the high-spin surface-peaked reactions). However, a possibly more serious objection with the arguments of Ref. 24 is that they would seem to predict the same analyzing power for all transitions which are high spin, whereas the strongly observed high-spin states with two units of angular momentum less than the stretched configuration have an $A_y(\Theta)$ behavior in general quite different from that of the stretched states.

In contrast to the arguments of Ref. 24, an evaluation of the coupling coefficients specific to the stretched states would seem to favor final state protons with $S_{pp} = 1$. Alternatively, and from a different perspective, it is also thought that π^- absorption on an $S_{pp} = 0$ pair is suppressed because intermediate Δ-N s-wave channels are then disallowed by simple conservation laws. (Were this the case, it is possible that the ($J_{max}-1$) state suppression tied to $^1S_0$ proton final-state coupling in Ref. 21, in fact results from other aspects of the pion production dynamics.) If the final protons were indeed coupled to a spin triplet, and assuming that a nucleon spin is flipped at the pion "emission" vertex but not at the pion "scattering" vertex, this would imply an initial proton-neutron coupling that is spin anti-parallel. If the effects of the pion distortions were stable over a wide range of target mass, this latter coupling would be consistent with the ground-state analyzing power being opposite in sense to that of the stretched state, as is in fact observed for $j_>$ neutrons.

There is uncertainty in either of the above largely schematic arguments, although it is clear that the starting assumptions in either case have a bearing on several other related physical phenomena. Suffice to say for the moment, that the resultant systematics are an very interesting experimental observation of high-spin stretched state analyzing power behavior for which there is at present no (simple) convincing explanation.

EXTENSION OF (p,π^-) SYSTEMATICS TO HEAVY NUCLEI

Momentum matching considerations suggest that the population of analogous (p,π^-) transitions on heavier mass targets, as alluded to earlier (and presented "prematurely" in Fig. 2), should also be possible. In particular, the samarium and lead regions, where (p,π^-) transitions involve the (full) $1h_{11/2}$ and $1i_{13/2}$ high-spin neutron orbitals, are clearly the next likely candidates for study, leading to the possibility of sampling high-spin $(1h_{11/2})^3$ and $(1i_{13/2})^3$ configuration 2p-1h states, with spins up to $31/2^-$ and $37/2^+$, respectively. Is it possible to produce such states with the (p,π^-) reaction? If we calculate the amount of angular momentum available in a peripheral collision of 175 MeV protons with samarium and lead, we find values of 16-22 and 17-25 units in the two cases, easily enough to match the large L_{trans} (15 and 18 units for the $(1h_{11/2})^3$ and $(1i_{13/2})^3$ configurations, respectively), involved in the high-spin excitations for these heavy targets.

Recently obtained angular distribution data are displayed in the form of double differential cross section spectra for the ^{144}Sm(p,π^-)^{145}Gd and ^{208}Pb(p,π^-)^{209}Po reactions in Fig. 8. We see that these heavy target spectra are qualitatively similar to the (p,π^-) spectra obtained during earlier survey studies[15] of the oxygen-calcium-zirconium regions. In both Sm and Pb cases, there is evidence for a strongly excited group of unresolved residual "discrete" states at relatively low excitation energy, and the <u>total</u> amount of such strength is fairly large. At forward angles, the "discrete" cross section strength, summed over a region on the order of 1 MeV, amounts to 15 and 25 nb/sr for ^{145}Gd and ^{209}Po,

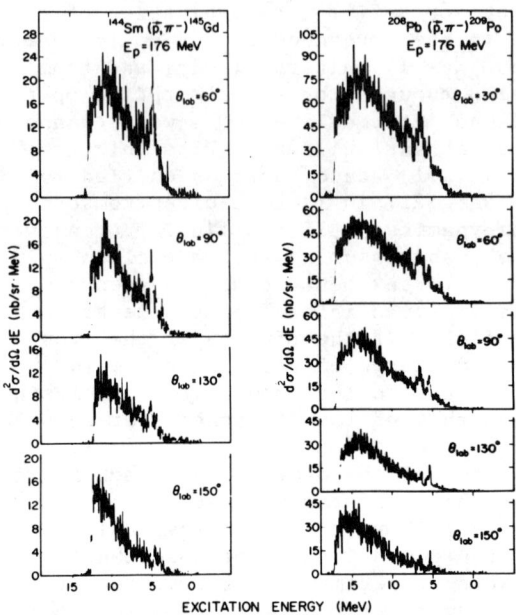

Fig. 8. Spectra for the ^{144}Sm(p,π^-)^{145}Gd and ^{208}Pb(p,π^-)^{209}Po reactions at E_p = 176 MeV as a function of θ_{lab} showing the angular dependence of low-lying (unresolved) discrete strength.

respectively, a value comparable to that observed for optimal lighter mass targets.[15] It is clear here, however, from the rapidly changing shape with angle of the concentrated strength excitation that for these heavy target cases a very large number of levels having quite different angular behavior are contributing to the cross section in this energy region. Some few of the discrete transitions involved -- presumably the highest-spin states -- still show some strength (although only of order a few nb/sr!) at the largest measured laboratory angles. We note here a complicating experimental feature -- in contrast to the lighter masses studied (Figs. 2 and 5), the continuous part of the spectrum extends well below the discrete-state region in excitation, providing a strong background for measurement of these relatively weak cross sections.

How can one attempt to distinguish the stretched-state configurations among this "forest" of discrete transitions, unresolved experimentally at the forward angles (or should we learn to have a better appreciation of the "forest")? It is already clear that this is somewhat more difficult than was the case for the lighter mass regions. We know that the high-spin stretched states are well matched near the nuclear surface, and in general they should also show a much greater strength at the largest angles, relative to transitions of lower total L, owing to a cross section distribution which is relatively flat with angle according to plane wave[21] expectations. From a different tack, the analyzing power distributions for the stretched configurations reached by (p,π^-) exhibit an experimentally determined universal behavior,[24,25] as we have seen in Fig. 7, corresponding to a stretched-state "signature": namely, small or negative A_y at forward angles, rising to a substantial positive A_y (.2-.4) beyond roughly 60-70 degrees.

In Fig. 9 are displayed spectra for the ^{144}Sm(p,π^-)^{145}Gd and ^{208}Pb(p,π^-)^{209}Po reactions obtained under conditions expected to emphasize the highest-spin transitions. A prediction[26] based on

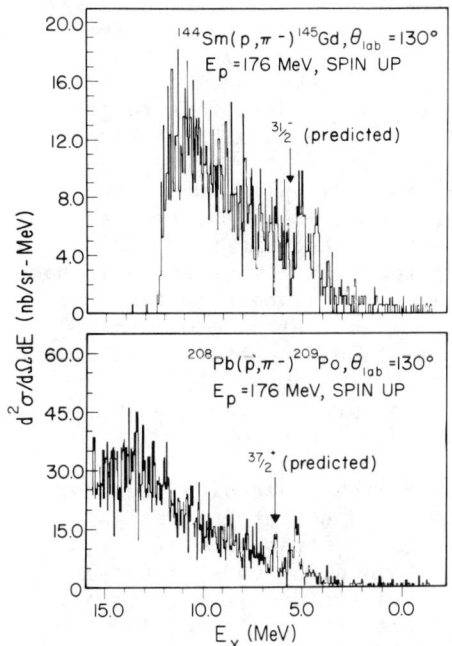

Fig. 9. Large angle spectra for the ^{144}Sm(p,π^-)^{145}Gd and ^{208}Pb(p,π^-)^{209}Po reactions taken with spin up protons. Arrows mark location of expected high-spin stretched states (see text).

shell model calculations gives the location of the $31/2^-$ stretched-state configuration, as indicated by the arrow in the top panel, for ^{145}Gd at 5.79 MeV. A much more crude estimate of the position[27] of the $37/2^+$ stretched state in ^{209}Po is indicated in the bottom panel of Fig. 9 at 6.4 MeV. Although the discrete transitions are less prominent relative to the continuum here, we do observe small peaks near the predicted[26,27] shell model excitation energies for the two stretched configurations. The highest-lying peaks in each case (E_x = 5.2 MeV in ^{145}Gd and E_x = 6.4 MeV in ^{209}Po) have a positive A_y at 130°, in accord with (p,π^-) stretched-state systematics. However, the situation becomes less clear when one considers the 150° spectra, for which different peaks then those mentioned above seem to be strong in the ^{145}Gd spectrum, while at least two fragmentary peaks remain in the ^{209}Po spectrum.

Detailed computer modeling of these large angle spectra in the spirit of the Brown-Scholten-Toki[21] model is the next step in order to determine whether or not what we are seeing at the largest angles is in reasonable accord with the expected high-spin state behavior. Nonetheless, it seems clear that this phenomenon is similar in many ways to that observed during the survey of lighter targets, and therefore we have at least qualitative evidence for the excitation of high-spin stretched or nearly stretched 2p-1h states with the (p,π^-) reaction spanning the entire periodic table. In fact, these nuclear excitations at very high momentum transfer (500-700 MeV/c), represent a fundamental and systematic response of the nucleus to this high-q probe.

HIGH-SPIN STATES IN THE (p,π^+) REACTION

Using the same general momentum sharing arguments we have been discussing with regard to the (p,π^-) reaction raises the question as to whether (p,π^+) also populates high-spin 2p-1h states. The answer is that it does, but not as selectively. Such high-spin states are indeed observed[28] for (p,π^+) reactions on light target nuclei, and presumably involve constraints on the reaction path that are similar to those we have just been talking about for (p,π^-) reactions. However, transitions involving lower angular momentum transfer and/or less optimal q sharing are often of comparable strength, largely because (as mentioned earlier) of the possible coherence of multiple reaction paths involving a variety of target nucleon orbitals for the several fundamental NN → NNπ pion production processes. For example, in an earlier study[28] we observed the strong population of the 2p-1h $J^\pi = 9/2^+$ state at 9.5 MeV excitation in ^{13}C with the ^{12}C(p,π^+) reaction. This state is not configured with form $[j_>]^3$, as we have talked about so far, but rather is the largest spin that one can make with "cross-shell" transitions involving the $p_{3/2}$, $p_{1/2}$, and $d_{5/2}$ orbitals. The point is that low spin states in the spectrum (see Refs. 15,28), have strength comparable to that of the high-spin stretched state, although a distinctly different cross section angular distributions.[28]

For heavier nuclei, in addition to the preceeding arguments about the relative enhancement of low spin states in (p,π^+), the neutron excess favors $0\hbar\omega$ $(\nu j_>) \to (\pi j_>)$ transitions [i.e., (p,π^-)] over the analogous $(\pi j_>) \to (\nu j_>)$ transitions [i.e., (p,π^+)]. This has the effect of further diminishing the importance of the high-spin $(j_>)^3$ configuration strength for (p,π^+) in heavy target nuclei. The situation is pictorially illustrated in Fig. 10, where spectra for both (p,π^+) and (p,π^-) reactions, populating states in the same residual nucleus, ^{49}Ti, are displayed. We see the strong, selective (and by now familiar), population of high-spin discrete strength for ^{48}Ca$(p,\pi^-)^{49}$Ti in the lower panel of Fig. 10, whereas these same states, which could be excited through processes (a) or (b) discussed earlier, by the interaction of the incident proton with an $f_{7/2}$ target proton are masked, at the very least, by other states of lower spin coherently excited in the (p,π^+) reaction.

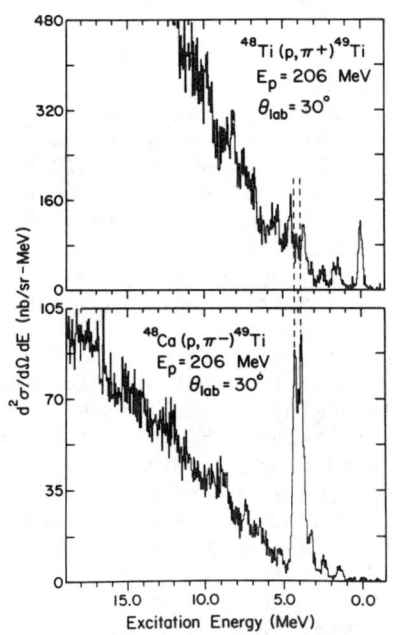

Fig. 10. A Comparison of (p,π^+) and (p,π^-) spectra, showing the striking difference in selectivity for population of states in the common residual nucleus ^{49}Ti.

The fact that no appreciable TNM (p,π^+) strength for these high-spin states is seen -- certainly less than half that of the absolute cross section observed for (p,π^-) -- may not be too surprising since there are only two $f_{7/2}$ protons with which to interact compared to eight $f_{7/2}$ neutrons.

ISOLATION OF FUNDAMENTAL NN → NNπ PROCESSES IN (p,π^+)

Now we ask the more general question -- is this TNM picture we have been discussing also dominant in $A(p,\pi^+)A+1$ reactions as it so strongly appears to be for (p,π^-)? -- and if so, how is the behavior we observe in nuclear pion production related to what we know about the fundamental free NN → NNπ processes? With regard to the latter question, we can begin by examining the results of the phase shift predictions displayed in Fig. 7 of VerWest and Arndt,[11] given in terms of the isospin reaction channels $\sigma_{II'}$, where I and I' are the incident and exit channel isospin of the nucleon pairs, repsectively. One notes immmediately that σ_{10}, which is a component of fundamental process (b) pp → $np\pi^+$ and drives the process (a) pp → $d\pi^+$, dominates other channels near threshold by almost an order of magnitude. On the other hand, pn → $nn\pi^+$ and pn → $pp\pi^-$ are given by a similar combination of isospin cross sections $1/2[\sigma_{11} + \sigma_{01}]$, and hence should be expected to be similar in magnitude in the nuclear medium and much weaker than processes containing the term σ_{10}. Experimentally, analyzing power measurements near threshold have been made only for the strong NN → NNπ channels -- $A_y[pp \to d\pi^+]$ is found typically to be large and negative at most angles.[3]

In comparative $(p,\pi^\pm)$ studies, made under similarly constrained conditions, one might exploit the properties of selected high-spin final states to isolate specific NN→NNπ channels in the nuclear medium, as well as investigate the influence of these fundamental channels on the observed $A_y(\Theta)$. In order to attack this problem we have recently obtained broad-range spectra (30°≤ Θ_{lab} ≤ 150°, T_p = 200 MeV) for the $^{13}C(p,\pi^+)$ and $^{13}C(p,\pi^-)$ reactions, populating the mirror nuclei ^{14}C and ^{14}O. The raw spectra (see Fig. 11) themselves already exhibit many interesting features. First of all, one notices that more strong states are seen in the (p,π^+) than in the (p,π^-) spectrum, and in general the yield is significantly larger for most discrete states excited by (p,π^+), in accord with expectations based on the coherence of multiple (p,π^+) amplitudes (vs. the highly restricted TNM path for all (p,π^-) transitions), and the relative strength (i.e., dominance of σ_{10} "inside" the nucleus) of the underlying fundamental NN → NNπ processes.

More specifically, this particular choice of $(p,\pi^+)-(p,\pi^-)$ comparison, where the population of mirror residual states is studied, may be used as a tool for identifying the high-spin nature of states populated in the (p,π^+) reaction. This is because the observed relative strength of residual states in (p,π^-) is dominated by momentum-matching considerations, thus apparent mirror peaks strong in both spectra probably indicate high-spin 2p-1h (with respect to ^{13}C) states accessible via the allowable NN→NNπ processes. We will concentrate here on the cross section and

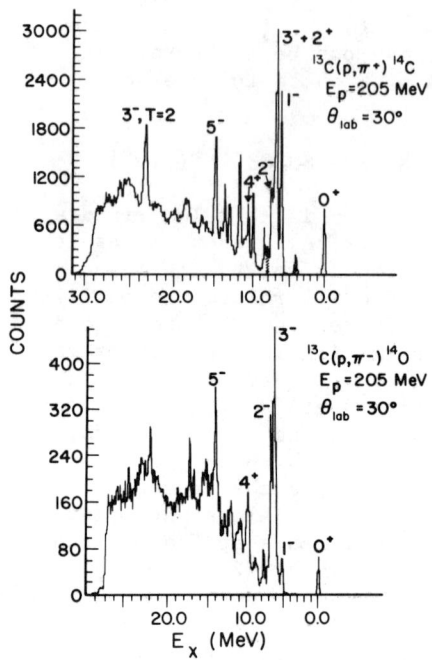

Fig. 11. A comparison of forward-angle $^{13}C(p,\pi^+)$ and $^{13}(p,\pi^-)$ spectra, populating states in the mirror nuclei ^{14}C and ^{14}O. J^π assignments are indicated for states known previously or to be discussed in the text.

analyzing power data for a few such discrete (p,π^+) transitions, and on the $A_y(\Theta)$ behavior for the continuum regions in both (p,π^+) and (p,π^-) spectra. Of the strongest (p,π^+) transitions observed in the ^{14}C spectrum (top panel Fig. 11) there are two states not previously identified: one at $E_x = 14.87$ MeV, to which we assign $J^\pi = 5^-$ based on various A = 14 data systematics, and a second strongly excited (high-lying and sharp) state at $E_x = 23.2$ MeV, for which I will present various speculations as to its structure.

Although there are 4^- states in ^{14}C, whose structures are of great interest, and which are known particularly from inelastic pion scattering studies,[29] a high-spin 5^- state with configuration $|^{13}C \times (\pi p_{3/2})^{-1}(\pi p_{1/2})(\nu d_{5/2})\rangle_{5^-}$ cannot be excited by inelastic scattering[30,31] on ^{14}C. A likely candidate because of the foregoing arguments is the strong peak at relatively high excitation ($E_x \cong 15$ MeV) in the (p,π^+) spectrum with a correspondingly strong peak observed for the mirror-like state in the (p,π^-) spectrum. It is plausible that this 15-MeV peak in fact corresponds to the anticipated 5^- state in ^{14}C reached via $pp \to np\pi^+$ through the $(\pi p_{3/2}) \to \{(\pi p_{1/2})(\nu d_{5/2})\}_{3^-}$ transition (where maximal coupling of the $p_{3/2}$ hole and $p_{1/2}$ spectator neutron outside the ^{12}C core leads to total spin of 5^-). Correspondingly, the mirror 5^- state in ^{14}O could be reached via $pn \to pp\pi^-$ by a $(\nu p_{3/2}) \to \{(\pi p_{1/2})(\pi d_{5/2})\}_{3^-}$ transition.

The same (p,π^+) transition without the spectator $p_{1/2}$ neutron, is just that leading to the stretched $9/2^+$ state at 9.5-MeV[28] excitation in ^{13}C we have just discussed. The cross section angular distributions for these two transitions are plotted together (with an appropriate relative normalization factor) in Fig. 12, and indeed show a very similar behavior. [Also plotted in this figure are data for the lower 4^- state in ^{14}C, identified as primarily a neutron

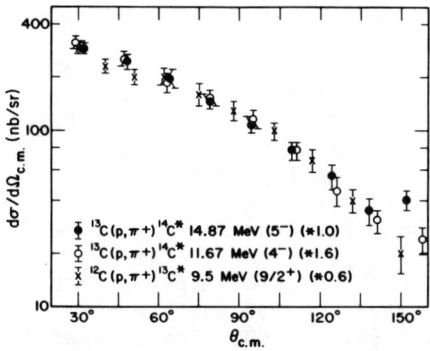

Fig. 12. Cross section angular distributions for several states of a similar nature excited by (p,π^+) on ^{12}C and ^{13}C targets as described in the text.

particle-hole excitation $|(\nu p_{3/2})^{-1}(\nu d_{5/2})>_{4^-}$ from inelastic pion scattering measurements[29] on ^{14}C. This state should be reached by the process $p+n \to nn\pi^+$ through interaction with a $p_{3/2}$ target neutron. Its unexpectedly large cross section suggests a more detailed consideration of the sensitivity of the various probes to the structure of these 4^- states.] Alternatively, a state at the excitation appropriate for the isobaric analog of the strong state near 15 MeV seen in the (p,π^+) spectra, has recently been observed and identified as 5^- in inelastic electron and pion scattering studies[30] on ^{14}N at $E_x = 16.9$ MeV. In sum, these observations lend circumstantial

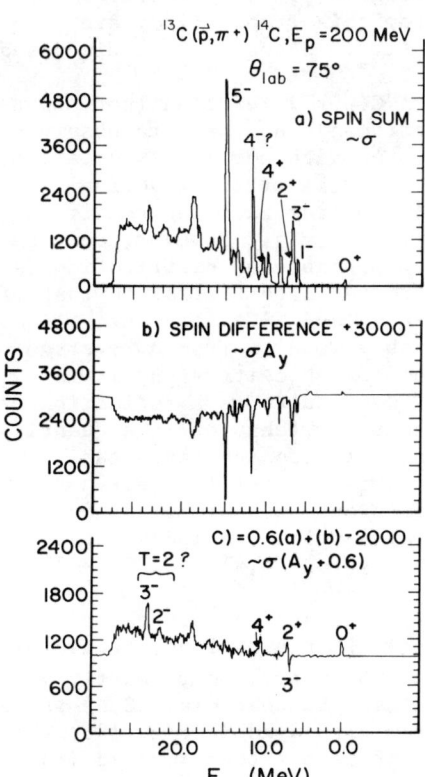

Fig. 13. Spin-dependent spectra for $^{13}C(p,\pi^+)^{14}C$ showing large negative analyzing powers for the continuum and most sharp states as evidenced by the mirror reflection of the spin sum $\sim\sigma$ spectrum a) by the spin difference $\sim\sigma A$ spectrum displayed in b). Transitions which do not share this behavior stand out in the "manufactured" spectrum c), where the effect of an average $A_y \sim -0.6$ has been removed from the σA spectrum of b).

support to the assignment of 5^- for the strongly excited state in $^{13}C(p,\pi^+)^{14}C$ at 14.87 MeV.

The most striking general feature of the observed analyzing powers in the (p,π^+) spectrum is the overall negative A_y for both the continuum and nearly all discrete states, illustrated in Fig. 13b by the spin-difference spectrum that is a nearly perfect reflection of the spin-sum spectrum shown in Fig 13a. This behavior persists with angle, as revealed by the strikingly similar $A_y(\theta)$ results for two ($\sim$ 1 MeV wide) representative continuum regions displayed in the lower portion of Fig. 14, and three strong representative discrete transitions in the uppermost panel of Fig. 15. This characteristic angular distribution for $A_y(\Theta)$ can be compared to that of the free[3] $pp \rightarrow d\pi^+$ (σ_{10}) $A_y(\Theta)$ also plotted in Fig. 14. The latter has been "transformed" to the proton-nucleus frame by first deducing the momentum of the struck target proton (assuming a head-on p-p collision) associated with a given pion momentum and scattering angle in the $13C(p,\pi^+)$ reaction, then transforming to the struck proton rest frame in to order determine the laboratory kinematics at which to look up the analyzing power for the free $pp \rightarrow d\pi^+$ reaction. Although there is not exact agreement, qualitatively the shape of the resultant $A_y(\Theta)$ is quite similar to that which is observed for the (p,π^+) reaction. Also of interest, however, are the few exceptional (p,π^+) transitions for which $A_y(\Theta)$ deviates significantly from this trend. They are highlighted in the lower portions of Figs. 13 and 15, and discussed below.

Spin-difference spectra for the $^{13}C(p,\pi^-)$ reaction (not shown) reveal much more state-to-state variations in A_y than are observed for (p,π^+). This is in general agreement with our TNM expectations and would arise from a sensitivity to details of the reaction process brought about by the critical momentum matching conditions and highly constrained reaction path for (p,π^-). However, continuum regions of the (p,π^-) spectrum do show a stable A_y distribution (see Fig. 14), and this behavior is furthermore quite similar to that of corresponding (p,π^-) data from ^{18}O, ^{26}Mg, and ^{48}Ca targets.[10,17] Since the continuum for (p,π^-) presumably results from an average over many state-dependent contributions of opposite sign, it is tempting to interpret these stable $A_y(\theta)$ results as a reflection of the intrinsic free $pn \rightarrow pp\pi^-$ analyzing power behavior -- a quantity as yet unmeasured. Were this later speculation the case, the observed sign difference in Fig. 14 between A_y for the $(p,\pi^\pm)$ continua might then arise from the different angular momentum coupling of the final-state nucleon pairs which accompany their opposite isospin couplings [I' = 0 mainly for (p,π^+) vs. I' = 1 for (p,π^-)].

Since the analyzing power of the dominant free process $pp \rightarrow d\pi^+$ appears to be reflected in most $A(p,\pi^+)A+1$ transtitions, states with anomalous $A_y(\theta)$ may signal transitions which proceed primarily via normally weak (i.e., I' = 1) $NN \rightarrow NN\pi$ isospin channels. Such states are evident in the "synthetic" spectrum displayed in Fig. 13c, where the overall negative analyzing power effect has been removed (the factor 0.6 was chosen empirically to give zero contribution from the

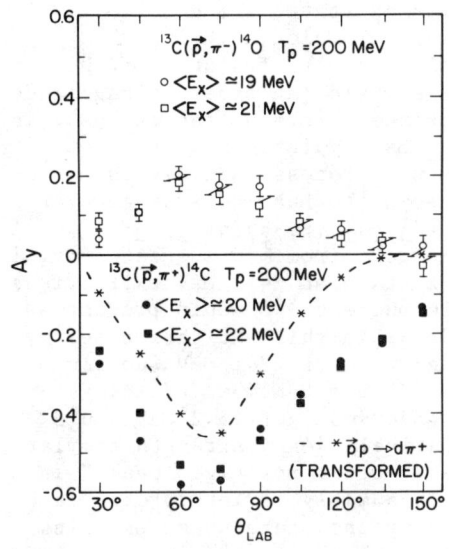

Fig. 14. Analyzing power distributions for regions of the (p,π⁺) and (p,π⁻) continuum. The negative $A_y(\Theta)$ for (p,π⁺) is similar to the strong NN → NNπ contributing processes (when transformed to the p-target frame) as described in the text.

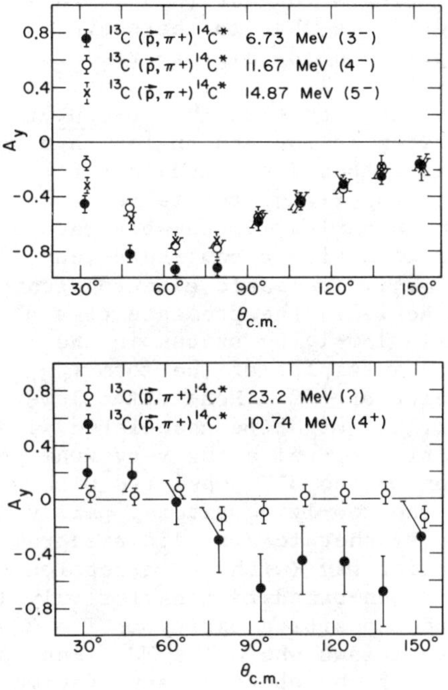

Fig. 15. Analyzing power distributions which follow the systematic σ_{10} analyzing power behavior (top), and those which deviate (bottom).

strong discrete states), and only states with a different analyzing power behavior appear. Several of interest are labeled in the figure, and angular distributions of the analyzing power for two of these states are displayed in the lower portion of Fig. 15. The anomalous behavior for the 10.74-MeV 4⁺ state is perhaps not surprising, since this state is believed to be predominantly a two-neutron excitation from the ^{14}C ground state,[32] hence accessible in (p,π⁺) only via pn → nnπ⁺ on the $p_{1/2}$ target neutron. The cross section observed for this state is indeed comparable in magnitude to that of its mirror counterpart in ^{13}C(p,π⁻), which would proceed by the charge-symmetric process pn → ppπ⁻. A comparison of these mirror 4⁺ transitions, excited by the mirror fundamental channels via $(\nu p_{1/2}) \to (2d_{5/2})^2_{4+}$, will allow us to study (p,π⁺) and (p,π⁻) under nearly identical conditions. Such investigations, combined with further near-threshold NN → NNπ $A_y(\Theta)$ measurements, may be very relevant for understanding possible modifications of fundamental NN→NNπ processes by the nuclear medium.

The strongest (p,π^+) transition with anomalous A_y behavior in Fig. 13 is that to the sharp E_x = 23.2-MeV state, while a coprresponding mirror-state candidate is visible in the (p,π^-) spectrum. A natural explanation for all observed features of the 23.2-MeV state in ^{14}C, except the large yield (an order of magnitude greater at forward angles than the strongest (p,π^-) transitions), is that it has T = 2 and therefore cannot be populated via pp → $(np)_{I'=0}\pi$, but rather through (p,π^+) processes involving σ_{11} or σ_{01} channels. Presumably such a state in ^{14}C has moderate angular momentum with a configuration $|^{13}C \times (p_{3/2})^{-1}(p_{1/2})(d_{5/2})>3^-$ [or 4^-,T=2]. Inelastic electron scattering data from ^{14}C suggest (Fig. 1 of Ref. 31) the presence of a 4^- T = 2 state at 24.3 MeV excitation, while low-lying states in the isobaric nucleus ^{14}B would predict a spin multiplet of the form $(p_{3/2})^{-1}(d_{5/2})$ in this excitation energy region of ^{14}C. Hence, the lower lying-state at 22.1 MeV seen in Ref. 31, might be associated with the 2^- T = 2 member of this multiplet, while the very weak state observed[31] at 23.2 MeV could then be the 3^-, populated with a (relatively weak) strength similar to the low-lying natural parity states. Other positive "peaks" in the synthethic Fig. 13c spectrum can presumably be interpreted in a similar way (with the exception of the ground state which exhibits this non-standard behavior only locally near 75°), namely, as being either positive parity or T = 2 states, the latter dominated by transitions where the TNM final-state nucleons are coupled to T = 1.

If the above interpretation of isospin nature of the high-lying sharp state in ^{14}C is correct, we are seeing evidence for an unexpectedly strong enhancement of normally weak NN → NNπ isospin channels inside the nucleus. An alternative explanation (also incomplete) is that the state at E_x = 23.2 MeV in ^{14}C is a T = 1 state with maximal spin coupling accessible in (p,π^+) -- namely, a configuration of the form $|^{13}C \times (\pi p_{3/2})^{-1}(\pi d_{5/2})(\nu d_{5/2})>7^+$ -- where we would then speculate that effects resulting from the strong constraints on the angular momentum coupling are at work to produce the observed state-dependent modification of the standard $A_y(\theta)$ pattern normally given by the dominant σ_{10} channel.

Additional evidence, however, comes from study of the $^{12}C(p,\pi^+)^{13}C$ reaction as shown by the spectrum in Fig. 16 where, as was the case for ^{14}C, a strongly excited sharp state appears at high excitation energy (E_x = 21.4 MeV). The excitation energy for this state is in fact in very good agreement with a state identified[31] with a M4 transition in $^{13}C(e,e')$, and a state also observed[33] relatively strongly in $^{13}C(\pi^-,\pi^{-'})$. Recent shell model calculations[34] have suggested a number of closely lying states near E_x = 21.5 MeV in ^{13}C with J^π = $7/2^+$ or $9/2^+$, and T = 3/2 or 1/2. Along with the experimental results this suggests that the 21.4-MeV state seen in our $^{12}C(p,\pi^+)^{13}C$ spectrum may be a moderately high-spin (7/2,9/2) T$_>$ state, much like our T$_>$ speculations regarding the strong high-lying state in ^{14}C. In fact, a comparison (Fig. 17) of the cross section and analyzing power angular distribution behavior for the $^{12}C(p,\pi^+)^{13}C^*$(21.4 MeV) and $^{13}C(p,\pi^+)^{14}C^*$(23.4 MeV) transitions shows a very comparable behavior, indicating similar processes involving similar states are likely involved. If indeed these transitions are to be associated with T$_>$ states, the relative

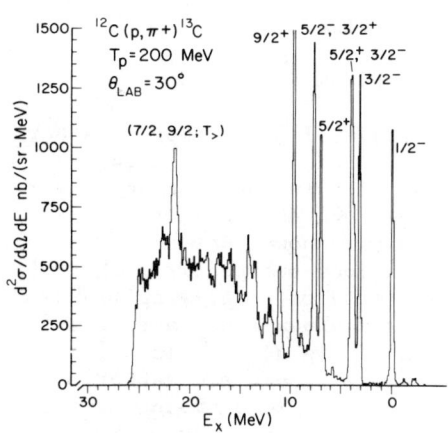
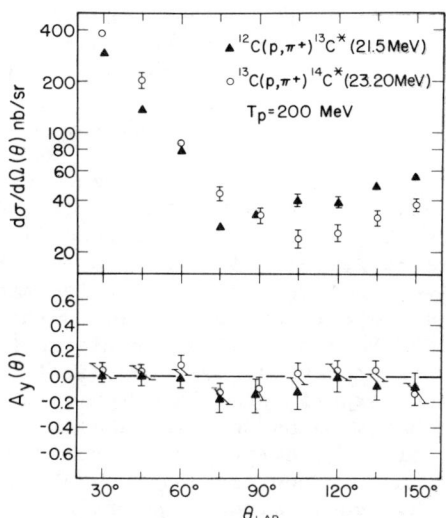

Fig. 16. Spectrum for the $^{12}C(p,\pi^+)^{13}C$ reaction at $E_p =$ 200 MeV showing in particular the strong excitation of a high-lying sharp state at an excitation energy of 21.5 MeV.

Fig. 17. Cross section and analyzing power distributions for the high-lying sharp states seen in both the $^{12}C(p,\pi^+)^{13}C$ and $^{13}C(p,\pi^+)^{14}C$ reactions.

narrowness of the ^{14}C state (compared to the experimental resolution < 200 KeV) might be explained since isospin conservation then restricts its particle decay channels. For the ^{13}C state, however, with width almost twice the experimental resolution, some particle decay channels may be open (for an initial configuration of the form $|^{12}C \times (p_{1/2})^{-1}(p_{3/2})(d_{5/2})\rangle$, for example, decay to the 1^+, T = 1 state at E_x = 15.11 MeV in ^{12}C by neutron emission is possible). Alternatively, if this high-lying ^{13}C state is T = 1/2, it could particle decay to the 1^+, T = 0 state at E_x = 12.71 MeV in ^{12}C.

It would be interesting to find other cases of such high-lying strongly excited sharp states in (p,π^+) spectra as an aid in solving what at the present has become somewhat of a mystery. In the meantime, we have thought of other reaction studies to test several of these hypotheses. For example we have initated conventional experiments with the $^{11}B(\alpha,p)^{14}C$ and $^{11}B(\alpha,d)^{13}C$ reactions for which we expect strong population of high-spin $T_<$ states in both residual nuclei. Should these high-lying states observed in the $^{12,13}C(p,\pi^+)^{13,14}C$ spectra not be strongly populated in the (α,d) and (α,p) reactions, and preliminary indications are that they are not, this would suggest once again their $T_>$ assignment. In that case, since the ^{13}C state may have the open particle decay modes described above, $^{12}C(p,\pi n)$ coincidence measurements could be used to investigate the isospin character of the state at E_x = 21.4 MeV in ^{13}C in order to pin down what we are observing here. The possibility of such studies is under investigation.

SUMMARY AND CONCLUDING REMARKS

In discussing our adventures with pions, I have taken a rather heuristic approach (similar to that which stimulated the actual investigations!) to the general concept of a two-nucleon model for near-threshold pion production. Pursuing such a view in our investigations of the reaction mechanism has shown that most features, either predicted or those which soon emerged from the data base, have a qualitative explanation in terms of underlying (TNM) NN → NNπ processes "immersed" in the nuclear medium. The fact that these experimental TNM signatures persist in (p,π) reactions probably means that fundamental NN → NNπ amplitudes are rather stable in their off-shell behavior. We spent a fair amount of time discussing how and why (p,π^-) transitions are relatively weak, highly constrained, and often very selective. A striking example of the latter is the systematic and selective excitation of high-spin 2p-1h states across the entire periodic table which can be viewed as a universal response of the nucleus to this high-q probe. While these constraints make the state-to-state variations in $A_y(\Theta)$ for (p,π^-) large, continuum $A_y(\Theta)$ are relatively stable with respect to target nucleus. The systematic stretched state Ay(Θ) behavior is yet to be simply explained (any explanation is clearly interwoven with the systematics of several other related phenomena), although in the meantime it may be used as an experimetal tool in the search for very high spin states produced by (p,π^-) in (heavy) target nuclei.

We found that in general, (p,π^+) transitions are strong, reflecting the strength of underlying pp → dπ$^+$ amplitudes (σ_{10}) plus coherence effects. In a spectropscopic comparison of the ^{13}C(p,π$^+$)^{14}C and ^{13}C(p,π$^-$)^{14}O reactions, strongly excited states populated in ^{14}O by the highly selective (p,π^-) reaction were used to help identify similar high-spin states in the mirror nucleus ^{14}C -- in this regard, circumstantial evidence was presented supporting a new $J^\pi = 5^-$ assignment for a state excited at E_x = 14.87 MeV in the (p,π^+) reaction. More generally, we find for (p,π^+) that the analyzing power for most discrete states and the continuum follow the negative analyzing power behavior of the underlying strong pp → dπ$^+$ process. The several states which don't follow this Ay(Θ) pattern are "interesting" and their behavior probably reflects contributions from other fundamental processes. Of particular note are high-lying states observed in the ^{12}C(p,π$^+$)^{13}C and ^{13}C(p,π$^+$)^{14}C reactions at E_x = 21.5 and 23.2 MeV, respectively. The suspected isospin T$_>$ nature of these states would point to the anomalous enhancement of normally weak NN → NNπ channels in the nuclear medium.

Finally, although I have stressed "minimal" theory in this presentation, certainly realistic TNM calculations which reproduce the (by now) relatively large number of (p,π) data systematics are highly desirable. This is the only way after all in which we can further test our TNM understanding of these processes while at the same time turn (p,π) into a probe which might truly compliment other more standard probes of nuclear structure. In pursuing such efforts, it seems that we should not lose sight of the remarkable success that relatively simple calculations have had in describing the recently observed experimental behavior of both discrete high-spin and continuum (p,π^-) excitations. Furthermore, the demonstrated dominance of fundamental NN $\to$ NNπ amplitudes for (p,π) reactions in general, suggests that certain NN $\to$ NNπ features (e.g., the negative $A_y(\Theta)$ for (p,π^+) from pp $\to$ dπ^+, etc.) must be installed in TNM codes at a <u>fundamental</u> TNM level if one is to have any hope of understanding (and calculating) the systematic features of the (p,π^+) data base or its exceptional components. In the later regard, it is interesting to note that the experimental situation for near-threshold NN $\to$ NNπ is also not very completely known — a planned program of experiments with the upcoming IUCF Cooler/Storage Ring may help to remedy this situation.

ACKNOWLEDGEMENTS

Contributions from many people are represented here. Prof. S.E. Vigdor, Dr. M.C. Green (Argonne), and Dr. T.G. Throwe have all been heavily involved in the pion production work since its early stages at IUCF. More recently the efforts of Prof. B.A. Brown (Michigan State), and Elie Korkmaz, whose Ph.D. thesis work is concerned with the ^{13}C$(p,\pi^{\pm})$ studies, are appreciated and acknowledged. Any errors or omissions in the manuscript are my responsibility!

Work supported in part by U.S. National Science Foundation PHY 81-14339.

REFERENCES

1. S. Dahlgren, B. Hoistad, and P. Grafstrom, Phys. Lett. <u>35B</u>, 219 (1971).
2. D.F. Measday and G.A. Miller , Ann. Rev. Nucl. Part. Sci. <u>29</u>, 121 (1979); B. Hoistad, in Advances in Nuclear Physics, eds. J. Negele and E. Vogt, Vol. 11 (Plenum Press, New York, 1979), p. 135, H.W. Fearing, in Progress in Particle and Nuclear Physics, ed. D.H. Wilkinson, Vol. 7 (Pergamon, Oxford, (1981), p. 113; AIP Conf. Proc. No. 79, Pion Production and Absorption in Nuclei, ed. R.D. Bent (American Institute of Physics, New York, 1982).
3. E.G. Auld, A. Haynes, R.R. Johnson, G. Jones, T. Masterson, E.L. Mathie, P. Ottewell, P. Walden, and B. Tatischeff, Phys. Rev. Lett. <u>41</u>, 462 (1978).
4. T.P. Sjoreen, P.H. Pile, R.E. Pollock, W.W. Jacobs, H.O. Meyer, R.D. Bent, M.C. Green, and F. Soga, Phys. Rev. C <u>24</u>, 1135 (1981); G.J. Lolos, E.L. Mathie, P.L. Walden, G. Jones, E.G. Auld, and R.B. Taylor, Phys. Rev. C <u>25</u>, 1086 (1982).
5. W.W. Jacobs, A.G. Drentje, P.H. Pile, P.P. Singh, T.P. Sjoreen, and S.E. Vigdor, Phys. Lett. <u>94B</u>, 319 (1980).

6. R.D. McKeown, S.J. Sanders, J.P. Schiffer, H.E. Jackson, M. Paul, J.R. Specht, E.J. Stephenson, R.P. Redwine, and R.E. Segel, Phys. Rev. Lett. 44, 1033 (1980).
7. M.J. Iqbal and G.E. Walker, Phys. Rev. C 32, 556 (1985).
8. J. Conte and M. Dillig, private communication.
9. M.C. Green in AIP Conf. Proc. No. 79, Pion Production and Absorption in Nuclei, ed. R.D. Bent (AIP, New York, 1982), p. 131.
10. T.G. Throwe, S.E. Vigdor, W.W. Jacobs, M.C. Green, C.W. Glover, T.E. Ward, and B.P. Hichwa, submitted to Phys. Rev C, and T.G. Throwe, Ph.D. Thesis, Indiana University, 1984 (unpublished).
11. B.J. VerWest and R.A. Arndt, Phys. Rev. C 25, 1979 (1982).
12. W.W. Jacobs, T.G. Throwe, S.E. Vigdor, M.C. Green, J.R. Hall, H.O. Meyer, W.K. Pitts, and M. Dillig, Phys. Rev. Lett. 49, 855 (1982).
13. S. Cohen and D. Kurath, Nucl. Phys. A101, 1 (1967).
14. K. Kume, Proc. 6th Int'l Symp. on Polarization Phenomena in Nucl. Phys. (Osaka, 1985), p. 189.
15. S.E. Vigdor, T.G. Throwe, M.C. Green, W.W. Jacobs, R.D. Bent, J.J. Kehayias, W.K. Pitts, and T.E. Ward, Phys. Rev. Lett. 49, 1314 (1982), and Nucl. Phys. A396, 61c (1983).
16. J.W. Watson, M. Ahmad, B.D. Anderson, A.R. Baldwin, A. Fazely, P.C. Tandy, and R. Madey, Phys. Rev. C 23, 2373 (1981).
17. J.J. Kehayias, R.D. Bent, M.C. Green, M. Hugi, H. Nann, and T.E. Ward, to be published in Phys. Rev. C, and J.J. Kehayias, Ph.D. Thesis, Indiana University, 1983 (unpublished).
18. M. Behar, A. Filevich, G. Garcia Bermudez, M.A.J. Mariscotti, and E. Ventura, Nucl. Phys. A366, 61 (1981).
19. H. Nann, W.W. Jacobs, A.D. Bacher, J.D. Brown, G.D. Cravens, G.T. Emery, W.P. Jones, E.J. Stephenson, B.A. Brown, Phys. Rev. C 30, 1509 (1984).
20. H. Nann, "Transfer Reactions to Stretched States", Proceedings of this Workshop.
21. A. Brown, O. Scholten, and H. Toki, Phys. Rev. Lett. 51, 1952 (1983).
22. M.C. Green, J. Brown, W.W. Jacobs, E. Korkmaz, T.G. Throwe, S.E. Vigdor, T.E. Ward, P.L. Jolivette, and B.A. Brown, Phys. Rev. Lett. 53, 1893, (1984).
23. O. Scholten, H. Toki, J. Aichelin, A. Bonasera, R. Kaps, and H. Safafin, Phys. Rev. C 32, 653 (1985).
24 H. Toki and K.-I. Kubo, Phys. Rev. Lett. 54, 1203 (1985).
25. S.E. Vigdor, W.W. Jacobs, T.G. Throwe, and M.C. Green, Phys. Rev. Lett. 54, 1204 (1985).
26. R. Lawson, private communication.
27. B.A. Brown, Private communication.
28. F. Soga, P.H. Pile, R.D. Bent, M.C. Green, W.W. Jacobs, T.P. Sjoreen, T.E. Ward, and A.G. Drentje, Phys. Rev. C 24, 570 (1981).
29. D.B. Holtkamp, S.J. Seestrom-Morris, S. Chakravarti, D. Dehnhard, H.W. Baer, C.L. Morris, S.J. Greene, and C.J. Harvey, Phys. Rev. Lett. 47, 216 (1981).

30. D.F. Geesaman, D. Kurath, G.C. Morrison, C. Olmer, B. Zeidman, R.E. Anderson, R.L. Boudrie, H.A. Thiessen, G.S. Blanpied, G.R. Burleson, R.E. Segel, and L.W. Swenson, Phys. Rev. C 27, 1134 (1983); J.C. Berstrom, R. Neuhausen, and G. Lahm, Phys. Rev. C 29, 1168 (19840; R.A. Lindgren, in Proc. Int'l Conf. on Highly Excited States and Nuclear Structure, (Orsay, 1983).
31. M.A. Plum, R.A. Lindgren, J. Dubach, R.S. Hicks, R.L. Huffman, B. Parker, G.A. Peterson, A. Alster, J. Lichtenstadt, M.A. Moinester, and H. Baer, Phys. Lett. 137B, 15 (1984).
32. F. Ajzenberg-Selove, Nucl. Phys. A360, 1 (1981).
33. S.J. Seestrom-Morris, D. Dehnhard, M.A. Franey, G.S. Kyle, C.L. Morris, R.L. Boudrie, J. Piffaretti, and H.A. Thiessen, Phys. Rev. C 26, 594 (1982).
34. T.S.H. Lee and D. Kurath, Phys. Rev. C 22, 1670 (1980).

SELECTIVE POPULATION OF STRETCHED HIGH-SPIN STATES IN THE (d,α) REACTION

H. Nann
Department of Physics, Indiana University
Bloomington, Indiana 47401, USA

ABSTRACT

Recent experimental investigations of the (d,α) reaction on target nuclei between A = 40 and 64 with 80 MeV vector polarized deuterons have shown an extremely strong selectivity for exciting [(target) x $(1f_{7/2})^{-2}_{J=7}$] configuration high-spin states in the final nuclei. The transitions to such states are characterised by L = 6 angular distributions of the differential cross section and typical J = 7 patterns of the vector analzying power. Their location and their transition strength in the (d,α) reaction on the even Ni-isotopes is used to critically test predictions from shell model calculations. On the even Ca-isotopes, this strong and systematic selectivity is employed to determine the extent of proton core excitations in their ground state wave functions.

INTRODUCTION

Transfer reaction studies at beam energies around 60 MeV and higher have attracted quite some interest since single-particle and cluster states with relatively high angular momenta are selectively populated as a result of the angular momentum mismatch between entrance and exit channels. This feature, especially for the (d,α) reaction, leads to new and exciting structure information about high-spin states not available at lower incident energies. Because of the simplicity and relative purity of these high-spin states, experiments aimed at determining their excitation energies and transition strength distributions are of great importance for comparison with theoretical predictions based, for example, on multi-configuration shell-model calculations. Such comparisons can be useful in defining the appropriate configuration space and in determining the proper residual interaction to be used in these calculations. On the other hand, the selective enhancement of high-spin states in the final nucleus can be employed to determine those small components in the target ground state wave function from which these stretched configuration high-spin states exclusively can be reached.

CHARACTERISTIC FEATURES OF THE (d,α) REACTION AT 80 MeV

In a two-nucleon transfer reaction, such as (d,α), the proton-neutron pair is preferentially transferred in a completely aligned configuration due to geometrical coefficients in the structure factor.[1] At low bombarding energies, this feature is inconspicuous, since small orbital angular momentum transfers are kinematically favored. However at higher bombarding energies, momentum matching requires that the transferred proton-neutron pair carries large values of the orbital angular momentum. Together, these features yield large enhancements in the cross sections of states which are formed with the maximum J within a given region. For example, within the $1f_{7/2}$ orbit space, residual states with the [(target) x $(1f_{7/2})^{-2}_{J=7,T=0}$] configuration will be strongly enhanced. Indeed, these features can be observed in a comparison of the ^{60}Ni(d,α)^{58}Co reaction at 80 (Ref. 2) and 17 (Ref. 3) MeV bombarding energies. By choosing angles giving approximately the same momentum transfer at the two incident beam energies, for example, $\Theta_{c.m.} \sim 6°$ at 80 MeV and $\Theta_{c.m.} \sim 35°$ at 17 MeV, the relative population of different states can be compared. The 80-MeV spectrum is dominated by the transition to the 7^+ state at 2.69 MeV in ^{58}Co, whereas in the 17-MeV spectrum, the transitions to this state and to the 5^+ state at 0.37 MeV show comparable yields. The ratio of the

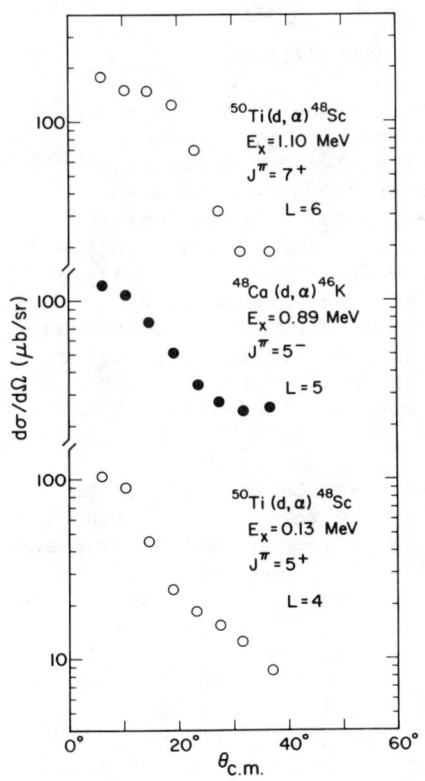

Fig. 1. Differential cross sections characteristic of various L-transfers.

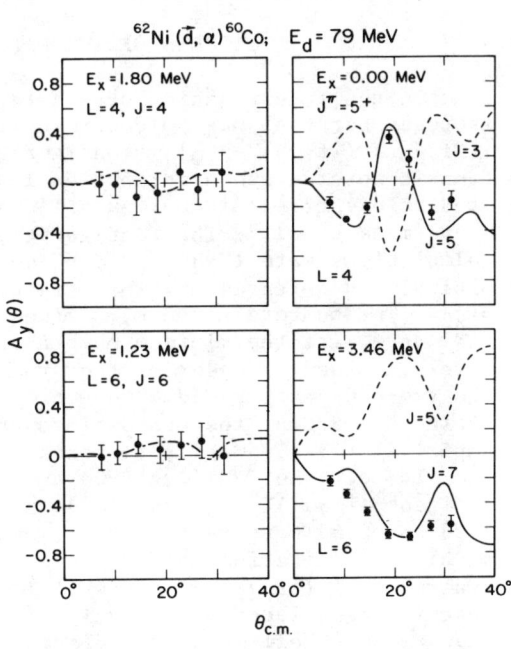

Fig. 2. Experimental and DWBA vector analyzing power angular distributions for various possible L- and J-combinations.

differential cross sections for the L=6 transition to the 2.64-MeV, $J^\pi=7^+$ state and the L=4 transition to the 0.37-MeV, $J^\pi=5^+$ state is about 3.5:1 at 80 MeV incident energy and about 1:1 at 17 MeV. This shows clearly the kinematic enhancement of the L=6 transition strength over the L=4 strength at 80 MeV.

The measured angular distributions of the differential cross section for different orbital angular momentum transfers are sufficiently distinct to allow unambiguous identification of the transferred L-value, as can be seen in Fig. 1, where samples of L = 6, 5 and 4 angular distributions are displayed. Once the orbital angular momentum transfer has been established, the spin of the transferred proton-neutron pair can be determined from the characteristic J-dependent features of the vector analyzing power (VAP) angular distributions. Experimental L=6, J=7 and J=6 as well as L=4, J=5 and J=4 vector analyzing power patterns obtained from the ^{62}Ni(d,α)^{60}Co reaction are presented in Fig. 2. Both the L-patterns of the differential cross section and the J-patterns of the vector analyzing power have been observed to be very stable across the $1f_{7/2}$ orbital (from A=40 to A=64).

RESULTS OF THE Ni(d,α)Co REACTION

In the investigation of the (d,α) reaction on the even nickel isotopes our goal was to determine the location of stretched $[(Ni)_{0^+} \times (1f_{7/2})^{-2}_{7,0}]_{7^+}$ configuration states and their (d,α) transition strength distribution in order to compare them to predictions of multi-configuration shell-model calculations by Glaudemans et al.[4] The configurations taken into account in these calculations were $f^{-n}q^m$ and $f^{-n-1}q^{m+1}$, where f denotes the $1f_{7/2}$ orbital and q represents the remaining fp-shell orbitals $p_{3/2}$, $f_{5/2}$, $p_{1/2}$. The numbers n and m, defined with respect to the doubly-magic ^{56}Ni core, are the minimum number of holes and particles, respectively, needed to describe the ground state of a specific nucleus. The two-body matrix elements employed were obtained from a mixture of the Kuo-Brown, the schematic surface delta interaction, and empirical $(f_{7/2})^2$ matrix elements.

In both the ^{58}Ni(d,α)^{56}Co and ^{60}Ni(d,α)^{58}Co reactions, the dominant $(1f_{7/2})^{-2}_{7,0}$ transition strength is to a single state at 2.28 and 2.69 MeV, respectively, with several much smaller fragments at higher excitation energy. Figure 3 shows a comparison of the experimental $(1f_{7/2})^{-2}_{7,0}$ transition strengths for the ^{58}Ni(d,α)^{56}Co reaction as a function of excitation energy and the predictions of Glaudemans' shell-model calculations. The overall agreement is quite good. In the case of the ^{62}Ni(d,α)^{60}Co reaction, however, the situation is quite different as can be seen in Fig. 4. Here the experimental $(1f_{7/2})^{-2}_{7,0}$ transition strength is distributed over many levels between 1.5 and 4.8 MeV, with a centroid of 3.9 MeV. The shell-model predictions completely fail to account for the observed strength distribution, although the summed strength agrees well within the experimental uncertainties. It is likely that the interpretation of the ^{62}Ni(d,α)^{60}Co results requires a much larger configuration space and most likely the use of a different residual interaction.

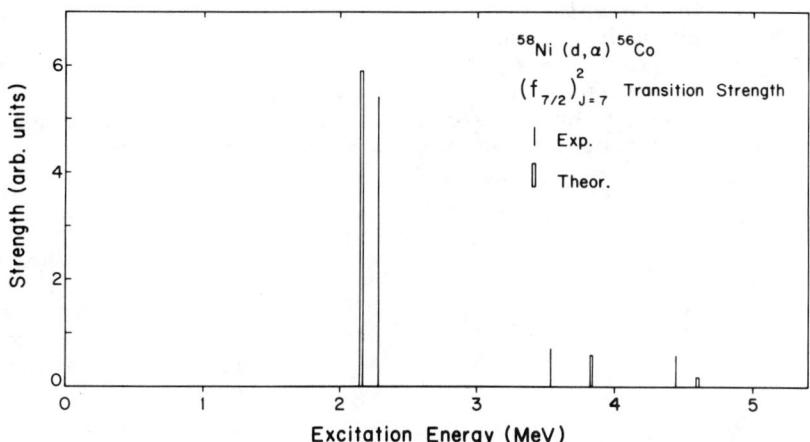

Fig. 3. Experimental and theoretical distributions of the $(f_{7/2})^2_{J=7}$ transition strength for the ^{58}Ni(d,α)^{56}Co reaction.

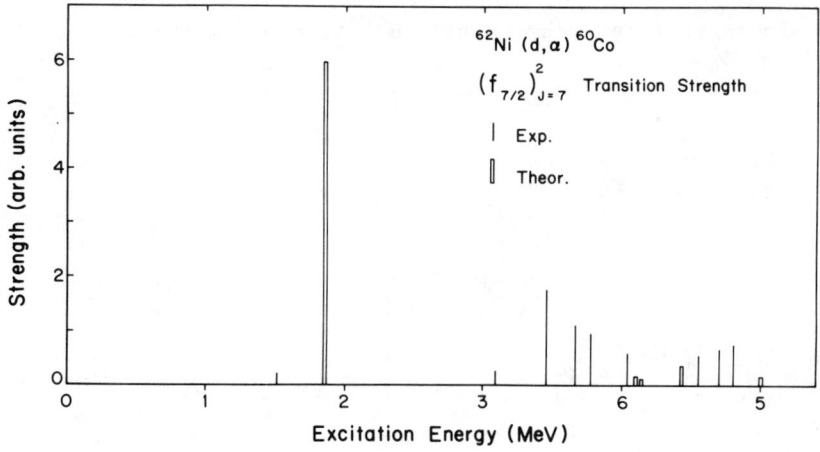

Fig. 4. Experimental and theoretical distributions of the $(f_{7/2})^2_{J=7}$ transition strength for the ^{62}Ni(d,α)^{60}Co reaction.

RESULTS OF THE Ca(d,α)K REACTION

The extremely strong selectivity of the (d,α) reaction on 1f-shell target nuclei at 80 MeV bombarding energy for picking up proton-neutron pairs in the completely aligned $(1f_{7/2})^2_{7,0}$ configuration was used in the ^{even}Ca(d,α)K reaction to determine the extend to which proton core excitations exist in the ground states of the even calcium isotopes. Although these proton core excitation components in the ground state wave functions are quite small, the strongest peak in the spin up (d,α) spectrum (the detection system being on the right side of the beam) on each Ca-isotope at $\Theta_{lab} = 14°$ belongs to a L = 6, J = 7 transition. This is depicted in Figs. 5 and 6. Transitions to states populated by the pickup of other configuration proton-neutron pairs from the major shell-model components of the target ground state wave function are weaker.

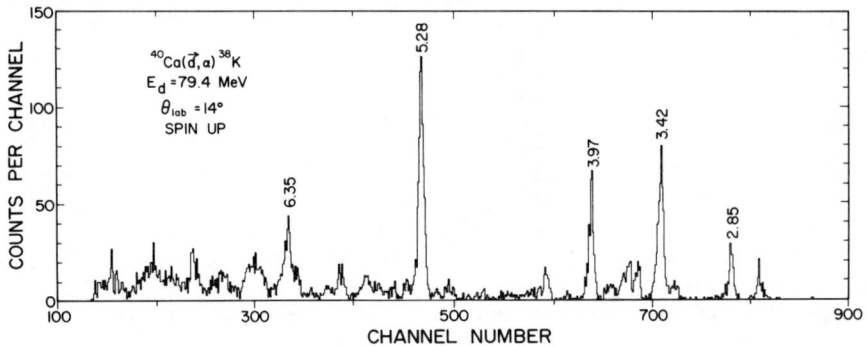

Fig. 5. Alpha-particle spectrum of the $^{40}Ca(d,\alpha)^{38}K$ reaction.

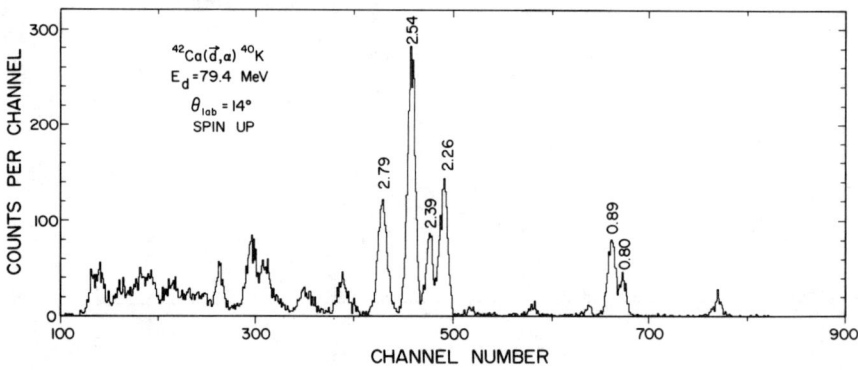

Fig. 6. Alpha-particle spectrum of the $^{42}Ca(d,\alpha)^{40}K$ reaction.

Unambiguous L = 6, J = 7 transitions were identified leading to the states at 4.54 and 5.95 MeV in ^{46}K, at 1.91 MeV in ^{42}K, at 2.54 MeV in ^{40}K, and at 5.28 MeV in ^{38}K. The measured angular distributions of the differential cross section are shown in the left-hand panels of Figs. 7 and 8 with the corresponding vector analyzing powers in the right-hand panel. It is interesting to note that in ^{46}K two $J\pi = 7^+$ states are populated but not in the other K-isotopes. In ^{38}K, there is another 7^+ state known at 3.46 MeV. However, this state is very weakly excited in the ^{40}Ca(d,α)^{38}K reaction. This is brought forth only from a channel by channel analysis of the vector analzying power of the peak around 3.4 MeV.

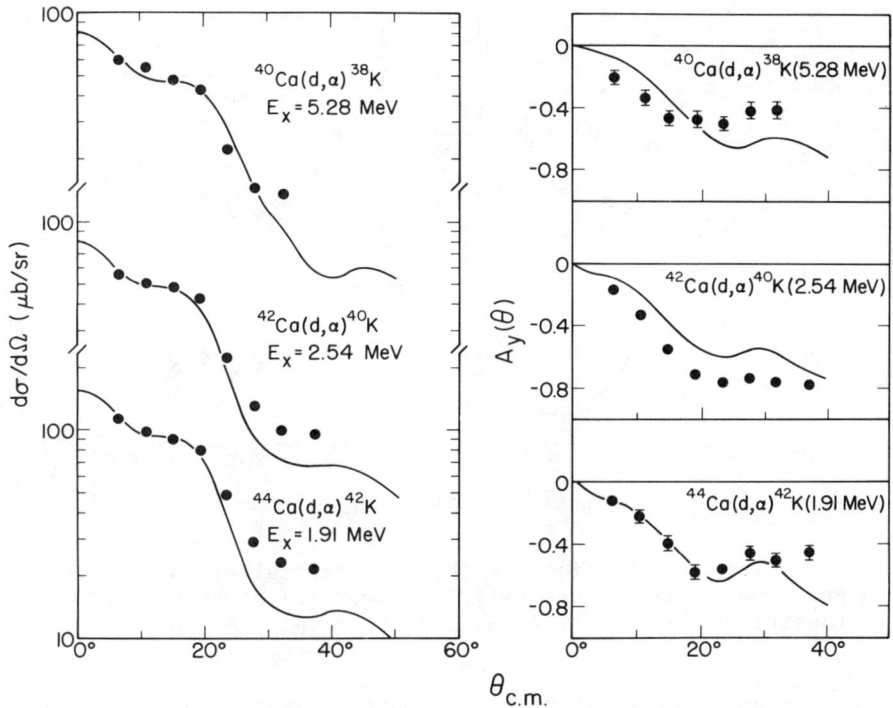

Fig. 7. Angular distributions of the differential cross section (left-hand panel) and vector analyzing power (right-hand panel) for L = 6 transitions in the ^{40,42,44}Ca(d,α) reactions.

From the expression for the spectroscopic amplitude[5] for the pickup of a proton-neutron pair in the stretched $(1f_{7/2})^2_{7,0}$ configuration, the proton occupation number n_p(Ca) in the ground states of the even Ca-isotopes can be linked to that of ^{50}Ti, n_p(Ti), by the following relation

$$n_p(\text{Ca}) = n_p(\text{Ti}) \frac{n_n(\text{Ti})}{n_n(\text{Ca})} \frac{\sigma_{\text{exp}}(\text{Ca})}{\sigma_{\text{exp}}(\text{Ti})} \frac{\sigma_{\text{DWBA}}(\text{Ti})}{\sigma_{\text{DWBA}}(\text{Ca})} \qquad (1)$$

Here the n_n are the neutron occupation numbers and σ_{exp} and σ_{DWBA} are the experimental and calculated (d,α) differential cross

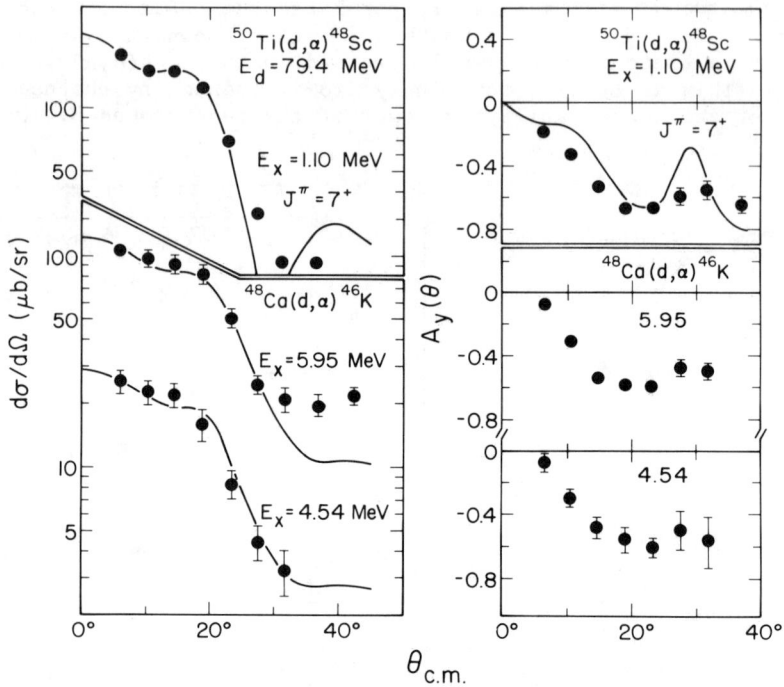

Fig. 8. Angular distributions of the differential cross section (left-hand panel) and vector analyzing power (right-hand panel) for L = 6 transitions in the ^{50}Ti(d,α) and ^{48}Ca(d,α) reactions.

sections, respectively. It should be noted that n_p(Ca) extracted from Eq. (1) does not depend on the absolute normalization of the distorted-wave Born approximation (DWBA) cross section for the (d,α) reaction. Furthermore, the dependence on optical-model parameters is greatly reduced as long as they are very nearly the same for all nuclei. In the analysis the following values were used: n_p(Ti) = 1.85 (Refs. 6 and 7) and n_n(Ti) = 6.7 (Ref. 8), n_n(^{48}Ca) = 6.8 (Ref. 9), n_n(^{44}Ca) = 2.9 (Ref. 9), n_n(^{42}Ca) = 1.7 (Ref. 9) and n_n(^{40}Ca) = n_p(^{40}Ca).

Microscopic DWBA calculations were performed with the code DWUCK4 using optical model parameters from the literature,[10] and standard nonlocality and finite range corrections were applied. These parameters have been used successfully in an extensive series of (d,α) analysis of 80 MeV incident energy for nuclei between ^{50}Ti and ^{64}Ni.

There remains the question of what form factor for the transferred proton-neutron pair to use in this comparison between the pickup from the Ca-isotopes and ^{50}Ti, since for the Ca-isotopes it involves not only "small" components of the ground state wave functions but also the final states are very far away from the $1f_{7/2}$ proton centroid. The latter situation is illustrated in Fig. 9 for ^{40}Ca. The shell-model $1f_{7/2}$ orbital lies about 7.3 MeV above the $1d_{3/2}$ orbital.

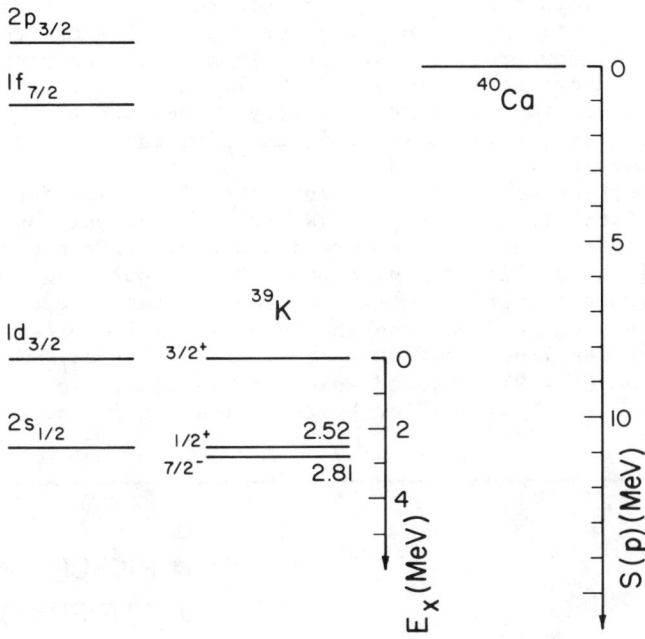

Fig. 9. Energy diagram for ^{39}K with respect to ^{40}Ca.

According to the standard prescription one uses a separation energy for each nucleon equal to one-half the deuteron separation energy for that particular transition. One then generates a radial wave function from a standard Woods-Saxon well (r_o = 1.25 fm, a = 0.65 fm, λ = 25) with the depth adjusted to match the separation energy. This prescription yields $1f_{7/2}$ proton occupation numbers of 1.8, 1.5, 1.9, and 1.6 for the masses 40, 42, 44, and 48, respectively. These values are clearly too large showing the deficiency of the standard form factor prescription.

As a first step toward a more realistic form factor, the prescription was modified in the following way. Recently Sick et al.[11] have shown that, for a few special cases, precise valence-nucleon radial wave functions can be obtained by combining information at large radii from low energy "sub-Coulomb" transfer reactions and at smaller radii from measurements of the highest allowed multipolarity nuclear magnetization density by elastic electron scattering. In a procedure which incorporates both informations, the magnetic form factors are fitted using phenomenological radial wave functions computed from a Woods-Saxon potential with its depth adjusted to the separation energy of the unpaired nucleon. The results of this analysis are then represented by the root-mean-square radius of the valence nucleon wave function.[12] It has been demonstrated[13] that the use of such radial wavefunctions in analyses of single-nucleon transfer reactions leads to a more realistic determination of absolute spectroscopic factors. If one uses the $1f_{7/2}$ rms radii[12] of 4.08 for neutrons (extracted from ^{49}Ti) and 4.10 for protons (extracted from ^{51}V), the proton occupation numbers reduce to 1.1, 1.0, 1.4 and 1.0 for the calcium isotopes 40, 42, 44 and 48 respectively. These numbers are still uncomfortably large indicating that the form factor is probably still not correctly described.

If an effective one-body potential well is used for generating a more realistic form factor, it is probably not only sufficient to change the radius but also to vary its depth. This means that the transferred nucleon is no longer bound by the observed separation energy. From a comparison of the proton separation energies of ^{A}Ca and $^{A+1}$Sc one deduces that the shell-model $1f_{7/2}$ orbital in the calcium-isotopes lies about 6 to 7 MeV above the $1d_{3/2}$-$2s_{1/2}$ orbitals (see Fig. 9). Use of an effective binding energy for a $1f_{7/2}$ proton, which is smaller by this amount than the actual

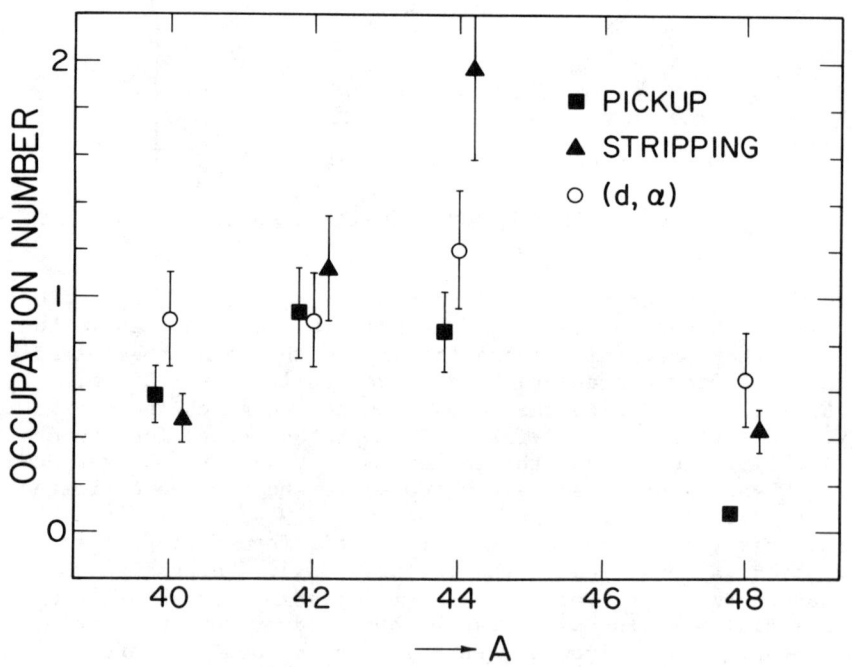

Fig. 10. Number of $1f_{7/2}$ protons in the ground states of the even calcium isotopes.

separation energy, yields $1f_{7/2}$ proton occupation numbers of 0.9 ± 0.2, 0.9 ± 0.2, 1.2 ± 0.3, and 0.7 ± 0.2 for the Ca-isotopes 40, 42, 44, and 48, respectively.

These last occupation numbers are believed to be the most realistic ones that presently can be derived from the (d,α) data at 80 MeV bombarding energy. In Fig. 10 they are compared to results from single-nucleon stripping and pickup reactions.[14-19] The agreement is quite good with exception of the results from the $^{48}Ca(d,^{3}He)^{47}K$ reaction[15] which yield a very small $1f_{7/2}$ proton occupancy in the ground state wave function of ^{48}Ca. At present, it is unclear how to explain it. One possiblity is that the $1f_{7/2}$ proton pickup strength in the $^{48}Ca(d,^{3}He)^{47}K$ reaction is much more fragmented than the $(1f_{7/2})^{2}_{7,0}$ proton-neutron pickup strength in the (d,α) reaction and has therefore escaped detection. This is corraborated by the fact that in the $^{48}Ca(d,^{3}He)$ study both the $2s_{1/2}$ and the $1d_{3/2}$ pickup strength were observed to be fragmented. It may be relevant that in a weak coupling picture $7/2^-$ states in ^{47}K can be constructed by coupling a $2s_{1/2}$ or $1d_{3/2}$ proton hole to the collective 3^- state in ^{48}Ca, while in ^{46}K one cannot make a 7^+ state by coupling a proton-neutron pair in the $2s_{1/2}$ and $1d_{3/2}$ orbitals to any low-lying collective state in ^{48}Ca. On the other hand, if a solution for this discrepancy is sought in terms of an interference between one- and two-step processes, it would have to be constructive in the (d,α) case and destructive for $(d,^{3}He)$. Moreover, this interference could not alter the shape of the one-step (d,α) VAP angular distributions, experimentally observed to be stable for the transitions to both 4.54 and 5.95 MeV states in ^{46}K and all other 7^+ transitions studied across the $1f_{7/2}$ shell (up to ^{64}Ni). Unfortunately, absolute multi-step calculations of this kind are presently not feasible.

These proton occupation numbers for the $1f_{7/2}$ orbital agree quite well with the results of shell-model calculations by Zucker[20] who obtains 0.57, 0.72, 0.83 and 0.64 for the calcium-isotopes 40, 42, 44 and 48, respectively. The agreement for ^{48}Ca is astonishing.

CONCLUSION

Investigations of the (d,α) reaction on target nuclei between ^{40}Ca and ^{64}Ni with 80 MeV vector polarized deuterons have shown an extremely strong selectivity for exciting residual nuclear states formed by picking up proton-neutron pairs in the completely aligned $(1f_{7/2})^{2}_{7,0}$ configuration. Such transitions are characterized by pure $L = 6$ angular distributions of the differential cross section and distinct $J = 7$ patterns of the vector analyzing power. This strong and systematic selectivity was successfully used to locate $[(^{even}Ni) \times (1f_{7/2})^{-2}_{7,0}]$ high-spin states in the even Co-isotopes and their transition strengths distributions. A comparison to predictions from multiconfiguration shell-model calculations yielded information about the used configuration space and residual interaction. Moreover, this selectivity was employed to determine the amount of $1f_{7/2}$ proton core-excitations in the even Ca-isotopes. In order to extract meaningful proton occupation numbers it was found that new avenues for a more realistic form factor for the picked up proton-neutron pair have to be explored. The method used in the present analysis can only be considered a first try to obtain a better form factor description.

ACKNOWLEDGEMENT

I would like to thank all my collaborators for their assistance and especially G.T. Emery and W.W. Jacobs for their stimulating discussions. This work has been supported in part by the U.S. National Science Foundation.

REFERENCES

1) N.K. Glendenning, Phys. Rev. B137, 102 (1965).
2) H. Nann, A.D. Bacher, W.W. Jacobs, W.P. Jones, and E.J. Stephenson, Phys. Rev. C24, 1984 (1981).
3) M.J. Schneider and W.W. Daehnick, Phys. Rev. C4, 1330 (1972).
4) R.B.M. Mooy, P.W.M. Glaudemans and A.G.M. van Hees, Phys. Lett. 104B, 251 (1981), and P.W.M. Glaudemans, private communication.
5) I.S. Towner and J.C. Hardy, Adv. in Phys. 18, 401 (1969).
6) E. Newman and J.C. Hiebert, Nucl. Phys. A110, 366 (1968).
7) P. Doll et al., J. Phys. (London) G5, 1421 (1979).
8) A. Moalem, J.F.A. van Hienen, and E. Kashy, Nucl. Phys. A307, 277 (1978).
9) P. Martin, M. Buenerd, Y. Dupont, and M. Chambre, Nucl. Phys. A185, 465 (1972).
10) R.M. DelVecchio and W.W. Daehnick, Phys. Rev. C6, 2095 (1972).
11) I. Sick et al., Phys. Rev. Lett. 38, 1259 (1977); I. Sick, Commun. Nucl. Part. Phys. 9, 55 (1980).
12) S.K. Platchkov et al., Phys. Rev. C25, 2318 (1982).
13) A.E.L. Dieperink and I. Sick, Phys. Lett. 109B, 1 (1982).
14) P. Doll et al., Nucl. Phys. A263, 210 (1976).
15) S.M. Banks et al., Z. Phys. A316, 241 (1984), and Nucl. Phys. A437, 381 (1985).
16) G. Bruge, H. Farragi, Ha Duc Long and P. Roussel, CEA-N-1232, p.124 (1970).
17) U. Lynen et al., Proc. Conf. on Direct Reactions with ^{3}He, Tokyo, 1967, ed. K. Matsuda and H. Kamitsubo.
18) J. Bommer et al., Nucl. Phys. A160, 577 (1971).
19) D.H. Youngblood, R.L. Kozub, R.A. Kenefick, and J.C. Hiebert, Phys. Rev. C2, 477 (1970).
20) A.P. Zucker, Proc. Top. Conf. on the Structure of $1f_{7/2}$ Nuclei, Legnaro (Padova) 1971, ed. R.A. Ricci (Editrice Compositori, Bologna, 1971), p. 94.

SESSION D

NUCLEAR STRUCTURE FROM TRANSFER AND KNOCKOUT REACTIONS

TOWARDS ABSOLUTE SPECTROSCOPIC FACTORS

Gerhard J. Wagner
Physikalisches Institut der Universität Tübingen
Morgenstelle, D-7400 Tübingen, FRG.

ABSTRACT

A brief review of previous attempts to determine absolute spectroscopic factors from transfer reactions or electron scattering alone shows that these approaches are too inaccurate or too indirect, respectively. Our approach towards absolute spectroscopic factors for 3s proton transfer in the Pb region combines precise relative spectroscopic factors from (d,^{3}He) reactions with absolute information from charge density differences.

INTRODUCTION

Knowledge of absolute spectroscopic factors is fundamental for our understanding of the nuclear many-body problem. Simple mean field theories allow for spectroscopic factors close to unity e.g. in $^{208}Pb \pm 1$ nucleon overlaps. We use the definition:

$$C^2 S_\rho := |<A-1,f|a_\rho|A,0>|^2 \qquad (1)$$

for nucleon removal processes. Here $\rho = (n,l,j,m,t_3)$ represent quantum numbers in a "natural orbit" basis which is assumed to be at hand. (The isospin coupling factor C^2 will be dropped for simplicity of notation in the following). The reactions we use are considered as direct in the sense that the effective operators may be represented by the annihilation operator a_ρ ("sudden approximation"). We note that spectroscopic factors of unity (with a (2j+1) weighing) require a completely occupied ρ-subshell in the target ground state $|A,0>$ and a pure hole-state $|A-1,f>$. Correlations of any type in one or the other state will reduce the value of S.

Interest in a precise determination of spectroscopic factors in the Pb region has been revived by recent theoretical developments[1,2] where depletions of normally occupied shells near the Fermi level by about 30 % were predicted to result from long-range correlations, tensor correlations and short-range correlations, respectively. This raises the question whether absolute factors may be determined to that accuracy. Therefore I shall briefly review relevant previous work in the next chapter. We shall find that neither electron scattering alone (such as in ref. 1) nor transfer reactions alone are sufficiently accurate. Our own approach[3] which then will be presented in some detail uses a combination of both. In the final chapter an extension of this method is proposed which avoids certain model dependencies, however at the price having to use a sum rule with its notorious experimental difficulties. Its application will provide at least a lower limit to the 3s proton occupancy in ^{208}Pb.

This work has been the incentive for a joint effort by groups from NIKHEF Amsterdam, Indiana University, CEN Saclay and Universität Tübingen to tackle the problem of absolute spectroscopic factors and it should be considered as a first step only. Peter de Witt Huberts in his talk will present the second step and dwell on the comparison of (d,^{3}He) and (e,e'p) reactions. This allows me to be quite brief on the questions of reaction mechanism which otherwise form a substantial part of our discussions.

PREVIOUS WORK

The attempt to determine occupation probabilities from elastic and inelastic electron scattering[1] alone depends on assumptions on the shell-model parity of the corresponding states. Only if a fragmentation of transition strength due to configuration mixing may be excluded can one attribute the observed "quenching" of the single-particle transitions to other effects such as tensor- and short-range correlations. For stretched states at relatively high excitation energies and level densities fragmented strength may be difficult to observe. For the difference of ground state charge densities of ^{206}Pb and ^{205}Tl we shall demonstrate below that low-energy configuration mixing leads to a "quenching" of the 3s-single proton contribution.

Given the definition eq. 1 transfer reactions offer a much more direct access to spectroscopic factors and occupation probabilities.

$$n_\rho = \sum S_\rho(0 \to f) \qquad (2)$$

The sum has to be taken over the full (A-1) spectrum. Hence only a lower limit of n_ρ can be obtained experimentally.

Of all nuclear transfer reactions the (e,e'p) reaction is least pestered by absorption in the hadronic channels. As a result its cross sections are more directly related to S than those from pick-up or stripping reactions which are sensitive to the asymptotic normalization of the single-particle wave functions; this leads to the well-known strong dependence of the deduced spectroscopic factor on the assumed shape of the single-particle wave function. The (e,e'p) reaction at 500 MeV has been performed[4] on ^{16}O using various kinematical arrangements to test the DWIA assumptions. The transition to the ground state of ^{15}N is a good candidate for a pure single-particle transition as one starts with a doubly-magic target and as this state is the only known 1/2$^-$ state in ^{15}N that carries strength. Yet, the analysis using several bound state and optical potentials consistently yielded a spectroscopic factor of 0.5 to 0.6. One probably has to await progress in the description of (e,e'p) reactions mechanisms which is expected to come from the activities at NIKHEF before one can appreciate the significance of this result. Does it indicate strong $1p_{1/2}$ depletions in ^{16}O?

In contrast to previous results an investigation of the

^{16}O(g.s.) - ^{17}O(g.s.) overlap by sub-Coulomb elastic scattering[5] yields a neutron spectroscopic factor exhausting 93 % of the single particle value. This is derived from the amplitudes in the oscillatory pattern of the angular distribution which results from the interference of Coulomb and neutron-transfer amplitudes. Exact finite range calculations have been used but the deduced asymptotic normalization is fairly stable against the assumptions due to the sub-Coulomb nature. By the same token the spectroscopic factor depends strongly on the assumed radius of the $d_{5/2}$ neutron orbit which is derived from magnetic electron scattering. Very recently[6] the reliability of this technique has been questioned. Also the role of meson exchange currents in magnetic electron scattering needs further clarification as pointed out by R. Lindgren during this workshop.

In this situation spin-dependent sum rules were highly welcomed as a means to test the reliability of absolute spectroscopic factors. They were applied[7,8,9] to open-shell nuclei where (nearly) complete sets of relative pick-up and stripping spectroscopic factors pertaining to the same subshell are available. Interestingly normalization factors close to unity were obtained[7] indicating both the reliability of the absolute spectroscopic factors and the validity of the simple shell model (without depletions by short-range correlations etc.). Unfortunately, it turns out that the set of inhomogeneous linear equations connecting the partial sums would be also satisfied if all transitions were quenched through such depletions by the same (i. e. spin-independent) amount and if all spectroscopic factors were correspondingly too large. This might appear as an improbable coincidence, but it is more likely than not that DWBA procedures historically were tailored to yield spectroscopic factors satisfying the sum rules when applied to the readily accessible final states. An investigation of the influence of short-range correlations[10] confirmed the spin-independence and led to an estimate of a 5 % effect. In the light of refs. 1,2 a much larger effect can not be excluded. In any case, in this situation the sum rules are of little help in solving the problem of absolute spectroscopic factors.

Our short list of previous approaches towards absolute spectroscopic factors ends with a reference to work[11,12] studying the impact of magnetic electron scattering on spectroscopic factors. There a reanalysis of existing spectroscopic factors from transfer reactions using radii from electron scattering led to a 20 to 30 % reduction of the spectroscopic factors. Therefore the sum rule values which were originally exhausted by the transitions in the accessible range of a few MeV were now also low. This indicated missing strengths at higher excitation energies, as expected from tensor-correlations and short-range correlations which should affect the spectrum at energies of 2hω and above. The results of refs. 11,12 of course have also to be viewed in the light of the criticism of magnetic electron scattering mentioned above. Yet it should be stressed that the combined input from electron scattering and transfer reactions used there signaled the way we are following.

THE DIFFERENCE OF CHARGE DENSITIES OF ^{206}Pb AND ^{205}Tl

By a combined analysis of elastic electron scattering and transfer reactions we approach the determination of 3s proton occupancies in ^{206,208}Pb. As input we use (i) the charge density difference $\Delta\rho$ between ^{206}Pb and ^{205}Tl as determined from elastic electron scattering[13,14] on ^{206}Pb and ^{208}Pb and (ii) precise relative spectroscopic factors from (d,^{3}He) reactions on ^{206}Pb and ^{208}Pb.

In refs. 13 and 14 the ratio of cross sections for elastic electron scattering on ^{205}Tl and ^{206}Pb has been carefully measured for momentum transfers $q \lesssim 2.7$ fm^{-1}. The most conspicuous feature of this ratio is a spike at $q \approx 2$ fm^{-1}. The ratio is sensitive to the difference $\Delta\rho$ of the charge densities. The spike results from the 3s proton contribution to $\Delta\rho$, because 2d and 1g contributions (fig. 8 of ref. 15) as well as the effects of core polarization (fig. 10 of ref. 15) behave smoothly near $q = 2$ fm^{-1}. The amplitude of the observed spike is smaller than expected for a pure 3s proton single-particle density (fig. 3 of ref. 15). In fact, Campi has obtained a good fit to the experimental ratio (fig. 11 of ref. 15) by assuming $\Delta\rho = 0.7 \rho_{3s} + 0.3 \rho_{2d}$ where ρ_{3s} and ρ_{2d} represent single-particle densities (normalized to unity) calculated with wave functions from mean field theory. We shall consequently in the following assume[15] that the contribution of the 3s proton charge density amounts to $z = 0.7 \pm 0.1$.

We are open to the possibility, however, that the ongoing discussion on the interpretation of the data will lead to another value of z. We warn, however, against the naive interpretation of the charge difference in coordinate space $\Delta\rho(r)$ as exhibiting near the origin exclusively the 3s proton contribution; for core polarization effects whose exact size is difficult to estimate are expected to contribute essentially near $r \approx 0$ (fig. 9 of ref. 15). It is also naive to celebrate the value of $z \approx 0.7$ as proof for the expected[1,2] 30 % "quenching" of the 3s proton occupancy in ^{206}Pb or even in ^{208}Pb. The spectroscopy with transfer reactions will reveal configuration mixing in the ground state of ^{205}Tl as the origin of the "quenching".

RELATIVE SPECTROSCOPIC FACTORS FOR PROTON REMOVAL FROM ^{206}Pb AND ^{208}Pb

Whereas the electron scattering experiment thus provides an absolute measure of the difference between the densities of ^{206}Pb and ^{205}Tl, such a quantity is not obtainable for the more fundamental pair ^{208}Pb and ^{207}Tl due to the unstability of ^{207}Tl. In order to relate the former information to the latter we have performed a precise relative determination of the spectroscopic factors for proton removal from ^{206}Pb and ^{208}Pb, respectively, leading to the ground states of the respective Tl-isotopes. Previous pick-up measurements[16,17] did not match the level of accuracy set by electron scattering.

We investigated the (d,³He) reactions on ²⁰⁶Pb and ²⁰⁸Pb using 52 MeV deuterons from the Karlsruhe cyclotron. High relative precision was achieved by frequently interchanging the targets leaving all other conditions unchanged. We used targets of ≈1 mg/cm² thickness whose relative thickness was determined to better than 0.5 % by measuring the small angle elastic deuteron scattering ($5° < \theta_{LAB} < 9°$) on both sides of the beam and subsequent optical model analysis. The isotopic composition of the isotopically enriched targets was known to better than 0.05 %. Reaction products were measured with ≈90 keV overall resolution with four ΔE-E telescopes, two on each side of the beam, consisting of Si surface barrier detectors. Fig. 1 shows that the ground states which we are mostly interested in are well separated from the first excited states in both reactions. The resolution was insufficient, however, to resolve known ¹⁶ 1/2⁺ states at 1.22 and 1.44 MeV in ²⁰⁵Tl.

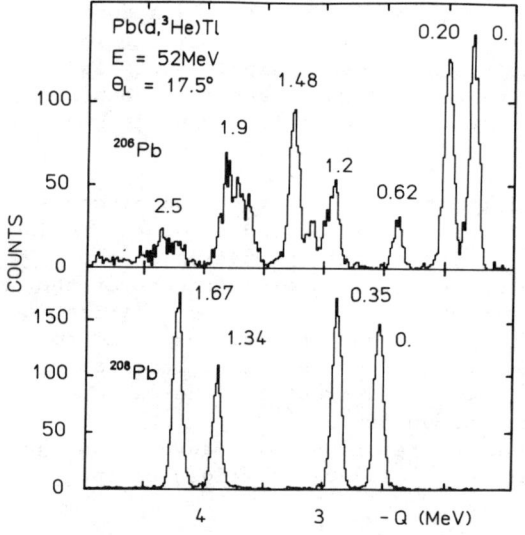

Fig. 1 Spectra of (d,³He) reactions on ²⁰⁶Pb (top) and ²⁰⁸Pb (bottom).

Measured angular distributions for the ground state transitions and their ratio are displayed in fig. 2 together with statistical error bars whenever these exceed the size of the data points. The curves are the results of local, zero-range DWBA calculations with finite-range corrections in the local energy approximation. The dash-dotted curve in the lower part of fig. 2 shows the result expected for equal spectroscopic factors in both reactions: the observed angular pattern of the cross section ratio is qualitatively reproduced; the predicted ratio exceeds unity on the average. This is merely the result of the slight Q-value difference of 761 keV and the change in

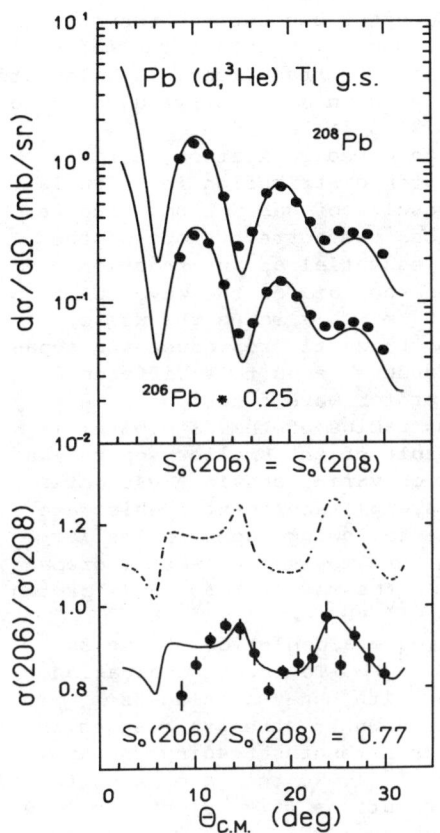

Fig. 2 Cross sections (top) and their ratios (bottom) for the ground state transitions of the $(d,^3He)$ reactions on ^{206}Pb and ^{208}Pb. Statistical errors are given whenever exceeding the size of the data points. Smooth curves represent the results of DWBA calculations. The dash-dotted curve assumes equal spectroscopic factors for both transitions; the corresponding solid curve is the result of a least-squares fit of $S_o(206)/S_o(208)$ to the data.

potential well radii which were assumed to scale as $A^{1/3}$. The data lie substantially below the ratio predicted for equal spectroscopic factors. The solid curve in the lower part of fig. 2 is the result of a least-square fit to the data. The ratio of ground state spectroscopic factors $R = S_o(206)/S_o(208)$ is obtained as 0.77 ± 0.01 (stat.) ± 0.015 (syst.). The estimated error includes uncertainties arising from the determination of relative target thicknesses and from long-time unstabilities of charge integration. It also includes part of the uncertainties arising from the DWBA analysis[17,18] namely those produced by common variations in the optical model or bound state potential[19] parameters of both isotopes. Indeed the deduced value of R is extremely stable against such variations even if these lead to severe deteriorations of the fit to the individual cross sections. By CCBA calculations we convinced ourselves that the presence of the low-lying 2_1^+ state in ^{206}Pb has absolutely no effect on the value of R.

However, the error given above does not account for the strong sensitivity of the result to modifications which affect the difference of the rms radii of the bound state wave functions. To calculate these we assumed a Woods-Saxon potential with a geometry adjusted to nuclear matter distributions[19] (r = 1.2675 fm, a = 0.81 fm, λ_{so} = 25) and with a depth adjusted to the observed separation energies. The resulting rms radius of the 3s proton distribution in ^{208}Pb is 5.357 fm in good agreement with the results of sub-Coulomb transfer measurements[20]. Our procedure guarantees the correct slope of the wave function at large radii which is essential as our reaction - according to the DWBA analysis - probes the tail of the wave function with a maximum sensitivity near r ≈ 10 fm. This makes the use of Hartree-Fock wave functions which do not exactly reproduce the separation energy futile. The adopted procedure leads to a difference $\delta_{3s} = <r^2>_{206}^{1/2} - <r^2>_{208}^{1/2}$ for the 3s proton wave functions of -0.14 fm. Note that the calculated rms radius of the 3s orbital is larger in ^{206}Pb than in ^{208}Pb as a result of the smaller separation energy in ^{206}Pb. The sensitivity of R on variations in δ was determined as dR/dδ ≈ 2.7 fm^{-1}. From the (e,e'p) experiments, which are discussed by P. de Witt Huberts in the following paper, a smaller value of R ≈ 0.65 was deduced. If one were to ascribe the discrepancy to this origin alone it would signify a rms-radius for the 3s proton wave function about 0.025 fm larger in ^{206}Pb than in ^{208}Pb.

Our result R ≈ 0.77 does not indicate a depletion of the 3s proton occupancy in ^{206}Pb relative to ^{208}Pb. To obtain occupation numbers one has to sum over all states with non-vanishing 3s strengths: n = $\sum$ S(3s). For ^{208}Pb the known 3s strength resides in the ground state. For ^{206}Pb we found 3s strengths feeding the known 1/2$^+$ states[16] at 1.22 and 1.44 MeV in ^{205}Tl. To this aim we performed a ℓ-decomposition of the angular distributions (fig. 3) of the 1.20 and 1.48 MeV groups and obtained $S_1(206)/S_o(206)$ = 0.16 ± 0.02 and $S_2(206)/S_o(206)$ = 0.05 ± 0.03. For a more accurate determination clearly high resolution experiments such as the ^{206}Pb($\vec{d},^3$He) measurement performed by H. Nann et al[21] with the QDDM spectrograph at the IUCF are desirable. Our values lead to a ratio of occupation numbers

$$n(206)/n(208) \approx \sum_{i=0}^{2} S_i(206)/S_o(208) = 0.93 \pm 0.04. \qquad (3)$$

In considering the ratio we avoid the difficulties of the determination of absolute spectroscopic factors and we minimize the effects of neglecting possible high-lying 3s strengths. These are supposed to result from short-range correlations and from 2 $\hbar\omega$ excitations due to tensor forces and should be quite similar for both target nuclei. We find equal occupation numbers within twice the experimental error.

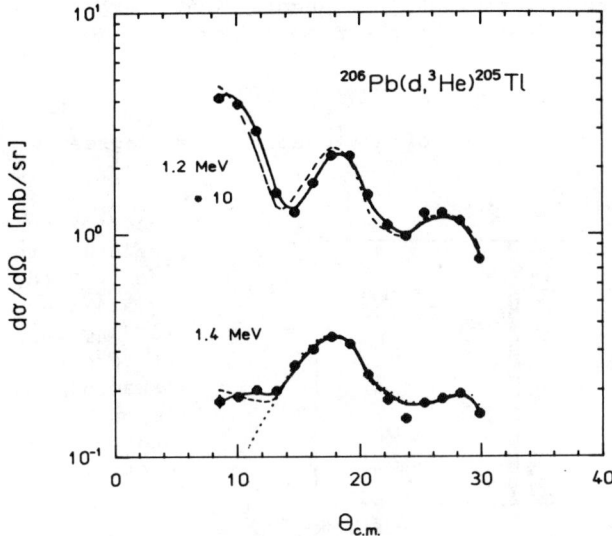

Fig. 3 Angular distributions for the ^{206}Pb(d,^{3}He) reactions leading to the unresolved groups at excitation energies of 1.2 and 1.48 MeV, respectively. Also shown are the results of a χ^2-fit based on empirical angular distributions. For the 1.20 MeV group a superposition of a $2d_{3/2}$ and $3s_{1/2}$ distributions is compared to a pure $2d_{3/2}$ distributions (dashed). For the 1.48 MeV group a $1h_{11/2}$ (dotted), a $1h_{11/2} + 2d_{3/2}$ superposition (dashed) and a $1h_{11/2} + 2d_{3/2} + 3s_{1/2}$ (full curve) superposition are shown.

A SIMPLE ACCOUNT FOR THE OBSERVATIONS

The situation as far as the 3s proton transfer is concerned is summarized in fig. 4. One notices that in the pick-up reaction from ^{206}Pb about the same total strength as in ^{208}Pb is spread among (at least) three states. The fragmentation of the strength in ^{205}Tl is readily explained by phonon-hole coupling models[22,23] where the ground state of ^{205}Tl is obtained as

$$|1/2^+> = \alpha|3s^{-1} \times 0^+> + \beta|2d_{3/2}^{-1} \times 2^+> + \gamma|2d_{5/2}^{-1} \times 2^+> \quad (4)$$

where $|0^+>$ and $|2^+>$ represent the ground and first excited states of ^{206}Pb. Depending on the assumed forces values of α^2 range[22,23] from 0.73 to 0.87. In ^{207}Tl no fragmentation of 3s strength has been observed, which in the framework of the model would signify $\alpha(207) =$

1. Then in zero-order, namely for an uncorrelated target ground state wave function $|0^+>$ one obtains the plausible prediction

$$R = S_o(206)/S_o(208) = \alpha^2 \tag{5}$$

which, with the quoted values of α^2, qualitatively agrees with the data.

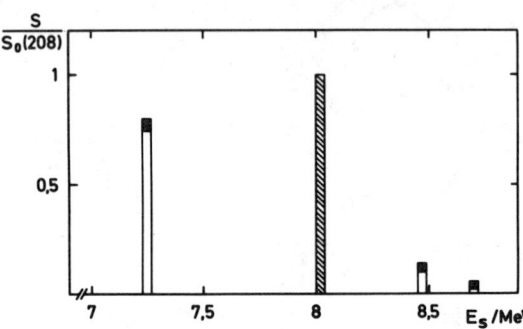

Fig. 4 Spectroscopic factors (relative to that of the $^{208}Pb(d,^3He)^{207}Tl$ (g.s.) transition) leading to $1/2^+$ states in ^{207}Tl (shaded) and ^{205}Tl (open bars) vs. proton separation energy.

Using the wave function of eq. 4 it is also straightforward to calculate the 3s proton contribution z to the $^{206}Pb-^{205}Tl$ charge difference. One finds simply

$$z = \alpha^2. \tag{6}$$

Again prediction and experiment are fully compatible.

The model also accounts for the relative strength of the first $3/2^+$ states at 0.35 MeV in ^{207}Tl and 0.20 MeV in ^{205}Tl. Fig. 5 shows our results leading to $S_1(206)/S_1(208) = 0.55$ for the first excited states. The calculations predict 0.52 (ref.22) and from 0.53 to 0.70 (ref.23) depending on the force chosen.

In ref. 3 we investigated the modifications to eqs. 5 and 6 arising from correlated 0^+ wave functions. If these correlations deplete the 3s-proton shell one has to take into account Pauli corrections when applying the 3s proton annihilation operator in eq. 4. As shown in ref.3 the Pauli correction factors are different for R and z. This prevents us from relating $S_o(206)$ directly to z without additional assumptions. But for our goal of determining absolute occupation numbers a model dependence should be avoided anyway. The phonon-hole model has been extremely useful, however, to demonstrate the common origin of the fragmentation of 3s strength in ^{205}Tl and the "quenching" of the 3s proton contribution to the charge difference $\Delta\rho$: both result from low-energy configuration mixing in ^{205}Tl.

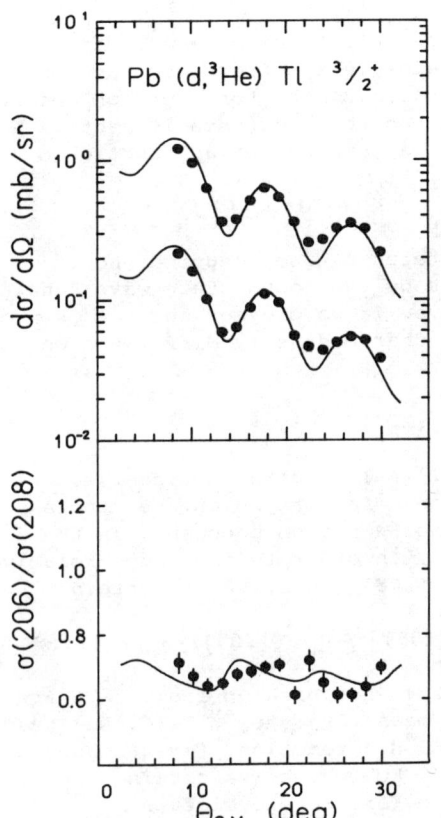

Fig. 5 Same as fig. 2 but for transitions leading to the first $3/2^+$ states in ^{207}Tl (upper distribution) and ^{205}Tl (lower distribution). The least-squares fit to the ratio of cross sections (bottom part) yields $R(3/2^+) = 0.55 \pm 0.03$ for the corresponding spectroscopic factors.

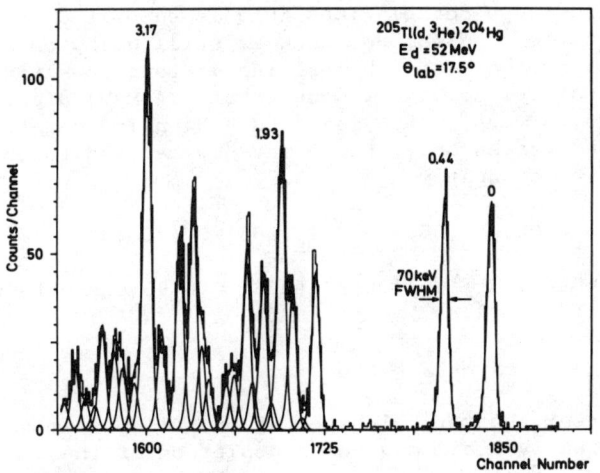

Fig. 6 Spectrum of the ^{205}Tl(d,^{3}He) reaction.

A NOVEL MODEL INDEPENDENT APPROACH

We exploit a well-known sum rule[24] to obtain the occupancy of the 3s proton shell in ^{208}Pb from relative spectroscopic factors and the absolute value of z. As shown in ref.24 the total proton density of J = 0 or 1/2 nuclei may be related to spectroscopic factors S for proton removal and the target-final state overlap function ϕ_{fo} by

$$\rho(\vec{r}) = (1/4\pi) \sum_{f,nlj} S_f(nlj) \cdot \phi_{fo}^2(\vec{r}) \quad (7)$$

where the sum runs over all final states f and occupied shells nlj. We approximate the overlap functions by the bound state wave functions of the DWBA procedure described above. If we project out the 3s proton contribution to the ^{206}Pb-^{205}Tl charge density difference we obtain

$$z = \sum S_f(206) - \sum S_f(205); \quad (nlj = 3s_{1/2}) \quad (8)$$

The sum should be taken over the full A-1 spectrum. Of course, the S_f now represent *absolute* spectroscopic factors which obey the sum rule $\sum S_f(A) = n(A)$, where n(A) is the 3s proton occupancy in the ground state of the target nucleus A. In order to introduce relative spectroscopic factors we divide by $n(208) = \sum S_f(208)$ to obtain

$$z/n(208) = [\sum S_f(206)/\sum S_f(208)] - [\sum S_f(205)/\sum S_f(208)]. \quad (9)$$

The relative spectroscopic factors for pick-up from ^{205}Tl are presently being determined by us by measuring the ^{205}Tl(d,^{3}He) reaction simultaneously with the ^{208}Pb(d,^{3}He) reaction. Fig. 6 shows a spectrum with energy resolution of 70 keV which was obtained in a period of optimum beam quality of the Karlsruhe cyclotron.

In applying eq. 9 we shall again assume that the percentage of strength lost in unobserved high-lying states is the same for all target nuclei. For Jastrow type correlations this is supported by ref. 10. The analysis of the ^{205}Tl(d,^{3}He) data is still going on. Yet, for the relative 3s proton strength residing in four low-lying states of ^{204}Hg (at 0 and 1.61 MeV and a doublet at 2.4 MeV) a preliminary value of $\sum' S_f(205)/S_o(208) \gtrsim 0.6$ is found. The prime denotes the partial sum over four states only. Eq. 9 may now be used to obtain a lower bound for the occupancy

$$n(208) > z[\sum S_f(206)/S_o(208) - \sum' S_f(205)/S_o(208)]^{-1}. \quad (10)$$

With the values z > 0.6 and $\sum S_f(206)/S_o(208) \approx 0.93$ one obtains the preliminary result n(208) > 1.8.

NEXT STEPS

In view of the current theoretical expectation this preliminary result indicates a remarkably small degree of depletion of the 3s proton shell in ^{208}Pb. Therefore we shall, as a next step, carefully check quantities and approximations entering into eq. 10.

For this purpose we plan to improve the spectroscopy of proton removal reactions from ^{205}Tl by (i) performing a $(\vec{d}, ^3\text{He})$ experiment with vector polarized deuterons at the Karlsruhe cyclotron; (ii) performing a ^{205}Tl(e,e'p) experiment, thereby exploiting the sensitivity of the (e,e'p) reaction at small missing momentum to s-strength; and (iii) attempting high resolution spectroscopy of the $(\vec{d}, ^3\text{He})$ reaction as soon as the K = 600 spectrometer at the IUCF will be in operation. The comparison of (i) and (ii) for well resolved final states should improve our understanding of the radial sensitivity of the $(d, ^3\text{He})$ and (e,e'p) reactions.

As eq. 10 shows n(208) scales with z. In view of the discussions on the interpretation of the ^{206}Pb-^{205}Tl charge density difference it therefore seems worthwhile to perform an independent analysis of the electron scattering data of refs. 13-15. Effective forces have been developed which lead to a fairly satisfactory reproduction[25] of the total charge densities in Hartree-Fock calculations for the Pb isotopes. It will be interesting to apply these forces to the present problem.

In summary, it has become clear that the approach towards absolute spectroscopic factors and occupation numbers is a tedious one. It is essential that a combination of electron scattering and transfer data be used. The present transfer data, for example, showed that the alleged "quenching" of the 3s proton contribution to the ^{206}Pb-^{205}Tl charge density difference largely results from configuration mixing. With eqs. 9,10 we demonstrated in a more constructive way the utility of this combination for a determination of occupation numbers.

ACKNOWLEDGEMENTS

This work is supported in part by the BMFT, by KfK Karlsruhe, and by a NATO travel grant. It was performed in collaboration with H. Clement, P. Grabmayr, S. Klein, K. Reiner, W. Reuter and G. Seegert. I like to thank the members of the Amsterdam-Indiana-Saclay-Tübingen-collaboration for many useful discussion. Comments and criticism by H. Müther, K.W. Schmid and I. Sick are also acknowledged. We are grateful to Dr. F. Begemann for the precise determination of the isotopic composition of the targets.

REFERENCES

1) V.R. Pandharipande, C.N. Papanicolas and J. Wambach, Phys. Rev. Lett. 53 (1984) 1133
2) M. Jaminon, C. Mahaux and H. Ngô, Nucl. Phys. A440 (1985) 228
3) P. Grabmayr, S. Klein, H. Clement, K. Reiner, W. Reuter, G.J. Wagner and G. Seegert, Phys. Lett. B, in press
4) M. Bernheim et al, Nucl. Phys. A375 (1982) 381
5) S. Burzynski et al, Nucl. Phys. A399 (1983) 230

6) T.E. Milliman et al, Phys. Rev. C32 (1985) 805
7) C.F. Clement and S.M. Perez, Nucl. Phys. A213 (1973) 510 and Nucl. Phys. A284 (1977) 469
8) A. Moalem and Z. Vardi, Nucl. Phys. A332 (1979) 195
9) A. Moalem and E. Friedman, Phys. Rev. Lett. 40 (1978) 1064
10) M. C. Birse and C.F. Clement, Nucl. Phys. A351 (1981) 112
11) A.E.L. Dieperink and I. Sick, Phys. Lett. 109B (1982) 1
12) S.K. Platchkov et al, Phys. Rev. C25 (1982) 2318
13) H. Euteneuer, J. Friedrich and N. Voegler, Nucl. Phys. A298 (1978) 452
14) J.M. Cavedon et al, Phys. Rev. Lett. 49 (1982) 978
15) B. Frois et al, Nucl. Phys. A396 (1983) 409c and priv. comm.
16) E.R. Flynn, R.A. Hardekopf, J.D. Sherman, J.W. Sunier and J.P. Coffin, Nucl. Phys. A279 (1977) 394
17) W.W. Daehnick, J.D. Childs and Z. Vrcelj, Phys. Rev. C21 (1980) 2253
18) H.J. Trost, Ph. D. thesis, Hamburg (1981)
19) J. Streets, B.A. Brown and P.E. Hodgson, J.Phys. G. 8 (1982) 839
20) P.W. Woods, R. Chapman, I.N.Mo, P. Skensved and J.A. Kuehner, Phys. Lett. 116B (1982) 320
21) H. Nann, priv. communication
22) N. Azziz and A. Covello, Nucl. Phys. A123 (1969) 681
23) L. Zamick, V. Klemt and J. Speth, Nucl. Phys. A245 (1975) 365
24) C.F. Clement, Nucl. Phys. 213 (1973) 493
25) J. Friedrich, priv. communication

COMPLEMENTARY ASPECTS OF THE QUASI-FREE (e,e'p) REACTION
AND THE (d,^{3}He) REACTION

P.K.A. de Witt Huberts
NIKHEF-K, P.O. Box 4395, 1009 AJ Amsterdam, The Netherlands

ABSTRACT

The quest for absolute spectroscopic factors is an old one but has not yet been successfully realized experimentally. Various uncertainties in the DWBA analysis of one-nucleon transfer reactions impair the precision of deduced spectroscopic factors. We argue that these uncertainties may be reduced if information from electron scattering is employed in the analysis of hadron-induced reactions. High-resolution data of the quasi-free proton knockout reaction (e,e'p) are presented and the spectroscopic information obtained is discussed. We indicate that a synthesizing approach to electromagnetic and hadronic reactions may lead towards the goal of absolute spectroscopic factors.

INTRODUCTION

Since the nuclear shell model plays a key role in our understanding of nuclear structure, spectroscopic factors, that are a measure of the overlap of the wave function of a physical state with a basis state of the shell model, constitute a natural and important interface between theory and experiment. The most prolific source of spectroscopic factors have been one nucleon transfer reactions (e.g. (d,p) and (d,^{3}He)). For these reactions an extensive apparatus of analysis involving exact finite range DWBA calculations has been developed. Since strongly interacting, strongly absorbed, ejectiles and projectiles are used in such reactions, the interaction is predominantly located in the stratospheric zone of the nucleus. The foremost consequence of this observation is that the spectroscopic factor, to be deduced from the form factor of the transferred nucleon, is strongly dependent on the Ansatz that one makes for the bound-state wave function. Consequently absolute spectroscopic factors cannot be obtained with acceptable precision. This and other subtleties involved in the analysis of one-nucleon transfer reactions have been extensively discussed some 15 years ago by Macfarlane.[1]

Recently it has been realized[2] that the traditional virtue of electron scattering, its ability to map out nulcear and single-nucleon densities, can be used to constrain the radial wave-function dependence of spectroscopic factors. Rather accurate information on the rms radius of the unpaired-nucleon wave function, obtained from high multipole (M7, M9) elastic electron scattering, has been used in a re-analysis of (d,^{3}He)-data. The resulting spectroscopic factors are ~30% smaller than the sumrule strength (2j+1). However, magnetic scattering data are available for only a few cases with stretched configuration (j=1+1/2) ground states that permit to extract accurate radial information.

In the quasi-free proton knockout reaction (e,e'p) the bound-state wave function can be mapped out in momentum, or equivalently, coordinate space. The (e,e'p) reaction has been exploited to some extent, notably so for the investigation of deeply-bound hole states.[3] Access to a variety of proton states near the Fermi surface is provided by the (e,e'p) reaction and weakly populated final states may be investigated with small interference by two-step processes.[4] Due to recent developments in instrumentation a missing mass resolution of (100-150) keV can now be achieved. One purpose of this talk is to present a selection of recent (e,e'p) data and to interpret these.

This will be done with a view to achieving progress towards the determination of absolute spectroscopic factors. Complementary aspects of (e,e'p) and (d,^{3}He) will be emphasized in the discussion.

RADIAL SENSITIVITY OF (e,e'p) AND (d,^{3}He) REACTIONS

In order to understand the dependence of the cross section on the Ansatz for the bound-state wave function (bswf) consider the expression for the quasi-free (e,e'p) cross section

$$\frac{d\sigma}{dp_o d\Omega_p de' d\Omega_e} = \kappa \, \sigma_{ep} \, S(E_m, p_m) \tag{1}$$

in which κ is a phase space factor and σ_{ep} is the electron-proton cross section. The spectral function $S(E_m, p_m)$, defined as the joint probability to have a proton with (E_m, p_m) in the nucleus, can be written as

$$S(E_m, p_m) = \sum_\alpha P_\alpha(E_m) |\phi_\alpha(p_m)|^2 \tag{2}$$

Here the overlap function $\langle \psi_{A-1} | \psi_A \rangle$ has been expanded in a single-particle basis $\{\alpha\} = \{nlj\}$. The functional $\phi_\alpha(p_m)$ is the Fourier transform of the coordinate representation $R_\alpha(r)$ of the bswf of the knocked-out proton. One has

$$R_\alpha(p_m) = \sqrt{\frac{2}{\pi}} \int_0^\infty j_{l\alpha}(p_m r) \, R_\alpha(r) r^2 \, dr \tag{3}$$

where $j_{l\alpha}$ is the spherical Bessel function of order l_α. The spectroscopic factor $C^2 S_\alpha$ is expressed in terms of the spectral function in the following way

$$C^2 S_\alpha = 4\pi \int_0^\infty S_\alpha \, p_m^2 \, dp_m$$

In a (e,e'p) experiment the spectral function is determined in a broad range of p_m-values (typically $p_m \simeq$ (0-250) MeV/c) and thus the bswf can be sampled in nearly the entire r-range of interest. For one-nucleon transfer reactions. e.g. the proton pickup reaction (d,^{3}He), the amplitude can be schematically written as

$$\sqrt{\sigma_\alpha(\Theta)} \propto \int_0^\infty f_s(r,\Theta) R_\alpha(r) \, r^2 dr \qquad (4)$$

In Table I we have summarized form factors pertinent to a variety of reactions, that probe single-particle aspects of nuclear structure. These include elastic magnetic electron scattering of high multipolarity[5] and the difference of charge scattering (e,e') cross sections from isotone pairs.[6] The region of the bound-state wave function that is being sampled in a given reaction can be assessed by tracing the cross section as a function of the lower limit of integration r_{min} in the expression for the reaction amplitude (3) and (4). This is illustrated in Fig. 1 for the 3s-proton knockout reaction Pb(e,e'p)Tl$_{gs}$ and in Fig. 2 for the pickup reaction ^{51}V(d,^{3}He)^{50}Ti at E_d = 52 MeV at the first maximum (Θ = 10°) of the ℓ=3 pickup angular distribution. The most notable feature of the (d,^{3}He) reaction is that, due to the strong absorption acting on both projectile and ejectile, the bswf is sampled mainly in the asymptotic tail. This property induces a strong dependence of the extracted spectroscopic factor on the shape of the bswf. For instance in a Woods-Saxon (WS) radial wave function basis, one finds $\Delta C^2 S / C^2 S \simeq 13 \, \Delta r_0/r_0$ (r_0 is the radius of the WS-potential). As is shown in Fig. 1 the r-space region being sampled depends on the value of p_m; the surface is probed at $p_m \sim$ 0 MeV/c, whereas at $p_m \simeq$ 200 MeV/c more of the nuclear interior is sampled. The regions of r-space, explored by either the (d,^{3}He) or the (e,e'p) reaction, are schematically indicated in Fig. 3 for ^{208}Pb, a case we will deal with in more detail in section 5. Because the bswf is mapped out with the (e,e'p) reaction, there is only weak dependence of the extracted spectroscopic factor on whether one uses harmonic oscillator, Woods-Saxon or Hartree-Fock wave functions in the analysis.

Table I Survey of the scattering amplitude expressions for various reactions, to illustrate the radial sensitivities involved.

Reaction	Scattering amplitude or form factor
high multipole magnetic (e,e') ($\Lambda = 2j_\alpha$)	$F_{M\Lambda}(q) \propto q \mu_n \int_0^\infty j_{\Lambda-1}(qr) R_\alpha^2(r) \, r^2 dr$
charge scattering from isotone pairs [$\Delta\rho = \rho(Z) - \rho(Z-1)$]	$F_{C0}(q) \propto \int_0^\infty j_0(qr) \, \Delta\rho(r) \, r^2 dr$
quasi free (e,e'p)	$\sqrt{S_\alpha(p_m)} \propto \int_0^\infty j_{1\alpha}(p_m r) R_\alpha(r) r^2 dr$
(d,^{3}He)	$\sqrt{\sigma_\alpha(\Theta)} \propto \int_0^\infty f_s(r,\Theta) R_\alpha(r) r^2 dr$

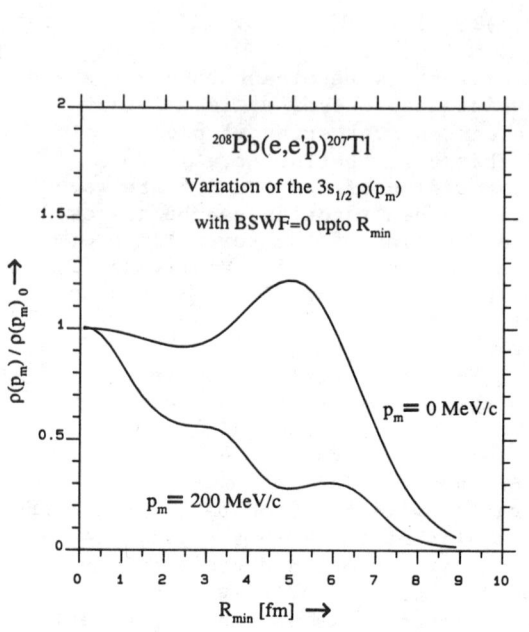

Fig. 1. R-space sensitivity of the (e,e'p) reaction.

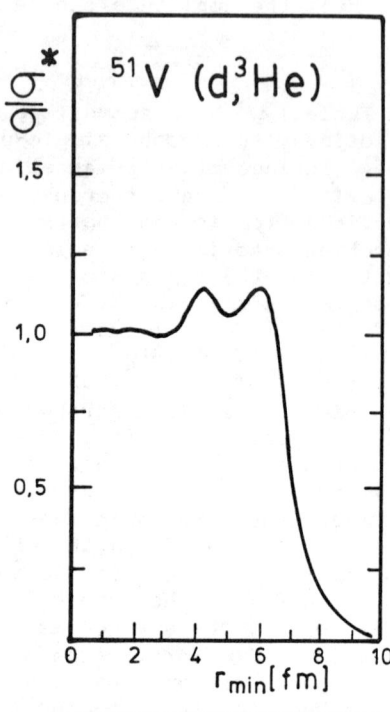

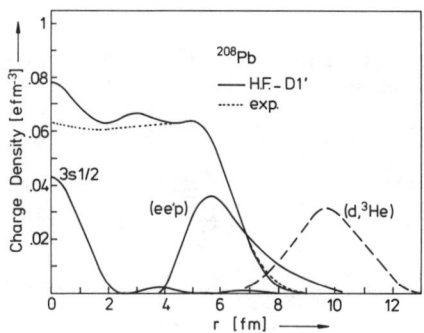

Fig. 3. Charge density of the ^{208}Pb and 3s proton density. The r-space sampling function of (e,e'p) at p_m=10 MeV/c and of (d,^{3}He) are also shown.

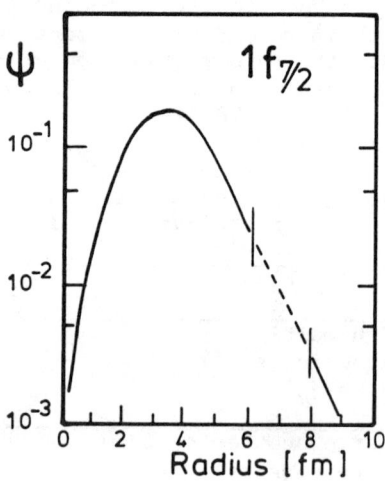

Fig. 2. R-space sensitivity curve of the ^{51}V(d,^{3}He)^{50}Ti reaction for ℓ=3 pickup.

SMALL WAVE-FUNCTION COMPONENTS AND TWO-STEP PROCESSES

The occupation probability of shell-model orbits at the Fermi surface could be assessed if methods were available to determine absolute spectroscopic factors. We are not that far yet, since the calibration of the (e,e'p) reaction depends on the theoretical treatment of the final-state interaction, that is rather schematic at present. Progress is being made in this field, however, and we discuss one example in section 4.

An alternative method to gauge correlations in the wave function would be to observe the knockout of particles from normally unoccupied states, located above the Fermi surface. Recently a weakly excited $1/2^+$-state has been observed in the reaction $^{12}C(e,e'p)^{11}B$.[7] The corresponding spectral function shows a p_m-dependence characteristic of $\ell=0$ knockout. This transitions comes about as a mixture of knockout from the $2s_{1/2}$ orbit and the coupling of the ^{12}C ground state with a $1s_{1/2}$-hole component of the $1/2^+$ state in ^{11}B.

The experimental spectral function is found to be in fair agreement with a calculation based on recent wave functions obtained in a large space ($2\hbar\omega$) shell-model calculation.[8] The notable feature of this calculation is the fairly large ($\simeq 20\%$) depletion of the 1p shell and a ten percent $1s_{1/2}$-hole probability of the ^{12}C ground state, i.e. the ^{4}He core is closed with a 90% probability. A similar observation on ^{4}He core non-closure has been made by Teeters and Kurath[9] in a classical shell-model study of non-normal parity states in the 1p-shell.

In order that weakly-populated final state be used as a quantitative gauge of wave function correlations, it must first be ascertained that the reaction proceeds via a direct one-step knockout mechanism. This aspect of the reaction mechanism can be investigated by study of the transition to the $5/2^-$-state at E_x = 4.45 MeV in the reaction $^{12}C(e,e'p)^{11}B$. In as much as the $1f_{5/2}$-shell is empty in the ^{12}C ground state the $5/2^-$ state should not be populated in a direct one-step process. The overlap $\langle ^{11}B(5/2^-)|^{12}C(0^+)\rangle$ calculated[8] in a $2\hbar\omega$ configuration space is indeed very small and thus the amount of spectroscopic strength at $E_x \simeq 4.45$ MeV may be used to set an upper limit to the amount of two-step processes in the reaction. Such processes might occur due to inelastic excitation of $^{11}B(3/2^- \to 5/2^-)$ in the final state. In the experiment protons were detected with relatively small kinetic energy T_p = 40 MeV with kinematics corresponding to a missing momentum p_m = 120 MeV/c, where the $\ell=1$ knockout probability is maximal. The value T_p = 40 MeV was chosen in order to enhance the probability of inelastic excitation in the final state. In the spectrum shown in Fig. 4, no strength is observed at the location of the $5/2^-$-state. From this data we infer an upper limit in terms of an equivalent spectral function, $S < 2 \cdot 10^{-10}$ (MeV/c)$^{-3}$MeV^{-1} to be compared with a typical weak transition strength $S \simeq 3 \cdot 10^{-9}$ (MeV/c)$^{-3}$MeV^{-1}. Hence, even when weak transitions are involved, the interpretation of the (e,e'p) reaction appears to be free to a large extent from complications due to two-step processes.

This observation may be compared with the results of a (p,2p) experiment, in which an incoming proton energy of 100 MeV was used[10] and the energy of the outgoing protons (40 MeV) was the same as in the present (e,e'p) experiment. The $5/2^-$ state has been found to be excited in the (p,2p) reaction with one third of the strength of the nearby $3/2^-$ state with an angular distribution characteristic for $\ell=1$ knockout. In this hadronic reaction two-step processes are expected to proceed through two main channels:
(i) inelastic excitation of the target nucleus $^{12}C(0^+ \to 2^+)$ and subsequent $\ell=1$ knockout;
(ii) $\ell=1$ knockout leading to the ^{11}B ground state followed by inelastic excitation of $^{11}B(3/2^- \to 5/2^-)$ by the outgoing protons.

Since the total reaction cross section for proton-nucleus scattering increases with decreasing T_p and two protons contribute to the inelastic excitation in the final state, it has been suggested[11] that the major contribution comes from the latter channel. Adopting the dominance of channel (ii) and assuming the two protons contribute coherently to the inelastic excitation, the probability of populating the $5/2^-$ state in (p,2p) should be a factor four larger than in the (e,e'p) reaction. However, from the present experiment, we infer a difference of at least a factor of ten. A more detailed investigation of the coupled-channel effects in the (p,2p) reaction is needed for a better understanding of this observation. The (e,e'p) results may also be useful to better pin down two-step processes in the (p,2p) reaction.

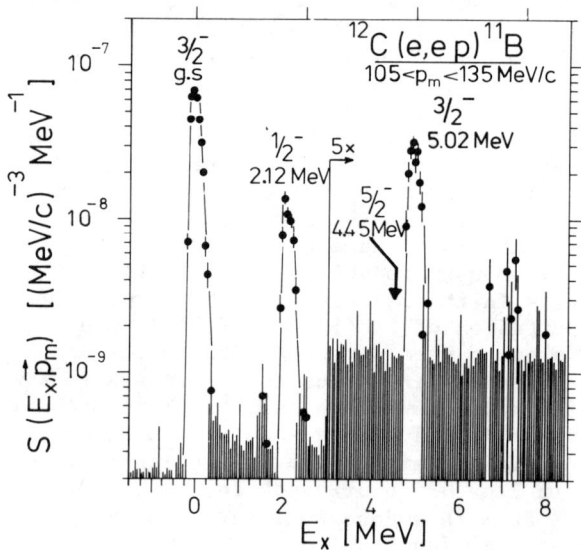

Fig. 4. Spectral function for the reaction $^{12}C(e,e'p)^{11}B$ at the maximum of the $\ell=1$ knockout process with an outgoing proton kinetic energy $T_p=40$ MeV. Note the break in the vertical scale at $E_x=3$ MeV.

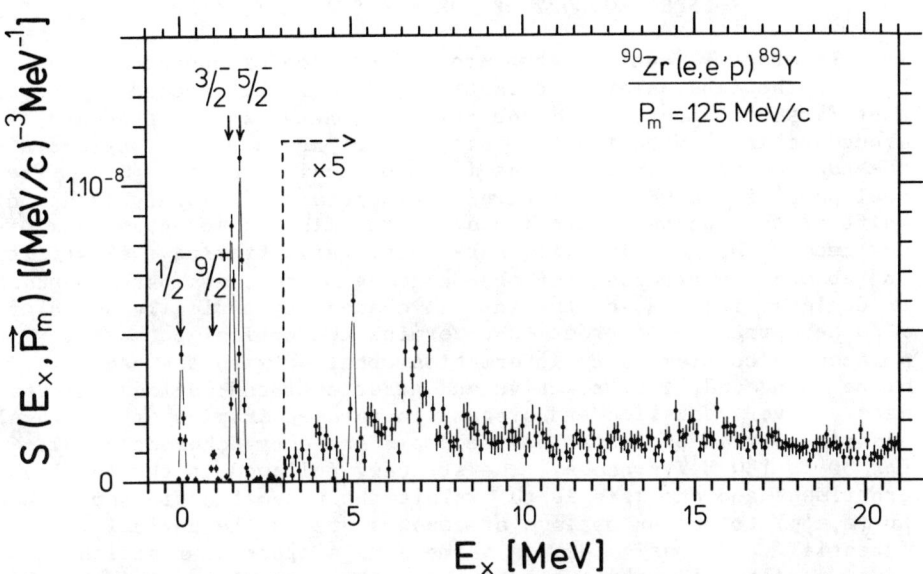

Fig. 5. High-resolution spectrum of the spectral function for the reaction ^{90}Zr(e,e'p)^{89}Y up to 20 MeV excitation energy in ^{89}Y.

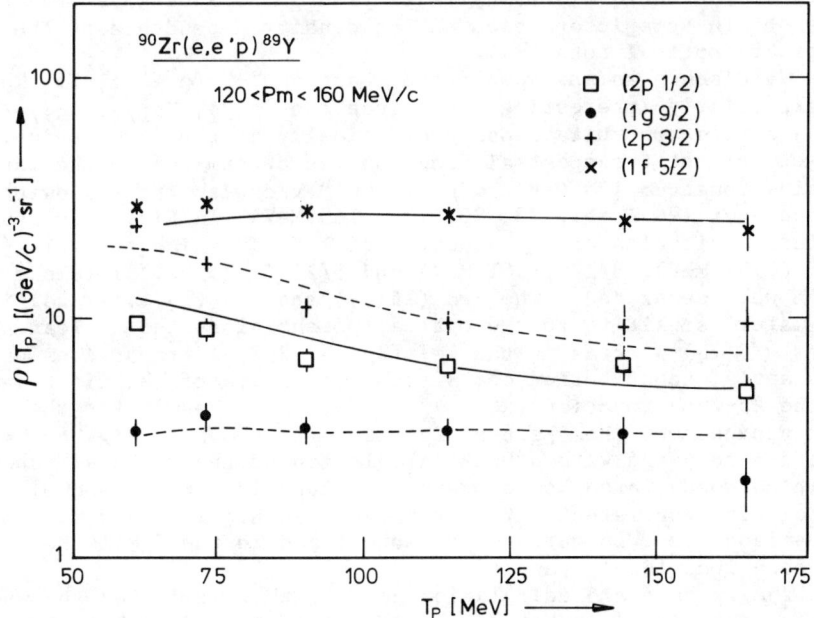

Fig. 6. Spectral function for various different quantum states at a fixed value of missing momentum p_m, as a function of outgoing proton energy T_p.

SPECTROSCOPY OF SURFACE STATES IN ^{90}Zr

In order to achieve absolute calibration of spectroscopic factors the final-state interaction (FSI) of the knocked-out proton must first be understood in quantitative detail. The standard procedure is to describe the fate of the proton on its way out of the nucleus as if it is moving in an optical potential (OP). The real part of the OP is relatively unimportant; it mainly leads to a shift of the spectral function along the missing momentum axis of a few tens of MeV/c. The imaginary part, parametrizing inelasticity and absorption processes of the outgoing proton, is most important; it depletes the flux of the initial channel by typically 30-50% for a 70 MeV proton. In order to determine the optical potential parameters complementary information obtained with the hadronic probe is needed, i.e. reactive and elastic proton-nucleus scattering data. A very detailed and systematic parametrization of the (p-A) elastic optical potential has been performed by Schwandt et al.[12] for 80 to 180 MeV protons. The strategy followed in the (e,e'p) experiment and analysis is to exploit the kinematic freedom at hand in (e,e'p) to put an optimum of constraints on the optical potential. The basic idea is to measure a spectral function in a given p_m-range for various outgoing proton energies. This allows a check on the prescribed energy-dependence of the OP parameters. In addition, by selecting different quantum states near the Fermi surface, the nuclear density is sampled in various ways and one may thus obtain some information on the density dependence of the imaginary optical potential.

We discuss as an example the reaction ^{90}Zr(e,e'p)^{89}Y. This is a particularly interesting case since the $2p_{3/2}$, $2p_{1/2}$, $1g_{9/2}$ and $1f_{5/2}$ orbits are quite close energetically in the A=90 region. In the experiment the spectral function was determined in the range of missing momentum 120 MeV/c $\leq p_m \leq$ 160 MeV/c with the following proton energies T_p=61, 73, 90, 114, 165 MeV. In Fig. 5 is shown a typical excitation energy spectrum of ^{89}Y, in which the 1/2$^-$ (gs), 9/2$^+$ (0.91 MeV), 3/2$^-$ (1.51 MeV) and 5/2$^-$ (1.75 MeV) levels are seen to be well separated. The ℓ=3 ($1f_{5/2}$) and ℓ=4 ($1g_{9/2}$) transitions are mainly sensitive to the optical potential at the nuclear surface (at $r \simeq 5$ fm) whereas in the ℓ=1 ($2p_{3/2}$, $2p_{1/2}$) transitions a substantial contribution comes from the region of the first maximum of the 2p-wave function (at $r \simeq 1.5$ fm). The result for the T_p-dependence of the spectral functions is shown in Fig. 6. The data are compared with a DWIA calculation of the spectral function in which Woods-Saxon bound state wave functions were used with a radial size consistent with Hartree-Fock orbitals. For ease of comparison the DWIA curves were normalized to the T_p=114 MeV point for each individual transition.

The 1g data and calculation agree rather well over the whole T_p range. This implies that the optical potential used has the correct energy dependence near the nuclear surface. For the $1f_{5/2}$ orbit, that peaks somewhat more inside the nucleus, slight deviations are observed at low energies. Here the absence of a surface term in the optical potential that was used may play a role. More pronounced effects are observed for the $\ell=1$ transitions.

The DWIA calculations neither reproduce the trend of the experimental data for the individual transitions, nor do they reproduce the energy dependence of the ratio $2p_{3/2}$ to $2p_{1/2}$ strength. This possibly indicates the inadequacy of the optical potential with a simple Woods-Saxon radial dependence in the nuclear interior. It may also indicate that the shape of the 2p wave function is different from the meanfield Woods-Saxon result. The use of Hartree-Fock Bogoliubov wave functions that include pairing degrees of freedom may cure this problem.

I will not pursue this particular question any further, except for noting that a more detailed investigation of the origins of the discrepancy is in progress.

Having found a consistent description of the $\ell=3$ knockout spectral function as to its proton-energy dependence, the question now arises how well the spectroscopic factor is determined on an absolute scale. We have investigated this matter of calibration using a data-set in which a broad range of p_m-values is covered (-50 MeV/c $< p_m <$ 250 MeV/c) and proton kinetic energies of 70-100 MeV were observed. Starting from the global optical potential paramenters of Schwandt et al., variations in the multiparameter space were allowed for, such that the χ^2 of the fit to elastic proton-Zr scattering data was constrained as $\chi^2 \leq 3\chi^2_{min}$. We thus admitted of a rather generous variation of parameters. For some parameters an additional χ^2 constraint was imposed by the shape of the spectral function. We divide the parameters in two distinct categories, i.e. the bound state wave function (bswf) space and the optical potential for the scattering state. The bswf was generated in either a local or non-local Woods-Saxon potential well while the well-depth V_o, for a given well radius r_o and diffuseness a_o, was varied to yield the proper separation energy. The rms radius of the bswf was varied in a range of 2% around the Hartree-Fock value.

For a given choice of bswf-shape the distorted spectral function was calculated, with an optical potential Ansatz. Then the calculated spectral function was normalized such as to yield a best fit to the (e,e'p) data. Thus a best-fit spectroscopic factor was obtained. The optical potential parameters, real V(r) and imaginary W(r) of a single or double humped Woods-Saxon form, and spin orbit term, were then varied within the above mentioned constraint $\chi^2 \leq 3\chi^2_{min}$. As was anticipated the imaginary potential is the most important source of uncertainty, affecting the spectroscopic factor. We find, within the framework of the optical potential description of the final state interaction that the spectroscopic factor for $1f_{5/2}$ knock-out is determined as $C^2S(1f_{5/2}) = 2.6 - 3.4$.

The choice of bound-state wave function, either Wood-Saxon type with a variation of r_o by $\pm$ 2% or Hartree-Fock type, influences the spectroscopic factor by less than 5%. The uncertainty due to the choice of OP parameters may be further reduced if theoretical constraints on the radial shape of the imaginary part of the optical potential are imposed. In addition, the analysis[13] of 65 MeV elastic proton scattering and analyzing power data in a Fourier-Bessel approach indicate that the shape of the OP may be

fairly well determined, also in the nucleons interior. These are
the next steps to be taken in order to further improve the accuracy
of the spectroscopic factors deduced from (e,e'p) data.

OCCUPANCY OF THE PROTON $3s_{1/2}$-ORBIT IN THE Pb-REGION

The concept of closed shells and a few active valence nucleons
is central to the shell-model description of nuclear structure.
Recent experimental observations from various different sources have
adduced suggestive evidence for a substantial depletion of shell-
model states near the Fermi-surface in the lead region.[14] The
experimental information consists mainly of i) the observed
quenching of the elastic magnetic response function of ^{207}Pb and
^{205}Tℓ, discussed in detail in the contribution of C. Papanicolas at
this conference and ii) the charge densities of ^{208}Pb, ^{206}Pb and
^{205}Tl. The charge densities $\rho(r)$ deduced from elastic electron
scattering at high momentum transfer show that the empirical density
at the center of the nucleus $\rho(r=0)$ is appreciably smaller[15] than
the value calculated in the closed-shell Hartree-Fock approximation.
Experiment and theory may be reconciled if an ad-hoc reduction of
30% of the occupation probability of the $3s_{1/2}$-orbit (that peaks at
r=0) is applied.

The difference of the charge density of ^{206}Pb and ^{205}Tl
$\Delta\rho(r) = \rho_{206}(r) - \rho_{205}(r)$ features the characteristic damped
oscillatory pattern of a $3_{1/2}$ orbital. The value of $\Delta\rho$ ar r=0 has
been shown[15] to be consistent with a difference of $3_{1/2}$ occupation
probability of 0.7 according to a density dependent Hartree-Fock
calculation. Within the framework of the mean field approximation
therefore the result is indicative of a 30% depletion of the
$3s_{1/2}$-orbit.

One would like to connect this observation to the doubly-magic
^{208}Pb, that presumably constitutes the best candidate for a truly
closed-shell nucleus. To this purpose we have performed a high
resolution study of the proton knockout reaction on both ^{206}Pb and
^{208}Pb. This is a collaborative effort of groups from Tubingen (G.J.
Wagner), Indiana (H. Nann), Saclay (B. Frois) and NIKHEF-K. The
main idea is to obtain information on relative spectroscopic factors
from two independent reactions, (e,e'p) and (d,^{3}He), that each
sample a totally different part of the bound-state wave function
(see Fig. 3). Recently precise relative spectroscopic factors for
3s-proton pickup have been measured[16] with the (d,^{3}He) reaction at
52 MeV.

The ^{206,208}Pb(e,e'p) experiment was performed at incident
electron energies of 410 and 350 MeV, with a missing mass resolution
of 130 keV. The ^{206}Pb (99.7% enriched) and ^{208}Pb (97% enriched)
targets were rotated in the 20 μA beam to avoid melting. The ratio
of target thicknesses was determined with an accuracy of one percent
from elastic and quasi-elastic cross sections. The stability of the
targets during the experiment was monitored by the singles rates in
the spectrometers. Taking account of this and other sources of
error the overall systematic error amounts to two percent in the
ratio of cross sections.

Coincidence cross sections were measured in three ranges of missing momentum p_m, $15 \pm \Delta p$, $100 \pm \Delta p$ and $200 \pm \Delta p$ MeV/c with $\Delta p = 25$ MeV/c. The knocked out proton energy was kept constant at $T_p = 100$ MeV in order to keep the FSI constant at each kinematics. In the given range of p_m-values the 3s wave function is sampled in a manner that is illustrated in Fig. 1.

At $p_m = 40$ MeV/c the characteristic selectivity for $\ell=0$ (i.e. 3s) knockout is manifest as is illustrated in Fig. 7. In all data taken the energy resolution was largely sufficient to separate the $1/2^+$ ground state from the first excited $3/2^+$ state at $E_x = 0.21(0.35)$MeV in ^{205}Tl(^{207}Tl).

In the data analysis two pieces of information were sought after i) precise relative spectroscopic factors for knockout leading to the ground state of ^{207}Tl and ^{205}Tl and ii) the relative occupation probability of the 3s-proton orbit in ^{206,208}Pb, that can be obtained by measuring all $\ell=0$ knockout strength up to $E_x \simeq 5$ MeV in the (A-1) system.

The ratio of ground-state transition cross sections is shown in Fig. 8. Assuming equal spectroscopic factors $S_o(206) = S_o(208)$ the distorted wave impulse approximation calculation yields the dashed curve, which deviates from a constant value of one quite slightly only. In the distorted wave calculation, the 3s wave function was generated in a Woods-Saxon well taking account of the proper separation energy and a well-radius dependence of $A^{1/3}$. Optical-potential parameters were taken from the Schwandt compilation. The ratio of spectroscopic factors $R_o = S_o(206)/S_o(208)$ by adjusting the calculated curve to the data points resulted in $R_o=0.69(3)$. There is no significant p_m-dependence manifest in the data. In the search for additional $\ell=0$ strength, the spectrum was integrated in $\Delta E_x = 1.0$ MeV bins up to $E_x = 5.5$ MeV and a multipole decomposition was performed of the spectral function of each bin. Distorted spectral functions for 3s, 2d, 1g and 1h knockout were employed in the procedure. Whereas the $\ell=0$ strength in ^{208}Pb $\rightarrow$ ^{207}Tl is found exclusively in the ground-state transition, appreciable fragmentation is observed in ^{206}Pb $\rightarrow$ ^{205}Tl. The two known excited $1/2^+$ states at $E_x = 1.22$ and 1.44 MeV in ^{205}Tl are populated with a summed strength relative to $S_o(208)$ of $(S_1(206)+S_2(206))/S_o(208)= 0.16(2)$. Adding up all strength in the fragmentation plot (see Fig. 9) up to $E_x = 5.5$ MeV we obtain for the ratio of $\ell=0$ strength $R=S_{total}(206)/S_o(208) = 0.84(9)$. To the extent that the fraction of $\ell=0$ strength located beyond the E_x interval considered is equal for the two lead isotopes, R represents the relative occupation probability of the 3s orbit.

On comparison of the (e,e'p) result with the values from the recent (d,^{3}He) experiment,[16] $R_o=0.77 \pm 0.01$(stat.) $\pm$ 0.015(syst.) and $R=0.93 \pm 0.04$, the following observation can be made. Whereas the results for R are in fair agreement, the precise values for R_o indicate that the (e,e'p) value $R_o = 0.69 \pm 0.03$(stat.) $\pm$ 0.02 (syst.) is somewhat smaller than the (d,^{3}He) result. This is considered not to be overly surprising since the bound-state wave function is sampled in such different manners in the two reactions. There is also the possibility that the standard prescription for the

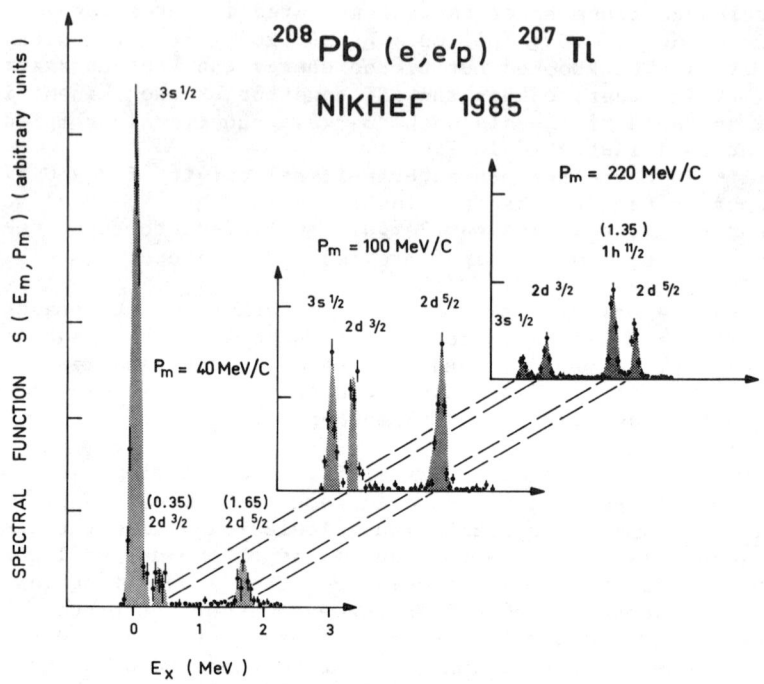

Fig. 7. Spectral function of ^{208}Pb(e,e'p)^{207}Tl for three different values of p_m.

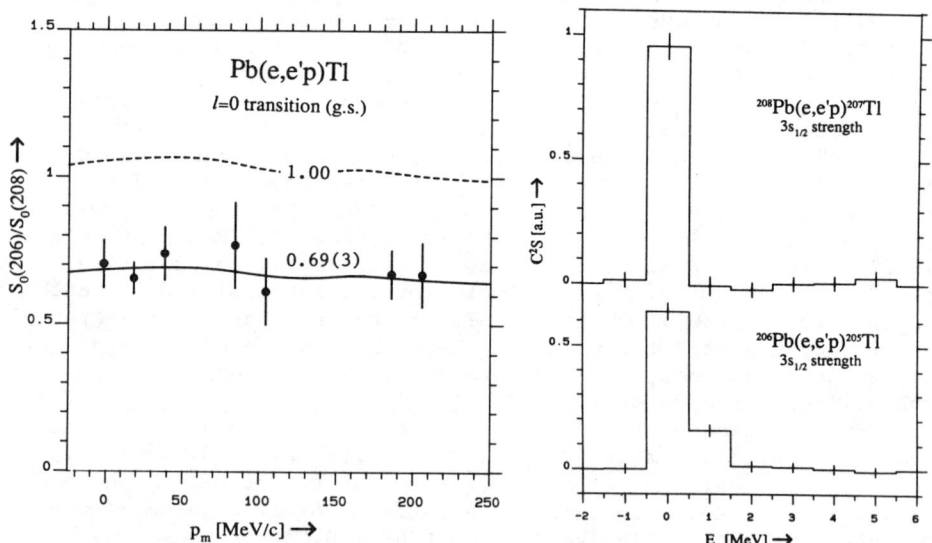

Fig. 8. Ratio of spectroscopic factors for the transitions to the ground state of ^{205,207}Tl.

Fig. 9. Fragmentation of $\ell=0$ strength in ^{207}Tl and ^{205}Tl.

change of the geometry of the mean field, in which the $3s_{1/2}$ proton is bound, in going from the open-shell nucleus ^{206}Pb to the closed-shell ^{208}Pb is not adequate. If, for instance, the Woods-Saxon well radius r_o is increased by one percent in ^{206}Pb the results of a reanalysis show that $R_o(e,e'p) = 0.67(3)$ and $R_o(d,^3\text{He}) = 0.68(1)$. Further detailed analysis and additional data on other nuclides are needed to clarify the observed difference of results of the (e,e'p) and (d,^{3}He) reactions.

In order to obtain absolute 3s occupation probabilities one has to link the spectroscopic data to the charge-density difference $\Delta\rho = \rho(206)-\rho(205)$. The relation between spectroscopic factors and $\Delta\rho$ is rather involved.[16,17] Further theroretical investigations of this matter are needed and are in fact underway.[18]

REFERENCES

1. M.H. Macfarlane, Proc. Intern. Conf. on properties of nuclear states (Montreal), 385 (1969).
2. S. Platchkov et al., Phys. Rev. C 25, 2318 (1982).
3. J. Mougey, Nucl. Phys. A396, 39c (1983).
4. G. v.d. Steenhoven et al., to be published in Rapid Communications Phys. Rev. C 34 (Nov. 1985).
5. T.W. Donnelly and I. Sick, Rev. Mod. Phys. 56, 461 (1984) and Reference 2.
6. B. Frois, Nucl. Phys. A396, 409c (1983).
7. G. v.d. Steenhoven et al., Phys. Lett. B, Vol. 156B, 146 (1985).
8. A.G.M. van Hees and P.W.M. Glaudemans, Z. Physik A315, 223 (1984).
9. W.D. Teeters and D. Kurath, Nucl. Phys. A275, 61 (1977).
10. D.W. Devins et al., Austr. J. Phys. 32, 323 (1979).
11. G.E. Walker, Proc. of the 3rd Workshop of the Bates User Theory Group (MIT, July 1984), Ed. G. Rawitscher and G.E. Walker, private communication.
12. P. Schwandt et al., Phys. Rev. C 26, 55 (1982).
13. H. Sagaguchi et al., Annual Report RCNP (1984). Osaka University, p. 12.
14. V.R. Pandharipande et al., Phys. Rev. Lett. 53, 1133 (1984).
15. J.M. Cavedon et al., Phys. Rev. Lett. 49, 978 (1982).
16. P. Grabmayr et al., to be published in Phys. Lett. B and G.J. Wagner, contribution to this conference.
17. M. Jaminon, C. Mahaux and H. Ngo, Phys. Lett. 158B, 103 (1985).
18. G. Co and J. Speth, to be published and G. Co, private communication.

QUASI-FREE KNOCKOUT REACTIONS INDUCED BY HADRONS

N. S. Chant
Department of Physics and Astronomy
University of Maryland, College Park, MD 20742

ABSTRACT

The experimental status of quasifree knockout reactions is discussed. Results for both nucleon and nucleon cluster removal are described and the use of pions, nucleons, and nucleon cluster projectiles is considered. An attempt is made to indicate open questions and possible directions for future work.

INTRODUCTION

With the new IUCF spectrometers it should be possible to take data for reactions of the type A(a,cd)B leading to three-body final states with much better energy resolution and statistics than has been possible in the past. If we select detection angles and energies close to the values for the free a+b → c+d reaction we can expect the mechanism suggested by the diagram in Fig. 1 to dominate the reaction. We refer to this process as quasifree knockout.

At the IUCF workshop three years ago[1] I discussed some of the considerations relevant to studying the upper vertex, specifically the nucleon-nucleon interaction in (p,2p) and (p,pn) reactions. Clearly there is a great deal of physics in the lower vertex as well, namely the spectroscopic factor distribution, or spectral function, and the individual nucleon momentum distributions. In addition, other reactions such as (p,pd) or (p,pα) can provide information on clustering and on nucleon-cluster interactions in nuclei. Finally, different probes such as alpha particles or pions can change the emphasis of the physics quite drastically.

I would like to review briefly some of the issues of interest for all of these reactions. More detailed information on some items can be found in the earlier talk as well as the proceedings of various workshops[2] and conferences.[3-5] In the process I will attempt to indicate some experimental possibilities for IUCF.

(p,2p) AND (p,pn) REACTIONS

Theoretical Background

Although distorted wave impulse approximation calculations have been published for (p,2p) using a local pseudo-potential, most current analyses of experimental data employ a factorized calculation.[1,2] Thus, if spin-orbit terms in the distorting potential are ignored the cross section for a (p,2p) reaction can be written in terms of a two-body interaction term and a spectral function which reduces, for discrete states, to a distorted momentum density for the struck nucleon.

For the reaction A(a,cd)B details are as follows:

PWIA

$$\frac{d^3\sigma}{d\Omega_c d\Omega_d dE_c} = K\, S_b \left|\phi(-\vec{p}_B)\right|^2 \frac{d\sigma}{d\Omega_{a+b}}$$

K - kinematic factor

S_b - spectroscopic factor $A \rightarrow B+b$

$\phi(-\vec{p}_B)$ - b-B momentum wave function

$\frac{d\sigma}{d\Omega_{a+b}}$ - half-shell two-body cross section for $a+b \rightarrow c+d$

DWIA

$$|\phi|^2 \rightarrow \sum_\Lambda |T^{\alpha L \Lambda}|^2$$

where

$$T^{\alpha L \Lambda} = \frac{1}{(2L+1)^{1/2}} \int \chi_d^{(-)*} \chi_c^{(-)*} \phi_{L\Lambda}^\alpha \chi_a^{(+)} d^3r$$

$\chi^{(\pm)}$ - incident/emitted distorted waves

$\phi_{L\Lambda}^\alpha$ - bound state wave function for b+B

DWIA with Spin-Orbit Distortions

Optical potentials:

$$V + iW \rightarrow V + iW + V_{so}\, \vec{\ell}\cdot\vec{s}$$

Distorted waves:

$$\chi^{(\pm)} \rightarrow \chi_{\rho\sigma}^{(\pm)} \qquad \rho,\sigma = \pm\tfrac{1}{2}$$

Hence:

$$T^{\alpha L \Lambda} \rightarrow T^{\alpha L \Lambda}_{\rho_a \sigma_a \rho_c \sigma_c \rho_d \sigma_d}$$

and (for p,2p)

$$\frac{d^3\sigma}{d\Omega_c d\Omega_d dE_c} = K'\, S_b \sum_{M \atop \rho_a \rho_c \rho_d} \left| \sum_{\Lambda \rho_b \atop \sigma_a \sigma_c \sigma_d} (L\,\Lambda\,\tfrac{1}{2}\,\rho_b | JM) \right.$$

$$\times (2L+1)^{1/2}\, T^{\alpha L \Lambda}_{\rho_a \sigma_a \rho_c \sigma_c \rho_d \sigma_d} \langle \sigma_c \sigma_d | t | \sigma_a \rho_b \rangle \Big|^2$$

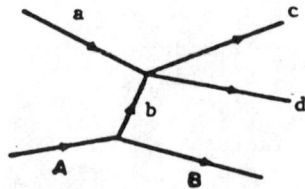

FIG. 1. First order diagram for quasifree A(a,cd)B reaction.

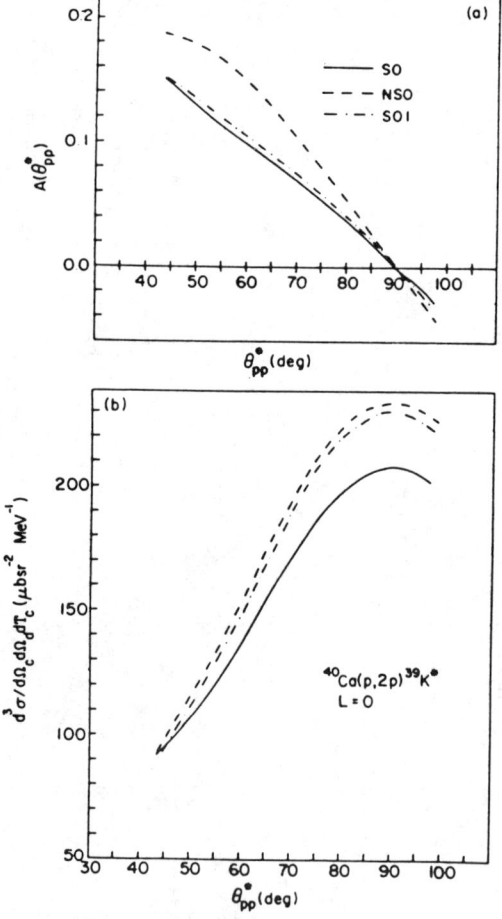

FIG. 2. Calculations for a quasifree angular distribution for $^{40}Ca(p,2p)^{39}K(2.52$ MeV) at an incident energy of 150 MeV as a function of the effective p+p scattering angle. The kinematics is chosen such that the residual nucleus is at rest. (a) Polarization analyzing powers; (b) differential cross sections.
SO - spin-orbit DWIA
NSO - no-spin-orbit DWIA
SO1 - spin-orbit terms for projectile only.

The quantity $\langle|t|\rangle$ is the p+p $\frac{1}{2}$-shell t-matrix evaluated for the asymptotic kinematics. It is frequently approximated by nearby on-shell values. Notice that, even in a factorized DWIA, a more complicated expression results if spin-orbit terms are included in the distorting optical potentials. Specifically, while there is still a factorization in amplitude, the cross section no longer factorizes.

Calculations including the effects of spin-orbit distortions have been carried out for a number of reactions.[6] In some cases effects are fairly small. An example is shown in Fig. 2 for ^{40}Ca(p,2p) at 150 MeV. Both the cross section and polarization analyzing power are reduced in the spin-orbit calculation but not changed greatly. In Fig. 3 predictions are shown for the (p,2p) reaction on ^{208}Pb in a coplanar geometry. For the $3S_{1/2}$ L=0 ground-state transition at 45.12° detection angles the no-spin-orbit analyzing power is approximately zero since the effective scattering angle is close to 90°. However, in the spin-orbit calculation the analyzing power exhibits a rapid swing between quite large positive and negative values at around 50 MeV. This behavior is found in many other cases and is associated with the minimum arising from a node in the struck proton momentum wave function. This is similar to behavior found in many other reactions where small differences in partial cross sections for spin up and spin down can, near a minimum, produce very large analyzing powers.

This, then, is my first proposed experiment. I believe such an effect has not been observed previously. The predicted analyzing powers are large, the measurement simple, and a proper description of the data should serve as a rather sensitive test of our treatment of the nucleon wave function and of the spin dependence of the distorting potentials.

Studies of the Nucleon-Nucleon Interaction

Despite the complications which can arise due to spin-orbit distortions it is still useful to use the factorized (no-spin-orbit) DWIA expression for the cross section as a guide to experimental design. This tells us that we can, at least in principle, determine the effective interaction from experiment provided we have a good description of the spectral function or distorted momentum distribution. Even better, we can attempt to hold this term fixed while varying the p+p kinematics. This is the basis of the so-called factorization test or quasifree angular distribution.

This is all discussed more extensively in my earlier IUCF talk. The main points to recall are:

a) By taking advantage of different detection geometries it is possible to study the nucleon-nucleon effective interaction in different density regions of the nucleus for a single transition. This is shown in Fig. 4 for the ^{40}Ca(p,2p)^{39}K reaction at 150 MeV incident energy. It is seen that at the ±41° detection geometry the yield arises mainly from the nuclear surface at about 10% of central density or less whereas at ±20° the cross section has an

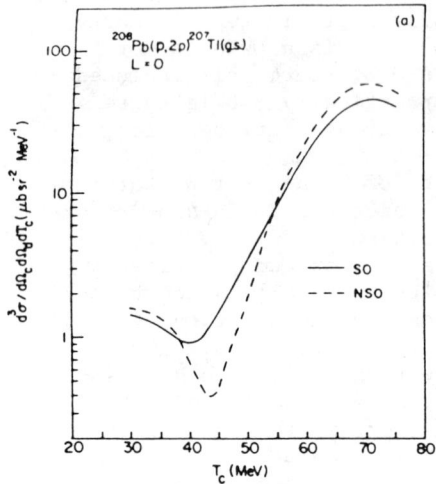
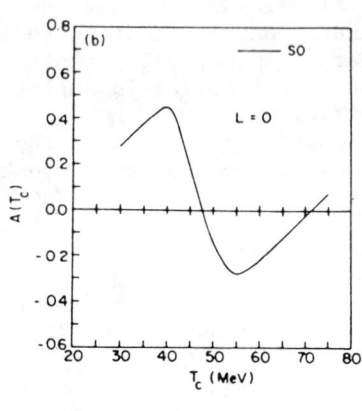

FIG. 3. Calculations of the L=0 ($3s_{1/2}$) energy sharing distribution for ^{208}Pb(p,2p)^{207}Tl (g.s.) at an incident energy of 150 MeV, $\theta_c = \theta_d = 45.12°$. (a) Differential cross section; (b) polarization analyzing power.

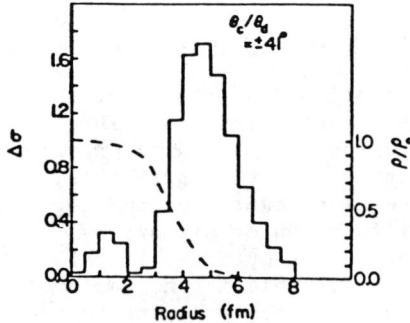
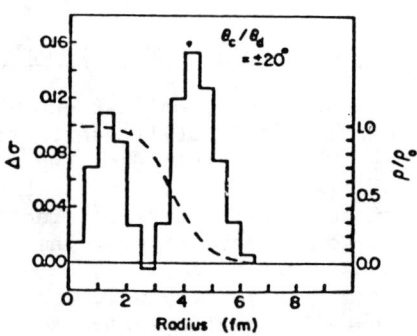

FIG. 4. Radial contributions to DWIA cross section for ^{40}Ca(p,2p)^{39}K (2.52 MeV, L=0) at 150 MeV. The detected angles are ±41° and ±20°, respectively. The detected energies are equal. The broken curve is the ^{40}Ca density distribution.

additional 40% contribution from the nuclear interior at essentially full central density. In a factorized DWIA the nucleon-nucleon matrix elements which enter for the ±41° data are half off-shell with a ratio of momenta K_{on}/K_{off} of 1.08 and thus should differ little from on-shell matrix elements. At ±20° the ratio increases to 1.49 and off-shell effects start to play a role. Further, one might expect half-shell matrix elements to be appropriate for the low density surface contributions while fully off-shell terms might be more important for the interior contribution. It is worth noting that, since the struck nucleon makes a transition from a bound to a scattering state, not only are different matrix elements involved in comparison with inelastic scattering but also the role of Pauli effects is likely to be different.

This then would be the basis of a second experiment in which (p,2p) measurements would be made over a range of opening angles from around ±45° to ±20° or less in order to examine the behavior of the nucleon-nucleon interaction over a range of densities and with different off-shell kinematics.

b) A caveat is that progress in such an experiment must be accompanied by improved theoretical calculations in which the factorization approximation is removed. Simply stated, if we design experiments in which we emphasise the role of the nucleon-nucleon interaction in the nuclear interior it makes no sense to try to understand the data in terms of a theory in which the interaction is assumed to be the free interaction. Experimental efforts to study this issue[1,7,8] for $^{40}Ca(p,2p)$ suggest that the factorization approximation works well at incident energies as low as 75 MeV. However, these studies were for a single 2s transition in geometries which maximized the surface localization of the reaction and were not far off-shell. Thus, generalization is likely to be extremely dangerous.

Attempts to improve on factorized DWIA calculations are currently underway and improved calculations were reported at the Osaka meeting by Kudo and Miyazaki[9] for the 200 MeV TRIUMF $^{16}O(p,2p)^{15}N$ data.[10] Their analysis can be best described as a density dependent factorised DWIA. That is the product of amplitudes

$$T^{\alpha L \Lambda}_{\rho_a \sigma_a \rho_c \sigma_c \rho_d \sigma_d} \langle \sigma_c \sigma_d | t | \sigma_a \sigma_b \rangle$$

is replaced by

$$T^{\alpha L \Lambda}_{\rho_a \sigma_a \rho_c \sigma_c \rho_d \sigma_d} = \int \chi^{(-)*}_{\rho_d \sigma_d}(\vec{r}) \chi^{(-)*}_{\rho_c \sigma_c}(\vec{r}) \langle \sigma_c \sigma_d | G(\rho(r)) | \sigma_a \sigma_b \rangle$$
$$\times \phi^{\alpha}_{L\Lambda}(\vec{r}) \chi^{(+)}_{\rho_a \sigma_a}(\vec{r}) d^3r$$

where $\langle | G(\rho(r)) | \rangle$ is an antisymmetrized off-shell transition matrix calculated using the density dependent Hamburg potential. We see

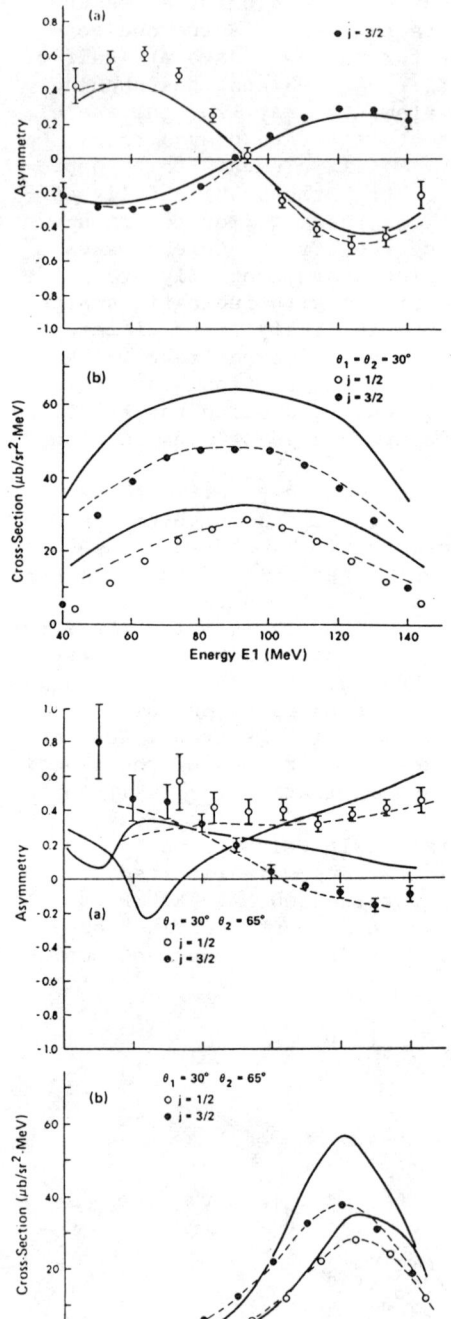

FIG. 5. $^{16}O(\vec{p},2p)^{15}N$ at 200 MeV. Analyzing powers (a) and cross sections (b). The continuous curves are a factorized DWIA prediction.[10] The broken curves result from using a density dependent interaction.[9]

that the factorization approximation is retained but a density
dependent t-matrix is used which differs at each radius. Results
for ^{16}O are shown in Fig. 5. It is seen that agreement with
experiment is somewhat better than the density independent calculations in the original paper. Thus, further studies along these
lines would be interesting.

c) Any experiment to study the effective interaction in
(p,2p) can clearly be enhanced by using polarized incident beams.
Not only will the incident polarization play a role but also the
(p,2p) reaction enables us to study the effective nucleon-nucleon
interaction for a polarized nucleon target. Specifically, for
asymmetric angles or energies, owing to differences in distortion
effects, the reaction can be preferentially localized on one side
or other of the nucleus. As a result there can be polarization in
orbital angular momentum for the struck nucleon and hence in
spin.[11] This can be seen easily in a semi-classical model and is
calculable in DWIA. Thus we have a polarized nucleon target.
Specifically, in the approximation that spin-orbit terms in the
optical potentials can be neglected the nucleon-nucleon cross
section appearing in the factorized DWIA may be written

$$\frac{d\sigma}{d\Omega}(\vec{P}_a) = \frac{d\sigma}{d\Omega}(0)\{1 + (\vec{P}_a+\vec{P}_b)\cdot\vec{A} + \vec{P}_a\cdot\vec{P}_b\ C_{nn}\}$$

where $\vec{P}_a$ is the incident beam polarization, $\vec{P}_b$ is the struck
nucleon polarization and A and C_{nn} are the (half-shell) nucleon-
nucleon polarization analyzing power and spin correlation coefficient, respectively.[12] The expression holds for polarization
normal to the plane in a coplanar experiment. For $\ell=0$ transitions
P_b is zero and hence, the (p,2p) analyzing power should simply
equal A. For $\ell>0$ there is a simple relationship between the values
of P_b for $j=\ell\pm 1/2$ members of a spin-orbit doublet which is exact in
the limit of degenerate states. Thus, at least in principle, cross
section and analyzing power measurements for spin-orbit partners
can be combined to extract A and C_{nn} independently of a DWIA model.

In practice DWIA calculations will be needed to evaluate the
data and spin-orbit terms in the distorting potentials can, in some
cases, confuse the issue.[6,13] However, it seems reasonable to
assume that a combination of $\ell=0$ and $j=\ell\pm 1/2$ cross section and
analysing power data will enable our proposed experiment to explore
the spin-dependence of the interaction with least ambiguity.
Clearly if one makes quantitative progress it might be worthwhile
adding a focal-plane polarimeter in order to add additional spin-
dependent data and thus delineate the nucleon-nucleon spin dependence in even more detail.

An interesting result which has been discussed pre-
viously[11,12,14] is the observation, for ^{16}O(p,2p) and ^{40}Ca(p,2p) at
200-300 MeV, that the nucleon-nucleon polarization analyzing power
may perhaps be reduced in the nuclear interior. Specifically, for
$\ell>0$ transitions, it is found that the value of A extracted from
spin-orbit doublet data is close to zero not only in symmetric
geometries for which the effective p+p scattering angle is close to
90° but also in asymmetric geometry measurements where the free

scattering value is around 0.3. This behavior is not explained as a spin-orbit distortion effect and is not observed for $\ell=0$ transitions for which the expected free scattering values are obtained. It has been suggested[12] that it is merely a distortion effect not properly included in factorized DWIA such that, as a result of refraction the effective p+p scattering angle becomes close to 90°. Again we see the need for DWIA calculations which do not make the factorization approximation and, thus, do not restrict the p+p kinematics to their asympototic values. In any case, it is clearly worth further investigation in a higher precision measurement and partially motivates the experiment proposed.

Studies of Momentum Wave Functions

In order to obtain information on the struck particle momentum distribution it is obviously desirable that distortion effects be fairly small so that uncertainties in the DWIA analysis do not play a major role. In Fig. 6 we show calculations for $^{40}Ca(p,2p)^{39}K$ (L=0, 2.52 MeV) at 400 MeV. Shown is the predicted experimental measurement in comparison with the actual momentum distribution assumed. Clearly, distortion effects are important and must be reliably calculated. However, the qualitative features of the momentum distribution are not obscured and useful information can be expected. Also shown is the predicted result of a 500 MeV (e,ep) measurement. Since there is now only one proton to deal with distortion effects are, as expected, less but must still be calculated.

As an example of an attempt to use the (p,2p) reaction to study the struck nucleon momentum wave function, 500 MeV data from the measurement of the $^4He(p,2p)^3H$ reaction carried out at TRIUMF[15,16] is shown in Fig. 7. Notice that the data extend to momenta of about 500 MeV/c. These data show rather nicely the sort of problems one must face in trying to obtain momentum distribution information. The single particle wave functions used were obtained from fits to electron-^{4}He scattering data. The dot-dash curve uses a conventional analysis by Lim in terms of a simple Eckart parametrization. For the continuous curve the electron scattering data were first corrected for meson exchange contributions which, as we see, leads to a wave function with reduced high momentum components. The broken curve uses the meson exchange corrected wave function but with the spin-orbit terms in the distorting potentials set to zero. We see that, though negligible below about 250 MeV/c, the spin-orbit terms do improve agreement with experiment at the largest momenta and are at least as important as getting the wave function right. Thus, even for this very light target the proper treatment of distortion effects is quite important to the analysis.

For these data we see that it is at around 400 MeV/c that things start to get interesting. In generalized Breuckner-Hartree-Fock calculations[17] for ^{4}He and ^{16}O it is at this point that two-body correlations start to become important. Thus, measurements at even higher momenta are desirable. These certainly should include additional (e,ep) measurements. However, as a third proposed

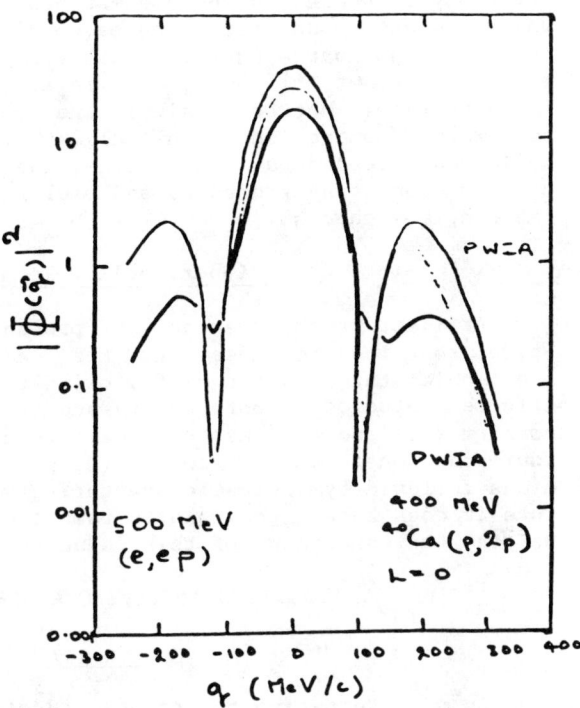

FIG. 6 Comparison of predicted experimental measurement (DWIA) and actual momentum distribution (PWIA) for $^{40}Ca(p,2p)^{39}K$ (L=0, 2.52 MeV) at 400 MeV. The broken curve is the predicted 500 MeV (e,ep) result.

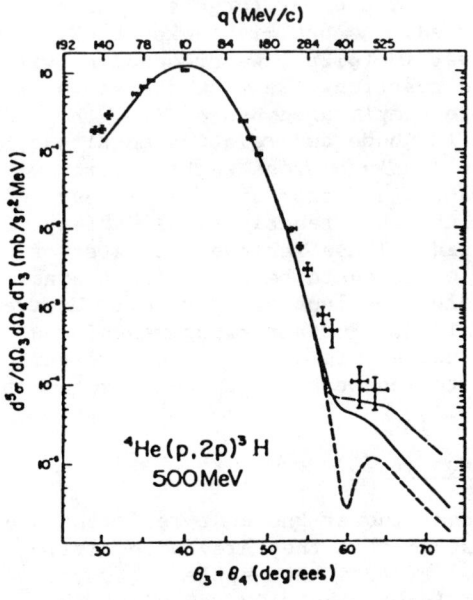

FIG. 7. Coplanar symmetric data; ——— meson exchange corrected, DWIA with spin-orbit terms; --- same with no spin-orbit terms; -·-· Lim Eckart function.

experiment it should be interesting to make (p,2p) measurements on
a variety of nuclei to the highest possible recoil momenta for
which adequate count rates can be obtained. For a 500 MeV incident
beam in a coplanar symmetric equal energy geometry 400 MeV/c is
reached at about ±61° and 800 MeV/c at ±78° with respect to the
incident beam. Whether a simple knockout mechanism will persist at
such recoil momenta seems unlikely. Nevertheless, particularly if
analogous (e,ep) data is available, one should be able to see the
onset of competing processes and explore possible corrections to
the simple mechanism.

Inner Shell Hole States

At least at the present IUCF proton energy of 200 MeV use of
(p,2p) reactions to investigate the location and width of inner
shell-hole states is not useful owing to multiple scattering
effects. Studies of both (p,2p) and (p,pn) reactions at IUCF[18-20]
confirm this and at least one analysis[21] reproduces most of the
four-body continuum in terms of (p,2p) transitions to low excitations followed by inelastic scattering of the emitted protons.
Once it does become possible to ramp the Cooler Ring to 400 or 500
MeV the current status of this issue should be reevaluated.

REACTIONS INVOLVING NUCLEON CLUSTERS

$(\alpha,\alpha p)$

As an alternative to protons and electrons, quasifree knockout
reactions can be induced by composite particles. Data for the
$(\alpha,\alpha p)$ reaction at 140 MeV on ^{2}H, ^{6}Li, and ^{19}F targets[22] and on ^{16}O
and ^{40}Ca targets[8] have been reported. As one might expect DWIA
calculations show that, in contrast to (p,2p), we are dealing with
a more strongly surface localized reaction. Results are shown in
Fig. 8 for ^{40}Ca. Despite the more complicated projectile, the
results are quite well described in shape and relative magnitude by
a DWIA calculation. However, the absolute cross sections are in
disagreement with experiment by about a factor of two, although
relative values are consistent with other reactions. A variety of
explanations have been investigated. These include estimates of
off-shell effects which might be expected to be significant since
only 35 MeV is available in the alpha-nucleon c.m. system. However
so far the discrepancy is unexplained. Further experimental and
theoretical studies would be of interest in order to resolve this
issue. One possibility might be to repeat some of the measurements
at higher incident alpha energies.

$(p,p\alpha)$ and $(\alpha,2\alpha)$

Although some electron induced cluster knockout reactions are
beginning to appear[23,24] I understand that they are quite difficult. In order to keep distortion effects manageable in, say,
$(e,e\alpha)$ it is desirable to have at least about 100 MeV kinetic

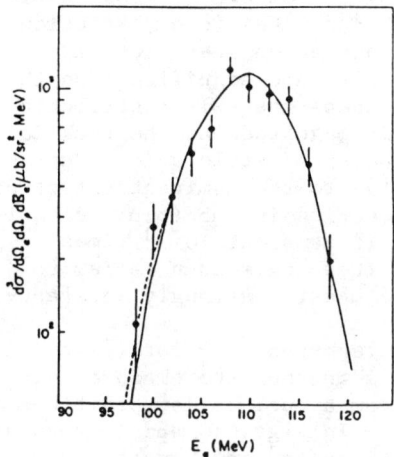

FIG. 8(a). Energy sharing distribution: $^{40}Ca(\alpha,\alpha p)^{39}K(1/2^+, 2.5$ MeV$)$ $\theta_\alpha/\theta_p = 90°/-51.41°$, $E_\alpha = 139.2$ MeV; (solid line) deep potential, (dashed line) shallow potential.

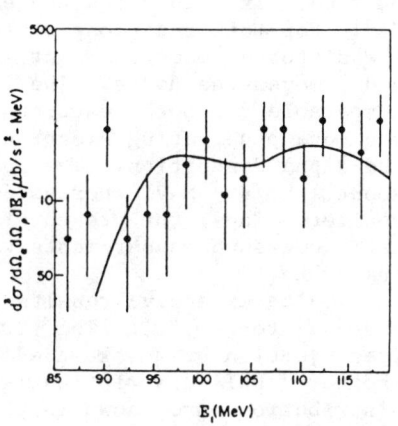

FIG. 8(b). Energy sharing distribution: $^{40}Ca(\alpha,\alpha p)^{39}K(3/2^+, 0.0$ MeV$)$ $\theta_\alpha/\theta_p = 90°/-51.41°$ at $E_\alpha = 139.2$ MeV.

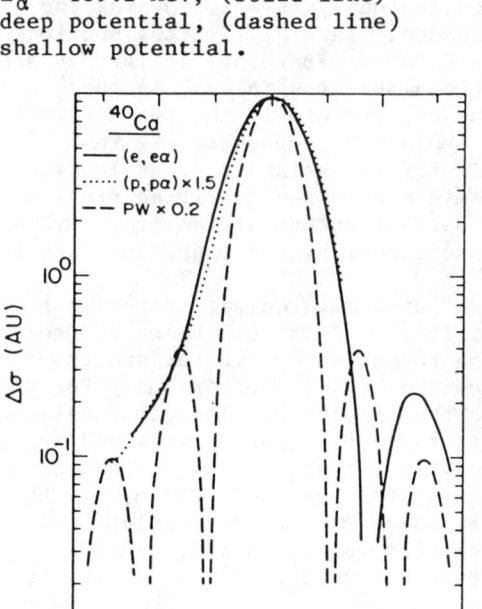

FIG. 9. Calculated distorted and undistorted momentum distributions for 1 GeV (e,eα) and 0.8 GeV (p,pα) vs P_{recoil}.

energy for the emitted alpha particle (i.e., a momentum transfer of 870 MeV/c). To get some idea of how things compare we show in Fig. 9 DWIA calculations[4] for 800 MeV (p,pα) and 1 GeV (e,eα) reactions on ^{40}Ca. In both cases the emitted alpha has an energy of 120 MeV. For both reactions distortion effects are significant so that the distorted momentum distribution broadens the PWIA distribution and removes the nodes. The reduction in magnitude at the peak is comparable for both reactions. Thus there is little gain in using the more penetrating electron probe owing to the dominant effect of the alpha distortion. The peak cross section in the (p,pα) case is about 1 μb/sr^2 MeV, whereas for (e,eα) it is about 10^{-11} times smaller. Thus, for present facilities it appears much easier to carry out such measurements with a more massive strongly interacting probe.

Quite extensive results have been reported[25-30] for (p,pα) and [31-33] for (α,2α). The first issue is whether the clean quasifree ejection of an extended cluster from a nucleus is possible and whether a DWIA description is adequate. In Figs. 10 and 11 angular distributions are shown for L=0 transitions for both reactions at the quasifree kinematic point. For the 140 MeV (α,2α) reaction agreement with the free α+α data is excellent, presumably owing to the strong surface localization of the reaction. For the 100 MeV (p,pα) reaction the case is not quite as convincing. Thus, either the reaction mechanism is more complicated than assumed or, for this less well localized reaction, an analysis which avoids the factorization approximation is needed. In Fig. 12 cross sections and polarization analyzing power data for ^{9}Be(p,pα) at 147 MeV are compared with theory.[30] Since the alpha is spinless, in the absence of spin-orbit effects one expects simply the p+α analyzing power. For the data shown both L=0 and 2 components are involved. We see that agreement is fairly satisfactory at low recoil momenta suggesting underlying dominance by the quasifree process. However spin-orbit terms in the optical potentials do play a role in other kinematic regions and additional higher precision data is needed.

In Fig. 13 DWIA predictions[34] are compared with experiment for heavier targets. Agreement is fairly good in both cases so one is encouraged to use these reactions to extract α-cluster spectroscopic factors. Results are shown in Fig. 14 for (p,pα). For the ground state transitions studied the results are largely consistent with (^{6}Li,d) studies from Rochester.[35,36] However, we would argue that the results are more stable with respect to uncertainties in the optical potentials than the transfer reaction, which, for these L=0 transitions, is mismatched by about 4 units of angular momentum. In contrast, the (p,pα) reaction is, of course, exactly matched and involves optical potentials somewhat less subject to uncertainty.

The absolute values obtained in the (p,pα) study are in agreement with shell model predictions for the 1p shell targets. In the sd shell the experimental values start at around 3 times shell model predictions, increasing to around 20 times for ^{40}Ca. Since the structure calculations in question[37] are limited to an sd basis

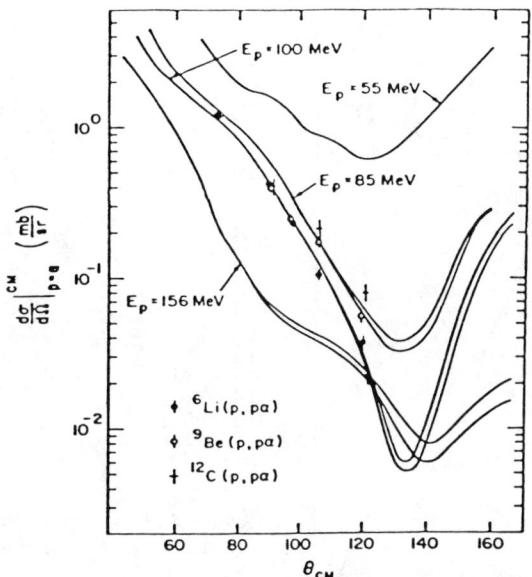

FIG. 10. Angular distribution for (p,pα) at 100 MeV. The curves represent free p+α scattering.

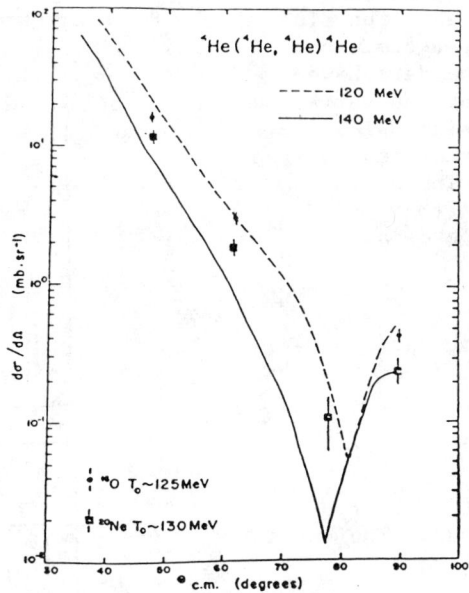

FIG. 11. Angular distribution for (α,2α) on ^{16}O and ^{20}Ne at 140 MeV. The curves represent free α+α scattering.

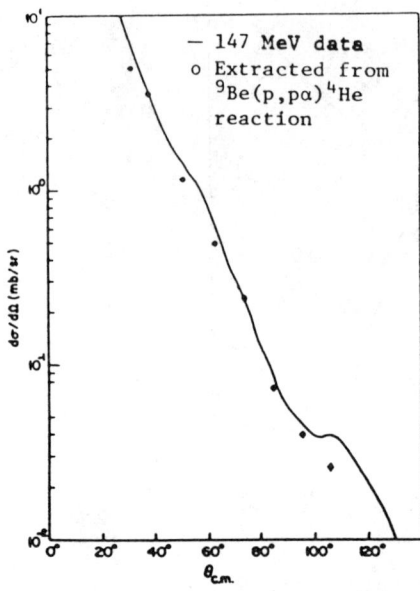

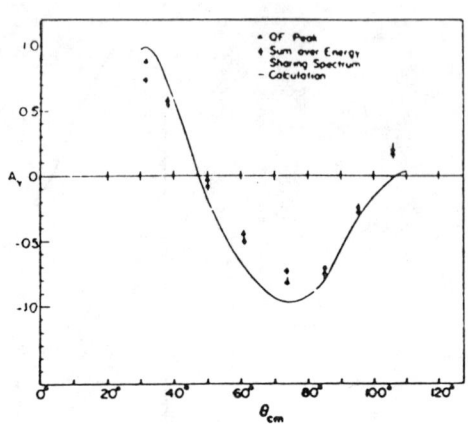

(b) $A(\theta)$. The curve is a fit to free p+α scattering.

(a) ^{9}Be(p,pα)^{5}He cross sections for the quasifree peak (zero recoil momentum) as a function of the two-body p-α c.m. scattering angle. The data have been corrected for the variation in distortions using DWIA calculations and normalized to free p-α scattering. The curve represents the 147 MeV p-α elastic scattering data.

FIG. 12. ^{9}Be(p,pα) at 150 MeV.

(c) Energy sharing distribution and analyzing power. The curves are DWIA calculations.

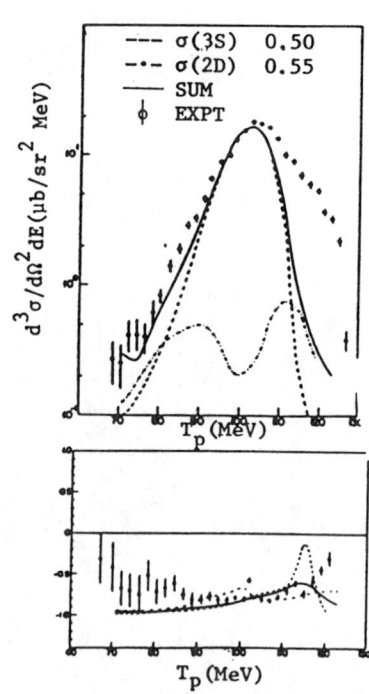

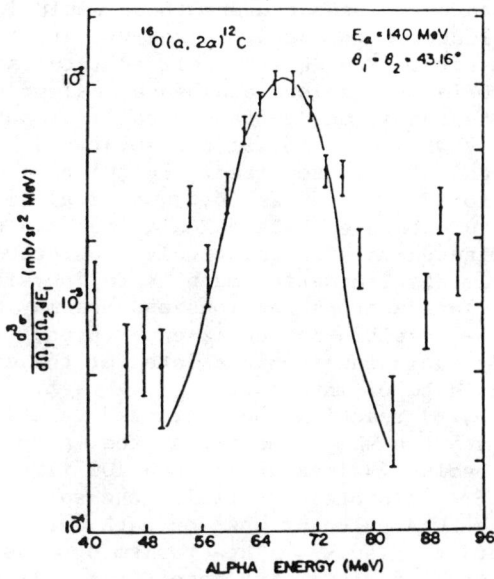

FIG. 13a. $^{16}O(\alpha,2\alpha)^{12}C$ (g.s.) at 140 MeV.

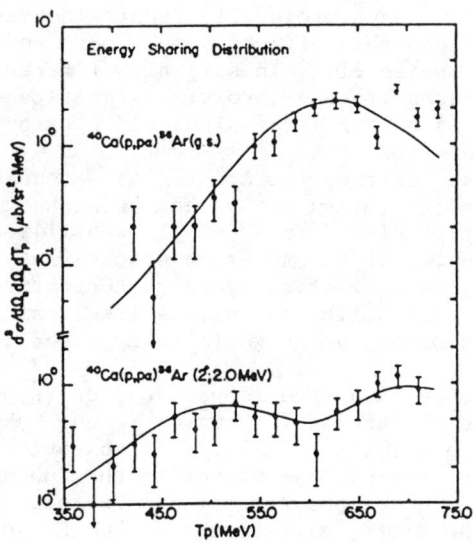

FIG. 13b. DWIA calculations for $^{40}Ca(p,p\alpha)$ at 100 MeV.

the problem at the upper end of the shell at least partially reflects omission of 2p-2h and 4p-4h components in the target wave functions. In an oscillator picture these would lead to 6S and 7S terms in the cluster form factor in addition to the 5S term arising from the $(sd)^4$ term. Even quite small admixtures could lead to much enhanced DWIA predictions and hence smaller spectroscopic factors owing to the strong surface localization of the added terms. Probably we should not be too surprised by the discrepacies in absolute magnitude I have cited. In the case of two nucleon transfer reactions[38,39] such as (p,t) or (d,α) discrepancies in absolute magnitude between DWBA calculations and experiment of 5 to 10 can be eliminated only in relatively sophisticated calculations. Despite this limitation much useful physics has been learned from relative cross sections and one may hope that similar progress will be possible in the case of (p,pα). Clearly, more extensive study including accurate data for the excited state transitions would be of interest.

For the (α,2α) reaction the extracted relative spectroscopic factors have much the same behavior as the (p,pα) results. However, the absolute values are around 100 times larger than the (p,pα) values.[31,32] Since essentially the same approximations are involved in the DWIA calculations for both reactions I find this discrepancy hard to dismiss. The problem does not seem to arise from a poor choice of distorting potentials. It can be remedied by a rather unphysical increase in the rms radiius of the α-cluster wave function. However in this case agreement with the (p,pα) data is lost. As shown in Fig. 15 the (α,2α) reaction is much more strongly surface localized than the (p,pα) case. Thus it seems reasonable to conclude either:

a) The incident projectile itself induces clustering in the surface region and this effect is much more severe for the more massive alpha projectile. In slightly different language this implies a two-step process involving significant excitation of the target to levels having large alpha width prior to the actual knockout process. Or:

b) In the extreme surface region (around 0.1% of central density) probed in (α,2α) the extent of alpha clustering is around 100 times that predicted in a "simple" shell model.

A calculation of the two-step process has yet to be carried out. However, in view of the good factorization of the (α,2α) reaction this possibility seems less likely since some broadening of the angular distribution would be expected in a multi-step process.

Clearly more extensive comparisons of (p,pα) and (α,2α) reactions would be of interest. However if one takes the present results at face value the (α,2α) spectroscopic factors imply that essentially all the nuclear matter in the low density region has "crystallized" into alpha particles. In this context it is worth noting that the $^{16}O(α,2α)$ reaction at 800 MeV does not exhibit the large discrepancy cited above.[40] Rather the measurement is essentially consistent with the (p,pα) spectroscopic factors. This behavior is consistent with the fact that, at the higher alpha

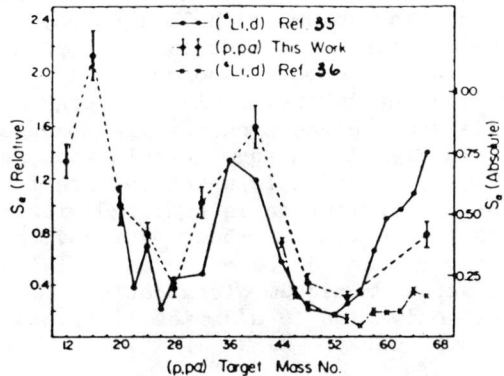

FIG. 14. Extracted spectroscopic factors for ground state transitions as a function of target mass. The lines merely guide the eye. The left scale indicates the relative value (normalized to unity at A=20) and the right scale the absolute value extracted in (p,pα).

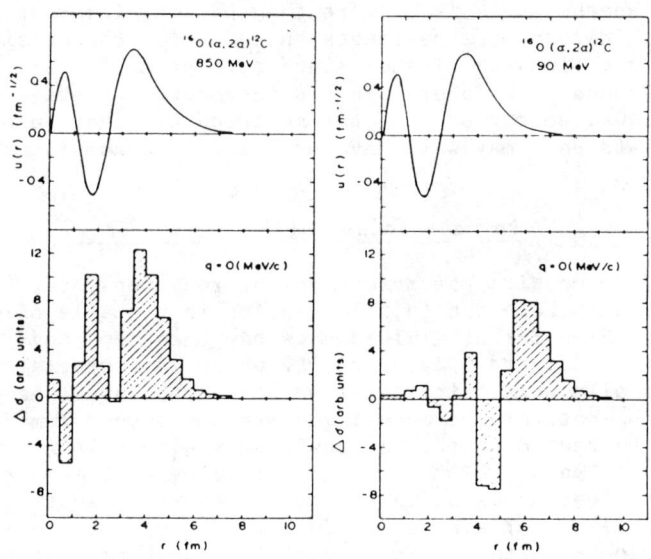

FIG. 15. Radial localization of the $^{16}O(\alpha,2\alpha)^{12}C$ reaction at 90 and 850 MeV.

energy, the (α,2α) reaction is no longer highly surface localized and samples the nuclear volume much like (p,pα) at around 200 MeV.

Despite the effects of distortion in (p,pα) reactions there is some sensitivity to the α-cluster momentum wave function. In Fig. 16 we see that, even at 100 MeV incident energy, relatively crude data[26,27] provide some guidance on the rms momentum width of the wave function. In Fig. 17 we show calculations for the ^{40}Ca(p,pα)^{36}Ar(g.s.) reaction in which the predicted experimental distorted momentum distribution is compared with the plane wave result. Despite a reduction of about 30 in overall magnitude, as at 800 MeV the experiment can be expected to follow roughly the envelope of the actual momentum distribution. This possibility provides further motivation to a series of (p,pα) measurements even at the present IUCF energy.

(p,d^{3}He) Quasifree Reactions

An interesting feature in the case of cluster knockout is that in addition to a quasifree elastic scattering of the projectile from the cluster a quasifree reaction is also possible. An example is the (p,d^{3}He) reaction which proceeds via a (p,d) reaction on an alpha cluster. This process has been observed[41] on 1p shell targets at 100 MeV and, for ^{9}Be, 200 MeV. Typical data is shown in Fig 18. As in the case of (p,pα) the energy sharing distributions reflect rather severely distorted momentum distributions and the extracted spectroscopic factors agree well with the (p,pα) values. Whether this is of more than passing interest is unclear. In counter experiments the data for the quasifree reaction can sometimes be cleaner since both particles are composite and backgrounds are lower. In a spectrometer experiment with roughly equal energy sharing the deuteron will have about beam rigidity and data may well have much lower random rates than (p,pα).

(p,pd), (d,dα), (^{3}He,^{3}Heα), Etc.

Other composite particle knockout reactions are, of course, possible. Of these the (p,pd) reaction is probably of greatest interest. Some rather limited data has been reported[18,42-44] for fairly light targets. Ignoring ^{6}Li which is a special case where things usually work fairly well owing to the low alpha-d binding energy, the data are of modest quality and some parameter tweaking seems to be needed to obtain consistency with DWIA calculations.

As pointed out[23,24] for (e,ed) this type of measurement is of interest since, at least in principle, there is sensitivity to nucleon-nucleon correlations. In this context, the measurement of interest would be to obtain a quasi-free angular distribution at fixed recoil momentum. This would reveal the extent to which the p-n correlation in the target differ from the correlation in a free deuteron.

While this type of measurement would certainly be of interest provided the data were of adequate quality the theoretical analysis

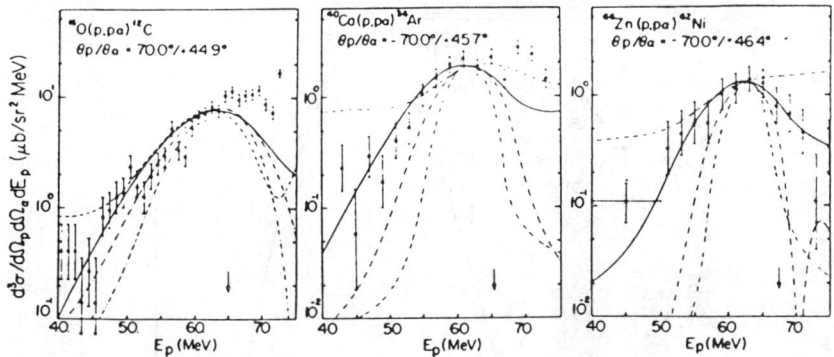

FIG. 16. Energy sharing distributions for (p,pα) ground state transitions. Curves are normalized DWIA calculations for different alpha particle bound state radius parameters r_0 (0.7, ----; 2.3, ———; 1.9, -·-·; 2.5, -··-).

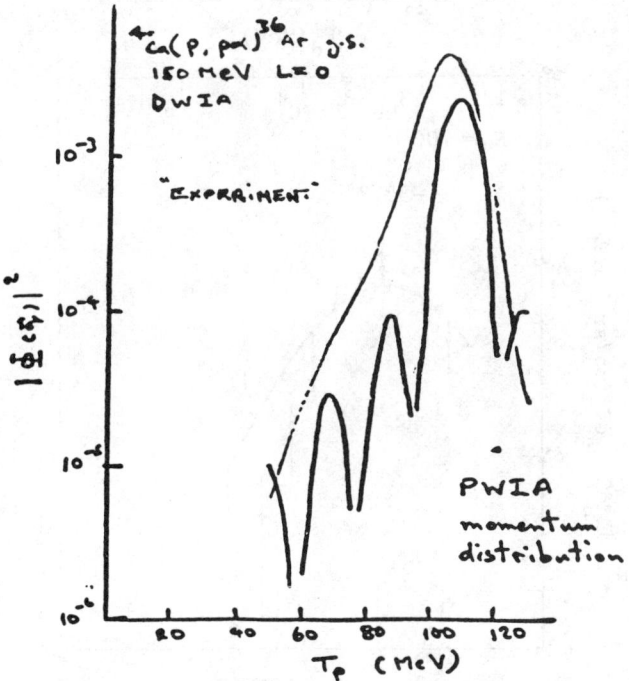

FIG. 17. Comparison of the assumed momentum distribution with the predicted experimental measurement for ^{40}Ca(p,pα)^{36}Ar (g.s.) at 150 MeV.

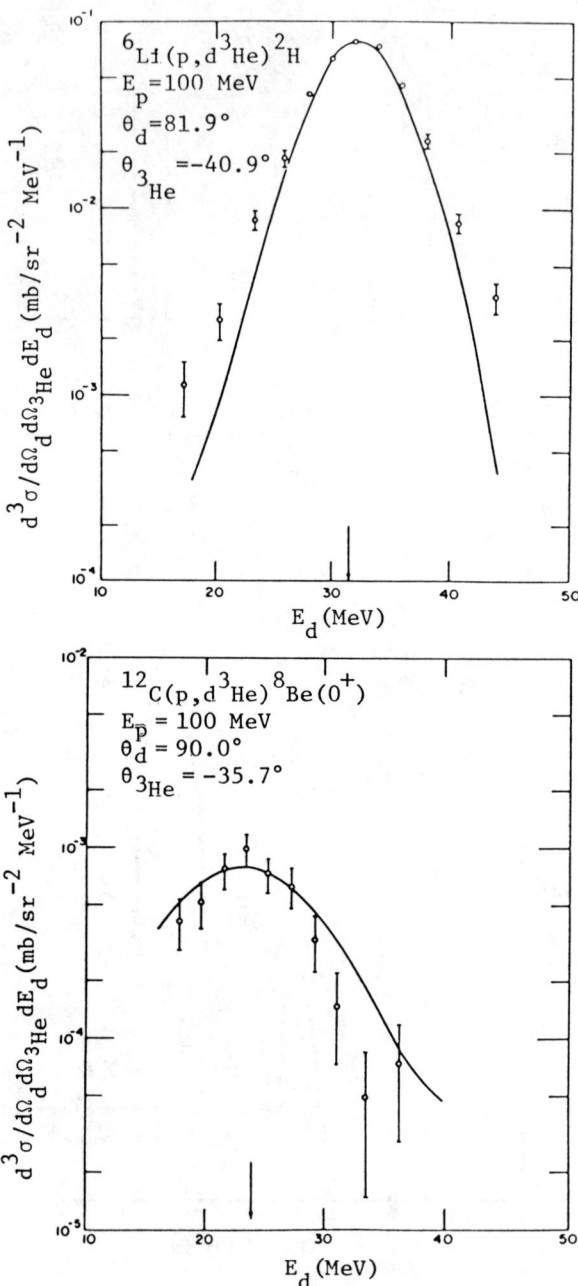

FIG. 18. The (p,d³He) reaction on ⁶Li and ¹²C at 100 MeV.

is fairly difficult. Owing to the large radius and low binding energy of the deuteron the simplifications which seem to work quite well for α knockout are unlikely to be reliable for (p,pd). Clearly a more microscopic treatment is needed including, as a minimum, some of the refinements found in current two-nucleon transfer reaction calculations.[38]

PION INDUCED PROCESSES

$(\pi,\pi p)$, $(\pi,\pi n)$, $(\pi^+,\pi^0 p)$

In addition to nucleon and nucleon cluster projectiles, quasi-free knockout can also be induced by pion beams. Results have been reported by several groups for $(\pi,\pi p)$ and $(\pi,\pi n)$ reactions[45-47] and there is also data for the quasifree charge exchange reaction $(\pi^+,\pi^0 p)$. Results for 199 MeV $^{12}C(\pi^+,\pi^+ p)^{11}B$ are shown in Fig. 19. The resolution was sufficient to separate the 1p shell transitions from the excitation region corresponding principally to 1s shell removal. Despite questions as to the applicability of a factorized DWIA in this case[48] the simple calculation[49,50] seems to describe the data rather well and the absolute spectroscopic factors obtained are in rather good agreement with the Cohen & Kurath wave functions.[51] The interesting feature here is whether some special feature pertaining to pion dynamics is revealed since clearly the extraction of spectroscopic factors or momentum densities can be better done using easier knockout reactions. With this in mind a measurement[46] was carried out at LAMPF in which the invariant mass of the emitted pion and proton was intentionally swept through the delta region. At present it appears that somewhat more extensive data of greater precision is really needed.[52]

Of similar interest is the 240 MeV SIN experiment[47] on ^{16}O comparing identical $(\pi^+,\pi^+ p)$ and $(\pi^-,\pi^- p)$ transitions. In the Δ resonance region the ratio of cross sections for these reactions should be around 9 arising from the intrinsic $\pi^\pm + p$ interaction. In fact, the ratio is as much as 5 times larger. This can be understood nicely in terms of destructive interference[53] between the relatively weak π^-+p amplitude and knockout of a nucleon by a Δ while its decay nucleon is recaptured in the nucleus. It is argued that this is a rather direct manifestation of a Δ-nucleus interaction and additional measurements for other targets would provide valuable confirmation of the proposed explanation.

$(\pi^+,2p)$

Measurements of pion absorption in nuclei are the subject of considerable interest at present. Much of the work is oriented toward understanding the global features of the reaction. One possible mechanism which can play a role is a simple quasifree reaction[54] in which the absorption takes place on a deuteron-like cluster in the nucleus. Some of the evidence in support of this interpretation was reviewed earlier by Roos.[4] In Fig. 20 results are shown[55,56] for the reaction $^{12}C(\pi^+,2p)^{10}B$. The data[57] are

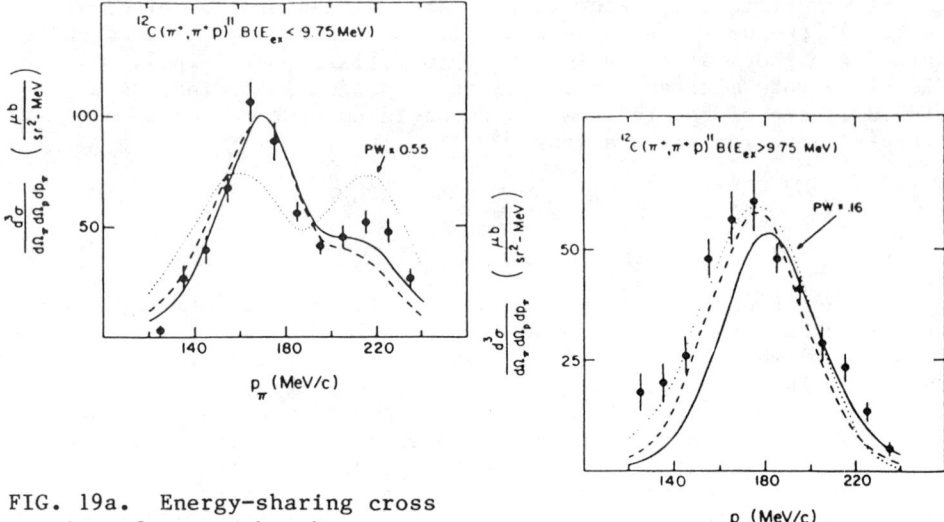

FIG. 19a. Energy-sharing cross sections for $p_{3/2}$ knockout at an incident energy of $T_\pi = 199$ MeV and angles of $\theta_\pi = -117.5°$ and $\theta_p = 30.0°$. The dotted curve is a PWIA calculation. The full (dashed) curves are DWIA calculations.

FIG. 19b. Energy-sharing cross sections for $1s_{1/2}$ knockout.

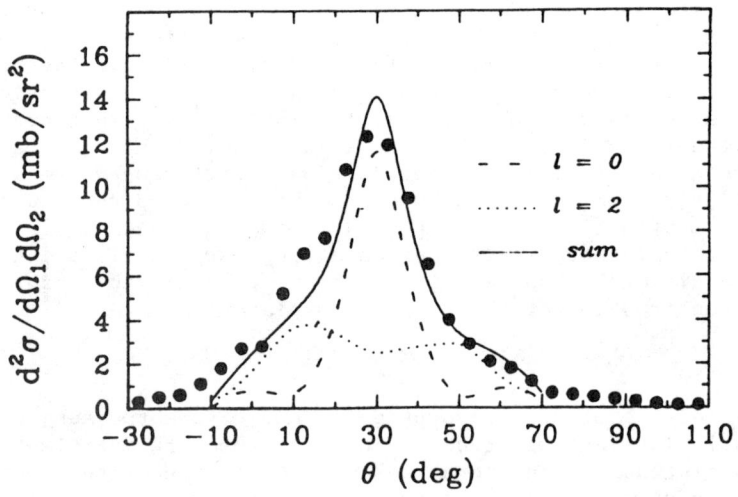

FIG. 20. Angular correlation for the $^{12}C(\pi^+,2p)$ reaction. The curves are DWIA quasideuteron calculations.

integrated over a fairly large range of energy so some guesses are needed in carrying out the DWIA calculation. However, there is pronounced peaking at the angle corresponding to $\pi+d \to 2p$ kinematics and we see that contributions from two strong low lying transitions, one L=0 and one L=2, are at least plausible candidates in explaining this data. It should be noted that more sophisticated (and much harder) calculations[58] suggest about 30% contributions from other configurations which we have not included in our simple cluster approach. Also there are arguments that the so-called L=2 component is simply a multiple scattering background. However, the approach looks encouraging.

It is worth remembering that in the pion absorption the nucleons are emitted almost back to back with 100 MeV or so of energy each. Consequently, high relative momenta could be involved and the process might be sensitive to details of the two-nucleon correlations. This is an old argument which is probably a little naive and, in any case, awaits data of considerably higher precision before it is worth exploring carefully.

SUMMARY AND CONCLUSIONS

While the simplifications afforded by an electromagnetic probe make (e,ep) the "classic" quasifree knockout reaction many interesting problems exist with hadronic probes which we have attempted to indicate. In addition to (p,2p) studies, cluster knockout reactions such as (p,pα) and (α,2α) pose interesting problems. Moreover reactions such as (e,eα) are only barely feasible at present. Data for quasifree knockout and quasifree reactions with pion beams are now being obtained and are providing new information concerning pion dynamics in nuclei.

ACKNOWLEDGMENTS

I would like to thank Professor P. G. Roos for his many invaluable contributions to this work. Thanks are due to Ms. Judy Myrick for assistance in preparing this manuscript. This work was supported in part by the National Science Foundation.

REFERENCES

1. N. S. Chant, in The Interaction Between Medium Energy Nucleons in Nuclei-1982, IUCF, AIP Conference Proceedings No. 97, edited by H. O. Meyer (AIP, New York, 1983), p. 205.
2. N. S. Chant, in Studying Nuclei with Medium Energy Protons, University of Alberta/TRIUMF Workshop, Edmonton, Alberta, Canada, edited by J. M. Greben (TRIUMF report TRI-83-3) p. 233 (note: pp 239 and 238 should be interchanged).
3. N. S. Chant, in Clustering Aspects of Nuclear Structure and Nuclear Reactions, Winnipeg-1978, AIP Conference Proceedings No. 47, edited by W. T. H. van Oers, J. P. Svenne, J. S. C. McKee, W. R. Falk (AIP, New York, 1978), p. 415.

4. P. G. Roos, in *Clustering Aspects of Nuclear Structure*, 4th International Conference on Clustering Aspects of Nuclear Structure and Nuclear Reactions, Chester-1984, edited by J. S. Lilley and M. A. Nagarajan (Reidel Publishing Company, Dordrecht, Holland, 1985), p. 279.
5. N. S. Chant, in *Momentum Wave Functions - 1982 (Adelaide, Australia)*, (AIP Conference Proceedings No. 86) edited by Erich Weigold (AIP, New York, 1982), p. 19.
6. N. S. Chant and P. G. Roos, Phys. Rev. C $\underline{27}$, 1060 (1983).
7. P. G. Roos et al., Phys. Rev. Lett. $\underline{40}$, 1439 (1978).
8. C. Samanta, Ph.D. thesis, University of Maryland, 1981.
9. Y. Kudo and K. Miyazaki, contributed paper, 6th International Symposium on Polarization Phenomena in Nuclear Physics, Osaka (Japan), 1985.
10. P. Kitching et al., Nucl. Phys. $\underline{A340}$, 423 (1980).
11. Th. A. Maris et al., in *Nuclear and Particle Physics at Intermediate Energies*, edited by J. B. Warren (Plenum Press, New York, 1976), p. 425.
12. C. A. Miller, Invited paper presented at the 9th International Conference on the Few Body Problem, Eugene, 1980; C. A. Miller, Common Problems in *Low-and Medium-Energy Nuclear Physics*, edited by B. Eastel, B. Goulard, and F. C. Khana (Plenum Press, New York, 1979), p. 513.
13. N. S. Chant et al., Phys. Rev. Lett. $\underline{43}$, 495 (1979).
14. P. Kitching et al., Preprint TRI-PP-84-34, May 1984 and Advances in Nuclear Physics, to be published.
15. M. B. Epstein et al., Phys. Rev. Lett. $\underline{44}$, 20 (1980).
16. W. T. H. van Oers et al., Phys. Rev. C $\underline{25}$, 390 (1982).
17. J. G. Zabolitzky and W. Ey, Phys. Lett. $\underline{76B}$, 527 (1978).
18. L. Rees, Ph. D. thesis, University of Maryland, 1983.
19. J. Watson et al., Phys. Rev. C $\underline{26}$, 961 (1982).
20. M. Ahmad et al., Nucl. Phys. $\underline{A424}$, 92 (1984).
21. G. Ciangaru et al., Phys. Rev. C $\underline{27}$, 1360 (1983).
22. A. Nadasen et al., Phys. Rev. C $\underline{19}$, 2099 (1979).
23. S. Frullani and Jean Mougey, *Advances in Nuclear Physics, 14* edited by J. W. Negele and Erich Vogt (Plenum Press, New York, 1984), p. 1.
24. J. Y. Mougey, Paper at CEBAF Workshop, 1985.
25. P. G. Roos et al., Phys. Rev. C $\underline{15}$, 69 (1977).
26. T. A. Carey et al., Phys. Rev. C $\underline{23}$, 576 (1981).
27. T. A. Carey et al., Phys. Rev. C $\underline{29}$, 1273 (1984).
28. A. Nadasen et al., Phys. Rev. C $\underline{22}$, 1394 (1980).
29. A. Nadasen et al., Phys. Rev. C $\underline{23}$, 2353 (1981).
30. C. W. Wang et al., Phys. Rev. C $\underline{31}$, 1662 (1985).
31. N. S. Chant et al., Phys. Rev. C $\underline{17}$, 8 (1978).
32. C. W. Wang et al., Phys. Rev. C $\underline{21}$, 1705 (1980).
33. J. D. Sherman et al., Phys. Rev. C $\underline{13}$, 20 (1976).
34. N. S. Chant and P. G. Roos, Phys. Rev. C $\underline{15}$, 57 (1977).
35. N. Anantaraman et al., Phys. Rev. Lett. $\underline{35}$, 1131 (1975).
36. H. Fulbright et al., Nucl. Phys. $\underline{A284}$, 329 (1977).
37. W. Chung et al., Phys. Lett. $\underline{79B}$, 381 (1978).
38. N. S. Chant, Nucl. Phys. $\underline{A211}$, 269 (1973).

39. N. Frascaria et al., Phys. Rev. C 16, 603 (1977).
40. N. Chirapatpimol et al., Nucl. Phys. A264, 379 (1976).
41. A. A. Cowley et al., Phys. Rev. C 15, 1650 (1977).
42. C. Samanta et al., Phys. Rev. C 26, 1379 (1982).
43. R. Warner et al., Nucl. Phys. A443, 64 (1985).
44. J. Y. Grossiord et al., Phys. Rev. C 15, 843 (1977).
45. H. J. Ziock et al., Phys. Rev. C 24, 2674 (1981).
46. H. J. Ziock et al., Phys. Rev. C 30, 650 (1984).
47. G. S. Kyle et al., Phys. Rev. Lett. 52, 974 (1984).
48. E. Levin and J. M. Eisenberg, Nucl. Phys. A355, 277 (1981).
49. L. Rees, N. S. Chant, and P. G. Roos, Phys. Rev. C 26, 1580 (1982).
50. N. S. Chant, L. Rees, and P. G. Roos, Phys. Rev. Lett. 48, 1784 (1982).
51. S. Cohen and D. Kurath, Nucl. Phys. A101, 1 (1967).
52. J. Cohen and J. V. Noble, Phys. Lett. 150B, 45 (1985).
53. M. Hirata, F. Lenz, and M. Thies, Phys. Rev. C 28, 785 (1983).
54. P. G. Roos, L. Rees, and N. S. Chant, Phys. Rev. C 24, 2647 (1981).
55. B. G. Ritchie, N. S. Chant, and P. G. Roos, Phys. Rev. C 30, 969 (1984).
56. B. G. Ritchie, N. S. Chant, and P. G. Roos, Phys. Rev. C 32, 334 (1985).
57. A. Altman et al., Phys. Rev. Lett. 50, 1187 (1983).
58. K. Ohta, M. Thies, and T.-S. H. Lee, Ann. Phys. (NY), to be published.

HIGH-SPIN INNER AND OUTER SUBSHELLS VIA TRANSFER REACTIONS
Sydney Galès
Institut de Physique Nucléaire, B.P. n°1, 91406 Orsay Cedex, France

Abstract : One nucleon transfer reactions, investigated at high incident energy, are particularly well suited to the study of high-spin inner and outer subshells. Single-particle modes at high excitation energies are observed in a wide variety of nuclear reactions. These "giant single-particle (hole) resonances" are the response of a nucleus to the external field acting on its single-particle degrees of freedom.

The advantages and the limitations of the transfer reaction approach will be discussed using the results from neutron and proton pick-up and stripping reactions. The strong selectivity in angular momentum transfer, the exploration of a large excitation energy range, the use of polarized beams and the new decay studies help us to establish the main features of the single-particle (hole) strength functions (identification, centroïd energy, spreading mechanisms). Stricking similarities are observed between single-hole and single-particle response functions such as the dependance of the strength distribution with the mass number, or its behaviour accross the transition regions. The problem raised by the subtraction of the underlying background, the method used to extract the strength distributions, the assumptions made to describe the reaction process(DWBA, form factors, unbound nature of the state) as well as the case of overlapping subshells will be examined.

I - INTRODUCTION

Nuclear structure aims at the understanding of the damping of simple modes of excitation such as the single-particle states and the collective vibrations. To day our knowledge of the single particle motion comes from the early experiments on nucleon scattering and knock-out reactions as well as from spectroscopic studies with transfer reactions. For the states located near the Fermi surface (valence subshells) the particle (hole) fragmentation can be studied via one nucleon pick-up or stripping processes such as (d,p), (p,d), (d,^{3}He), (^{3}He,d) or (^{3}He,α). Up to a few MeV excitation energy (0-3 MeV), the comparison of the results from these various processes as well as high energy resolution in the exit channel yield rather precise informations on the fragmentation of the single-particle strengths for both low and high-spin valence subshells. Well above the Fermi surface, the particle states are unbound and they can be studied directly by the direct scattering of nucleons.

Until recently, the information about the deeply-bound states was rather scare. The results from (p,2p)[1] and later on from (e,e'p)[2] knock-out reactions have established the usefulness of the shell model picture up to mass number A = 40.

During the last decade, significant progresses have been made in the study of the single-particle modes, located at intermediate excitation energies (4 - 20 MeV) well below (inner-hole) or above (outer subshells) the Fermi sea. With the help of a wide variety of transfer reactions, the experimental results span a large domain of the nuclear mass chart. In the mean time, the increasing amount of empirical systematics has induced the development of theoretical nuclear models and the key role played by the coupling between single-particle motion and surface vibrations has been pointed out.

In the residual energy spectrum from a transfer reaction, these high-lying modes appear as a "giant resonance like structure" superimposed on a substantial background of more complicated states. From the raw data to the experimental single-particle strength function a number of steps have to be taken, each of them being valid under certain assumptions.

In this paper I would like to present our present knowledge of the single-particle strength function from transfer reactions, the advantages and limitations of such approach, the assumption made in the data analysis and the remaining open questions. All along the discussion, the empirical systematics will be compared to the predictions of the nuclear models.

II - HIGH-LYING SINGLE-PARTICLE STATES VIA TRANSFER REACTIONS

A - THE INDEPENDANT PARTICLE MODEL, THE QUASIPARTICLE APPROXIMATION AND THE DAMPING MECHANISM

As mentionned in the introduction, high-lying single-particle (hole) states do not appear as simple as would be bare particle (hole) in a Hartree-Fock nucleus. Within the independent particle model (IPM), the spectral function is represented by a delta function, as schematically sketched in fig. 1a. In a real nucleus this result is no more valid since one has to take into account the residual interaction between nucleons. If the IPM is a good approximation, the spectral function has a peak around the single-particle energy but has a finite value at each energy (fig. 1b). The quasiparticle approximation consists of replacing the spectral function of fig.1b by a lorenztian probability distribution centered around a quasiparticle energy E_{qp} with a width at half maximum Γ. It is rather important to point out that both quantities, E_{qp} and Γ have a physical significance only if the spectral function has a pronounced peak. The tails of the distribution (or the underlying background of fig.1b) are not related to the quasiparticle concept[3]. The simplest quasiparticle approximation, namely a smooth strength function with symmetric distribution around the quasiparticle energy (dashed line fig. 2b) could be improved by introducing a smooth energy dependance of the width $\Gamma(E)$ and of the energy E_{qp}. The resulting line has an asymmetric shape as shown in fig. 1c.

For closed shell nuclei and for the valence states located near the Fermi surface the phenomenological IPM (Woods-Saxon + spin orbit potential) appears to be a very good approximation. For example the experimental single-particle energies for both weakly bound neutron and proton states in ^{208}Pb are well reproduced by static wells (Woods-Saxon for example) with appropriate parameters. Whereas this effect would suggest an effective mass m* of about unity, there are several effects which indicate that the problems are not so simple. Both the velocity dependance and the energy dependance of effective interactions give large deviations of m*/m from unity[4]. The effect of the energy dependance drops off rapidly as one goes away from the Fermi surface[5], so that m*/m has a value of about 0.7 for either deep-hole or high-lying states. Calculations with Skyrme-type interaction seem to describe nucleon vibrations rather well, yielding effective masses m*/m ≅ 0.6 to 0.8. Such interaction however, gives substantially too large single-particle spacing in the region of the Fermi surface[6].

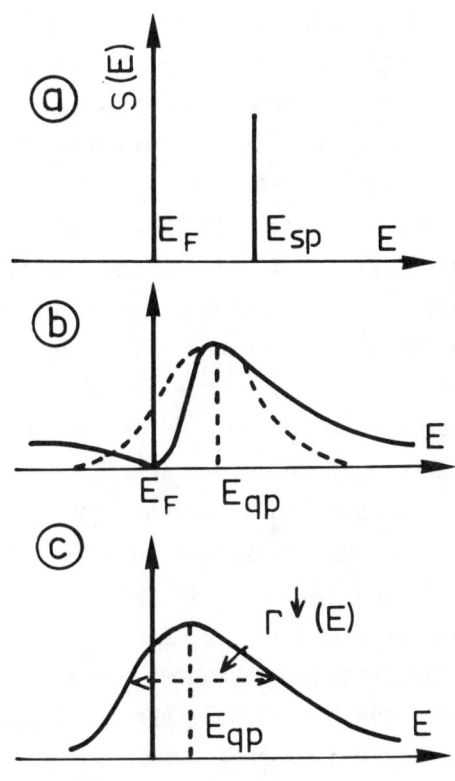

Fig.1 a) Spectral function in the case of the IPM.
b) modifications due to the residual interaction (full curve) and the result of the quasiparticle approximation (dashed curve).
c) Spectral function for a quasiparticle state with an energy dependance of the width and of the quasiparticle energy Eqp.

A dynamical theory has been proposed[7] which suggests that the energy dependance of the effective interaction produces reduction in m*/m as one goes away from the Fermi surface, namely, that the compression of single-particle levels due to the coupling to surface vibrations dies off rather rapidly with energy increasing or decreasing away from the Fermi energy. Another property, affected by the coupling to collective vibrations is the distribution of spectroscopic strength of the single particle states. In the HF approximation, all spectroscopic factors are equal to unity. The coupling of single-particle states to vibrations implies that the single-particle strength will be spread, giving rise to spectroscopic factors less than unity[6,8]. The understanding of the damping mechanism of deep-hole (high-lying) states in nuclei has been outlined in several theoretical review papers[9-11]. Such knowledge will yield information on the single-hole (particle) response function in the nuclear medium. The spreading is believed to occur in two stages as shown in fig. 2. In the first stage of fragmentation, the bare single-particle state is spread over several doorway states. As a second step the doorway states themselves have their strengths damped over the other many degrees of freedom. The amount of statistical damping is strongly dependent on the level density of the adjacent complicated states

i) if the underlying level density is low, then sharp peaks may be observed, characterizing the first step stated above (see. fig. 2).

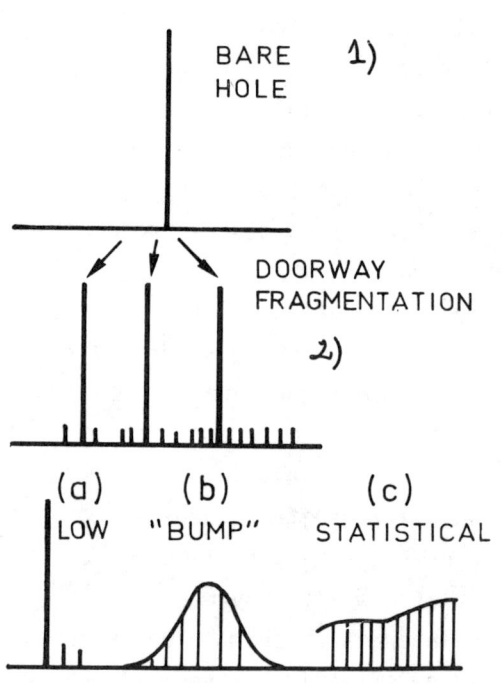

Fig.2 - 1) simple particle state, 2) Doorway fragmentation, 3) Resulting strength distributions

ii) If the underlying level density is moderate, then the single-particle state appears as "a giant" structure.
iii) If the level density is high and the coupling with the grass states is strong, the single-particle strength spreads over a large energy range and may be observed as an almost structureless continuum spectrum (fig. 2).

This mechanism is responsible for the observation of overlapping structures corresponding to the closely spaced inner (outer) sub-shells of a medium-heavy nucleus.

B - THE TRANSFER REACTION APPROACH - ADVANTAGES AND LIMITATIONS

The main features of a spectrum from the transfer reaction A(a,b)B are schematically illustrated in fig. 3. The observation of the "giant" structures depends critically on the selectivity in angular momentum ℓ of the reaction. The extraction of valuable information on a particular single-particle strength distribution is strongly related to a good description of the pick-up process.

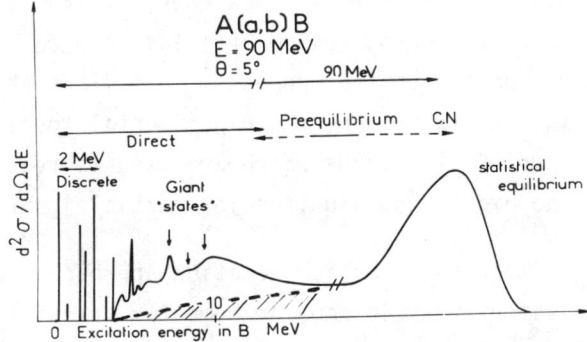

Fig. 3 - Schematic representation of the excitation energy region of a typical spectrum from the reaction A(a,b)B.

One of the first example of deep-hole states using the transfer reaction approach, namely the neutron pick-up reaction, is the work reported by Sakaï and Kubo on the $^{118}Sn(^{3}He,\alpha)^{117}Sn$ reaction [12]. The residual energy spectra is shown in fig. 4 and a weak enhancement of the cross section is

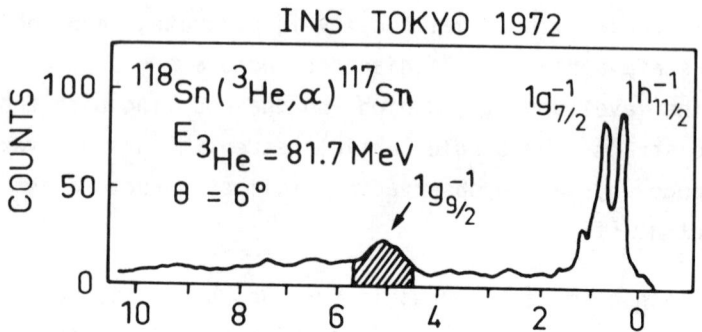

Fig. 4 - Residual energy spectra from the $^{118}Sn(^{3}He,\alpha)^{117}Sn$ reaction at 81.7 MeV. An enhancement of cross-section is clearly seen around 5.2 MeV in ^{117}Sn.

observed around 5.2 MeV in ^{117}Sn. According to the authors, this excitation was attributed to the pick-up of a neutron from the next lower shell, e.g $1g_{9/2}$, 2p subshells. Very rapidly a growing amount of experimental data has been accumulated and high-lying hole strengths have been observed in (p,d), (d,t), (d,^{3}He) and (^{3}He,α) reactions at various energies from mass number A = 90 to A = 208. In recent review articles, a detailed information has been reported on the subject [13-16]. Through selected examples, I would like to present the experimental tools, the reaction model and the empirical methods which are used to reach a detailed knowledge on the high-lying single hole (particle) strength functions.

Within the transfer reaction approach, various factors such as the selectivity in angular momentum transfer ℓ, the characteristic patterns of the angular distributions (ℓ dependance of the cross-sections), the energy resolution in the exit channel, the explored energy range, the use of polarized incident beams, the nature and the magnitude of the "true" underlying background and the correct description of the reaction process play key roles in the identification, localisation and determination of the amount of strength for a given subshell.

Let me emphasize that all the best conditions cannot be achieved in a single experiment but the comparison between different reactions carried out at different incident energies may give the most precise answer to the addressed questions.

For example the strong selectivity for large angular momentum transfer is clearly observed in the energy spectrum from the ^{116}Sn(^{3}He,α) reaction displayed in fig.5. The large mismatching suppress the small ℓ contributions ($\ell = 3$ but mainly $\ell = 1$). The well known concentration of $\ell = 4$, $1g_{9/2}$ hole strength[12-17] at 5.3 MeV dominates the spectrum. The width of the peak is less than 1 MeV and, the total amount of strength is of about 20-30% of the sum rule limit.

Another remarkable feature of the displayed spectrum is the continuous decrease of the cross-section beyond 20 MeV at forward angles, which was not observed with lower energy projectiles. This is one of the first results where an excitation energy range of 50 MeV has been explored. Therefore, the total neutron-hole strength of all orbits down to the bottom of the well, may be involved.

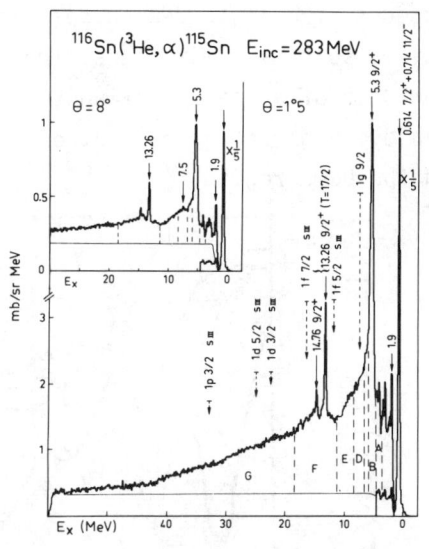

Fig. 5 - Excitation energy spectrum from the reaction ^{116}Sn(^{3}He,α) ^{115}Sn at 283 MeV (taken from Ref.17). The full line represents the background assumption. Dashed arrows indicate the positions of some subshells as given by Hartree-Fock calculations.

Around 50 MeV excitation energy, the pick-up contribution is most probably very small and the measured cross-section of the residual "non pick-up" background is then extrapolated towards lower excitation energies. Under this assumption it is found that the missing $1g_{9/2}$ strength is spread over a wide energy interval (6.6 to 11.6 MeV). In addition two new gross structure "peaks" are identified at 7.5 and

15 MeV respectively. The results of the analysis show that a large fraction of $1f_{5/2}$ strength is located between 8.6 and 11.6 MeV in agreement with the previous result of Siemssen et al [18]. The main contribution to the higher energy bump (11.5-18.5 MeV) is thus attributed to the $1f_{7/2}$ deep-hole state. High-energy transfer reactions have also some limitations. The angular distributions are not characteristic of a unique ℓ transfer and it is also very difficult to achieve high energy resolution to look for possible "fine structure" in the bump region.

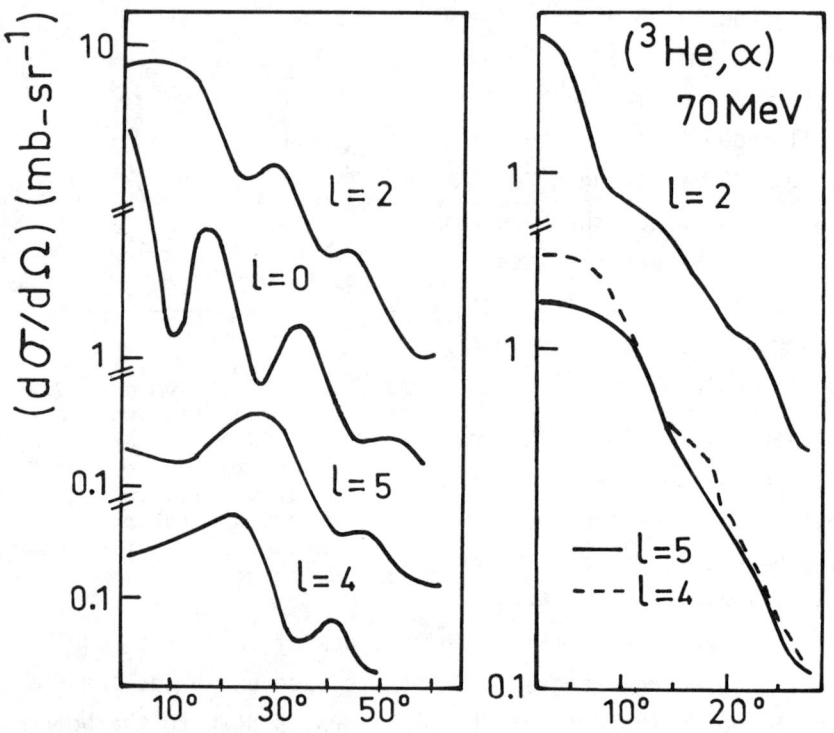

Fig. 6 - *Typical angular distribution patterns for various ℓ transfer from high and low incident energy transfer reactions.*

In fig. 6 are shown the typical features of the ℓ-dependance of the cross-sections for transfer reactions. Low energy proton induced transfer reactions lead to very characteristic pattern for the angular distributions, very high energy (^{3}He,α) or (α,^{3}He) reactions give rise to large cross-sections for large angular momentum transfer ($\ell=5,6$), but from the shape it is very difficult to distinguish between ℓ and $\ell \pm 1$ transfer. The importance of high energy resolution experiment is demonstrated in fig. 7.

Above the well resolved low-lying states, both the ^{116}Sn(d,t) ^{115}Sn and ^{116}Sn(^{3}He,α) spectra exhibit rather strong enhancements of group of states located in ^{115}Sn around 5 MeV excitation energy. Moreover some "fine structure" peaks could be observed in that region and the strength distributions of the $1g_{9/2}$ and 2p subshells are found to overlap in the explored energy range. These measurements have also clearly evidenced the change of the fragmentation of the $1g_{9/2}$ hole-strength as the mass number varies for the Sn isotopic chain [19].

The main limitations of the transfer reaction approache come from the model dependance of the extracted strength distributions and from the uncertainties introduced in the analysis by the shape and the magnitude of the subtracted continuum cross-sections which are not yet calculable.

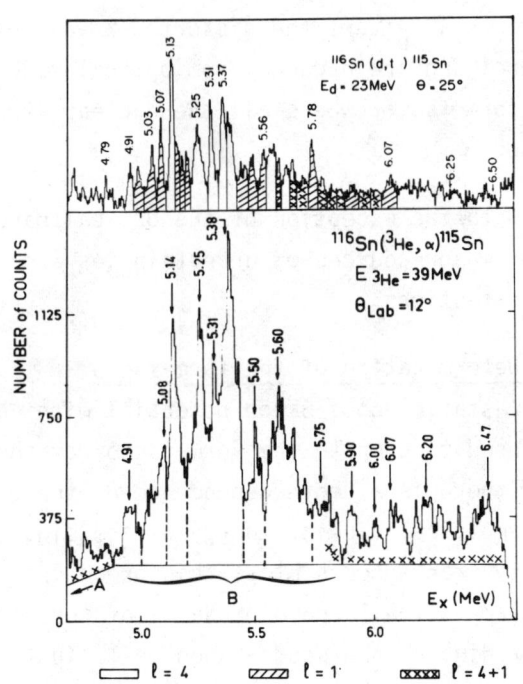

Fig. 7 - Excitation energy spectra from the reactions ^{116}Sn(^{3}He,α)^{115}Sn [19] at 39 MeV and the reaction ^{116}Sn(d,t)^{115}Sn [20] at 23 MeV. The solid line in the lower part of the figure represent the background assumption.

For a single-nucleon transfer reaction $A + a \rightarrow b + B$, the reaction model is based on the following assumptions.

Direct one-step process

In the framework of the Distorted Wave Born Approximation (DWBA) the transition amplitude can be written as [21]

$$T_{RA} = \int d\vec{r}_{aA} \, d\vec{r}_{bB} \, \chi_b^{(-)*}(\vec{k}_b, \vec{r}_{bB}) \, F(r_{aA}, r_{bB}) \, \chi_a^{(+)}(\vec{k}_a, \vec{r}_{aA}) \tag{1}$$

where χ_a, χ_b are the distorted waves for the entrance and exit channel and are deduced from optical model parameters and F is the form factor where all the nuclear structure information is contained.

With the exception of the optical parameter ambiguities, two other major sources of uncertainties are present in the calculated cross-sections.

Determination of the form factor

A static Woods-Saxon potential with ad hoc geometry parameters is used to compute the form factor of the transfered nucleon. Moreover there is a depth dependence of the potential in order to fit the empirical binding energy of the single-particle state. For the highly fragmented states, this procedure may be inadequate. Another problem is the large dependance of the computed cross-sections with the radius of the Woods-Saxon well. In the case of deeply-bound states, the tail of the form factor in the surface region where the reaction occurs is very small, whereas for quasibound and unbound states, the resonance method, introduced by Vincent and Fortune [22] is used to obtain a correct description of the tail of the form factor.

Exact Finite Range and Zero-Range Approximation

If we are able to compute the 6 dimensions integral mentionned above T_{BA} (see formula 1) then the experimental transfer cross-section $(d\sigma/d\Omega)_{exp.}$ is directly proportional to the spectroscopic strength C^2S of a nuclear single-particle (hole)state through the relation

$$\left(\frac{d\sigma}{d\Omega}\right)_{exp} = C^2 S \left(\frac{d\sigma}{d\Omega}\right)_{DW}^{EFR} \quad (2)$$

where

$$\left(\frac{d\sigma}{d\Omega}\right)_{DW}^{EFR} = K \, |T_{BA}|^2 \quad (3)$$

$\left(\frac{d\sigma}{d\Omega}\right)_{DW}^{EFR}$ is the computed Exact Finite Range (EFR) distorted wave cross-section and K a constant factor which contains the kinematics and spin statistics for a given transition T_{BA}.

However, most the calculations are made using the Zero Range approximation $\left(\frac{d\sigma}{d\Omega}\right)_{DW}^{ZR}$ where one assumes that the interaction between the ejectile b and the transfered nucleon x can be approximated by a delta function with a strength D_0

$$V_{bx} \cong D_0 \, \delta(r_{bx}) \, \delta_{\ell,0} \quad (4)$$

The zero range approximation leads to the following relation between the computed cross-section and the experimental one

$$\left(\frac{d\sigma}{d\Omega}\right)_{exp} = N \, C^2 S \left(\frac{d\sigma}{d\Omega}\right)_{DW}^{ZR} \quad (4)$$

where N is the normalization constant and is proportional to D_0^2. For transfer reactions induced by complex projectiles (^{3}He, α etc...) the values of this normalization constant are not well established and if EFR calculations are not carried out this approximation may be a source of large uncertainties.

In conclusion to this description of the model dependence of the deduced strength distributions, it may be stated that for rather pure single-particle states located at low excitation energy, the uncertainty is of the order of ± 20% whereas for highly excited states distributed over a large energy range, the errors may be of the order of ± 40%.

The analysis procedure of the data is schematically presented in fig.8. After the substraction of an empirical background (hatched area in fig. 8), the enhancements of cross-sections observed at

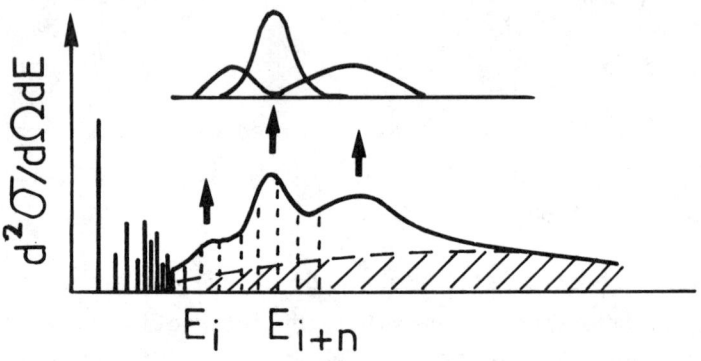

Fig. 8 - Schematic representation of the analysis procedure of the transfer reaction cross-section data for high-lying states (see text).

high excitation energy are fitted by a number of gaussian peaks having different widths and/or the whole range of excitation energy is divided into adjacent energy bins (0.2 to 1 MeV large). The measured angular dependance of the differential cross-sections $d\sigma/d\Omega dE$ compared to DWBA calculations yields to a ℓ identification and a J assumption is made on the basis of shell model and sum-rule arguments. Then for each fragment i with quantum number $n\ell J$ a spectroscopic factor $C^2 S_i$ is extracted. The first and second moment of the distribution are identified with the centroid energy $\tilde{E}_{n\ell j}$ and the damping with $\Gamma^{\downarrow} = 2.35\sigma$, respectively, whereas the sum of the spectroscopic factors $\Sigma\, C^2 S_i$ tell us to what extend the sum-rule is exhausted. The different steps discussed above can be summarized as follows

$$\frac{d^2\sigma}{d\Omega dE}\bigg|_{exp} \xrightarrow{\text{REACTION MODEL}} ;\ \text{J assumption} \rightarrow \frac{d^2\sigma}{d\Omega dE}\bigg|_{DW} \rightarrow S^{\ell J}(E)$$

In fig. 9 is displayed a typical result which summarizes our present knowledge on the $1g_{9/2}$ neutron inner-hole strength in the Cd, Sn, Te isotopes and in the ^{96}Zr and ^{144}Sm nuclei. The $g_{9/2}$ orbit has been studied with different reactions in many nuclei with neutron number ranging from 50 to 82 and more than 50% of the sum-rule is concentrated in a rather narrow energy range (1 to 4 MeV). The plotted quantities E_x and Γ represent the behaviour of the mean energy and of the spreading width of the $1g_{9/2}$ hole strength when the neutron number varies from N = 50 to N = 82. A smooth increase with mass number (from 3.1 MeV in ^{105}Cd to 7.5 MeV in ^{143}Sm) is observed for the centroid energy of the $1g_{9/2}$ hole state. This result indicates the usefulness of the quasiparticle concept even for such broad and fragmented structures. Turning now to the spreading width, a transition from line fragmentation to line broadening is observed for the $1g_{9/2}$ hole state in the Cd and Sn isotopes. On the contrary in the Te and ^{144}Sm nuclei only a broad peak ($\Gamma \geq 2$ MeV) is excited. An increasing spreading with neutron number N characterizes the trend displayed in fig. 9.

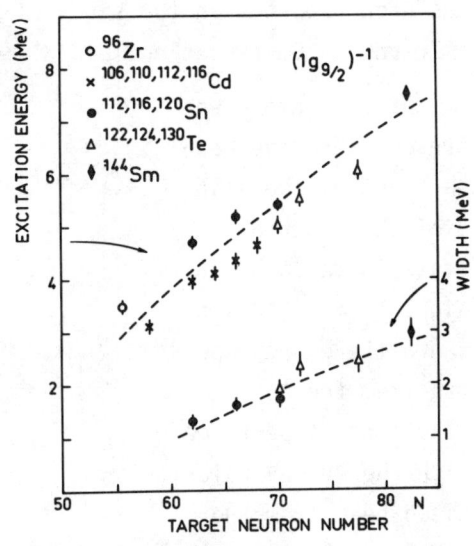

Fig. 9 - Excitation energies and widths of the $1g_{9/2}$ hole state observed in (p,d) and (^{3}He,α) reactions on ^{96}Zr, Cd, Sn, Te and ^{144}Sm nuclei versus neutron number.

C - TRANSFER REACTION TO INNER-HOLE STATES USING POLARIZED BEAMS

In single-nucleon transfer reactions, the angular distributions allow to determine the transfered orbital angular momentum, but not the total spin J, which has been infered using shell-model arguments. However, DWBA calculations and ($\vec{p}$,d), ($\vec{d}$,t) measurements to low-lying states have shown some years ago that analysing power Ay(θ) could exhibit strong J-dependence.

This property has been recently tested and proven to be very useful for spin assignments to fragmented inner-hole states.

The first spin determination of deeply-bound proton hole states has been reported in the study of the ($\vec{d}$,³He) reaction at 52 MeV on ⁹⁰Zr and ¹⁴⁴Sm target nuclei[23,24].

As an example we present here the measurement of the spin J of the inner-hole states in the tin region. In fig. 10 are shown the energy spectra from the ¹¹²Sn($\vec{d}$,t) reaction [25] performed with the 40 MeV polarized beam from the ISN Grenoble isochronous cyclotron. The good energy resolution allows us to study into detail the fine structure "peaks" observed in the energy range 3.6-6.0 MeV where are located the $1g_{9/2}$ and 2p inner-hole strengths.

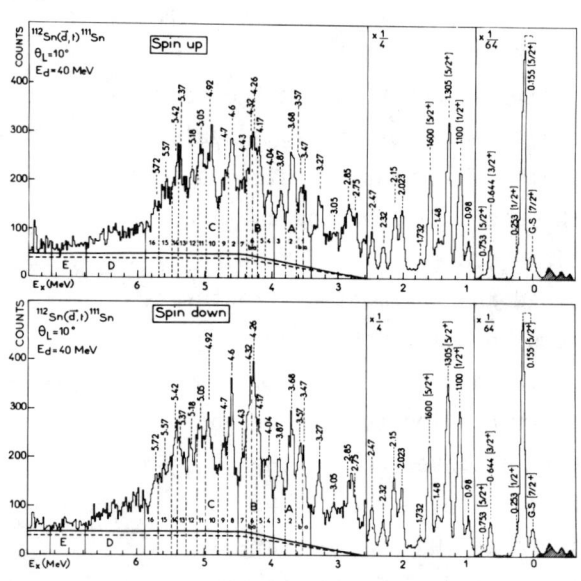

Fig. 10 - Triton energy spectra from the ¹¹²Sn($\vec{d}$,t)¹¹¹Sn reaction at θ = 10° lab angle and for the two spin directions. The solid and dashed horizontal lines show the assumed backgrounds. Above 3.5 MeV the spectra were divided into adjacent energy bins.

The angular distribution and the analyzing power for the state at 4.17 MeV are displayed in fig. 11.

The angular distribution could be only reproduced by a mixing of ℓ values (ℓ = 4+1) having different weights. This mixing being fixed, the analyzing power data is also well reproduced assuming the following spins, J = 9/2 + 1/2 without any further normalization. The detailed analysis of σ(θ) and Ay(θ) for each substructure leads to a very precise determination of the $1g_{9/2}$, $2p_{1/2}$ and $2p_{3/2}$ hole-strength up to 7 MeV in ^{111}Sn as shown in fig. 12. A large spreading of the $2p_{1/2}$ strength is observed and some $2p_{3/2}$ components are excited around 4.5 to 5.5 MeV (see fig. 12). A large spreading of the

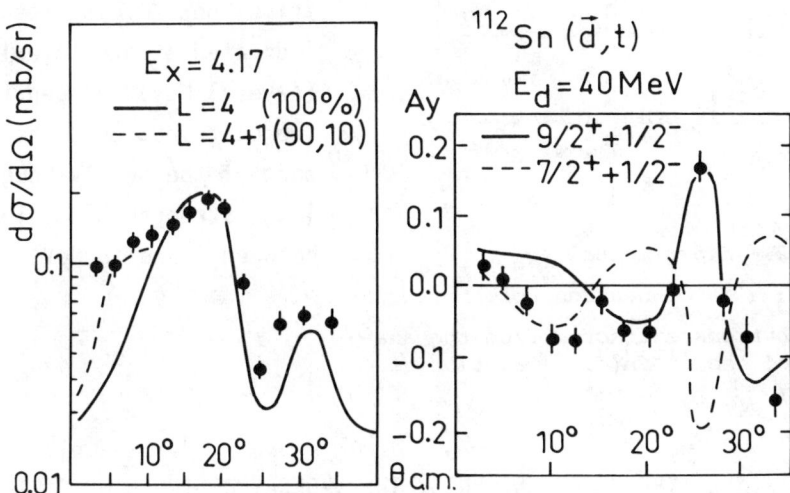

Fig. 11 - Experimental differential cross-section and analyzing power for the $^{112}Sn(\vec{d},t)^{111}Sn$ reaction to inner-hole states. The solid, dashed and dotted curves are the DWBA predictions for the indicated ℓ and J values.

$2p_{1/2}$ strength is observed and some $2p_{3/2}$ components are excited around 4.5 to 5.5 MeV (see fig. 12).

Similarly, the $^{116}Sn(\vec{d},t)^{115}Sn$ reaction at 40 MeV [26] and the $^{120}Sn(\vec{p},d)$ reaction at 90 MeV [27] have been investigated up to 8 MeV excitation to establish the spin J of the hole state. Recently, analysing power measurements were carried out up to rather high excitation energy (20 MeV) in ^{115}Sn and ^{89}Zr.

In the first case, the ^{116}Sn($\vec{d}$,t)^{115}Sn reaction at 52 MeV [28] was employed and pointed out a concentration of $1f_{7/2}$ strength around 15 MeV excitation energy in ^{115}Sn in agreement with angular distribution measurements of Refs 17 and 18.

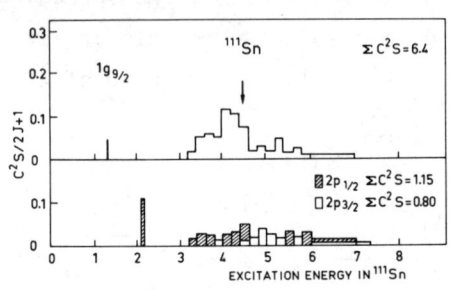

Fig. 12 - Experimental $1g_{9/2}$, $2p_{1/2}$ and $2p_{3/2}$ inner-neutron hole strength distributions extracted from the analysis of the ^{112}Sn($\vec{d}$,t) reaction at 40 MeV.

The ^{90}Zr($\vec{p}$,d) reaction was studied at 90 MeV in order to tentatively unravel the $1f_{5/2}$ and $1f_{7/2}$ hole strength distributions [29]. The results indicated a large overlap (several MeV) between the two distributions, the main part of the deeply-bound $1f_{7/2}$ strength being located between 7 and 16 MeV.

D - NEW TYPE OF EXPERIMENTS. THE γ-DECAY OF INNER-HOLE STATES

Up to now the experimental studies of the neutron inner-hole states were carried out via inclusive measurements. The main component of the $1g_{9/2}$ hole state is located below the neutron threshold in the Pd and Sn mass regions. Consequently the excited states will decay by emitting γ-rays. More information can be obtained on the damping mechanism via a decay study, the γ-ray transition probability being very sensitive to the admixture of collective and/or non collective components to the hole states.

H. Sakaï et al have recently studied the γ-decay of the $1g_{9/2}$ inner-hole in the Pd isotopes via the reaction (^{3}He,αγ) at 70 MeV [30]. As discussed in the introduction of subsect II.A, the degree of fragmentation of the hole state may be roughly divided into three classes;

sharp, bump or flat structures. A good example of sharp peak is the single state ($J^\pi = 9/2^+$, Ex = 2.4 MeV) in ^{101}Pd. Intermediate structures are observed around 4 MeV excitation energy in ^{101,105,107}Pd and show intermediate features. A M_1 transition is still present but it is no longer the major transition, and statistical decays take place at the same time. The high and flat energy regions of the spectra from neutron pick-up reaction show a purely statistical decay. Thus at first, the γ-decay seems to depend on the degree of spreading of the inner-hole state. These conclusions are schematically summarized in fig.13.

More complete measurements were obtained recently at Orsay on the γ-decay through an investigation of the ^{112}Sn(^{3}He,αγ) reaction [31] at 32 MeV.

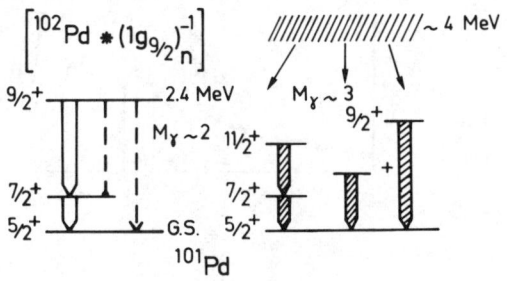

Fig. 13 - Schematic representation of the γ-decay of the sharp (2.4 MeV) and broad peaks (4 MeV) in ^{101}Pd. Taken from Ref. 30.

The α-particles emitted near zero degree (2°3-8°9) were focused by means of the large solid angle superconducting spectrometer SOLENO [32] onto a Si detector whereas the incident ^{3}He beam from the Orsay MP Tandem was stopped in a Faraday which has a half opening angle of 1°.

The γ-rays emitted in coïncidence with the α-particles were detected in two Ge(Li) counters located in the horizontal plane at 90° and 142° with respect to the beam axis. Since the α-particles are detected around the beam direction with a perfect azimuthal symmetry, the α-γ angular correlation depends only on the angle θ between the beam direction and the γ-ray detector axis (geometry II of Litherland and Ferguson [33]. In this work the α-γ angular correlation W (θ) was measured at two angles leading to a determination of the anisotropy ratio $R = W(90°)/W(142°)$.

The coincident α-particle spectrum with γ-rays emitted at 90° and 142° lab angles is presented in fig. 14. The magnetic field has

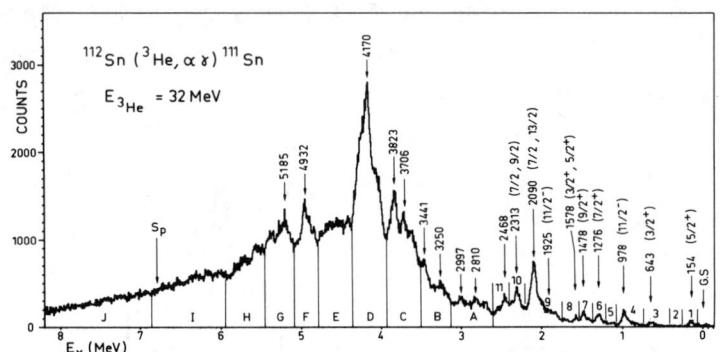

Fig. 14 - α-spectrum in coincidence with γ-rays emitted at 90° and 142° lab angles from the $^{112}Sn(^{3}He,\alpha\gamma)^{111}Sn$ reaction at 32 MeV.

been set to obtain the maximum solid angle (70 msr) for the 4.2 MeV $1g_{9/2}$ hole state in ^{111}Sn.

Above the low-lying states ($Ex \geq 2.6$ MeV) the spectrum has been divided into adjacent slices. In fig. 15 is presented the γ-decay pattern of the inner-hole states in ^{111}Sn. From the γ-spectrum associated with the excitation energy region 2.6 - 4.4 MeV (slices A, B, C, D) one may conclude that about 20-30% of the decay intensity proceeds through a ground-state transition. On the other hand more than 50% of the decay goes through a number of levels located between 1 and 2 MeV which are assumed to be weak coupling states (see fig. 15). In the γ-decay associated with the high-energy tail (Ex = 4.8 - 8.2 MeV, slice F, G, H, I, J in fig. 14) no γ-rays corresponding to the ground state transition are observable.

In summary with the exception of the decay to the ground and to the few first excited states, the main part of the decay occurs

through the feeding of weak coupling states. This result may be taken
as the first and direct experimental evidence of the importance of
surface vibrations mixing
with quasiparticle states
in the spreading mechanism
of the inner-hole states.

Since the high-spin inner
hole states (J = 9/2) decay
preferentially to high-spin
final states (J = 7/2, 9/2,
11/2) whereas the opposite
is true for the low spin
components (J = 1/2, 3/2)
of the distribution, the
measured anisotropy ratio
for each energy bin and
each secondary transition
has been used to determine
the high or low spin na-
ture of the decaying state.
It was then possible to
demonstrate that the $1g_{9/2}$
hole state decay involves
the secondary transitions

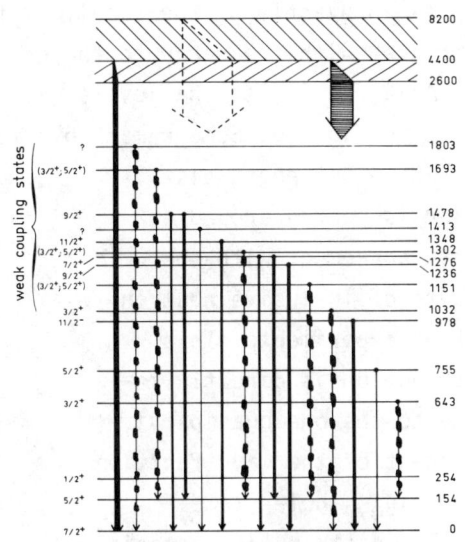

Fig. 15 - The γ-decay pattern of inner-
hole states in ^{111}Sn. Two decay schemes
are displayed in the figure. One corresponds to the energy region (2.6-4.4 MeV)
and the other one to the high-energy
tail of the hole strength distribution
(4.4-8.2 MeV). The wide arrows corresponds to the inobserved primary transitions. The secondary transitions are
shown in the fig.

represented by solid arrows in fig. 15. A second family of secondary
transitions (dashed arrows in fig. 15) characterizes the decay of the
low-spin (J = 1/2, 3/2) part of the strength distribution.

Using the two families of secondary transitions we could build
up the strength distributions of the $1g_{9/2}$ and 2p inner-hole states.
For each secondary transition, the two γ-coincidence spectra have
been extracted and the contribution of each γ detector has been added.
After the corrections due to the energy dependence of the (^{3}He,α)
cross-sections for ℓ = 4 and ℓ = 1 transfers (taken from DWBA predictions) and due to the variation of the solid angle of SOLENO with

α-kinetic energy, one obtains a distribution directly proportional to the strength function.

As an example of the results of this analysis, the hole-strength associated with the decay to the gs, to high spin states (J = 7/2, 9/2, 11/2, 13/2), to the levels with J = 5/2 and J = 3/2 are presented in fig. 16. Some remarkable features have to be outlined.

i) The γ_0 strength distribution is peaked around 4.2 MeV in agreement with the results of inclusive neutron pick-up experiments. This distribution is directly related to the one quasiparticle component of the wave function of the $1g_{9/2}$ hole state.

ii) The distribution associated with high-spin secondary transitions is very similar up to 5.5 MeV but a large amount of strength is spread up to 8 MeV. The distribution corresponding to the J = 5/2 states displays a broad secondary enhancement around 6.8 MeV.

iii) The strength function deduced from the decay to low-spin states (J = 3/2) is strongly fragmented and no concentration of strength is observed.

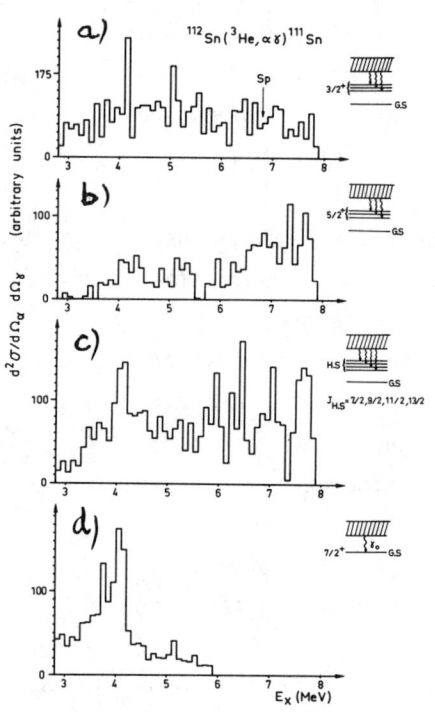

Fig. 16 - Coincident α-spectra with secondary γ-ray transitions with spins a) J_C = 3/2 b) J_C = 5/2 c) J = 7/2, 9/2, 11/2 ... and d) same as a), b), c) but for the ground state transition γ_0.

We would like to emphasize that the method discussed here to build up strength functions is very powerful.

All the distributions are free from any assumption on the background and almost entirely independent from the reaction process used to populate these states. Moreover the method allows us to obtain the partial strength functions associated with the collective components in the wave function of the $1g_{9/2}$ inner-hole state.

To obtain the full inner-hole strength distributions, the spectra labelled b), c), and d) have been added up to produce the $1g_{9/2}$ hole distribution whereas the spectrum a) is assumed to represent the 2p hole strength.

In fig. 17 are presented the histograms of the theoretical and experimental distributions for the $1g_{9/2}$ inner-hole state in ^{111}Sn. The experimental histogram has been normalized to the theoretical one at the maximum of the distribution (E_x = 4.2 MeV). The same comparison is made with the strength function deduced from a previous ($\vec{d}$,t) experiment [25] and the maximum of the ($\vec{d}$,t) distribution has been used for the normalization of our result.

The agreement both in shape and magnitude between the predictions of Soloviev et al [10] and our experimental results is remarkable. The secondary bump predicted by the nuclear model calculations is observed for the first time around 6.8 MeV.

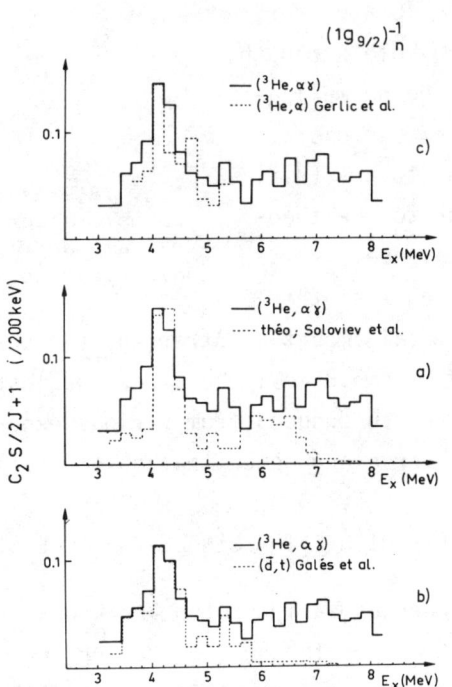

Fig.17 - a) Deduced experimental $1g_{9/2}$ strength function in ^{111}Sn (solid lines) is compared to the predicted one from the qs-ph model (dashed lines) [10,34]. b) same as a) expect the comparison made with the strength function deduced from the ^{112}Sn($\vec{d}$,t)^{111}Sn experiment [25].

The results from the ($\vec{d}$,t) experiment [25] are also in good agreement with the new data up to 5.8 MeV. Inclusive experiments were not able to detect the high energy part of the strength function.

The same type of comparison is made in fig.18 a,b, for the 2p hole strength. Here the normalization of our distribution to the ($\vec{d}$,t) one or to the theoretical one, is made assuming that the

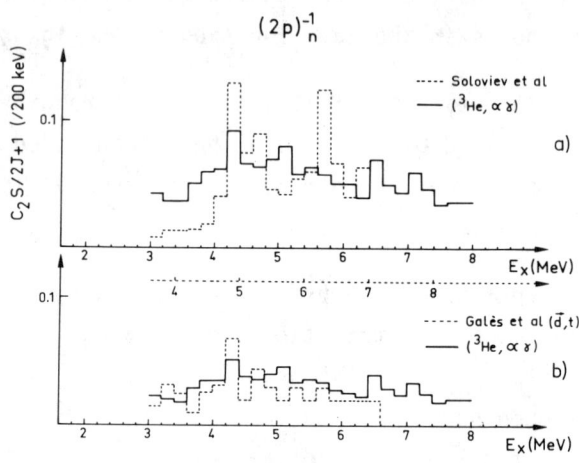

Fig. 18 - a) Same as fig. 17a for the 2p hole strength distribution, b) Same as fig. 17b for the 2p hole strength distribution.

same total amount of strength is observed in the energy range 4.5 < Ex < 5.8 MeV. One clear conclusion from this comparison of the 2p strength deduced from various experimental works is that one does not observe any concentration of strength.

E - HIGH-LYING SINGLE-PARTICLE STATES VIA STRIPPING REACTIONS

Whereas early proton knock-out reactions have been successively used to probe the hole structure in nuclei, pratically no information is available on high-lying "particle" states. The properties of the $T_>$ part (or IAS) of the single-particle strengths for subshells located well above the Fermi sea are well established [15]. On the other hand the $T_<$ part, where the bulk of strength is concentrated, is poorly known. Therefore one nucleon stripping reactions appear as a powerful tool to reach the properties of high-lying proton and neutron states in a way similar to the one employed to investigate the hole response function.

1. The (α,t) and (³He,d) reactions to high-lying proton states

The first experimental evidence of high-lying proton excitation in heavy nuclei was obtained while studying the (α,t) reaction at 80 MeV on ^{144}Sm target nucleus [15,35]. These experiments were complemented later on by an investigation of the (³He,d) process at 240 MeV.

The (α,t) experiment was performed with α-particle beams delivered by the Grenoble cyclotron. Outgoing tritons were detected at the focal plane of the QD spectrometer. Both high (50 keV) and low (230 keV) energy resolution spectra were recorded up to 22 MeV excitation energy.

Typical triton residual energy spectrum from the ^{144}Sm(α,t) reaction is displayed in fig. 19a. At low excitation energy, one observes a selective population of the $1h_{11/2}$ proton state due to the known selectivity of this reaction for large ℓ values (ℓ =5,6). Before this study the distribution of the proton strength above 2.5 MeV was unknown. At intermediate excitation energy (4 < E_x < 15 MeV) two broad bumps

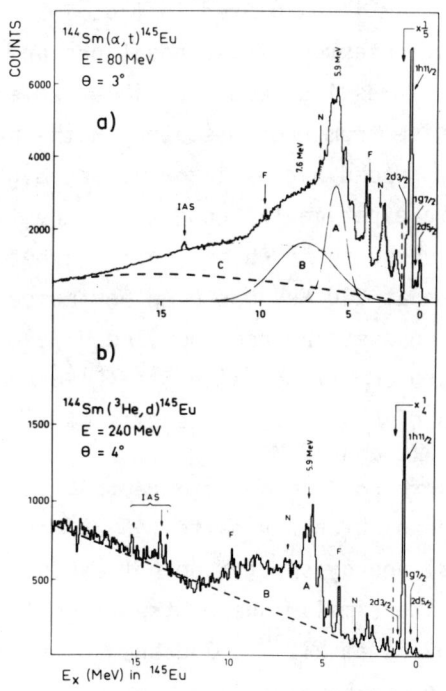

Fig. 19 - Triton (deuteron) residual energy spectra from the reactions ^{144}Sm(α,t)^{145}Eu (a) and ^{144}Sm(³He,d)^{145}Eu (b). The dashed lines are the assumed background line shapes from the PWBU model.

are excited above a substantial background. Their energy spacing, their respective widths, and in particular the fact the second bump

(B) appears as a shoulder of the main peak (A) recalls the similar features observed in the study of the deeply-bound $1g_{9/2}$ and 2p hole states in the tin isotopes [12,13,15-20].

Around 14 MeV a sharp peak corresponding to the $1h_{11/2}$ and $1i_{13/2}$ IAS is observed for the first time.

In order to gain further insight in the study of this phenomena, the same range of excitation energy in ^{145}Eu was investigated at 240 MeV using the ^{3}He beam from the Orsay synchrocyclotron and the large magnetic spectrometer Montpellier. The resulting residual energy spectrum is displayed in fig. 19b. The same structures A and B are also observed. Their positions and widths being consistent with the (α,t) results. One plausible explanation for these peaks is that they arise from proton states in the next major shell consisting of closely spaced high-spin orbitals, e.g $2f_{7/2}$, $1h_{9/2}$ and $1i_{13/2}$. Under the same experimental conditions the (α,t) reaction at 80 MeV has been studied on ^{208}Pb and ^{116}Sn target nucleus whereas the (^{3}He,d) reaction at 240 MeV has been performed on ^{120}Sn and ^{208}Pb nuclei. The resulting spectra from the (^{3}He,d) experiment are shown in fig.20. High-lying proton strengths in $^{117, 121}$Sb are observed around 11.0 MeV with a width of 5.5 MeV. In the case of ^{209}Bi two structures are excited around 7.2 and 10 MeV.

2. *Background line shape, data reduction and DWBA analysis*

The high-energy parts (15-25 MeV) of the triton and deuteron residual energy spectra do not exhibit pronounced structures and have quite similar shapes (see figs 19 a,b and 20). It has been shown by Wu et al.[36], Budzanowski et al.[37] and Matsuoka et al.[38], that the break-up processes give an important contribution to the reaction cross-sections induced by fast α or ^{3}He-particles (10-40 MeV/a.m.u) scattered on medium-heavy weight target nuclei.

An attempt to describe the underlying continuum using a simple Plane Wave Break-Up model (PWBU) has been attempted. The method developed by Matsuoka et al[38] was employed. The theoretical predictions were normalized to the data at forward angles ($\theta < 6°$) and at high excitation energy ($E_x \simeq 22$ MeV). The results are shown as dashed lines in figs. 19 a,b and 20.

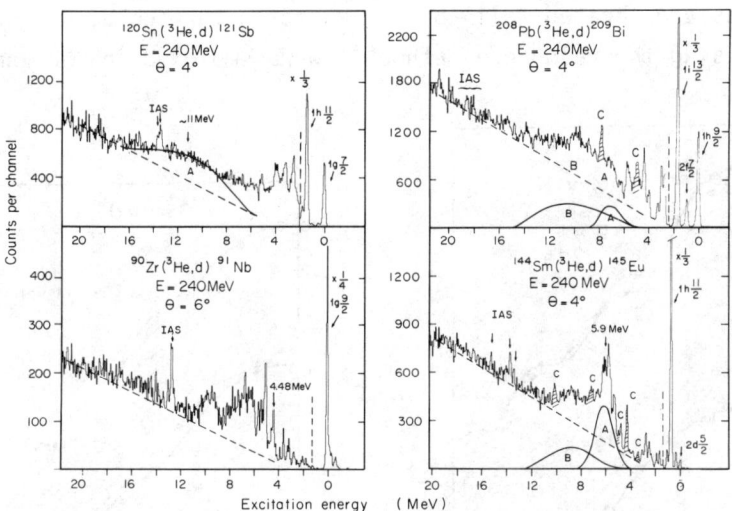

Fig. 20 - Residual energy spectra from the (^{3}He,d) reaction at 240 MeV on ^{90}Zr, ^{120}Sn, ^{144}Sm and ^{208}Pb target nuclei. The dashed lines are the predictions from the PWBU model for the background line shapes.

In the case of the (α,t) reaction, a large amount of cross-section located between 10 and 20 MeV excitation energy cannot be explained by the elastic break-up. Our estimate of the background cross-section does not take into account the inelastic or the absorptive break-up processes which are known to be important[39]. This failure of the simple PWBU model in the case of the (α,t) reaction leads us to use the so-called "empirical background line shape" which is shown as a solid line in fig. 19 a. The predictions of the PWBU Model were however in good agreement with the empirical systematics in the case of the (^{3}He,d) reaction (see figs. 19b and 20). Under these assumptions, the remaining part of the cross-sections was fitted by one (^{117,121}Sb) or two gaussian peaks (^{145}Eu, ^{209}Bi A and B) having different widths. Another method has been used, namely dividing the excitation energy range of interest into adjacent slices, 1 MeV wide.

The parameters used in the DWBA analysis were successfully tested on known low-lying states[35]. Moreover since the observed gross structures are located well above the proton threshold, unbound form factors using the resonance method[22] were employed in the analysis.

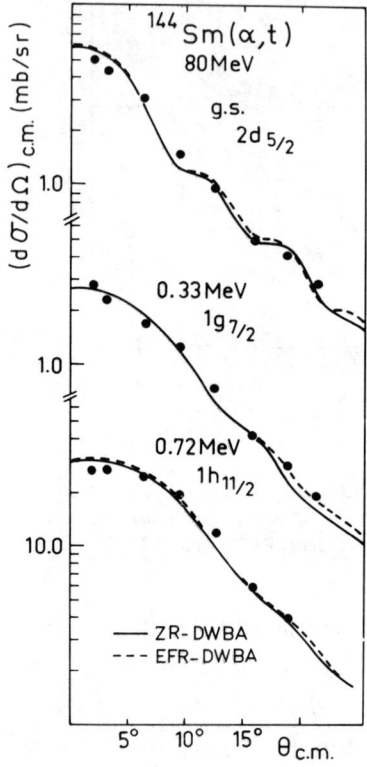

Fig. 21- Experimental data and theoretical DWBA curves from the $^{144}Sm(\alpha,t)$ reaction to low-lying states.

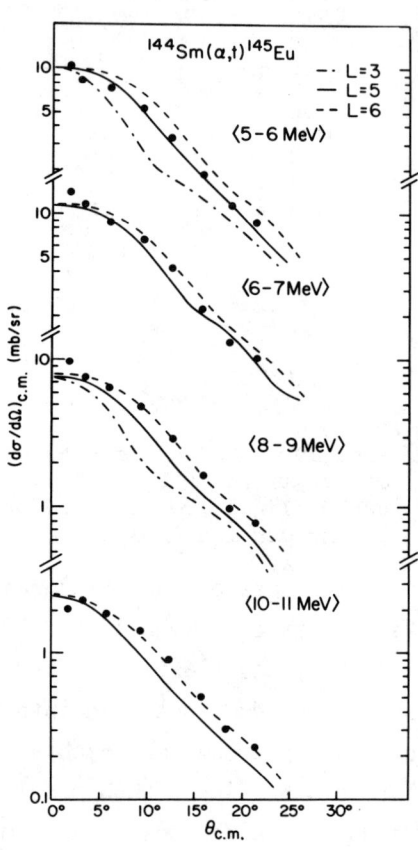

Fig. 22 - Angular distributions from the reaction $^{144}Sm(\alpha,t)$ to high-lying states. Solid, dashed and dot-dashed lines are DWBA calculations for the indicated ℓ values.

Typical experimental angular distributions and theoretical DWBA curves for both well-known low-lying states ($2d_{5/2}$, $1g_{7/2}$ and $1h_{11/2}$ levels) and the high-lying "peaks" (5-11 MeV) in ^{145}Eu are shown in figs 21 and 22, respectively.

A very good agreement is found between the data and DWBA calculations if one assumes an $\ell = 5$ transfer for the region 5-7 MeV and an $\ell = 6$ transfer for the region 8-11 MeV. For completeness one should mention that an $\ell = 3$ transfer does not reproduce the trend of the experimental data but 20% mixing of $\ell = 3$ in the bump A will not change significantly the quality of the fit. As regard to the deduced strengths, the results of the DWBA analysis indicates that the full $\ell = 5$ $1h_{9/2}$ proton strength is found in the bump A with a possible 20% admixture of $\ell = 3$, $2f_{7/2}$ strength and that the $\ell = 6$ $1i_{13/2}$ proton strength is concentrated in region B. For the ^{117}Sb and ^{209}Bi nuclei, a detailed analysis has been reported in ref.35.

3. Comparison to nuclear models

Soon after the publication of our first results on high-lying protons strengths, theoretical predictions were made by the Dubna group using the quasiparticle-phonon coupling nuclear model[40] and by the Orsay group[41] using a self-consistent approach (HF + RPA).

In fig.23 are presented the theoretical strength functions (unit of strength per unit energy interval for the high-lying $2f_{7/2}$, $1h_{9/2}$ and $1i_{13/2}$ outer subshells in ^{145}Eu calculated within the quasiparticle-phonon model[40]. The results of the calculations are compared to the experimental values deduced from the analysis of the (α,t) and (^{3}He,d) reactions. In the

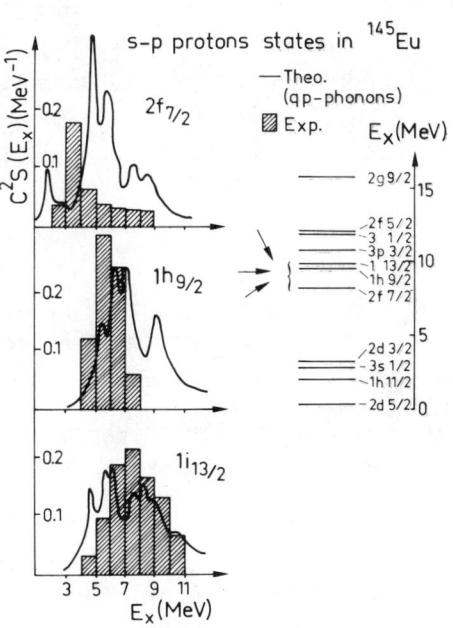

Fig.23 - Comparison between theoretical [40] and experimental proton strength distributions for high-lying subshells in ^{145}Eu.

figure is also indicated the level scheme of proton particle states calculed in a Woods-Saxon potential. The theoretical strength functions differ noticeably from each other. However the values of the second moment σ of the three distributions are close.

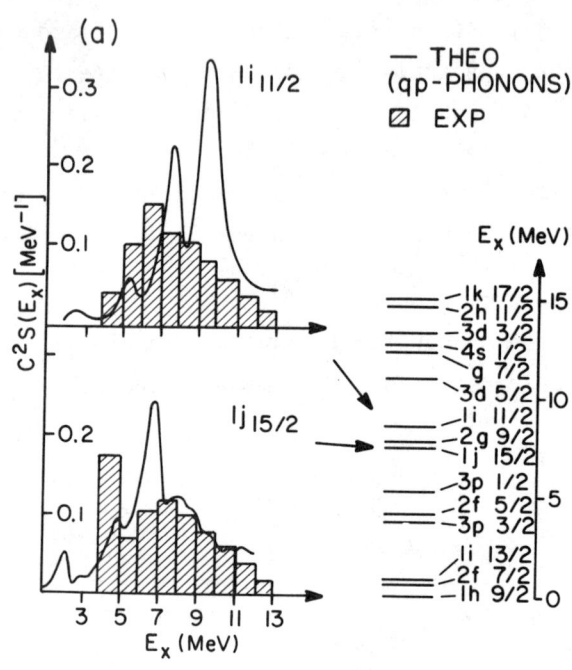

Fig. 24 - Same as Fig. 23 for the case of the $1i_{11/2}$ and $1j_{15/2}$ proton strengths in ^{209}Bi. Ref.40.

The agreement with experiment is quite good for the $1i_{13/2}$ subshell whereas the damping of the $2f_{7/2}$ and $1h_{9/2}$ strengths is rather large compared to the experimental ones. One may notice also an energy difference of about 1.0 to 1.5 MeV between the centroïd energies of the theoretical and experimental strength distributions. Such desagreement may be explained by a less accurate determination of the experimental strengths in the overlapping regions.

In Fig. 24, similar comparison is made in the case of the outer $1i_{11/2}$ and $1j_{15/2}$ subshells in ^{209}Bi. Here again a qualitative agreement is achieved. The theoretical calculations confirm the strong overlap between the $1i_{11/2}$ and $1j_{15/2}$ strength functions in ^{209}Bi. For the same nucleus, e.g. ^{209}Bi calculations in the framework of the single-particle vibrations coupling nuclear model have been carried out by N. Van Giai et al [41].

The theoretical predictions are made for both the low-lying and

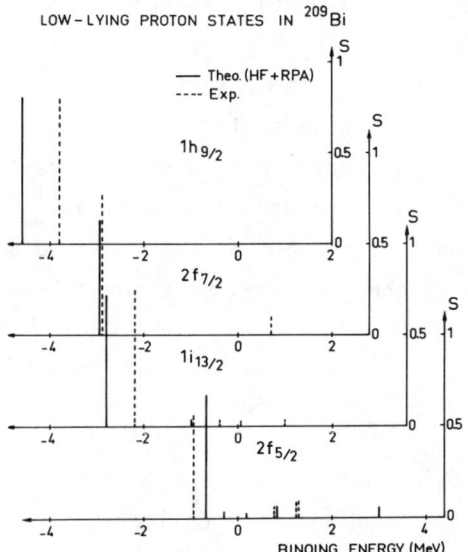

Fig.25 - Comparison between theoretical[41] and experimental strengths for the low-lying single-particle states in ^{209}Bi.

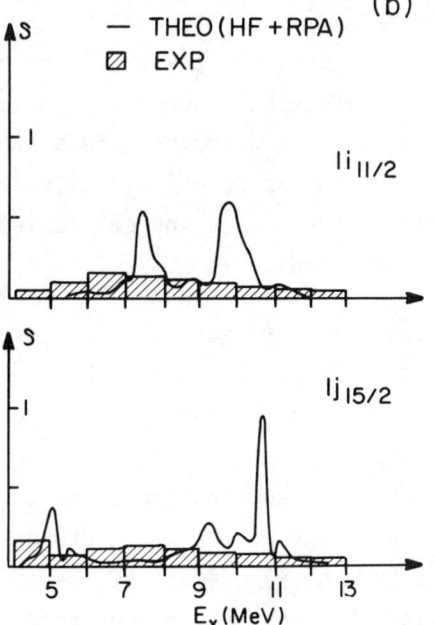

Fig.26 - Experimental and theoretical strength distributions for high-lying $1i_{11/2}$ and $1j_{15/2}$ proton states in ^{209}Bi.

high-lying proton subshells, giving an additional interest to such approach since the whole single-particle proton spectrum can be compared to the experimental results.

In fig. 25 the theoretical predictions for the strength distributions of the low lying $1h_{9/2}$, $2f_{7/2}$ and $1i_{13/2}$ subshells are compared to the experimental results. It is worth pointing out that there is no free parameter in such calculations, the single particle states being generated from the H.F. potential using the Skyrme III force. The agreement is rather satisfactory both for the position and the strength located in the main fragment of the low-lying $1h_{9/2}$, $2f_{7/2}$ and $1i_{13/2}$ proton states in ^{209}Bi.

Turning now to the high-lying $1i_{11/2}$ and $1j_{15/2}$ strength distributions the comparison between the predicted spectral functions and the experimental ones is displayed in fig. 26. The damping of these high-lying subshells is not reproduced by the calculations. The concentration of the high-lying proton strengths is correctly predicted as well as the observed strong overlap of the two strength distributions. The theoretical strength functions have much more structure than the experiment, a general feature already noted in the case of inner-hole strength distributions [11]. Surface vibrations mixing can account for 40-60% of the observed damping but the mixing with states of higher complexity is not taken into account in such calculations.

4. Proton pick-up and stripping to high-lying states in the transition region : The Sm isotopes

In order to test if the mechanism responsible for the damping of the high-lying proton strengths is linked with the collectivity of the low-lying states, one could look for a system which is known to show a strong variation of its collective properties with mass number. The Heidelberg group has investigated proton hole states in the Sm isotopes[23]. The spectra from the reaction ^{144,148,152}Sm(d,^{3}He) taken from Ref. 23 and displayed in fig. 27 exhibit broad peaks around 4 to 5 MeV excitation energy. The centroïd energy of the bumps decreases from ^{143}Pm to ^{151}Pm whereas its width increases. These peaks correspond to the population of the $1g_{9/2}$ proton hole state. The behaviour of the strength distribution has been qualitatively explained by the authors as a result of the increasing collectivity of the core (^{144}Sm spherical; ^{152}Sm, deformed).

The existing empirical systematics on inner-neutron hole states has been also extended across the N = 82 neutron closed shell through the study of the (^{3}He,α) and (p,d) reactions on the Sm and Nd isotopes [13,15,42]. In fig. 28 are plotted the centroid energy of the high-energy component of the $1h_{11/2}$ neutron-hole strength in the Sm-isotopes[42], the measured mean energy of the $1g_{9/2}$ proton-hole strength in the Sm isotopes[23], and the β_2 values of the lowest 2^+ state for the same isotopic chain. The trend of the centroïd energy curves (for both neutron and proton hole states) presents a striking similarity and seems to be strongly correlated with the increase of the values of β_2 in the heavier Sm isotopes.

The features described above were the main motivation of a recent study of the (α,t) reaction on the Sm isotopes[43] (A = 144 to A = 154) where an onset of deformation occurs around ^{152}Sm.

In fig. 29 are presented the spectra obtained in that study, investigated with the 100 MeV α-particles beam delivered by the K500 MSU cyclotron, and the S320 broad range spectrograph.

For the spherical ^{144}Sm target one observes the population of the $1h_{9/2}$ and $1i_{13/2}$ high-lying states and clear gap between the low-lying valence states and the "giant" structures. When one goes to the heavier Sm isotopes, the energy spacing between valence and high-lying states decreases strongly and disappears in ^{153}Eu and ^{155}Eu. This observation is

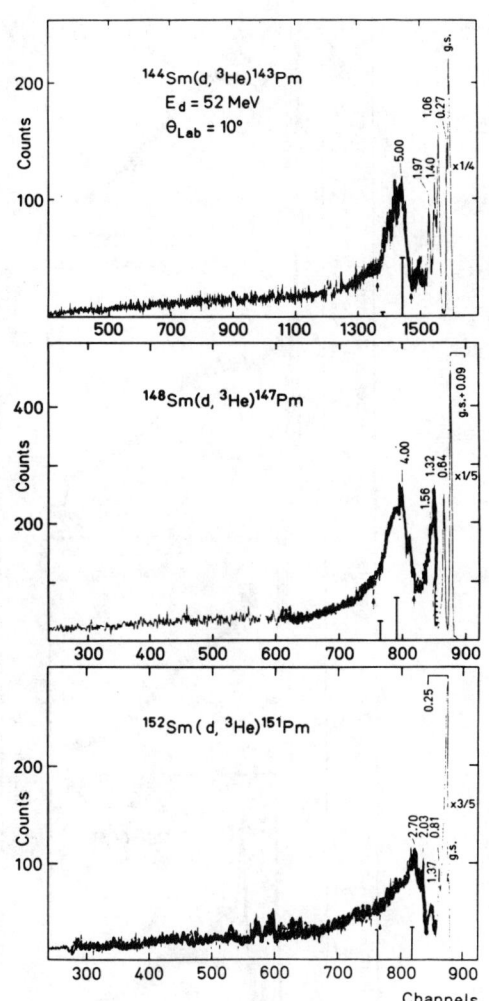

Fig. 27 - Taken from Ref. 23
^{143,147,151}Pm excitation energy spectra from the (d,^{3}He) reaction at 52 MeV.

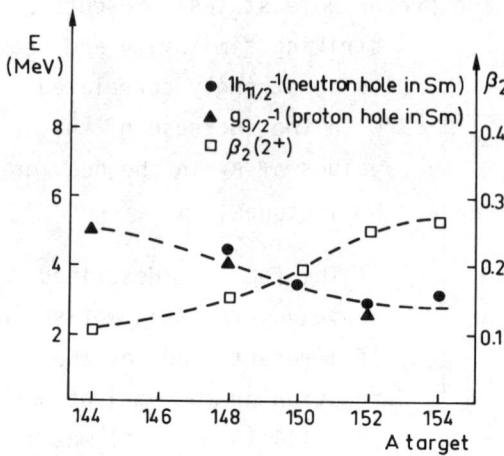

Fig. 28 - Measured centroïd energies full circle, $(1h_{11/2})_n^{-1}$ and full triangles, $(1g_{9/2})_p^{-1}$ of the inner-hole states in the Sm isotopes. The β_2 values for the lowest 2^+ state of the same target are indicated by empty squares.

consistent with the existence of a shape transition from spherical to deformed around N = 90. As regards to the high-lying structures, the broadening of the bumps follows the same behaviour, whereas sharp peaks appears at rather low energies in the deformed Eu isotopes and may result from the splitting into Nilsson orbitals of the high-spin $1h_{9/2}$ and $1i_{13/2}$ outer subshells.

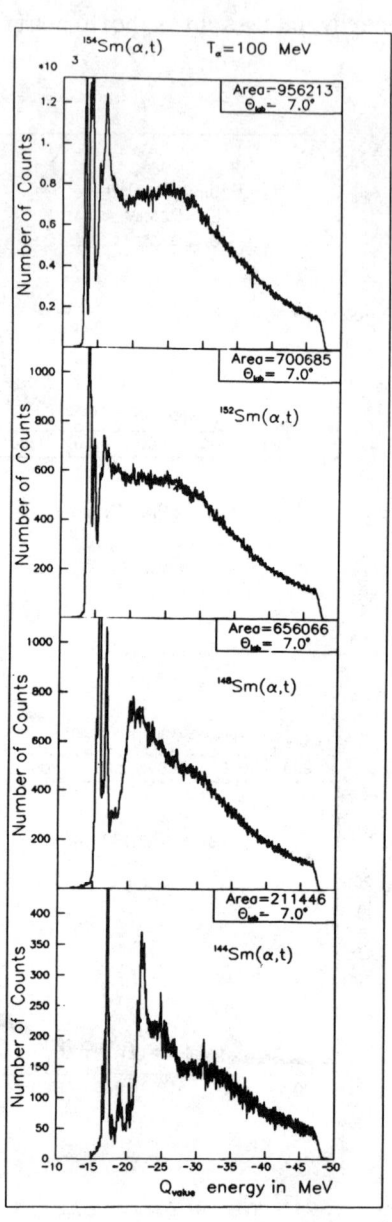

Fig. 29 - Triton energy spectra from the reaction $^{144,148,152,154}Sm(\alpha,t)$ at 100 MeV incident energy.

6. The (α, ³He) reaction to high-lying neutron states

Following the transfer reaction approach which has been quite successful in probing high-lying proton strength distributions, we report here on the study of quasi-bound or unbound high-spin neutron states using the (α,³He) reaction on the ^{208}Pb target nuclei and on the tin isotopes (A=112 to 124) at 183 MeV incident energy.

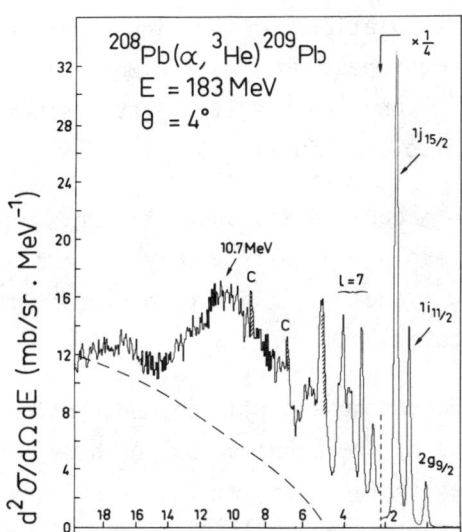

Fig.30 - Residual energy spectrum from the (α,³He) reaction at 183 MeV on ^{208}Pb target. The dashed lines are the predictions of the PWBU model for the background.

The α-particle beam was delivered by the K220 Orsay synchrocyclotron. The (α,³He) process has been chosen due to its known selectivity in angular momentum transfer for large ℓ values. In fig.30 is displayed the residual energy spectrum from (α,³He) reaction on ^{208}Pb target.

First, let me comment on the strong selectivity of this process for high-spin single-particle states. The first low-lying neutron states in ^{209}Pb have a large spectroscopic strength (0.8 to 0.6) but in the displayed ^{208}Pb(α,³He)^{209}Pb spectrum, one can see that the $1j_{15/2}$ (ℓ= 7) state has a cross section 10 times larger than the one of the $2g_{9/2}$ (ℓ = 4) ground-state. The fragmentation of the $1j_{15/2}$ subshell has been a long outstanding problem in nuclear structure. Here for the first time the fragmented components are clearly seen in the energy range 3-6 MeV.

At higher excitation energy, the observed spectrum is dominated by a wide "bump" centered around 10.7 MeV in ^{209}Pb. It may be already assumed that this large enhancement of cross-section arises from neutron stripping to high-spin outer subshells. To assess such assumptions, HF calculations of the single-particle neutron spectrum in ^{209}Pb were carried out[44]) and the resulting level scheme is displayed in fig. 31 a.

At about 10 MeV above the $2g_{9/2}$ g.s. high-spin subshells namely, the $2h_{11/2}$, $1k_{17/2}$ and $1j_{13/2}$ are predicted by the calculations.

Recent quasiparticle-phonon calculations by the Dubna group for hole strength functions of the high-lying neutron states in ^{209}Pb[45]) are displayed in fig. 31 b. The centroïd energies and the damping widths of the strenghts are in overall agreement with our previous assumptions.

Another recent investigation of the $(\alpha,^3\text{He})$ reaction has been undertaken on the tin isotopic chain[46]). This was made in close connection with the known properties of the $1g_{9/2}$ inner-neutron hole states in the Sn isotopes. In that case a transition from line fragmentation to line broadening has been established when the mass number varies from A = 113 to A = 125.

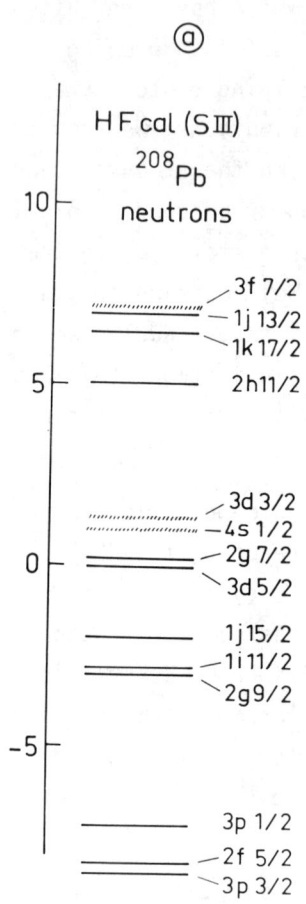

Fig.31 - a) Single-particle level scheme for neutron states in ^{208}Pb (ref. 44).

The ^{113,117,119,121,123,125}Sn energy spectra from the $(\alpha,^3\text{He})$ reaction are presented in fig. 32. Before this study pratically no information was available for neutron states located above 3 MeV.

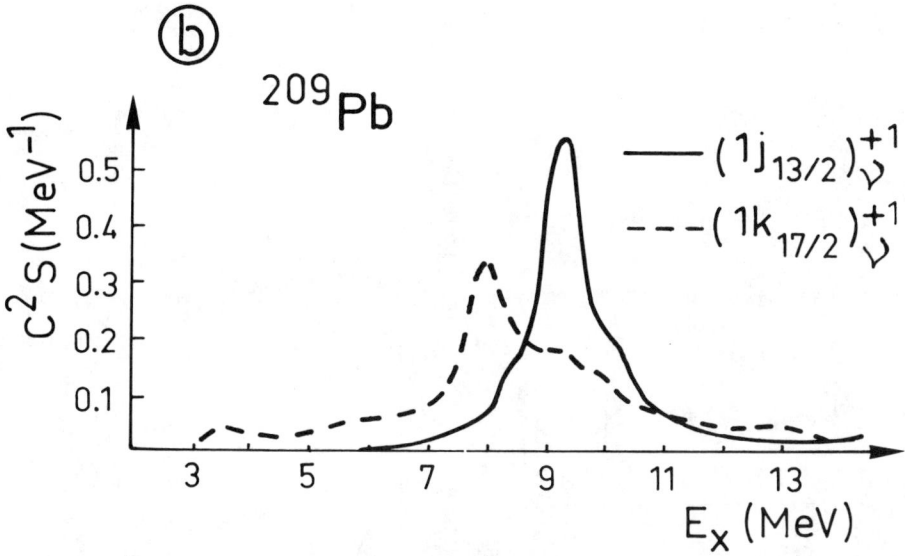

Fig. 31 - b) Strength functions of the high-lying neutron subshells $1j_{13/2}$ and $1k_{17/2}$ in ^{209}Pb (taken from ref.45).

In fig. 32 one observed the strong population of the $1h_{11/2}$ low-lying states. At intermediate excitation energies (4-12 MeV) a sudden onset of neutron single-particle strengths dominates the residual energy spectra in all Sn insotopes. In ^{121}Sn, this complex structure consists of a sharp peak with a width of less than 1 MeV and additionnal broader structures which appear as a shoulder of the main peak. We would like to stress that the characteristics of these high-lying excitations display striking similarities with the well known behaviour of the $1g_{9/2}$ and 2p inner-hole strengths in the same tin isotopes. The centroïd energies of the peaks appear at the same Q-values in the six isotopes investigated here. The narrow part of the bumps is located around 5 MeV in ^{125,123,121}Sn. A broadening is clearly observed when the mass number varies from A = 125 to A = 113.

The high-energy part of the spectra (Q = - 25 to - 35 MeV) does not exhibit pronounced structures and has quite similar shapes. It is likely that the main part of that cross sections is due to the break-up of the α particle. The PWBU model was not able to explain

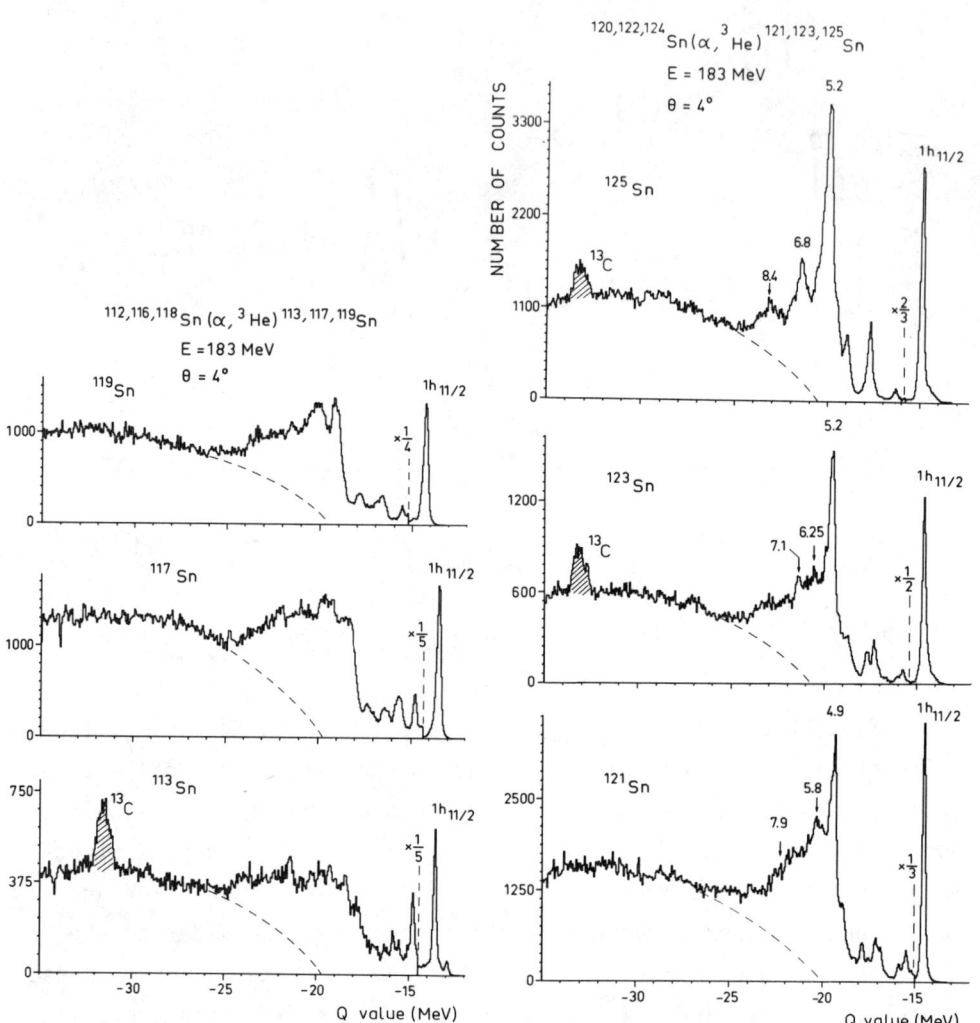

Fig. 32 - ^{3}He residual energy spectra from the reaction ^{112,116,118,120,122,124}Sn $(\alpha, ^3$He$)$ at 183 MeV and at lab. angle of 4°. On the horizontal scales are plotted the Q-values. The number at the top of the peaks correspond to their mean excitation energy. The dashed curves are the "empirical background line shapes".

this high-energy part of the spectra. A detailed analysis of the data has been made for the ^{117,119,121}Sn isotopes up to 12-15 MeV excitation energy. Assuming the background line shape shown in fig. 32, the deduced angular distributions in the regions of the bumps are well reproduced by DWBA calculations assuming an $1h_{9/2}$ and or a $1i_{13/2}$

neutron transfer. More than 50% of the $1h_{9/2}$ strength is located in the peak around 5 MeV in ^{125,123,121}Sn whereas the remaining part of that strength is strongly mixed with the $\ell = 6$ $1i_{13/2}$ strength in the broader structures. The calculations of Vdovin and Stoyanov[45] predict a strong overlap of the $i_{13/2}$ and $h_{9/2}$ neutron subshells and some increase of the damping widths in the Sn isotopes. The detail of these theoretical strength functions which are displayed in fig. 33 are not in a good agreement with the experimental data.

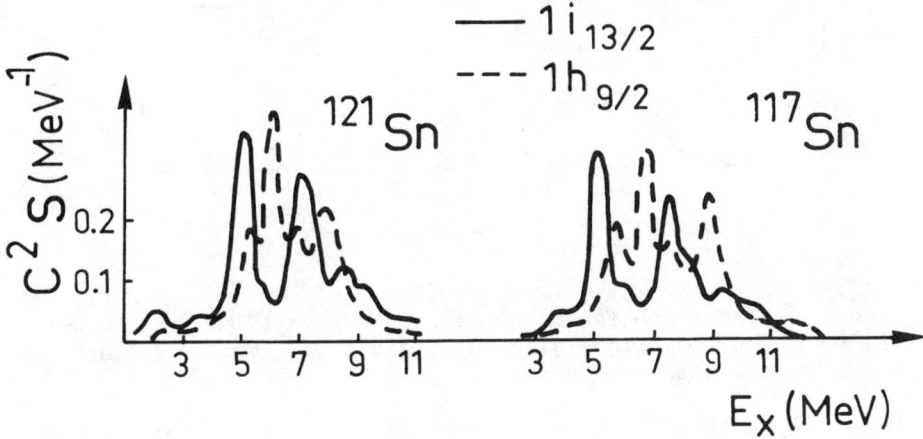

Fig. 33 - Neutron particle strength distributions for the $1i_{13/2}$ and $1h_{9/2}$ orbital in ^{127,121}Sn from ref. 45.

The main disagreement being the rather large fragmentation of the $1i_{13/2}$ subshell as tentatively observed experimentally. In the tin isotopes as one goes away from the shell closure (N = 82) the centroid and the spreading of the strength distribution increase. This result provides experimental evidence for a similar damping of both neutron hole and particle states in heavy nuclei.

Another impressive feature of the damping mechanism of high-lying single-particle (hole) states is the difference of the fragmentation of proton and neutron subshells in single closed-shell nuclei. In fig. 34 we display the calculated strength functions[47] of the $1h_{9/2}$

and $1i_{13/2}$ orbitals in ^{121}Sn and ^{121}Sb. In the tin isotopes, the peaks of the strength functions are much more pronounced than in the antimony isotopes. The width of the distributions in Sn isotopes are 1.5-2.0 MeV

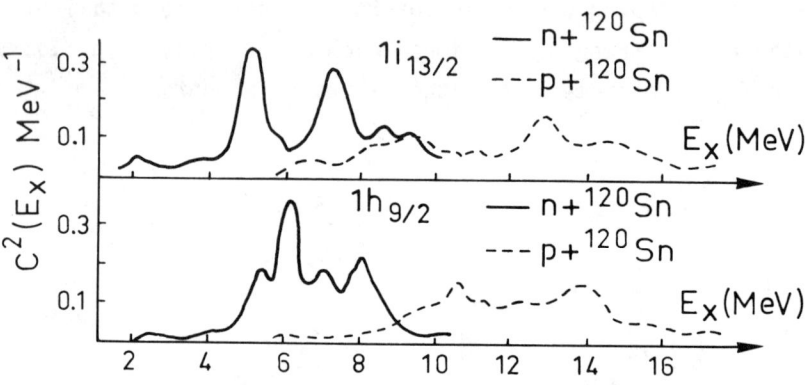

Fig.34 - The strength functions of the neutron (full lines) and proton (dashed curves) high-lying $1h_{9/2}$ and $1i_{13/2}$ subshells in ^{121}Sn and ^{121}Sb. Taken from Ref.47.

whereas in Sb isotopes they are of the order of 3.0 MeV. These differences are in qualitative agreement with our experimental data[35,46]. The reason for such behaviour is the considerable difference between the core-coupling matrix elements Γ_λ for the quasineutron in odd mass tin isotopes and for the proton in odd-mass Sb isotopes. Therefore the proton single-particle state interacts with even-even Sn core much more strongly than the neutron state. The same effect can explain why the neutron deep-hole $1f_{7/2}$ in ^{89}Zr becomes more fragmented than the proton hole $1f_{7/2}$ in ^{89}Y in agreement with the data of Refs 24 and 29. So we may conclude that in single closed shell nuclei the coupling between quasiparticle states and phonons leads to specific isotopic effect in the damping of single-particle states.

III - CONCLUSION AND OUTLOOKS

During the last few years, through more detailed experimental studies and refined theoretical approaches, we have reached a better understanding of the validity and the limitation of the independent single-particle nuclear model.

With the help of transfer reactions investigated at high incident energies, our empirical knowledge of the single particle response function has been greatly enhanced. Inner-neutron hole strengths have been observed in a variety of one nucleon pick-up reactions for a wide range of medium and heavy nuclei. New spectroscopic tools, like the use of polarized beams or the investigation of (p,pn) reactions, have or will proved their validity. More decisive progress is expected in the near future by the development of the decay studies. Strength functions of high-lying states could be directly deduced from exclusive experiments without any assumption on the underlying background or on the reaction mechanism. Extremely good energy resolution is already obtained with the new electron facilities like the MEA (NIKHEF-K) and (e,e'p) experiments on heavier targets like ^{208}Pb are now under our reach.

However the deduced strength distribution from the transfer reaction data are model dependent. In the framework of the nuclear reaction theory, the correct description of the process, as well as the calculation of realistic form factors for high-lying states are highly desirable. In addition the explanation of the underlying background, and the estimation of the magnitude of the multi-step processes are still to come.

Both the quasiparticle-phonon nuclear model and the self-consistent approaches have reached a good qualitative agreement with the empirical systematics. Moreover the link between single-particle and collective degrees of freedom in the nucleus has been established in order to explain the damping of high-lying single-particle strengths.

ACKNOWLEDGEMENTS

Many results from IPN Orsay have been used in this paper to illustrate the deep-hole and high-lying state properties. I would like to thank my colleagues from Orsay : Drs F. Azaiez, E. Gerlic, J. Guillot, H. Langevin-Joliot, S. Fortier, J.M. Maison, E. Hourani, J.P. Schapira and Dr. C.P. Massolo from University of La Plata who takes a large part in the analysis of the stripping reactions while she was at Orsay.

The following physicists have also participate at various stages in the work reported here, Drs G. Duhamel, G. Perrin, V. Comparat and P. Martin from ISN Grenoble, Pr. G.M. Crawley and J. Duffy from Michigan State University, Dr. D. Friesel from Indiana University (IUCF). I would like to acknowledge, Pr. V.G. Soloviev, Drs Ch. Stoyanov and A.I. Vdovin for communication of their results prior to publication and for helpful discussions and Dr. Nguyen Van Giai for the supply of his recent single-particle vibrations coupling calculations.

REFERENCES

1) G. Jacobs, Th. A.J. Maris, Rev. Mod. Phys. 45, 6(1973)
2) J. Mougey, Nucl. Phys. A355, 35 (1980) and Nucl. Phys. A396, 39C (1983)
3) H.A. Weidenmüller, in "Nuclear Structure Physics", U. Smilansky, I. Talmi and H.A. Weidenmüller, ed., Spinger Verlag (1973)
4) G. Bertsch and T.T.S. Kuo, Nucl. Phys. A112, 204 (1968)
5) J.P. Jeukenne, A. Lejeune and C. Mahaux, Phys. Reports 25C, 83 (1976)
6) V. Bernard and N. Guyen Van Giai, Nucl. Phys. A348, 75 (1980)
7) G.E. Brown, J.S. Dehesa and J. Speth, Nucl. Phys. A330, 290(1979)
8) Li Chu- sia and V. Klemt, Nucl. Phys. A364, 93 (1981)
9) N. Van Giai, Proc. Int. Symp. on Highly Excites States in Nuclear Reaction, ed. by H. Ikegami and M. Muruoka, RCNP, Osaka (Japan) 180 p. 682
10) V.G. Soloviev, Proc. Int. Symp. HESANS 83, ed. by N. Van Giai and N. Marty, Journal de Phys. C4, 69 (1984)
11) G.F. Bertsh, P.F. Bortignon and R.A. Broglia, Rev. Mod. Physics 55, 253 (1983)
12) M. Sakaï and K. Kubo, Nucl. Phys. A185, 217 (1972)
13) G.M. Crawley, Proc. Int. Conf. on the structure of medium-heavy nuclei. The Institute of Physics, 1979, Rhodos p.127 and ibid ref 9, p. 590
14) G.T. Wagner, ibid ref 9 p.465 and ibid ref.10, p.85
15) S. Gales, Proc. Int. Conf. on Nuclear Structure, Berkeley 1980, Nucl. Phys. A354, 193 C (1981)

    S. Gales, Lectures at the XII Masurian Summer School, Lake Mikolajki (Poland), Nucleonika, 27, 181 (1982)

    S. Gales, ibid ref.9, p.425

    S. Gales, ibid ref. 10, p.39

    S. Gales, Proc. of Niels Bohr Symposium on Nuclear Structure, ed by R. Broglia, G.B. Hagemann and B. Herskind, Elsevier Science Publishers, B.V., 1985, p.57
16) H. Langevin-Joliot, Lectures at the XIV Masurian Summer School, Lake Mikolajki, Poland (1983) Internal report Orsay IPNO-DRE-83/36
17) H. Langevin-Joliot et al, Phys. Lett., 114B, 103 (1982)
18) R.H. Siemssen et al, Phys. Lett. 103B, 323 (1981)
19) E. Gerlic et al, Phys. Rev. C21, 124 (1980)
20) G. Berrier-Ronsin et al, Phys. Lett. 67B, 16 (1977)

21) G.R. Satchler, Nucl. Phys. 55, 1(1964)
22) C.M. Vincent and H.T. Fortune, Phys. Rev. C2, 182 (1970)
23) P. Doll et al, Phys. Lett., 82B, 357 (1979)
24) A. Stuirbuick et al, Z. Phys. A, Atoms and Nuclei, 297, 307 (1980)
25) S. Gales et al, Nucl. Phys. A381, 40 (1982)
26) G. Perrin et al, Nucl. Phys., A356, 61 (1981)
27) G.M. Crawley et al, Phys. Rev. C23, 1818 (1981)
28) R.H. Siemssen et al, Nucl. Phys. A405, 205 (1983)
29) J. Kasagi et al, Phys. Rev. C28, 1065 (1983)
30) H. Sakaï et al, Phys. Lett. 103B, 309 (1981) and Nucl. Phys. A441, 640 (1985)
31) F. Azaiez et al, Nucl. Phys. A444, 373 (1985)
32) J.P. Schapira et al, Nucl. Inst. Methods, 224, 337 (1984)
33) A.E. Litherland and A.J. Ferguson, Can. J. Phys., 39, 788, (1961)
34) V.G. Soloviev, Ch. Stoyanov and A.I. Vdovin, Nucl. Phys. A342, 261 (1980)
35) S. Gales et al, Phys. Rev. Lett., 48, 1593 (1982) and Phys. Rev. C31, 31 (1985)
36) J.R. Wu, C.C. Chang and H.D. Holmgren, Phys. Rev. Lett., 40,1013, (1978)
37) A. Budzanowski et al, Phys. Rev. Lett. 41, 635, (1978)
38) N. Matsuoka et al, Nucl. Phys. A311, 173 (1978)
39) R.J. de Meijer and R. Kamermand, Rev. Mod. Phys.,57, 147, (1985)
40) Ch. Stoyanov and A.I. Vdovin, Phys. Lett., B130, 134 (1983)
41) Nguyen Van Giai and Pham Van Thieu, (private communication)
42) S. Gales et al, Nucl. Phys. A398, 19 (1983)
43) J. Duffy, G.M. Crawley et al, private communication
44) Nguyen Van Giai, private communication
45) A.I. Vdovin and Ch. Stoyanov, preprint JINR, Dubna E4-85-232
46) S. Gales et al, Phys. Lett. 144B, 323 (1985)
47) Ch. Stoyanov, A.I. Vdovin and V.V. Voronov, private communication.

SESSION E

SHELL MODEL CALCULATIONS

EFFECTS OF PARTICLE-HOLE EXCITATIONS IN LIGHT NUCLEI

P.W.M. Glaudemans
Fysisch Laboratorium, Rijksuniversiteit Utrecht,
P.O. Box 80.000, 3508 Utrecht, The Netherlands

ABSTRACT

It has become clear that the p-shell space is too restricted to further investigate a number of processes which are of present interest in light nuclei. With the advent of a new generation of very large computers (supercomputers) the effects of $N\hbar\omega$ excitations can be investigated in some detail. The energies, multipole moments and radii, calculated with several interactions in a complete $2\hbar\omega$ model space, are discussed and compared with experimental data.

INTRODUCTION

It is quite remarkable that so many processes observed in nuclei, which are in fact complex many-quark systems, can be reasonably well described in a rather simple shell-model picture. The explanation must be that although the shell model in a heavily truncated basis is in principle quite naive, it apparently contains enough physics to reproduce the main features of many experimental data. On the other hand it should not be surprising that more detailed measurements, such as carefully measured electromagnetic properties or electron scattering form factors, require a more sophisticated treatment of both the shell-model wave functions and the corresponding operators.

Observables can be expressed in terms of matrix elements $\langle \psi'_{real} | O_{real} | \psi_{real} \rangle \approx \langle \psi'_{sm} | O_{sm} | \psi_{sm} \rangle$, where ψ_{real} and ψ_{sm} denote realistic and shell-model wave functions, respectively, and O the operator for the process of interest. Here it is assumed that, with a proper choice for an effective operator O_{sm}, the shell-model matrix element does not deviate too much from the real matrix element. It thus follows that statements about the operator O_{sm} are closely related to the chosen configuration space for ψ_{sm}. One may thus expect to obtain more reliable information about processes of interest when for ψ_{sm} less naive assumptions are made.

In this paper we will focuss the discussion on some of the problems that one encounters when one tries to obtain better wave functions for light nuclei. This concerns mainly the difficulties in obtaining a suitable effective nucleon-nucleon interaction as well as problems involved in describing intruder states in a large shell-model basis. In our study of the p-shell nuclei a no-core model space is used, i.e. the hamiltonian contains only two-body matrix elements and no single-particle energies, see also Van Hees and Glaudemans[1,2].

THE EFFECTIVE NUCLEON-NUCLEON INTERACTION

Calculations on normal-parity states in p-shell nuclei in which the active nucleons are restricted to the 1p shell only, have been performed by Cohen and Kurath[3] many years ago. An extension of the model space such that also nonnormal-parity states are taken into account has been given recently by Van Hees and Glaudemans[1,2]. An improvement in the description of normal-parity requires the inclusion of $2\hbar\omega$ excitations. This is illustrated in fig. 1 with an example for A = 7. It should be remarked that the exact removal of

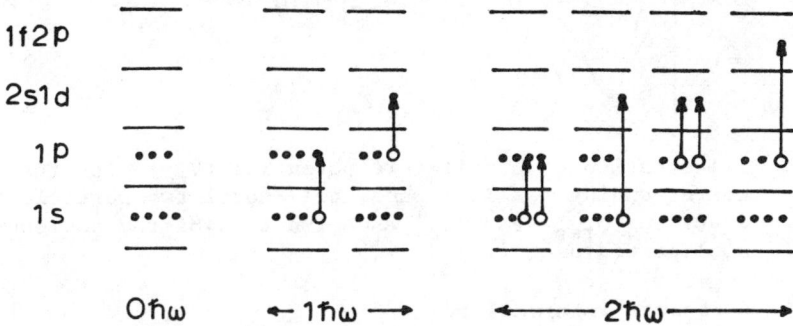

Fig. 1. Schematic representation of the excitations up to $2\hbar\omega$ for an A = 7 nucleus.

spurious-state effects requires
(i) a complete $2\hbar\omega$ space,
(ii) the use of harmonic-oscillator single-particle wave functions and
(iii) a translationally invariant interaction.

In the $2\hbar\omega$ space the hamiltonian is defined by 671 two-body matrix elements, even when one assumes a mass-independent interaction for the A = 4-16 nuclei. It is clear that with such a large number of two-body matrix elements it is not feasible to consider them as free parameters, whose values can be determined empirically from the observed p-shell spectra. One thus has to follow another approach. The requirement of translational invariance reduces the number of parameters from 671 to 51. A possibility to further reduce the number of parameters is to take a realistic interaction, which is derived from nucleon-nucleon scattering data and properties of the deuteron. In this paper some experiences with the Reid soft-core interaction and the Sussex matrix elements are discussed.

THE EVALUATION OF TWO-BODY MATRIX ELEMENTS

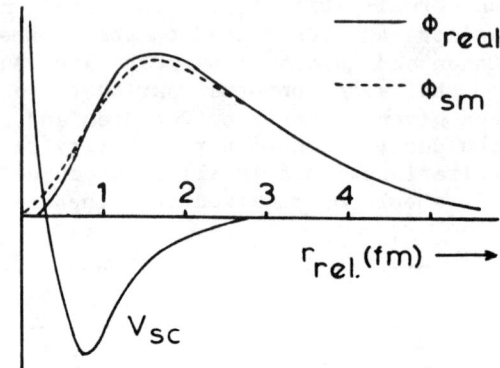

Fig. 2. Illustration of a soft-core potential (V_{sc}) with the corresponding realistic and shell-model two-particle wave functions ϕ_{real} and ϕ_{sm}, as a function of the nucleon-nucleon distance (r_{rel}).

The Reid soft-core interaction

It is not possible to use the Reid soft-core (RSC) interaction in a shell-model calculation without further modifications. The RSC potential V is strongly repulsive at short internucleon distance. This makes the intrinsic radial two-nucleon wave function approximately zero for small r. The relative wave functions derived from harmonic oscillator states do not satisfy this requirement, however, as is indicated schematically in fig. 2, because one works in a finite shell-model configuration space. Therefore, one has to modify V to obtain an effective interaction, which will be denoted by G.

We would like to find an interaction G such that it satisfies as well as possible the condition

$$\langle \phi_{sm} | G | \phi_{sm} \rangle = \langle \phi_{real} | V | \phi_{real} \rangle , \qquad (1)$$

where ϕ_{sm} and ϕ_{real} denote the shell-model and realistic two-particle state, respectively. Considering the identity

$$G = V + (G-V), \qquad (2)$$

one should realize that the correction (G-V) mainly affects the short-range behaviour of V, as follows from fig. 2. The two-body matrix elements of (G-V) can be expanded in terms of Talmi integrals I_p as follows[4]

$$\langle \phi_{sm} | G-V | \phi_{sm} \rangle = C \sum_p I_p , \quad p = 0,1,2,\ldots \qquad (3)$$

where C absorbs all numerical coefficients, while I_p is a function of r, depends on isospin and contains spin, spin-orbit and tensor components, for more details see e.g. ref.[4,5]. It is expected that I_p strongly decreases for increasing p. This stems from the fact that I_p contains a factor $(r/b/2)^{2p+2}$, where b is the harmonic oscillator size parameter. This factor becomes much less than unity for larger p, since I_p is important mainly for the correction at small internucleon distance, i.e. $r \ll b/2 \simeq 2$ fm.

Making the assumption

$$I_p = 0 \text{ for } p \geqslant 2, \qquad (4)$$

one can show[5] that one has only nine non-zero values I_p. In the present approach these nine Talmi integrals I_p are assumed to be mass-independent. The optimal values of I_p could be determined from a fit to p-shell spectra. It followed that the values I_p for p = 1 are about an order of magnitude smaller than those for p = 0 which further supports our assumption (4).

Summarizing we can say that with this approach the number of 671 two-body matrix elements are first reduced to 51 relative two-body matrix elements, which in turn can be expressed in terms of only nine Talmi integrals. The value of the size parameter b is considered as the tenth parameter.

The contribution of the Talmi integrals I_p to the two-body matrix elements is not a minor correction, however. This follows most clearly from the binding energies, which can be written as a sum of two terms

$$E_b = E_b(RSC) + E_b(I_p), \qquad (5)$$

where E_b (a negative value for bound states) denotes the calculated binding energy, while $E_b(RSC)$ and $E_b(I_p)$ denote the bare Reid soft-core and Talmi-integral contribution, respectively. It turns out that $E_b(RSC)$ is strongly positive, however, because the short-range repulsion dominates the matrix elements. Hence, a large negative contribution from $E_b(I_p)$ is needed to obtain the required negative total binding energy. The following relation is found to hold approximately in a $2\hbar\omega$ model space

$$E_b \approx \frac{1}{10} E_b(I_p) \approx -\frac{1}{10} E_b(RSC). \qquad (6)$$

As a typical illustration we give the numbers obtained from the b = 1.7 fm set of matrix elements for the ^{8}Li ground state. One finds $E_b(RSC) = +578$ MeV and $E_b(I_p) = -621$ MeV. From eq. (5) one thus obtains $E_b = -43$ MeV. This compares well with the experimental value $E_b = -42.7$ MeV, where it should be remarked that the Coulomb energy contribution is removed from both the experimental and theoretical value.

The I_p contribution derives mainly from the $I_{p=0}$ terms. In the example given above for ^{8}Li one finds: $E_b(I_{p=0}) = -643$ MeV and $E_b(I_{p=1}) = +22$ MeV.

The Sussex matrix elements

Well known sets of matrix elements for various values of b have been derived many years ago by the Sussex group[6] from a phase-shift analysis of nucleon-nucleon scattering data. We have investigated two sets i.e. those for b = 1.4 fm and b = 1.6 fm.

OBSERVABLES

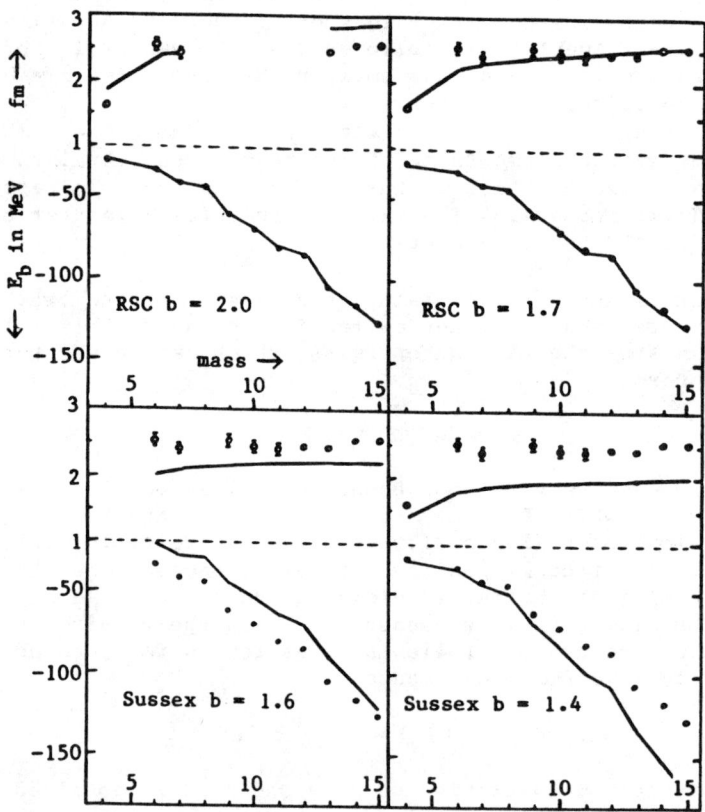

Fig. 3. R.m.s. radii and binding energies of ground states in A = 4-16 nuclei, calculated in a $2\hbar\omega$ space, are compared with experimental data (dots). The Reid soft-core interaction (RSC) and Sussex matrix elements, each for two different b values, have been used.

Spectra

It turns out that the RSC approach both with b = 2.0 and b = 1.7 fm as well as the Sussex matrix elements for b = 1.6 fm produce quite reasonable spectra. This does not hold for the b = 1.4 fm set of the Sussex matrix elements, which gives too expanded level schemes.

Binding energies

The binding energies of the four sets of matrix elements presently discussed are compared with the data in fig. 3, together with the nuclear radii. It is seen that the Reid soft core approach does reproduce the experimental binding energies quite accurately. For both sets (b = 2.0 and 1.7 fm) the deviations are generally less than 1 MeV. The results for binding energies of exotic p-shell nuclei with the RCS interaction are discussed in ref.[7]. For the Sussex interaction, however, the binding energies increase too strongly with mass number, see fig. 3. The b = 1.6 fm set is quite reasonable near A = 16, but not for lighter nuclei, whereas the b = 1.4 fm set reproduces the A ~ 5 region only.

Nuclear radii

The nuclear radii are very well reproduced by the RSC approach with b = 1.7 fm except for ^{6}Li, whereas the b = 2.0 fm set produces too large radii except for A ≤ 6, see fig. 3. The Sussex interactions for b = 1.6 fm and in particular for b = 1.4 fm yield too small values. It should be remarked that the theoretical numbers are mass radii, whereas the data represent charge radii. The difference between these two is expected to be very small, however, as follows from the values found for T_z = 0 nuclei, where both radii should be identical.

Magnetic dipole and electric quadrupole moments

The magnetic dipole moments and electric quadrupole moments in general agree quite well with the experimental data except for a number of notable discrepancies.

As examples of good and poor agreement we present in fig. 4 the g-factor for the ground states of ^{10}B and ^{12}B and the quadrupole moments for ^{11}B and ^{8}Li, respectively, obtained with a number of different interactions and model spaces.

It is no surprise that the g-factor of ^{10}B is very well reproduced by all calculations, because for a T_z = 0 nucleus only the isoscalar part contributes. The isoscalar contribution is known to be insensitive to the precise structure of the nuclear state, see e.g. ref.[8]. From fig. 4 it follows that the g-factor of ^{12}B depends strongly on the assumptions made in the theory. Note that the RSC interaction for b = 2.0 fm even produces the wrong sign, while Sussex for b = 1.6 fm yields a much too small value. These large differences for ^{12}B are all due to the isovector contribution, because again the isoscalar components are the same for all theories (g_{is} = 0.65). Some other cases where the g-factors are found to be quite sensitive to the interaction are ^{8}B, ^{9}Be,

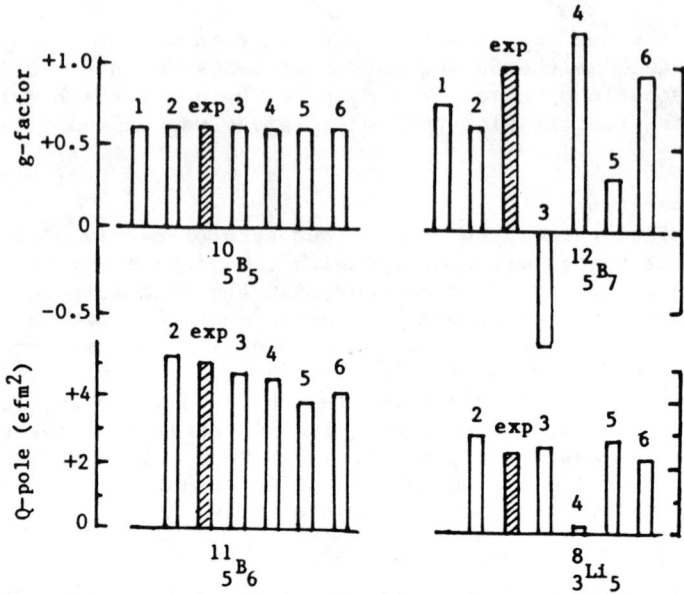

Fig. 4. Typical examples for a comparison between experimental values and those from several theoretical approaches for some g-factors and quadrupole moments. Theory 1,2 ($0\hbar\omega$ space, see refs. 3 and 2, respectively); theory 3-6 (present calculations in the $2\hbar\omega$ space with RSC for b = 2.0 and 1.7 fm and Sussex for b = 1.6 and 1.4 fm, respectively).

^{11}C, ^{12}N, ^{13}C and ^{13}N.

The quadrupole moment of ^{11}B is an example for little sensitivity to the details of the wave function, in contrast to ^{8}Li, where the RSC interaction for b = 1.7 fm fails completely. Some other cases where the Q-pole moments depend strongly on the interaction are ^{8}B, ^{12}B and ^{14}N.

It should be remarked that for the $0\hbar\omega$ space (fig. 4, set 2) an effective charge of 0.35e has been used. For the $2\hbar\omega$ calculations a much smaller value can be taken. The RSC sets with b = 2.0 and 1.7 fm require a value of $\Delta e \approx 0.1e$. For the Sussex matrix elements a different approach has been followed. In this case the size parameter for sets 5 and 6 (see fig. 4) has been increased from b = 1.6 and 1.4 fm, respectively, to b = 1.8 fm. With the latter value of b the rms charge radii can be quite well reproduced, see also fig. 3. Since the quadrupole moment is proportional with b no effective charge is found to be necessary to reproduce the experimental values of these moments.

THE PARTICLE-HOLE STRUCTURE OF p-SHELL WAVE FUNCTIONS

In figs. 5 and 6 the nature of the wave functions for the Reid

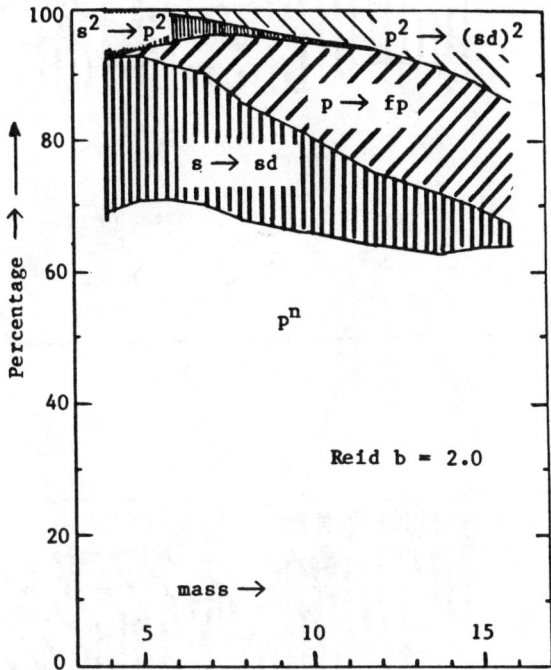

Fig. 5. Structure of the ground state of A = 4-16 nuclei in a $2\hbar\omega$ space with the RSC interaction for b = 2 fm.

soft core interaction with b = 2.0 fm and the Sussex interaction for b = 1.6 and 1.4 fm is depicted. It follows that in all cases the admixtures of the $2\hbar\omega$ components are large (20-40%) and roughly independent of the mass number. This mass independence may explain why earlier calculations in the p^n space were so successful. An analysis of the $2\hbar\omega$ components in terms of one particle - one hole excitations i.e. 1s -> 2s1d and 1p -> 1f2p reveals that the importance of each of these 1p-1h excitations varies gradually with A, but that the total 1p-1h contribution is again quite stable for A = 6-14. The same holds for the total 2p-2h admixtures which are given for $(1s)^2$ -> $(1p)^2$ and $(1p)^2$ -> $(2s1d)^2$ excitations separately. Note that for Sussex with b = 1.4 fm the 2p-2h contributions are even stronger than the 1p-1h intensities. It should also be remarked that in the middle of the p-shell mass region the 1p → 1f excitations in general contribute more than those of 1p → 2p.

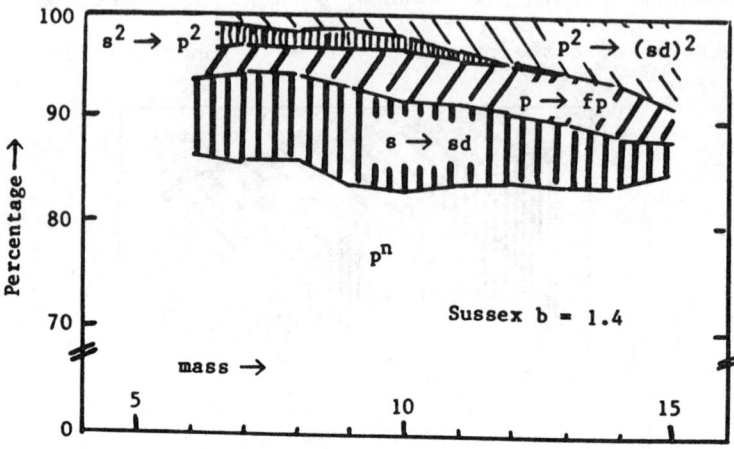

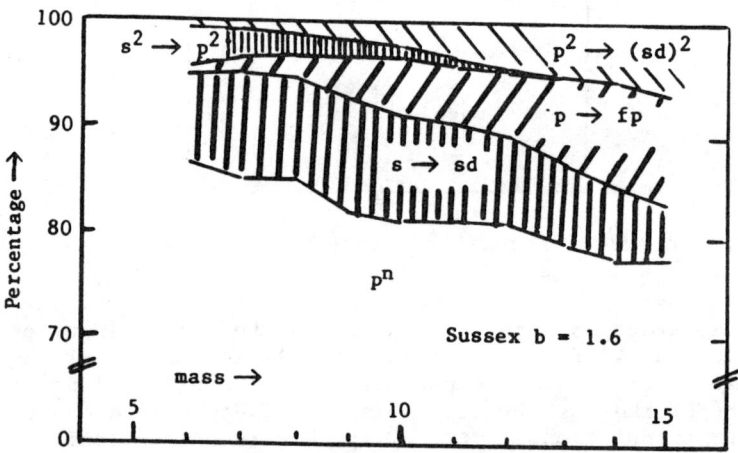

Fig. 6. Structure of the ground state of A = 6-15 nuclei in a $2\hbar\omega$ space with the Sussex interaction for b = 1.4 and 1.6 fm.

INTRUDER STATES

A major problem when enlarging the model space is a consistent description of intruder states, i.e. states that have a dominant particle-hole character. As typical examples we will discuss some states in ^{16}O.

Let us restrict ourselves to the 4^-(T=0) doublet, see fig. 7. In the $1\hbar\omega$ space just one 4^-(T=0) state can be formed by the

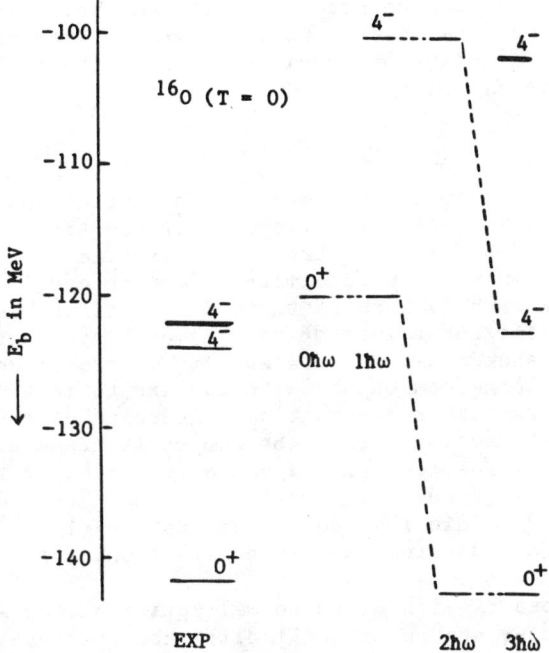

Fig. 7. Calculated binding energies of the 0^+ and 4^- states in ^{16}O as a function of included $N\hbar\omega$ excitations with $N = 0-3$.

coupling of a $p_{3/2}$ hole to a $d_{5/2}$ particle. In the $(0+1)\hbar\omega$ space the excitation energy of the yrast 4^- state can be well reproduced. An increase of the model space such that $2\hbar\omega$ excitations are included affects only the 0^+ ground state and pushes it downwards by about 20 MeV, see fig. 7. The negative-parity states can not be affected by $2\hbar\omega$ excitations and thus their excitation energy is increased by about 20 MeV. The inclusion of $3\hbar\omega$ excitations increases the binding energy of the 4^- by about 20 MeV such that again the correct excitation energy is obtained. In the $3\hbar\omega$ model space one obtains many additional 4^- states. Only the lowest member of this set is shown here. Its binding energy (and thus its excitation energy) lacks the correction of 20 MeV for which $N\hbar\omega$ excitations with $N \geq 5$ should be included. It is not possible to reproduce the experimentally observed 4^- doublet (see fig. 7) with the present approach, because the dimensions for $N \geq 5$ become too large without further truncations. It should be remarked that the shifts of about 20 MeV discussed above are found for all interactions discussed in this paper.

The situation for nonnormal-parity states in other nuclei is not always as serious as in ^{16}O. In many cases the low-lying nonnormal-parity states can be produced with $1\hbar\omega$ excitations and thus can be described consistently in a $3\hbar\omega$ space. With present-day computer facilities it might be possible to treat the p-shell nuclei in a complete $3\hbar\omega$ space.

SUMMARY

In this paper a no-core shell-model treatment of the A = 4-16 nuclei is discussed. The effective two-body interaction has been derived from the Reid soft-core potential (corrected for short-range effects with empirically determined Talmi integrals) and from the Sussex matrix elements. Each of these interactions has been applied with two different values of the harmonic oscillator size parameter b. It is shown that the states with a predominant p^n configuration have large components with $2\hbar\omega$ excitations. The latter admixtures are rather constant as a function of mass number. Depending on the interaction chosen the $2\hbar\omega$ excitations are found to contribute 20-40%. The resulting spectra for three of the four interactions are similar to those obtained in the p^n space (for the latter see refs. 1-3). This also holds for most multipole moments, although large deviations from experiment are found for a few cases.

A consistent description of nonnormal-parity states and normal-parity intruder states is still difficult, because it requires probably large admixtures of $N\hbar\omega$ excitations with $N \geq 3$. With present-day computing facilities it might be possible to treat all p-shell nuclei in the complete $3\hbar\omega$ model space. This would make a consistent description of most $0\hbar\omega$ and $1\hbar\omega$ states possible. A search for a better interaction and the study of the effects of $N\hbar\omega$ admixtures on critical observables such as particle transfer, (e,e') form factors, β-decay rates and other electromagnetic processes will be continued.

I would very much like to thank my collaborators Fons van Hees, Nico Poppelier and Lex Wolters, whose contributions to this project are indispensable.

References

1. A.G.M. van Hees and P.W.M. Glaudemans, Z. Phys. <u>A314</u>, 323 (1983).
2. A.G.M. van Hees and P.W.M. Glaudemans, Z. Phys. <u>A315</u>, 223 (1984).
3. S. Cohen and D. Kurath, Nucl. Phys. <u>73</u>, 1 (1965).
4. Y.S. Koh and P. Goldhammer, Phys. Rev. <u>C19</u>, 487 (1979).
5. D. Zwarts, thesis Utrecht University, 1984.
6. J.P. Elliott, A.D. Jackson, H.A. Mavromatis, E.A. Sanderson and B. Singh, Nucl. Phys. <u>A121</u>, 241 (1968).
7. N.A.F.M. Poppelier, L.D. Wood and P.W.M. Glaudemans, Phys. Lett. <u>157B</u>, 120 (1985).
8. P.C. Zalm, J.F.A. van Hienen and P.W.M. Glaudemans, Z. Phys. <u>A287</u>, 255 (1978).

LARGE SCALE CALCULATIONS OF THE NUCLEAR SPECTRUM

K. W. Schmid, E. Hammarén
Institut für Theoretische Physik, Universität Tübingen
D-7400 Tübingen, FRG

F. Grümmer
Institut für Kernphysik, Kernforschungsanlage Jülich
D-5170 Jülich, FRG

ABSTRACT

Three models are presented, which with increasing accuracy attempt to approximate the in large modelspaces numerically inaccessible complete shell-model expansion of the nuclear wave functions. All of these use angular momentum and particle number projected Hartree-Fock-Bogoliubov quasiparticle determinants as basic building blocks of the theory but differ in the degree of sophistication of the variational principles, which are used to determine the configuration mixing as well as the quasiparticle degrees of freedom. The performance of the three models is illustrated by a couple of numericaal examples.

INTRODUCTION

The microscopic description of many nuclear structure phenomena requires the use of single particle basis systems, which are far too large to allow for a complete diagonalization of an appropriate effective many nucleon Hamiltonian in the way of the well-known shell-model configuration-mixing (SCM) approach[1]. One is therefore forced to truncate the complete SCM expansion of the nuclear wave functions to a numerically-manageable number of A-nucleon configurations, which nevertheless should account for as many as possible degrees of freedom being relevant for the particular problem under consideration.

This can be achieved by expanding the nuclear wave function not with respect to a somehow arbitrary ad hoc chosen potential as it is done in the SCM method, but with respect to the optimal mean field each of the nucleons feels due to its interactions with all the others. The main advantage of these so-called "quasiparticle" approaches is that the mean field is extracted directly from the Hamiltonian via a variational procedure and thus the residual interaction between the resulting A-nucleon configurations and the corresponding reference determinant (the "quasiparticle vacuum") is minimized. Hence one may hope to obtain a rather good approximation to the exact SCM solutions with relatively few quasiparticle configurations, at least for that particular class of nuclear excitations to which the underlying variational principle is adjusted.

The price to be paid for this attempt to account for as many degrees of freedom as possible via as few as possible configurations is that the quasiparticle determinants usually break some of the symmetries required by the Hamiltonian. So, for example, the Hartree-

Fock-Bogoliubov (HFB)[2] determinants, which we are going to use as basic building blocks of our theory in the following, are in general neither eigenstates of the particle number nor of the angular momentum operators. Thus complicated projection techniques have to be used in order to obtain physical configurations with the desired quantum numbers from these "intrinsic" determinants.

In a recent paper[3] we have outlined the general structures of such spin and number-conserving quasiparticle approaches based on HFB-theory and presented the mathematical apparatus needed for practical calculations. Furthermore, in this paper approximation schemes of various sophistication have been developed yielding a hierarchy of various nuclear structure models. Three of these will be discussed in the present lecture.

THE MONSTER (HFB) APPROACH

The numerically simplest of these models is the MONSTER (HFB) approach[4] (MOdel handling many Number and Spin projected Two quasiparticle Excitations with Realistic interactions and modelspaces). Here we start by performing a standard selfconsistent HFB calculation using the (even A) vacuum

$$|q\rangle \equiv (\prod_{a=1}^{n} a_\alpha)|0\rangle \tag{1}$$

as test wave function and not worrying about possible symmetry violations except for requiring conservation of the average particle numbers. This yields an "intrinsic" HFB transformation

$$F \equiv \begin{pmatrix} A^T & B^T \\ B^+ & A^+ \end{pmatrix} \quad ; \quad F^+F = FF^+ = \mathbb{1} \tag{2}$$

connecting the quasiparticle creators a_α^+ and annihilators a_α to the corresponding particle operators c_i^+ and c_i of the chosen n-dimensional single particle basis:

$$a_\alpha^+ \equiv \sum_{i=1}^{n} (A_{i\alpha} c_i^+ + B_{i\alpha} c_i)$$

$$a_\alpha \equiv \sum_{i=1}^{n} (B_{i\alpha}^* c_i^+ + A_{i\alpha}^* c_i) \tag{3}$$

The corresponding mean field is obviously an average over usually many angular momenta and several different particle numbers.

Using the operators (3) then truncated intrinsic configuration spaces are constructed. For the considered even A system they consist out of the vacuum (1) and the two quasiparticle (2qp) excitations with respect to it:

$$\mathcal{C}_e(\text{intrinsic}) \equiv \{|q\rangle, |\alpha\beta\rangle \equiv a_\alpha^+ a_\beta^+ |q\rangle \quad ; \quad \alpha<\beta = 1,\ldots,n\} \tag{4}$$

For the neighbouring odd A systems only the 1qp determinants are included:

$$\mathcal{C}_0(\text{intrinsic}) \equiv \{|\alpha\rangle \equiv a_\alpha^+ |q\rangle \quad ; \quad \alpha=1,\ldots,n\} \tag{5}$$

The configurations (4), (5) are in general neither eigenstates of the angular momentum nor of the particle number operators. In the next step therefore these symmetries are restored with the help of the integral operator[5,6]

$$\hat{P}(Z_0 N_0 IM;K) \equiv \frac{2I+1}{32\pi^4} \int d\Omega D_{MK}^{I*}(\Omega) \hat{R}(\Omega) \int d\varphi_n e^{i\varphi_n(N_0-\hat{N})} \int d\varphi_p e^{i\varphi_p(Z_0-\hat{Z})} \tag{6}$$

where $R(\Omega)$ is the usual rotation operator, $D_{MK}^I(\Omega)$ its representation in angular momentum eigenstates, and $\hat{Z}$ and $\hat{N}$ are the proton and neutron number operators, respectively. Application of (6) to the configurations (4), (5) then yields the "physical" configuration spaces

$$\mathcal{C}_e(Z_0 N_0 IM) \equiv \{|qK;Z_0 N_0 IM\rangle, |\alpha\beta K;Z_0 N_0 IM\rangle; \alpha<\beta=1,\ldots,n\} \tag{7}$$

for the even A and

$$\mathcal{C}_0(Z_0 N_0 IM) \equiv \{|\alpha K;Z_0 N_0 IM\rangle \equiv \hat{P}(Z_0 N_0 IM;K)|\alpha\rangle; \alpha=1,\ldots,n\} \tag{8}$$

for the neighbouring odd A systems, respectively. Finally, for given Z_0, N_0 and I, the Hamiltonian is diagonalized in the corresponding non-orthogonal basis systems (7) and/or (8):

$$(H^{Z_0 N_0 I} - E^{Z_0 N_0 I} N^{Z_0 N_0 I}) f^{Z_0 N_0 I} = 0 \tag{9}$$

with

$$(f^{Z_0 N_0 I})^+ N^{Z_0 N_0 I} f^{Z_0 N_0 I} = 1 \tag{10}$$

and the overlap matrix $N^{Z_0 N_0 I}$ taking care of the non-orthogonality.

Using parity- and isospin-projection-conserving, axially symmetric quasiparticle operators (3) the MONSTER (HFB) approach was first applied in ref.[4] to several nuclei reaching from ^{20}Ne to ^{164}Er and later on used in more systematic studies in the $A \approx 130$[7,8] and $A \approx 80$[9] mass regions. In all the cases considered rather good agreement with the experimental data of many states, and, in small modelspaces, where these are available, also with the results of complete SCM diagonalizations was achieved.

As an example we shall present here some results from the $A \approx 130$ mass region. There as single particle basis for both protons and neutrons the $2s_{1/2}$, $1d_{3/2}$, $1d_{5/2}$, $0g_{7/2}$, $1f_{7/2}$, $0h_{9/2}$ and $0h_{11/2}$ orbits had been used, and as effective interaction a slightly renormalized microscopically calculated[10] nuclear matter G-matrix on the basis of the Bonn-potential[11] had been taken. Details of the renormalization as well as of the choice of single particle energies can be found in ref. 7 and will hence not be repeated here. Using one and the same Hamiltonian in refs.[7,8] more than 50 different nuclei in this mass region had been studied.

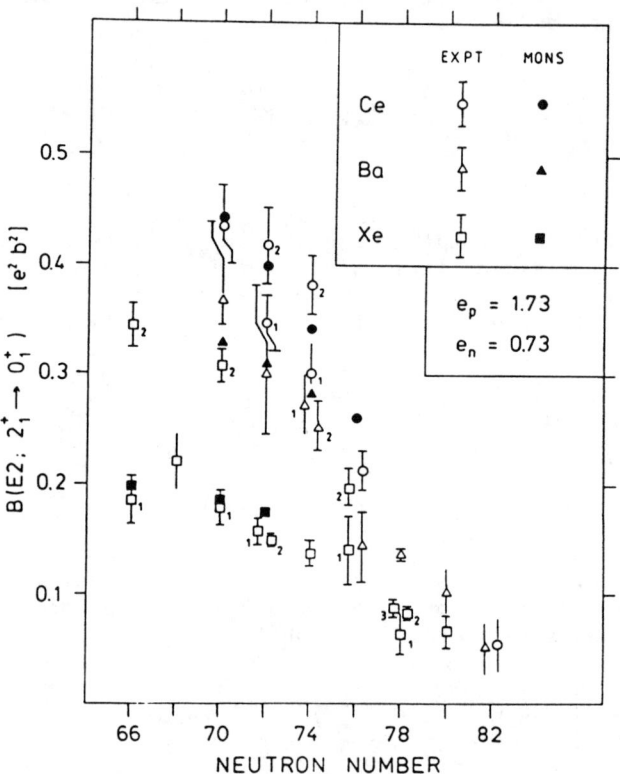

Fig. 1. $B(E2; 2_1^+ \rightarrow 0_1^+)$ transitions in some A 130 nuclei

Fig. 1 presents the systematics of the $B(E2; 2_1^+ \rightarrow 0_1^+)$ transition probabilities in some doubly even Ce, Ba and Xe isotopes. The experimental data (EXPT) are compiled from refs. 12-27. The theoretical results (MONS) are calculated with fixed effective charges $e_p = (1+\chi)e$ and $e_n = \chi e$ where $\chi = 0.73$ was obtained by a least square fit to the seven considered Ce and Ba isotopes. For the N=72 and N=74 Ce isotopes the experimental values indexed by "2" were used in the fit.

The theory describes quantitatively well the decreasing B(E2) probabilities towards the spherical N=82 and Z=50 (from Ce to Xe) shell closures. For the N=66 and N=70 Xe isotopes the theory seems to prefer the results from the earlier measurements (indexed by "1"). However, the surprisingly high values from recent measurements[18,22] (index "2"), where the estimated contributions of the side feeding were substracted, seem to indicate that the effective deformation of the lighter mass Xe isotopes should be considerably higher than theoretically predicted. This indication is supported by the calculated 2^+_1-0^+_1 energy differences in these Xe nuclei yielding a slightly smaller moment of inertia than experimentally observed. One obvious reason for this deficiency of the calculation could be the small number of valence protons (only four) of the Xe isotopes in our present modelspace. We shall therefore repeat some of our calculations within a larger modelspace including the ten $0g_{9/2}$ protons in the very near future. On the other hand it would also be of great interest to perform careful revised experiments for the B(E2) values of several other nuclei in this region to remove the impression of an uncertain experimental situation, which Fig. 1 definitely suggests.

In Fig. 2 we compare experimental and predicted properties of the yrast bands in some Ce isotopes. We have labelled the different parts of these bands indicating their dominant intrinsic structure.

The model accounts well for the changing energy behaviour of the ground and the crossing 2qp bands: 1) The steepening slopes of the $E_{ex}[I(I+1)]$ curves towards the more spherical heavier isotopes, 2) the non rotational behaviour of the γ-ray energies E_γ as a function of the angular momentum and 3) up to details the larger moments of inertia of the crossing 2qp bands.

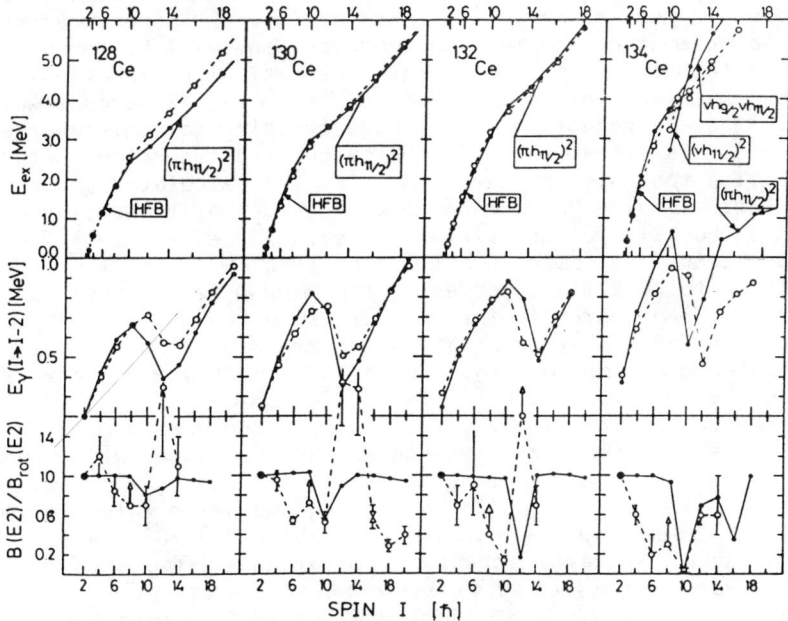

Fig. 2. Yrast properties of some Ce isotopes.

Up to ^{132}Ce the model predicts in agreement with experimental arguments[12,14] that the $h_{11/2}$ protons form the crossing 2qp band. In ^{134}Ce, however, the lowest 10^+ state is calculated to have a distinct $h_{11/2}$ neutron origin in agreement with the g-factor measurement[27] (see also Fig. 3). The calculated band is a K-isomer with an effective K-value $K_{eff}=10$. Consequently the level sequence associated with this isomer has odd and even spin members. Furthermore the moment of inertia of the calculated band is even samller than that of the ground band and hence much smaller than the moment of inertia of the band associated with the 10_1^+ isomer at 3208.4 keV in ref.[28]. The 11^+ level at 3973.2 keV[28] could, however, well be the first member of the isomeric band as we suggest in Fig. 2. For the 10_2^+ state at 3719.3 keV[28] and for the $\Delta I=1$ band associated with it, the model yields a $\nu h_{9/2}\, \nu h_{11/2}$ structure up to spin 14^+. At spin 16^+ the $(\pi h_{11/2})^2$ proton band with only even spin values becomes the yrast one in the calculation. Even though the negative g-factor[29] of the 10_2 state (see Fig. 3) and the energy behaviour (see $E\gamma$ for ^{134}Ce) support this interpretation, we consider it as tentative, only, because the position of the $\nu h_{9/2}$ orbit in the present basis was found to be about 1 MeV too low[7].

The experimental B(E2) transition probabilities of the ground bands in the Ce isotopes show a prominent reduction of the transition rates from the ones calculated for a rigid rotor. This so-called "prealignment anomaly" has been observed (see refs. for Fig. 1) also in Ba and Xe isotopes. The theoretical B(E2)'s in Fig. 2 as well as those for the Ba's and Xe's show an almost pure rotational behaviour up to the band crossing region, though in all the nuclei studied here large progressively with spin increasing 2qp admixtures were obtained. If the "prealignment anomaly" will persist future DSA measurements[22], where the problematic side feeding contributions will be removed with coincidence techniques, more elaborate theoretical explanations may have to be invented. The calculated B(E2)'s in Fig. 2 account well for the increasing reduction in the band crossing region. towards the heavier Ce isotopes. In ^{134}Ce the calculated reductions occurs one transition too late, however. In ^{134}Ce the calculated and experimental values at spin 10^+ correspond to the $10_2^+ \to 8_1^+$ transition and at spin 12^+ to the $12_1^+ \to 10_2^+$ transition. At spin values 14^+ to 18^+ the yrast transitions are shown. A still open problem remain the extremely high experimental B(E2)'s observed in the lighter Ce isotopes for the transition feeding states in the band crossing region. From the microscopic point of view these high values are difficult to understand. The calculated g-factors of Fig. 3 clearly demonstrate the nature of the ground and the crossing bands in the Ce isotopes. Up to the band crossing the values are slightly reduced from the purely collective $g \sim Z/A \sim 0.4$. At the band crossing and beyond the states with predominant proton character have large positive g-factors, whereas the $(\nu h_{11/2})^2\, K_{eff}=10^+$ band in ^{134}Ce has a large negative value in good agreement with experiment[27]. As discussed above the 10^+ state of the $\nu h_{9/2}\, \nu h_{11/2}$ band can nearly account for the negative g-factor of the 10_2^+ state[29], but we must consider this assignment as tentative.

Finally, as an example that the MONSTER (HFB) approach works also in even larger modelspaces, we quote a result of an earlier cal-

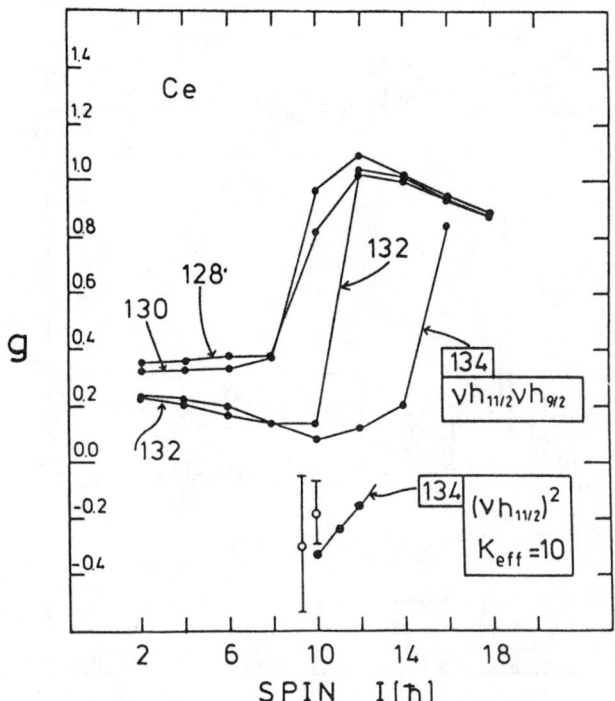

Fig. 3. g-factors in some Ce-isotopes.

culation[30] being based on the no-pairing limit of the present approach. Since, for realistic forces, in intrinsic HFB calculations for ^{20}Ne no pairing correlations are found, this calculation can be considered as a MONSTER (HFB) calculation. In ref.[30] all orbits from $0s_{1/2}$ up to the $0g_{9/2}$ have been used as single particle basis, and the kinetic energy operator together with the G-matrix of Barett et al.[31] being based on the Hamada-Johnstone potential served as effective Hamiltonian.

Figs. 4 and 5 show the B(E2)-strength distributions for the isoscalar electric quadrupole giant resonance as obtained in these calculations for ^{20}Ne and ^{28}Si, respectively. In ^{20}Ne inelastic α-scattering experiments[32] find 35(+25;-15)% of the classical energy weighted sumrule (CEWSR) for the isoscalar E2-mode between 18 MeV and 28 MeV excitation energy, while in ^{28}Si, also by inelastic α-scattering, 27(±6)% of the CEWSR are detected between 15.5 MeV and 23 MeV[33]. The theoretical predictions of ref.[30] are 24% and 26% for ^{20}Ne and ^{28}Si in these energy intervals, respectively. Furthermore, in both cases also the predicted spreading of the E2-strength is in nice agreement with the experimental findings.

However, in spite of its impressive applicability to a rich variety of nuclear structure problems all over the mass table, the MONSTER (HFB) approach is restricted to a rather special class of nuclear excitations.

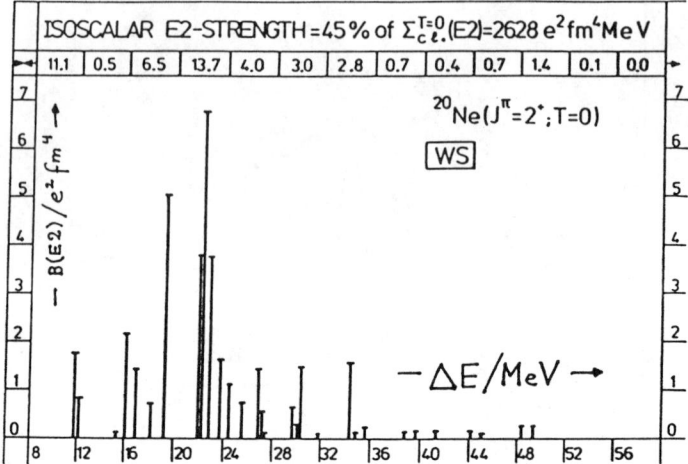

Fig. 4. The isoscalar E2-resonance in ^{20}Ne.

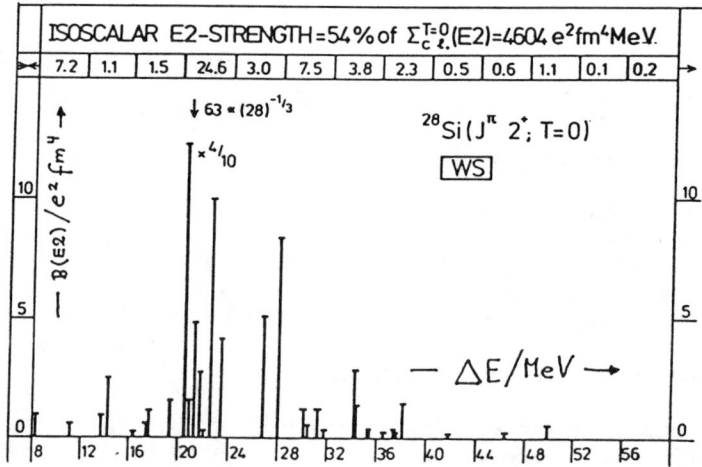

Fig. 5. The isoscalar E2 resonance in 28 Si.

This limitation is caused by the fact that due to the use of a spin-independent mean field all the dynamic changes in the structure of the wave functions with increasing spin have to be accounted for entirely by the configuration mixing of the included spin- and number-projected 2qp excitations. It is obvious that with such a configuration space drastic changes of, e.g., the deformation with growing angular momentum ("shape transitions") can hardly be described. Similarly the method becomes inadequate at the second backbend, where the yrast states adopt a predominantly 4qp nature with respect to the intrinsic vacuum, and, furthermore it is clear that the quality of the description of the excited states has to deteriorate with increasing spin. Finally, because the intrinsic mean field averages over both angular momenta and particle numbers, the MONSTER (HFB) approach should also not perform too well if drastic changes with the nucleon numbers are to be described as they occur for example in the neighbourhood of shell closures.

In order to cure these deficiencies the concept of a fixed intrinsic mean field had to be given up. Instead one has to optimize the HFB transformation for each spin separately, i.e. one has to solve the HFB variational problem with spin and number projection before the variation. This leads to the second model discussed in the present lecture.

THE MONSTER (VAMPIR) APPROACH

In this model one starts again from a variational calculation. However instead of using the intrinsic vacuum (1) as test wave function, now the angular momentum and particle number projected HFB vacuum

$$|\Psi_1\rangle \equiv |q(I)0;Z_0N_0IM\rangle \beta_{11} = \hat{P}(Z_0N_0IM;K=0)|q_1(I)\rangle \beta_{11} \tag{11}$$

defines the variational space. Here

$$\beta_{11} \equiv \langle q_1(I)0;Z_0N_0IM|q_1(I)0;Z_0N_0IM\rangle^{-1/2} \tag{12}$$

ensures the normalization and, for simplicity, axially symmetric quasiparticles (yielding K=0 for the vacuum) have been assumed. Furthermore, writing $|q_1(I)\rangle$ instead of $|q\rangle$ for the unprojected vacuum, we want to indicate its dependence on the actual considered spin value. Variation of the quasiparticle degrees of freedom in (11) then yields for each spin value a different HFB-transformation

$$F(I) = \begin{pmatrix} A^T(I) & B^T(I) \\ B^+(I) & A^+(I) \end{pmatrix} \quad ; \quad F^+(I)F(I) = F(I)F^+(I) = 1 \tag{13}$$

in terms of which the quasiparticle operators (3) have to be redefined as

$$\begin{pmatrix} a^+(I) \\ a(I) \end{pmatrix} = F(I) \begin{pmatrix} c^+ \\ c \end{pmatrix} \tag{14}$$

Consequently, already the intrinsic content (4) and (5) of the MONSTER configuration spaces (7) and (8) does now depend on the angular momentum, too.

The above described VAMPIR procedure (Variation After Mean field

Projection In Realistic modelspaces) as well as its combination with the MONSTER configuration spaces yielding the MONSTER (VAMPIR) model has been discussed in detail in ref.[34]. Later applications have then been reported in refs.[35,36]. The essential advantage of this method is that it yields always a 0qp state as dominant component for the yrast states, no matter how high the actual spin may be. Thus the spin limitation of the original MONSTER (HFB) approach vanishes and even drastic changes of the yrast structures with increasing spin can be described in a natural way. Because of the variation after projection furthermore obviously also the performance of the model near shell closures is drastically improved and, last but not least, since the configuration spaces (7) and (8) are now constructed with respect to the optimal transformation for each spin value separately, the quality of the description of the excited states does not depend on the considered angular momentum anymore.

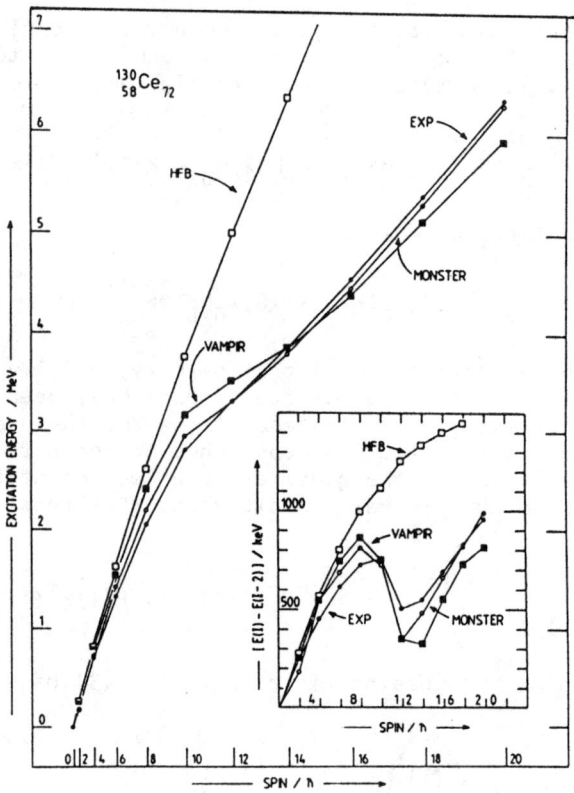

Fig. 6. Yrast band in ^{130}Ce.

Fig. 6 shows, as an example, the yrast band of ^{130}Ce as obtained by projection after the variation from the intrinsic HFB solution (HFB), by the MONSTER (HFB)-approach (same as in Fig. 2, here labelled as MONSTER) and by the projected vacua (11) obtained by variation after projection (VAMPIR), and compares these results[35] to the experimental data[14]. It is seen that the one determinantal VAMPIR approach fits the data equally well as the original MONSTER (HFB) approach, which needs about 130 projected quasiparticle determinants in order to account for the dynamical changes of the yrast states with increasing spin. This becomes evident from the large difference between the HFB and MONSTER (HFB) spectra.

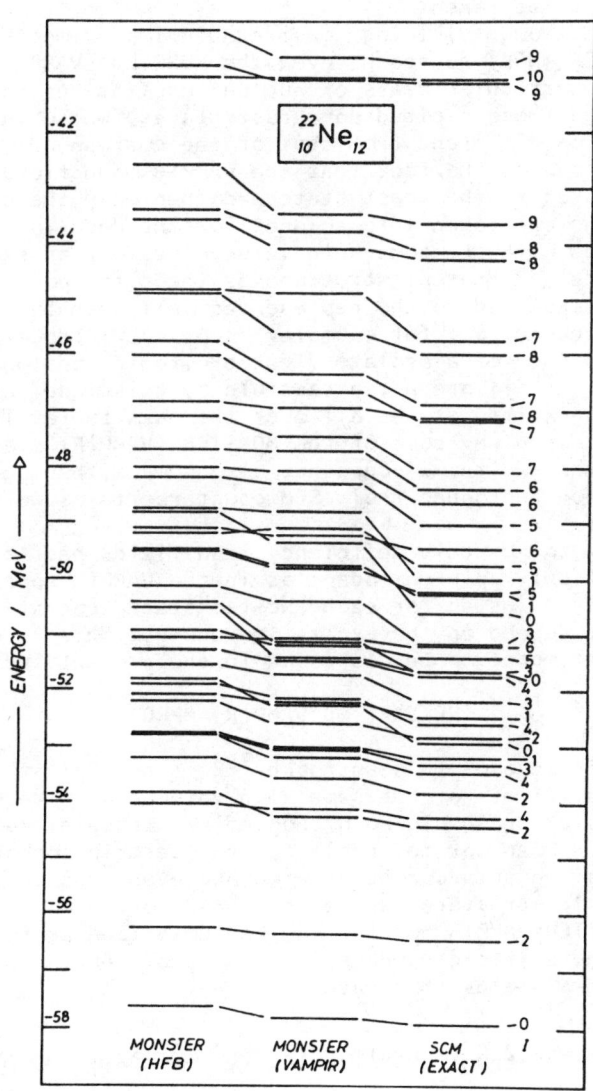

Fig. 7. Low energy spectrum of ^{22}Ne.

That the quality of the description of the excited states is considerably improved, if the spin-dependent VAMPIR mean fields are used for the construction of the MONSTER configuration spaces, can be read from Fig. 7, where the results of such a calculation[34] for the nucleus ^{22}Ne are compared to the MONSTER (HFB) results as well as to those of a complete SCM diagonalisation of the chosen MSDI-Hamiltonian, which could here be calculated since only the 1s0d single particle basis was taken.

However, though yielding considerable improvements with respect to the MONSTER (HFB) approach, even the MONSTER (VAMPIR) model is limited to a particular class of nuclear excitations, namely the yrast band and those excited non-yrast states, whose internal structure is not too different from that of the corresponding yrast states. This is due to the fact that the VAMPIR mean field for a given spin is adjusted to the yrast state and then only the projected 2qp configurations are taken into account for the description of the excited states with the same spin value. Thus, if an excited state has a completely different structure as the corresponding yrast one, it cannot be expected to the represented well even by the MONSTER (VAMPIR) procedure. So, for example, shape coexistence phenomena like the occurence of a prolate first excited 0^+ on top of an oblate groundstate in ^{28}Si, are not accessible by this model as well as many other similar excited states all over the mass table. This deficiency also is the reason why some of the MONSTER (VAMPIR) states in Fig. 7 (especially the excited 0^+-ones but also some with higher angular momenta) do not reproduce their SCM counterparts as well as most of the states do.

In order to cure this deficiency, mean fields had to be used, which are not only spin-dependent as in the VAMPIR approach but which also are newly adjusted for each excited state separately, i.e., which do also depend on the excitation energy. This is done in the third and last model to be discussed in the present lecture.

THE EXCITED VAMPIR APPROACH

The basic idea of this approach[37,3] is to apply the VAMPIR procedure several times for the same spin value always requiring that the actual solution should be orthogonal to those already obtained. Thus successively an optimal basis for the description of the lowest states of a given spin can be created and even drastic changes in their intrinsic structure can be accounted for.

The EXCITED VAMPIR model[38] may be summarized as follows: For any given spin and particle numbers first a usual VAMPIR calculation is performed. This yields the state (11)

$$|\Psi_1\rangle = \hat{P}(Z_0 N_0 IM; K=0)|q_1(1)\rangle \beta_{11} \equiv |\phi_1\rangle \beta_{11} \tag{15}$$

Next

$$|\psi_2\rangle = |\phi_1\rangle \beta_{12} + |\phi_2\rangle \beta_{22} \tag{16}$$

with

$$|\phi_2\rangle \equiv \hat{P}(Z_0 N_0 IM; K=0)|q_2(I)\rangle \tag{17}$$

is used as test wave function for the first excited state. Here β_{22} and β_{12} are determined by requiring normalization and orthognality of $|\Psi_2^{12}\rangle$ with respect to $|\Psi_1\rangle$ (Schmidt-orthogonalization). This procedure is now repeated up to

$$|\Psi_m\rangle = \sum_{i=1}^{m-1} |\Psi_i\rangle \beta_{im} \tag{18}$$

is determined. Finally, then the residual interaction is diagonalized in these configurations:

$$(H-E\mathbb{1})g=0 \quad ; \quad g^+g = gg^+ = \mathbb{1} \tag{19}$$

where

$$H_{ij} \equiv \langle\Psi_i|\hat{H}|\Psi_j\rangle \quad ; \quad i,j=1,\ldots,m \tag{19}$$

Obviously, this approach has a couple of advantages with respect to all other methods developed up to now in order to approximate the complete SCM expansion of the nuclear wave functions. First of all, even states with a completely different structure as the corresponding yrast state are described in a rather natural way. Second, the description of any state can always be improved by increasing the number of successive solutions m. It is trivial to prove that for larger m the prescription will always converge to the exact SCM solution. Third, the use of the variational principle also for the excited states minimizing the residual interaction between the different m solutions. Thus, successively an optimal basis of A-nucleon configurations is created. Diagonalizing the residual interaction within the lowest m solutions $|\Psi_i\rangle$ one obtains the best possible description of the m lowest states of the considered spin value, which can be achieved by m projected HFB type quasiparticle determinants. Fourth, and last but not least, the EXCITED VAMPIR model automatically selects the essential degrees of freedom, which are relevant for a particular state under consideration, no matter, whether they are of collective or single particle nature. Thus no a priori assumption about the structure of the states to be considered need to be made.

The numerical feasibility of the EXCITED VAMPIR approach has been shown in ref.[38] by applying it to the four energetically lowest 0^+-states in ^{50}Ti using the 1p0f-shell as basis and McGrory's[39] version of the Kuo-Brown-matrixelements[40] as effective interaction. As can be seen from Fig. 8, in this nucleus the original MONSTER (HFB) approach performs very bad. Because of the neutron $f_{7/2}$ subshell closure in the intrinsic HFB cacuum, this approach yields for ^{50}Ti only four linearly independent 0^+ states, which do not in the least

reproduce the experimental[41] situation. The MONSTER (VAMPIR) model
already improves the situation drastically. Here the subshell closure as well as the spherical symmetry of the intrinsic HFB mean
field are broken, and consequently now 39 linearly independent 0^+-
states are obtained. Furthermore, ground, first and second excited
state are already in much better agreement with experiment. However,
the EXCITED VAMPIR approach does even better. Though only m=4 was
used in these calculations, after diagonalization of the residual
interaction the MONSTER (VAMPIR) groundstate is lowered by 131 keV,
the first exited 0^+ by 350 keV and the second excited 0^+ by even
750 keV yielding rather good agreement with the experimental data.

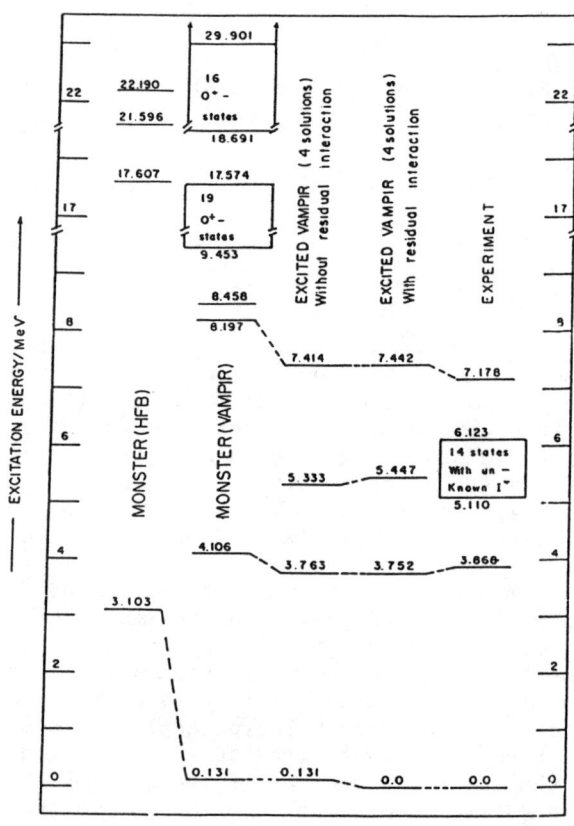

Fig. 8. 0^+-states in ^{50}Ti.

Furthermore, at 5.447 MeV an oblate intruder 0^+ state is obtained as
new second excited 0^+, which is not occuring at all in the MONSTER
(VAMPIR) spectrum being based on a prolate mean field. Unfortunately,
there is no experimental 0^+ state in the predicted energy region.
However, between 5.110 MeV and 6.123 MeV 14 states with unknown I^π
assignment are experimentally reported[41]. We are pretty sure that one
of these must be the counterpart of our second excited 0^+ state.

OUTLOOK

Instead of once more summarizing our results at the end of the present lecture, we would like to give a rough and incomplete outlook what we are doing with the above described methods at the moment and what we intend to do with them in the near future.

First of all, the systematic analyses of many nuclei in various mass regions will continue. So, e.g., at the moment the $A \approx 130$ region is reanalyzed in a larger, more appropriate basis system than used up to now and also the $A \approx 80$ region is reinvestigated. Simultaneously, calculations in a couple of rare earth nuclei and in some light sd-shell nuclei have been started. Besides aiming at a consistent parameter-free description of many nuclei, the main goal of these investigations is to obtain information about the effective interaction in different mass regions and modelspaces and how this interaction may be obtained from microscopically calculated G-matrices. In this context the study of the light nuclei is of special importance, since here easily 5 major shells can be included in the basis systems, so that, e.g., the effects of the giant quadrupole mode on the low excited states can be studied explicitly, while in the heavier systems, at least at the moment, such effects have still to be accounted for by appropriate renormalizations of the Hamiltonian.

Second, a number of special problems can be studied. So, e.g., VAMPIR wave functions have been used for investigations on the neutrinoless double β-decay[42]. Other items being high on our priority list are excited 0^+ states near shell closures (^{16}O, ^{90}Zr etc.), the double γ-decay, the Gamov-Teller resonances and many other problems.

Third, the methods discussed here, should still be improved. So, e.g., we are thinking at the moment about the possibility to use isospin-projection mixing HFB-transformations in order to incorporate proton-neutron pairing and also to break parity conservation on the quasiparticle level in order to obtain optimized mean fields also for negative parity states. Furthermore, on the long run also the assumption of axially symmetric quasiparticles should be given up and, last but not least, using many determinant test wave functions (MAD VAMPIR) would improve the results for odd mass nuclei considerably.

Fourth, in the meantime a microscopic reaction theory on the basis of earlier ideas[43] has been developed which embeds the above described large scale nuclear structure calculations into a full continuum approach. The necessary computer codes for this MERLIN model (Monster description of Electromagnetic Reactions and Low energy Inelastic Nucleon scattering) have been developed and tested. However, since up to now no results of realistic calculations with this magus were obtained, no further details will be given in the present lecture.

Summarizing, we may say that things look indeed rather encouraging at the moment. However, we have just started and even the few items on top of our priority test will still need a lot of time and careful analysis. Furthermore, though having developed a consistent theory with an amazingly large range of applicability, we should not forget that even the most refined version of the EXCITED VAMPIR model will still be only an approximation to the dream of a complete diagonalization in, e.g., a five major shell basis. It is out of this reason mainly that the fear of our Italian friend Gianpaolo Cô, who once translated MONSTER as "Many Obscure Numbers Stop Totally Every Research", will never become true.

ACKNOWLEDGEMENT

We would like to thank B. Fladt, T. Tomoda, T. Horibata, Zheng Ren-rong and M. Kyotoku, who contributed to some of the results discussed in the present lecture, and A. Faessler, who sponsored and supported this work from its very beginning.

REFERENCES

1. E. Halbert, J. B. McGrory, B. H. Wildenthal and S. R. Pandhya, Adv. Nucl. Phys. 4, 316 (1971);
   R. R. Whitehead, A. Watt, B. J. Cole and I. Morrison, Adv. Nucl. Phys. 9, 123 (1977);
   P. J. Brussard and P. W. M. Glaudemans, Shell model applications in nuclear spectroscopy (North-Holland, Amsterdam 1977).
2. N. N. Bogoliubov, ZhETF 34, 58 (1958) [JETP (Sov. Phys.) 7, 41 (1958)];
   N. N. Bogoliubov and V. G. Soloviev, Dokl. Akad. Nauk. SSSR 124, 1011 (1959) [Sov. Phys. Dokl. 4, 143 (1959)];
   J. G. Valatin, Phys. Rev. 122, 1012 (1969).
3. K. W. Schmid, F. Grümmer and A. Faessler, Phys. Rev. C29, 291 (1984).
4. K. W. Schmid, F. Grümmer and A. Faessler, Phys. Rev. C29, 308 (1984).
5. B. F. Bayman, Nucl. Phys. 15, 33 (1960).
6. R. E. Peierls and I. Yoccoz, Proc. Roy. Soc. A70 381 (1957);
   F. Villars, Varenna Lectures, Vol. 36, ed. C. Bloch (Academic, New York, 1966) p. 1.
7. E. Hammarén, K. W. Schmid, F. Grümmer, A. Faessler and B. Fladt, Nucl. Phys. A437, 1 (1985).
8. E. Hammarén, K. W. Schmid, F. Grümmer, A. Faessler and B. Fladt, submitted to Nucl. Phys. A.
9. T. Horibata, K. W. Schmid, E. Hammarén, A. Faessler and F. Grümmer, submitted to Nucl. Phys. A.
10. W. H. Dickhoff, Nucl. Phys. A399, 287 (1983).
11. K. Holinde, K. Erkelenz and R. Alzetta, Nucl. Phys. A194, 161 (1972).
12. J. C. Wells, N. R. Johnson, J. Hattula, M. P. Fewell, D. R. Haenni, I. Y. Lee, F. K. McGowan, J. W. Johnson and L. L. Riedinger, Phys. Rev. C30, 1532 (1984).
13. D. Husan, S. J. Mills, H. Gräf, U. Neumann, D. Pelte and G. Sailer-Clar, Nucl. Phys. A292, 267 (1977).
14. P. J. Nolan, R. Aryaeinejad, D. J. Love, A. H. Nelson, P. J. Smith, D. M. Todd, P. J. Twin, J. D. Garett, G. B. Hagemann and B. Herskind, Phys. Scripta T5, 153 (1983).
15. C. M. Lederer and V. S. Shirley, Table of isotopes, 7th ed., (1978).

16. G. Sailer-Clar, D. Husan, R. Novotny, H. Gräf and D. Pelte, Phys. Lett. 80B, 345 (1979).
17. M. Kitao, M. Kanbe and Z. Mazumoto, NDS 38, 191 (1983).
18. S. Xiangfu et al., preprint 1984;
    P. von Brentano, private communication 1984.
19. H. R. Hiddlestone and C. P. Browne, NDS 17, 225 (1976).
20. Yu. V. Sergeenkov and V. J. Sigalov, NDS 34, 475 (1981).
21. D. C. Kocher, NDS 17, 39 (1976).
22. A. Gelberg and P. von Brentano, private communication 1984.
23. T. Tamura, K. Miyano and S. Ohya, NDS 41, 413 (1984).
24. H. Hanewinkel, W. Gast, V. Kamp, H. Harter, A. Dewald, A. Gelberg, R. Reinhardt, P. von Brentano, A. Zemel, C. E. Alonso and J. M. Arias, Phys. Lett. 133B, 9 (1983).
25. T. Tamura, K. Miyano and S. Ohya, NDS 36, 227 (1982).
26. H. R. Hiddlestone and C. P. Browne, NDS 13, 133 (1974).
27. M. B. Goldberg, C. Broude, E. Dafni, A. Gelberg, J. Guber, G. J. Kumbartzki, Y. Niv, K.-H. Speidel and A. Zemel, Phys. Lett. 97B, 351 (1980).
28. M. Müller-Veggian, H. Beuscher, D. R. Haenni, R. M. Lieder and A. Neskakis, Nucl. Phys. A417, 189 (1984).
29. A. Zemel, C. Broude, E. Dafni, A. Gelberg, M. B. Goldberg, J. Gerber, G. J. Kumbartzki and K.-H. Speidel, Nucl. Phys. A383, 165 (1982).
30. K. W. Schmid, Phys. Rev. C24, 24 (1981).
31. B. R. Barett, R. G. L. Hewitt and R. J. McCarthy, Phys. Rev. C2, 1199 (1970);
    ibid C3, 1137 (1991).
32. K. T. Knöpfle, G. J. Wagner, A. Kiss, M. Rogge, C. Mayer-Böricke and Th. Bauer, Phys. Lett. 64B, 263 (1976)
33. K. van den Borg, M. N. Harakeh, S. Y. van der Werf, A. van der Woude and F. E. Bertrand, Phys. Lett. 67B, 4M5 (1977).
34. K. W. Schmid, F. Grümmer and A. Faessler, Nucl. Phys. A431, 205 (1984).
35. K. W. Schmid, F. Grümmer, E. Hammarén and A. Faessler, Nucl. Phys. A436, 417 (1985).
36. K. W. Schmid, F. Grümmer, E. Hammarén, M. Kyotoku and A. Faessler, submitted to Nucl. Phys. A.
37. K. W. Schmid and F. Grümmer, Z. Phys. A292, 15 (1979).
38. K. W. Schmid, F. Grümmer, M. Kyotoku and A. Faessler, submitted to Nucl. Phys. A.
39. J. B. McGrory, Phys. Rev. C8, 693 (1973).
40. T. T. S. Kuo and G. E. Brown, Nucl. Phys. A114, 241 (1968)
41. D. E. Alberger, NDS 42, 369 (1984) (compilation).
42. T. Tomoda, A. Faessler, K.W. Schmid and F. Grümmer, Phys. Lett. 157B, 4 (1985) and submitted to Nucl. Phys. A.
43. K.W. Schmid and G. Do Dang, Phys. Rev. C15, 1515 (1977);
    ibid C18, 1003 (1978).

GENERAL PROPERTIES OF THE RESIDUAL INTERACTION AS MEASURED NEAR CLOSED SHELLS

W.W. Daehnick

Nuclear Physics Laboratory, University of Pittsburgh
Pittsburgh, Pennsylvania 15260

Abstract

Residual interaction matrix elements for shell model calculations in a specified configuration space are unique to the extent that the nuclear interaction can be represented as a sum of two-body interactions. Many diagonal matrix elements have been obtained from two-nucleon multiplets near closed shells. The limited purity of such multiplets was taken into consideration by weighting the observed fractions of the two-nucleon configurations by their spectroscopic strengths and by using the resulting energy centroids. The occasional simultaneous knowledge of empirical particle-particle and particle-hole multiplets for the same configuration permits estimates of the uncertainty of the extraction procedure and indicates that diagonal matrix elements are often known to better than 10% or 100 keV, whichever is larger. Evidence is increasing that matrix elements for spaces of a major shell (or larger) have remarkable regularities along the lines initially pointed out by Schiffer. A significant but smooth A dependence has been found. The available matrix elements primarily depend on T and A and the parity of the multiplet. They depend in a general way on the coupling of j_1 and j_2, but not noticeably on the major quantum numbers l and n. These systematic features are also found in the matrix elements of Kuo and Brown and in some short range model forces (MDSI), although agreement of experiment with these theoretically derived matrix elements varies. The empirical data suggest that scaling of known matrix elements with $A^{-0.75}$ may be used to predict residual matrix elements for regions away from closed shells or in other major shells.

INTRODUCTION

The extraction of empirical matrix elements for shell model calculations from the spectra of "simple" nuclei goes back to the work of I. Talmi and collaborators, or possibly further. Such matrix elements can be defined even for small spaces (e.g. shell model sub-shells), but in this case they tend to have predictive power only for immediate neighbors with a basically similar structure. Even then, observables apart from level energies, such as electromagnetic transitions or stripping cross sections, are predicted with very limited accuracy.

In 1965, S. Cohen and D. Kurath published matrix elements extracted in a similar spirit, but derived from a simultaneous fit to many p-shell spectra[1]. The success of this effort is well known, and Cohen-Kurath p-shell wave functions are still in use. Until very recently it was impractical to extend such comprehensive shell model fits to the higher and much larger shells, which were only partially known. Historically, some of the most successful matrix elements for the higher shells came from the theoretical work of Kuo and Brown[2]. New insight was gained when J.P. Schiffer and collaborators focussed on unusually pure multiplets near closed shells and found surprisingly general features in empirical matrix elements derived from two-nucleon spectra[3,4].

In a way, Schiffer's observation of the simple universal shape of selected $|j,j,J>$ multiplets should not have been so unexpected, as his empirical systematics have a theoretical parallel in the matrix elements of Kuo and Brown, which were known then, although not presented in the same way. However, the prevalent view at the time was that the two sets of matrix elements - and hence the two approaches - did not agree very well. This happened, in my view, because people compared individual matrix elements, rather than systematic features, and put more stress on the precision and uniqueness of the numerical values of the derived and calculated matrix elements than appears justified in retrospect.

The theoretical work was criticized, because there was some evidence that the convergence of the perturbation expansion was not assured, at least for the Hamada-Johnston nucleon-nucleon force used at the time. A criticism of Schiffer's empirical matrix elements has been that two-nucleon multiplets are never totally without configuration admixtures, even near closed shells[5]. Consequently, some of his deduced values may be more appropriate for sub-shell spaces, and for this reason may not be directly comparable to those of Kuo and Brown, or others which were calculated for a full major shell.

DEFINITION OF THE EFFECTIVE INTERACTION

The Schroedinger equation for a system of N nucleons is given by Eq. 1:

(1) $$H_N \psi(1,\ldots,N) = \left(\sum_{i=1}^{N} T_i + \sum_{i<j}^{N} V_{ij}(r_i, r_j) \right) \psi = E\psi.$$

This equation is exact for the case that the nuclear interaction can be expressed in terms of two-body forces V_{ij}. In practice an exact solution of eq. 1 is developed as a series

(2) $$\psi_p = \sum_{k=1}^{g} a_{kp} \phi_{0k}.$$

where the choice of g depends on the accuracy desired. The wave functions ϕ_{0k} constitute a complete set of solutions of a simpler

Schroedinger equation, e.g. that of the independent particle model where the two-body interactions have been replaced by an average shell model potential $U(r_i)$:

(3) $$H_0\phi_0 = \left(\sum T_i + U(r_i)\right)\phi_0 = E_0\phi_0.$$

The difference between the exact Hamiltonian H_N and that for the independent particle model H_0 is called the residual interaction H':

(4) $$H_N = H_0 + H'.$$

H_0 must generate the correct single-particle states and energies, although it may not be fully defined in this way. In practice $U(r)$ is a harmonic oscillator or Woods-Saxon potential with realistic geometric parameters.

By inserting (2) into eq. 1, multiplying from the left with a basis state ϕ_{0l} and integrating, one transforms eq. 1 into the familiar matrix equation

(5a) $$\sum_{k=1}^{g} \langle\phi_{0l}|H_0 + H'|\phi_{0k}\rangle a_{kp} = E_p a_{lp}.$$

The g integrals in expression (5) are called the matrix elements H_{lk}, which can be calculated, in principle, if the nuclear two-body force is known. For two-particle states we have

(5b) $$H_{lk} = \langle\phi_{0l}|H_0|\phi_{0k}\rangle + \langle\phi_{0l}|H'|\phi_{0k}\rangle = E_k \delta_{lk} + H'_{lk}.$$

where E_ℓ is the unperturbed energy of the two-particle multiplet and $H'_{\ell k}$ is the matrix element of the residual nucleon-nucleon interaction.

Eq. 5a is solved by matrix inversion. The number of terms in eq. 5 depends on the expansion cut-off g in (2) and defines the configuration space which is included. It is clear that the residual interaction H' will differ for small and large values of g, i.e. for small and large configuration spaces. Only if g includes all possible configurations for all N nucleons is the solution Ψ_N exact in the physical sense.

The details of a calculation of H' from realistic or model forces are outside the scope of this talk: however, we note that once a truncated configuration space is chosen through g, an 'exact' theoretical solution is defined, and for such states one has the corresponding relation

(6) $\quad H'_{\ell m} + E_m \delta_{\ell m} = \Sigma_p <\phi_{o\ell}|\psi_p> E_p <\psi_p|\phi_{om}>$

This relation is derived from eq.5a by multiplying with a_{mp}, summing over p and using unitarity. It is eq. 6 which can be used for the deduction of empirical matrix elements, since the overlap integrals under the sum sign can often be measured. Eq.6 becomes particularly simple for diagonal matrix elements, where $\ell=m$. The product $S'(m,p)$ of the two identifcal overlaps is then proportional to the spectroscopic factor S measured in single-nucleon transfer reactions, and we may write

(7) $\quad H'_{mm} + E_m = \Sigma_p S'_{pm} E_p$

We refer to the sum in (7) as the centroid energy and note that the overlap integrals in eq. 6 add up to 1. Hence as long as all significant fractions of a particular configuration are found, only the relative values of S need to be accurately determined. If theory and experiment use identical configuration spaces and single-particle energies, and provided that the ground state of the target used in the transfer experiment can be viewed as a good single-particle state, identical matrix elements should be obtained.

EMPIRICAL SYSTEMATICS

In the early work of Schiffer the sum in eq.7 often was taken over just one term, i.e. only the dominant term was considered if the spectroscopic factors in a multiplet seemed large enough. In more recent work we have taken pains to find and include higher-lying spectroscopic strength in order to justify direct comparison with theoretical values for a major shell. For the multiplets previously used by Schiffer this change made a difference primarily for 0^+ states, but in addition, it has permitted the inclusion of many two-nucleon multiplets not pure enough for the simpler treatment.

One example of a less than pure multiplet and of our treatment is given in the table of Fig. 1. ^{88}Sr has a good inert core, i.e. a major shell closure for neutrons, and ^{87}Sr appears to be a good $g_{9/2}$ hole state. The low-lying states populated in the reaction ^{87}Sr(d,t)^{86}Sr may then be viewed as two-neutron-hole configurations. But they are not all of pure $(g_{9/2})^2$ nature.

As is seen in the table, four 0^+ states are known that contain some $(g_{9/2})^2$ strength. Their centroid energy is 632 keV, i.e. considerably higher than what would be deduced from the dominant state alone. As a consequence the extracted 0^+ matrix element is 632 keV less attractive than it would be in the simplest analysis. In

Ref. 5 this type of refinement has been applied to all known data on two-nucleon multiplets. The result is a broader base of evidence for empirical systematics which in essence, although not in every detail, supports the earlier systematics found by Schiffer.

It is interesting to check the consistency of such empirically deduced matrix elements by inspecting detailed results for the $(f_{7/2})^2$ configuration where results are available from five different nuclei, ranging from ^{42}Ca to ^{54}Co. Figures 2 and 3 give graphic representations of all diagonal matrix elements. The lines in Fig. 2 connect the subgroups for T=0 and T=1, respectively.

Table I

T=1, $(g_{9/2})^2$ diagonal matrix elements from ^{87}Sr(d,t)^{86}Sr. Configuration: $(\nu g_{9/2})^{-2}$, $\ell=4$, $E_0=2.685$ MeV=(unperturbed multiplet energy)

J^π	E_x^a (MeV)	C^2S (extracted)	C^2S (expected)	Matrix elements[b,c] $H'_{9/2}$	
0^+	0.000 2.102 2.203 3.101	0.130 0.013 0.022 0.012 } 0.18	0.2	-2.05	(+0.40-0.04)
2^+	1.077 1.854 2.642 2.789	0.58 0.21 0.11 0.030 } 0.93	1.0	-1.19	(+0.21-0.04)
4^+	2.230 3.362 3.481	1.28 0.02 0.37 } 1.67	1.8	-0.16	(+0.12-0.04)
$(6)^+$	2.857	2.40	2.6	0.17	(+0.11)
$(8)^+$	2.956	3.46	3.4	0.27	

b. The positive errors in this column mainly are due to missing strength which is assumed to be at 4 MeV and the negative errors contain random and fitting errors.

c. E_0 is based on the ground state energies of ^{86}Sr, ^{87}Sr, and ^{88}Sr (See Ref. 5).

The data shown stem not only from the relatively pure particle-hole multiplet of ^{48}Sc, but also from the strongly mixed pn and nn states of A=42 nuclei, and the two-hole states of mass 54 nuclei. For typical interaction energies of 1 to 2 MeV, data of varying precision seem to differ by a few hundred keV. Actually some of this difference is a systematic A dependence. The weakening of the interaction becomes noticeable as A increases (by 29%) in the $f_{7/2}$ shell.

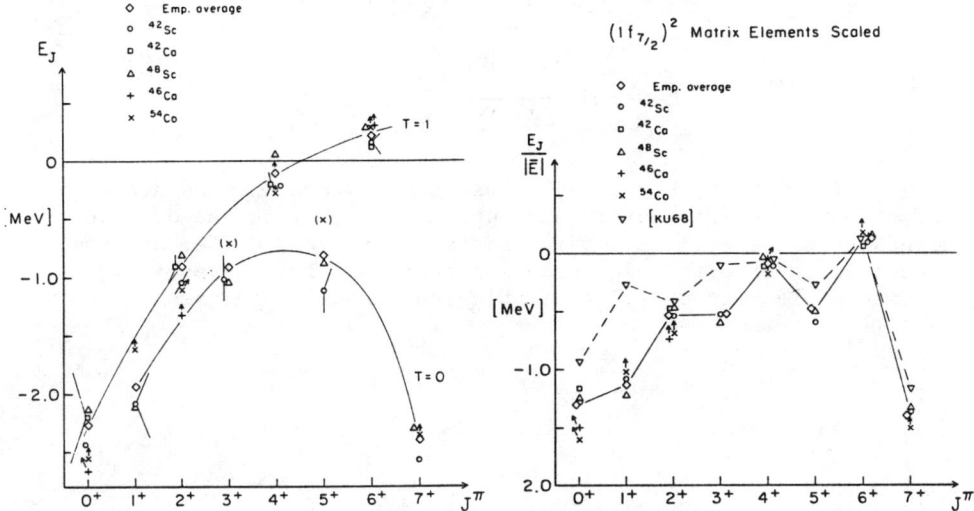

Fig.2. Diagonal matrix elements for known $(1f_{7/2})^2$ multiplets.

Fig.3. Comparison with Kuo-Brown ME, after mass scaling.

In Fig. 3 the matrix elements are replotted after scaling with $E \sim A^{-0.75}$, a factor which is suggested by a broader range of data. The differences are reduced and are no longer systematic. The dashed line connects corresponding theoretical values from Kuo and Brown and allows a first comparison of the two approaches.

It is important to realize that matrix elements which are derived from particle-particle spectra have systematic errors that differ in sign from those derived from particle-hole spectra. In both cases, the experimentally undetected spectroscopic strength lies at higher excitation energies, since one usually best resolves and deals with ground state multiplets. For particle-particle matrix elements errors of this kind produce excessive attraction, as was pointed out for the ^{86}Sr example. However, if comparisons are to be made, particle-hole spectra have to be transformed by a Pandya transformation, where the deduced particle-hole attraction is changed to particle-particle repulsion, and vice versa. Hence the missing strength causes erroneous repulsion, i.e. a systematic error of opposite sign. It is interesting that no systematic variance is seen for the ^{48}Sc values outside the roughly 100 keV scatter.

Available empirical matrix elements range over the entire periodic table and over many $|j_1,j_2,J\rangle$ combinations. Hence it is advisable to represent the matrix elements E_{JT} in a way that permits a comparison of very different configurations. For this purpose it is customary to introduce the concept of the average, or monopole interaction $\bar{E}_T$, and one defines

(8) $$\bar{E}_T = \frac{\sum_J (2J+1) E_{JT}}{\sum_J (2J+1)},$$

A look at $\bar{E}_T$, e.g., may expose any systematic mass dependence of the average interaction strength. In surveying the data one finds that the average T=0 interactions are very different from, and over an order of magnitude stronger than T=1 values. Fig. 4 presents a survey of all available T=0 monopole energies.

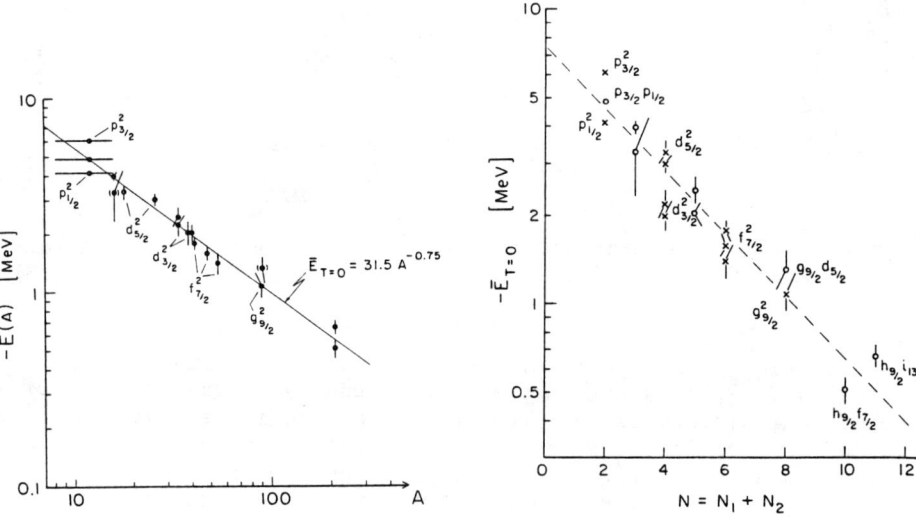

Fig.4 Empirical av. interaction energies as function of mass.

Fig.5 Average interaction vs. harm. osc. quantum number N.

We note that for multiplets near closed shells there is no significant exception to the simple power dependence proposed in Fig. 4. This behavior was unexpected, but it seems that any deviation from this smooth dependence cannot be much larger than about 10%. We note in particular the consistent fall-off of several $d_{5/2}$, $d_{3/2}$, and $f_{7/2}$ points, which were deduced at the beginning and the end of the shells. Points for $j_1 \ne j_2$ are not labeled individually. They also agree with the trend. A similar, but less precisely determined slope is seen for the much weaker T=1 interaction.

Historically[4] it was assumed that the average interaction energy was correlated with the (harmonic oscillator) energy quantum number N. This relation is shown in Fig. 5 for comparison. There is in fact some correlation with N, but the χ^2 value of a fit with a function of N is very much higher than that relative to A in Fig. 4. We here consider $\bar{E}_T$ and $\bar{E}$ as empirically known and use $\bar{E} \sim A^{-0.75}$ as the general scaling factor in future comparisons. (We note that because of the imperfect correlation with N, Schiffer used no quantitative prediction for the monopole interaction and in comparisons normalized his matrix elements with empirical values $\bar{E}$ which were calculated individually from each observed multiplet.)

A second parameter useful for graphical comparisons is the semi-classical coupling angle θ_{12}. The coupling angle is defined by

$$(9) \qquad \cos\theta_{12} = \frac{J(J+1) - j_1(j_1+1) - j_2(j_2+1)}{2[j_1(j_1+1)j_2(j_2+1)]^{1/2}}.$$

As pointed out by Molinari et al.[6] this parameter can approximate the dependence of the matrix elements on 6-j symbols to better than 15%, provided $J \neq 0$ and $j \geq 3/2$. For example, for even J

$$(10) \qquad \begin{pmatrix} j & j & J \\ +\tfrac{1}{2} & -\tfrac{1}{2} & 0 \end{pmatrix}^2 \approx [2/\pi(2j+1)^2] \tan\tfrac{1}{2}\theta_{12} \qquad [J \text{ even}]$$

It is important to keep these restrictions on J and j in mind later on. We make use of this representation in Fig. 6 which presents in graphical form all empirical matrix elements of the type $|j,j,J\rangle$.

As before the vertical coordinate indicates the mass-normalized matrix elements while the J coordinate has been replaced by the semi-classical coupling angle. The largest J values as before are on the extreme right of the graph, whereas J=0 is plotted for 180° (although J=0 does not have a unique representation in terms of θ_{12}). The T=0 and T=1 matrix elements are displayed separately. While not central to our discussion it might be of interest to mention that the solid curves represent a two-parameter fit to the data with a delta plus monopole force. Of the 8 multiplets, only the $(h_{9/2})^2$ values stray systematically from the curve. For this particle-hole multiplet, only the dominant states are known. Higher-lying strength should exist and, when found, will lower the (gamma) marks. It is conceivable that the individual interactions for these j^2 configurations in first order depend only on the mass of the nuclear core and a j-coupling coefficient.

The next figure presents a different set of data, but the solid curves of Fig. 6 have been retained to facilitate a comparison.

The dashed lines show the range of the data on the previous figure. The symbols shown in Fig. 7 represent theoretical matrix elements. Most symbols denote work by Kuo and Brown and hence 'a priori' calculations. All open circles indicate shell model fits to large sets of data. The larger circles are matrix elements from

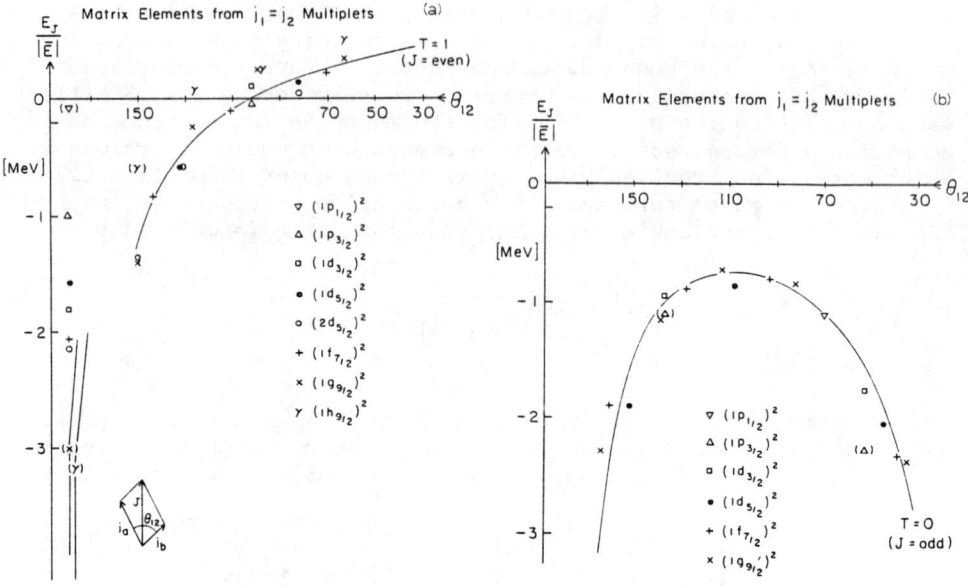

Fig. 6. Empirical j^2 matrix elements. The curves represent a two-parameter fit with a delta force plus monopole term.

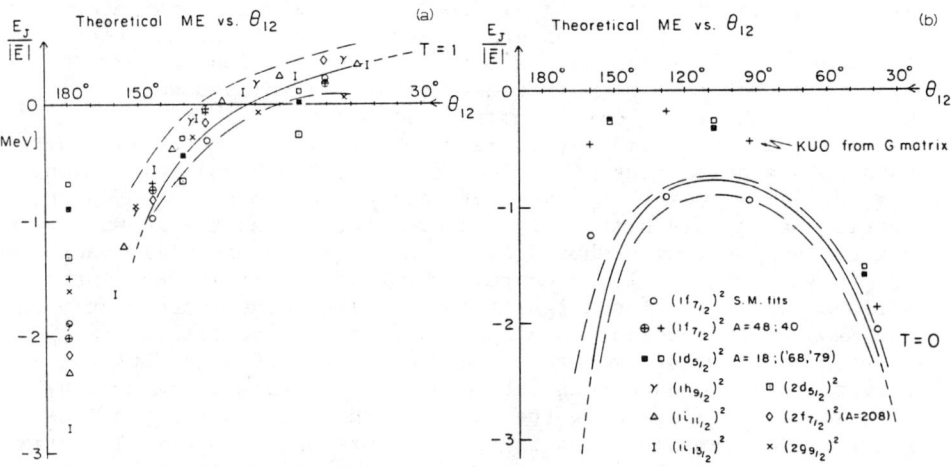

Fig. 7. Comparison of mass-scaled theoretical j^2 matrix elements.

Wildenthal for A=28. It is perhaps to be expected that the latter are within or very close to the empirical regions of Fig. 6. What comes as a bit of a surprise is the close agrement with Kuo's T=1 matrix elements. Of 36 T=1 predictions, including those for $(h_{9/2})^2$, all but one are inside the empirical region. We also note that the scatter after mass scaling is no greater than the scatter of values for the same matrix element calculated at different times or by different authors.

On the other hand, the Kuo-Brown predictions available for T=0 disagree with the empirical data, although they do describe an inverted U curve, similar to the data, and seem to obey mass scaling. As the $|j,j,J_{odd}>$ values are about a factor of three to five too small for low spin values, it is not surprising that the success of the Kuo-Brown matrix elements in shell model calculations has varied. Our systematics suggest that empirical modifications should be made primarily to their T=0 values.

In Fig. 8 we see a comparison of empirical j^2 data with the modified surface delta interaction. This interaction, which was fixed independently, does remarkably well for the T=0 interaction, although less well than Kuo-Brown for T=1.

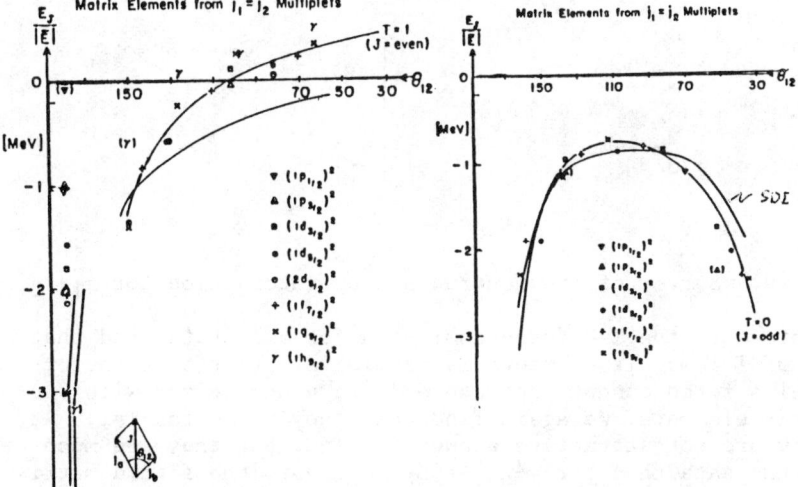

Fig. 8. Empirical ME of Fig. 6 and the MSDI model force.

The data available for $j_1 \neq j_2$ matrix elements are much less complete and often less precise than those for $j_1 = j_2$ discussed above, and much work remains to be done. Nevertheless, enough is known to draw some general conclusions. Fig. 9 presents an A-scaled compilation of the relatively large number of known n-p interactions.

While not consistent with any single curve, the points do cluster in a systematic way. It appears that there must be a grouping for Σj=even and Σj=odd. Actually, it is expected that $j_1 \neq j_2$ matrix elements must be represented by not just two, but four of eight distinct curves[6], depending on the combination of T, π, the evenness or oddness of J and of $\Sigma_j = j_1 + j_2 + J$. It may be useful to focus on configurations of one type, e.g. those like $(h_{9/2} i_{13/2})$. Figures 10a and 10b represent the T=0 and T=1 systematics, respectively.

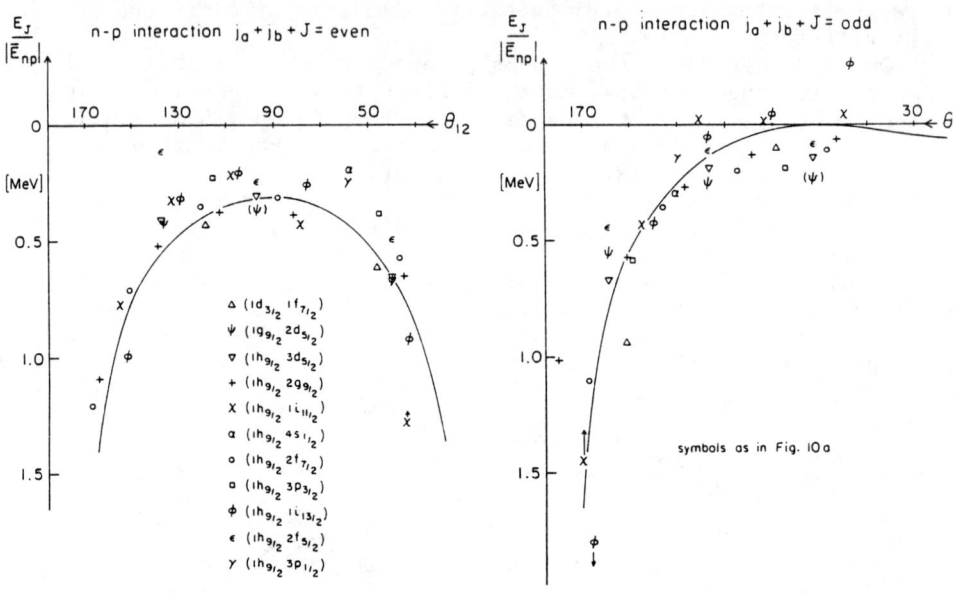

Fig. 9. Survey of the neutron-proton interaction for $j_1 \neq j_2$.

We note that the T=0 force dominates for all states and that one class of T=1 matrix elements is repulsive. (For this latter case, a delta force cannot contribute.) In a comparison with Kuo-Brown matrix elements, we again find that they agree for T=1. As above, they are not attractive enough for T=0, but they are much closer to the data than for j^2_{odd}. By contrast, the fitted matrix of Wildenthal[7], which will be discussed in the next talk, are slightly more attractive for this case than closed shell systematics.

CONCLUSION

In conclusion, I would like to reiterate some points that seem worthy of particular attention:
1) Diagonal matrix elements have been extracted from transfer reactions on "one-particle" targets near closed shells. Systematic errors can be tracked by the simultaneous derivation from particle-particle and particle-hole spectra.

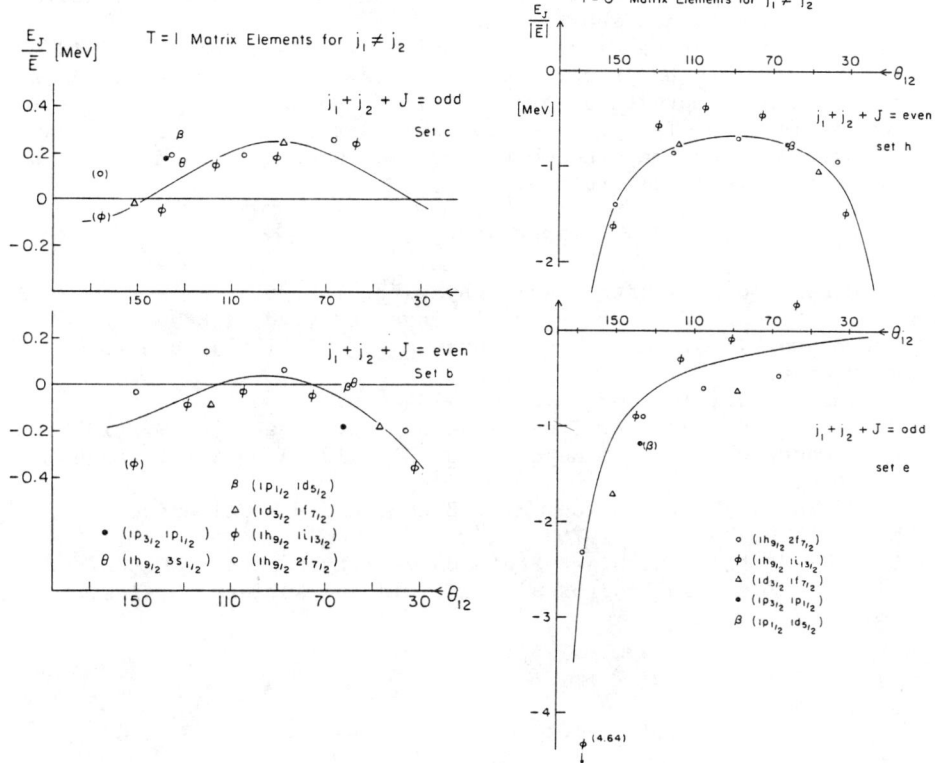

Fig. 10. Subset of $j_1 \neq j_2$ ME expected to have the same dependence on θ_{12}.

2) The measured matrix elements change by a factor of 9 as one moves through the periodic table from the p-shell to ^{208}Pb, in agreement with the calculations of Kuo and Brown. This overall mass dependence is fit by $\bar{E} \sim A^{-0.75}$.

3) All derived monopole energies are fit to within their uncertainties (typically 10% or better) by $\bar{E}(J=0) = 31.5 \, A^{-0.75}$. Monopole energies and individual matrix elements for the same configuration derived from the beginning <u>and</u> the end of the shell support this scaling factor, although the uncertainty of the individual points would permit larger or smaller slopes within a single shell.

4) If plotted against θ_{12} after mass scaling, the empirical diagonal matrix elements seem to trace out a few simple, characteristic curves. These curves and the observed scaling feature are consistent with the assumption that the residual interaction of bound nucleons is adequately represented by a delta plus monopole plus quadrupole force. Evidence for other

multipoles, beyond the component already contained in the delta force, has not been substantiated.
5) Shell model searches for large bodies of data generally yield matrix elements that approximately agree with the directly determined ones. Impressive agreement is found for the Kuo-Brown predictions for T=1, but the T=0 calculations are in significant disagreement so that the method or the input data used should be subject to renewed scrutiny.

REFERENCES

1. S. Cohen and D. Kurath, Nucl. Phys. 73, 1 (1965).
2. T.T.S. Kuo and G.E. Brown, Nucl. Phys. A85, 40 (1966), G.E. Brown and T.T.S. Kuo, Nucl. Phys. A92, 481 (1967), and subsequent papers.
3. J.P. Schiffer, Ann. of Phys. 66, 798 (1971).
4. J.P. Schiffer and W.W. True, Rev. of Mod. Phys. 48, 191 (1976).
5. W.W. Daehnick, Physics Reports 96, 319 (1983), and references therein.
6. A. Molinari, M.B. Johnson, H.A. Bethe, and W.M. Alberico, Nucl. Phys A239, 45 (1975).
7. B.H. Wildenthal, Michigan State University Report MSUCL-427 (1983). See also Progress in Particle and Nuclear Physics, Vol. 11.

ANALYSIS OF FEATURES IN THE A=34-48 REGION
IN TERMS OF $d_{3/2}$ and $f_{7/2}$ DEGREES OF FREEDOM

S. T. Hsieh*, X. Ji, R. Mooy and B. H. Wildenthal
Drexel University, Philadelphia, PA 19104

ABSTRACT

Schematic aspects of the energy spectra of A=33-48 (N,Z=16-28) nuclei are analyzed with calculations in a model space comprised of all configurations spanned by (A-32) particles in the $d_{3/2}$ and $f_{7/2}$ orbits. The analysis emphasizes the ground-state masses of both near-stable and neutron-rich nuclei, the energy separations of the $d_{3/2}$ and $f_{7/2}$ "single particle" (or "single hole") states, the excitation energies of nominally simple particle-particle and particle-hole states, and the energies of states characterized by "stretched" configurations of $f_{7/2}$ particles. An issue of particular interest is whether these "single-particle", "two-particle", and "three-particle" features are consistent with the existance of the low-lying "multiparticle-multihole", or "deformed", states which are striking features of the spectra of doubly-even nuclei both below and above N,Z=20. Results obtained with an interaction determined from a fit to a collection of level-energy data from this region are compared with experimental values. Some differences between these predictions and those based on a Hamiltonian deduced from the analysis of nominal "two-nucleon" states with "two-particle" spaces are noted.

INTRODUCTORY REMARKS

Nuclei in the upper sd-shell and in the lower fp shell exhibit a mixture of simple and complex features. Single-nucleon transfer data have established that the lowest $3/2^+$ state of each odd-mass nucleus has a $d_{3/2}$ single-particle or single-hole nature and that the lowest $7/2^-$ state has an $f_{7/2}$ single-particle nature. This is the rationale for our choice in this study of a $d_{3/2}$-$f_{7/2}$ model basis. (The role of the adjacent $1s_{1/2}$ and $1p_{3/2}$ orbits will be discussed below.) The efficacy of an

$(f_{7/2})^n$ interpretation of many of the dominant features of nuclei between A=42 and A=54 has long been known.[1-3] Between A=34 and A=38, $(d_{3/2})^n$ configurations dominate a smaller domain of low-lying states.[4] In addition, the low-lying states of nuclei which have one or three proton holes in the $d_{3/2}$ shell and an odd number of neutron particles in the $f_{7/2}$ shell have been shown to have the negative parity and the spins J = 2, 3, 4 and 5 characteristic of simple $f_{7/2}$-$d_{3/2}$ particle-particle and particle-hole structures.[5,6] Finally, in the heavier sd-shell nuclei, high-spin states have been identified which can be interpreted as simple "stretched" two-particle excitations into the $f_{7/2}$ orbit.[7,8] These excitations also can be tracked into the region above N,Z=20.[9,10]

In combination, knowledge about states which have simple $d_{3/2}$-$d_{3/2}$, $f_{7/2}$-$f_{7/2}$ and $d_{3/2}$-$f_{7/2}$ structures can lead directly to the empirical determination of the values of almost all of the diagonal two-body energy matrix elements of these orbits.[11,12] In the abstract, these values are the necessary and defining input for a shell-model calculation in the $d_{3/2}$-$f_{7/2}$ model space. Shell-model calculations in this d-f space which allow all possible particle-hole excitations between the two orbits are of interest on several grounds beyond merely systematizing the observed features of the simple two-nucleon spectra. Along with the simple, "shell-model", features just enumerated, nuclei in this region also exhibit striking collective, "deformed", features at low excitation energy.[13,14] In shell-model terms, these deformed states must arise predominantly from $d_{3/2}$-$f_{7/2}$ multiparticle-multihole excitations.

How consistent are these deformed features with the two-nucleon and single-particle spectra? What are the implications of the observed energy separations of the $d_{3/2}$ and $f_{7/2}$ single-particle (hole) states and of the coexistence of the two-nucleon with the deformed states for the ground state correlations in the heavy sd and light fp nuclei? Are the masses of the very-neutron-rich nuclei in this region consistent with the stable and near-stable masses in this model context?

In the present study we attempt to answer these questions by determining if all of these varied phenomena can be imbedded within a single, internally unified and consistent, shell-model treatment.

Simultaneously, we try to illustrate the differences between using a Hamiltonian based purely on deductions from "simple" spectra, as in references 11 and 12, and analogous values determined from a more extensive analysis of energy-level data as described here.

The present shell-model study[15] follows along the lines of investigation laid out by previous shell-model studies of the mixing of sd and fp orbits in the upper sd shell and Ca regions.[16-23] A number of complementary studies have employed the technique of mixing spherical and deformed shell-model configurations.[24-27] The technical procedures of constructing the shell-model matrices were accomplished in the present work with the aid of new computer codes developed at the University of Utrecht.[28]

GENERAL APPROACH

In this work we model states of A = 33-48 in terms of A-32 particles distributed without restriction over the $0d_{3/2}$ and $0f_{7/2}$ orbits. As judged by evaluation of the experimentally observed spectra in this region, the most serious omissions from this orbit space are the $1s_{1/2}$ orbit, which lies only 1 MeV below the $d_{3/2}$ orbit, and the $1p_{3/2}$ orbit, which lies only 2 MeV above the $f_{7/2}$. The next most serious omissions are those of the $0d_{5/2}$ and $1p_{1/2}$ orbits. In omitting these degrees of freedom from our study we eliminate, from the outset, our ability to deal with states which are dominated by the single-particle degrees of freedom associated with these orbits. Hence we can say nothing about $s_{1/2}$ hole states, $p_{3/2}$ particle states, or more complicated states which are dominated by configurations involving such excitations.

In compensation for these restrictions to our field of discourse we are able to generate a very deep array of particle-hole structures within the two-orbit space, since for the region around ^{40}Ca the dimensions of our vector space are quite large. These calculations are thus to be regarded more as an exploration of specific selected features of nuclear structure than as an attempt to give a complete model recapitulation of the low-lying nuclear spectra. We know from the outset that "single-particle" $1/2^+$, $3/2^-$ states and assorted other classes of states will not be encompassed within our results. We aim, rather, to account systematically for the simple $d_{3/2}$ and $f_{7/2}$ excitations and to discover which complicated excitations are built predominately out of these two orbits. Obviously, in nature the "$d_{3/2}$, $f_{7/2}$" states will be significantly affected by the nearby $s_{1/2}$ and $p_{3/2}$ orbits as well as by the more distant orbits. With the present model treatment we are attempting to identify a "$d_{3/2}$-$f_{7/2}$" subset of the total array of observed states that is affected by omitted configurations in a simple, uniform way.

Given the technology for constructing and diagonalizing shell-model matrices, all that is needed to generate predictions for level energies and wave functions from an assumed orbit space is the set of one-body and two-body energies for these orbits. Our study commences by taking as the single-particle energies for the $d_{3/2}$ and $f_{7/2}$ orbits the binding energies relative to ^{32}S of the lowest $3/2^+$ and $7/2^-$ levels of ^{33}S. There are 24 two-body matrix elements in the $d_{3/2}$-$f_{7/2}$ orbit space. We take as initial values for the 20 diagonal two-body matrix elements the mass-averaged results[12] of Daehnick. We use the values of Ref. 20 for the four off-diagonal matrix elements. We will refer to this Hamiltonian as the the "D_{av}" Hamiltonian.

With the "D_{av}" Hamiltonian we constructed and diagonalized matrices for 150 states in the A=33-48 region, concentrating on ground, low-lying and stretched levels as noted above. The multiparticle shell-model eigenvalues obtained in these diagonalizations should, if our assumptions are valid, correspond to experimentally observed nuclear levels in this region.[30,31] Such correspondences did emerge quite clearly when the theoretical and experimental spectra for the various nuclei were compared. The values of the 26 single-particle energy and two-body matrix element Hamiltonian parameters were then adjusted through two iteration cycles with the technique of Chung[29] to obtain values for this space that provide the better agreement with a selected set of level-energy data.

The level-energy data were considered in the form of binding energies relative to ^{32}S. The Coulomb correction for each group of isotopes was assumed to be constant within the group. The values of the corrections were based on comparisons of mirror levels in A = 34, 35, 38, 39, 42 and 43. These Coulomb corrections, along with the total binding energy of ^{32}S, were then subtracted from the total experimental binding energies[32] (for both ground and excited states) of the various nuclear levels considered in order to provide the actual data to which the

Hamiltonian parameters were fitted. We will refer to the values of the Hamiltonian parameters obtained in this fitting procedure as the "df_t" Hamiltonian. The values of the single-particle energies and two-body matrix elements of the "D_{av}" and "df_t" Hamiltonians, together with theoretical values, are shown in Table I. The two sets of empirical values can be seen to be very similar to each other. In the following sections we will discuss the consequences of using the "df_t" interaction, and in some instances the "D_{av}" interaction, for multiparticle shell-model calculations.

SPECTRA OF THE NOMINAL "TWO-NUCLEON" SYSTEMS

In the space and mass range we work with, the only states which are literally two-body in structure are those of A=34. Unlike the simple one-orbit $d_{3/2}$-model analysis of A=34, the two-orbit model incorporates mixing between the two 0^+ states that can be formed in this space, and likewise between the two 2^+ states. This mixing has an appreciable effect upon energies, but its effect upon wave functions would be very difficult to detect in analyses such as those of references 11 and 12. In the context of one-orbit $d_{3/2}$ and $f_{7/2}$ models, the states of ^{38}K(^{38}Ar) would correspond to two $d_{3/2}$ holes, those of ^{42}Sc(^{42}Ca) to two $f_{7/2}$ particles and those of ^{48}Sc to $f_{7/2}$ particle-$f_{7/2}$ hole combinations. From the standpoint of a shell closure at ^{40}Ca, the lowest quartet of levels in the nuclei ^{38}Cl, ^{40}K and ^{46}K correspond to $d_{3/2}$-$f_{7/2}$ particle-particle, hole-particle and hole-hole combinations, respectively. In our space these A>34 systems are not simply two-nucleon in structure, but it might be expected that the simple two-nucleon structures just enumerated would dominate the multiparticle wave functions. In Table II we list the experimental excitation energies for these states in comparison with the results obtained with the "df_t" Hamiltonian. The results obtained with the "D_{av}" Hamiltonian are quite similar. Inspections of the multiparticle shell-model eigenfunctions of the "two-nucleon" states around A=40 reveal that the expected simple particle-particle or particle-hole configurations are dominant, but that they and the background configurations together yield a net excitation from the $d_{3/2}$ to the $f_{7/2}$

Table I. Values of two-body matrix elements and single-particle energies for the $d_{3/2}$-$f_{7/2}$ orbit space.

matrix element						"ST"[a]	"D_{av}"[b]	"df_t"[c]
j_1	j_2	j_3	j_4	J	T			
d	d	d	d	0	1	-2.0	-2.51	-2.228
d	d	d	d	1	0	-1.5	-1.31	-0.925
d	d	d	d	2	1	-0.4	+0.15	+0.177
d	d	d	d	3	0	-2.8	-2.45	-2.798
d	f	d	f	2	0	-5.4	-3.71	-3.484
d	f	d	f	2	1	-0.4	+0.04	+0.325
d	f	d	f	3	0	-2.8	-1.66	-1.113
d	f	d	f	3	1	-0.2	-0.19	-0.289
d	f	d	f	4	0	-1.4	-1.36	-2.156
d	f	d	f	4	1	+0.3	+0.51	+1.177
d	f	d	f	5	0	-3.5	-2.29	-2.010
d	f	d	f	5	1	-0.5	-0.38	-0.243
f	f	f	f	0	1	-1.78	-2.11	-2.520
f	f	f	f	1	0	-1.45	-1.94	-2.673
f	f	f	f	2	1	-0.51	-0.86	-1.586
f	f	f	f	3	0	-0.93	-0.92	-0.997
f	f	f	f	4	1	0.00	-0.11	-0.135
f	f	f	f	5	0	-1.09	-0.83	-0.731
f	f	f	f	6	1	+0.24	+0.23	+0.395
f	f	f	f	7	0	-2.65	-2.40	-2.722
d	d	f	f	0	1	+2.62	+2.35	+1.605
d	d	f	f	1	0	-0.45	-1.14	-1.252
d	d	f	f	2	1	+0.45	+0.44	+0.633
d	d	f	f	3	0	-0.24	-0.31	-0.461
single particle energies								
$d_{3/2}$						-8.642[d]	-8.756[e]	-8.617
$f_{7/2}$						-5.708[d]	-5.460[e]	-5.836

[a] diagonal m.e. from Ref. 11, off-diagonal m.e. from Ref. 24
[b] diagonal m.e. from Ref. 12, off-diagonal m.e. from Ref. 20
[c] present results
[d] Ref. 30 and Ref. 32
[e] Ref. 12

Table II. Energies of spectra characterizable as "particle-particle", "particle-hole", or "hole-hole" in the $d_{3/2}$-$f_{7/2}$ space.

Energy (MeV)[a]		2J	#[b]	p[c]	
Exp.	Theo.				

"$d_{3/2})^{+1}$- $d_{3/2})^{+1}$", "$d_{3/2})^{+1}$- $f_{7/2})^{+1}$" and "$f_{7/2})^{+1}$- $f_{7/2})^{+1}$" states; A = 34, T = 0 and 1 (Cl,S)

0.146	-20.060	6	1	1	
-20.059	0.148	0	1	1	(T=1)
1.030*	1.526	2	1	1	
2.722	2.120	4	1	-1	
2.900*	2.901	4	1	1	(T=1)
4.075	3.448	8	1	-1	
3.632	3.594	10	1	-1	
3.982	4.491	6	1	-1	
5.318	5.315	6	2	-1	(T=1)
5.689	5.361	10	2	-1	(T=1)
5.280	5.660	14	1	1	
5.318	5.929	4	2	-1	(T=1)
	6.083	2	2	1	
	6.313	0	2	1	(T=1)
6.639	6.781	8	2	-1	(T=1)
	6.898	4	2	1	(T=1)
	7.413	6	2	1	
	7.650	10	1	1	
	8.246	8	1	1	(T=1)
	8.777	12	1	1	(T=1)

"$d_{3/2})^{-1}$- $d_{3/2})^{-1}$" states; A = 38, T = 0 (K)

-68.714	-69.002	6	1	1	
2.190*	1.519	2	1	1	

"$d_{3/2})^{-1} - d_{3/2})^{-1}$" states; A = 38, 2T = 2 (Ar)

-68.584	-68.849	0	1	1
2.280	2.653	4	1	1

"$d_{3/2})^{+1} - f_{7/2})^{+1}$" states; A = 38, T = 2 (Cl)

-57.703	-57.849	4	1	-1
0.671	0.676	10	1	-1
0.755	0.841	6	1	-1
1.309	0.985	8	1	-1

"$d_{3/2})^{-1} - f_{7/2})^{+1}$" states; A = 40, T = 1 (K)

-89.590	-89.956	8	1	-1
0.030	0.267	6	1	-1
0.800	0.603	4	1	-1
0.892	0.782	10	1	-1

"$f_{7/2})^{+1} - f_{7/2})^{+1}$" states; A = 42, T = 0 (Sc)

-117.276	-117.658	0	1	1	(T=1)
0.611	0.333	2	1	1	
0.617	0.501	14	1	1	
1.491	1.590	6	1	1	
1.511	2.086	10	1	1	

"$f_{7/2})^{+1} - f_{7/2})^{+1}$" states; A = 42, T = 1 (Ca)

-117.276	-117.658	0	1	1
1.525	1.479	4	1	1
2.752	2.984	8	1	1
3.189	3.463	12	1	1

"$f_{7/2})^{-1} - f_{7/2})^{-1}$" states; A = 46, T = 3 (Ca)

-154.151	-154.500	0	1	1
1.346	1.646	4	1	1
2.575	3.137	8	1	1
2.974	3.618	12	1	1

"$d_{3/2})^{-1} - f_{7/2})^{-1}$" states; A = 46, T = 4 (K)

-139.903	-139.930	4	1	-1
0.886	0.679	10	1	-1
0.587	0.901	6	1	-1
0.691	1.017	8	1	-1

"$f_{7/2})^{+1} - f_{7/2})^{-1}$" states; A = 48, T = 3 (Sc)

0.131	-178.279	10	1	1	
-178.074	0.027	12	1	1	
0.251	0.254	8	1	1	
0.622	0.543	6	1	1	
1.142	1.245	4	1	1	
1.096	1.616	14	1	1	
2.516	3.134	2	1	1	
6.702	7.019	0	1	1	(T=4)

* Centroid energies
a) binding energies are listed for ground states, excitation energies for excited states
b) the symbol \# indicates the ordering number of the state in the spectrum, the first state of a given J and P having \# = 1, etc.
c) parity P is indicated by +1 and -1

orbit of about 0.5 to 1.0 particles. Within a given multiplet, the amount of this $d_{3/2}$ to $f_{7/2}$ excitation is relatively constant.

ENERGY SPLITTINGS OF THE $d_{3/2}$ and $f_{7/2}$ "SINGLE-NUCLEON" STATES

In the same sense as for the "two-nucleon" states, the only pure "single-nucleon" states in our model are the $3/2^+$ and $7/2^-$ levels of ^{33}S. Other systems conventionally considered as "single-particle" or "single-hole", such as ^{39}K, ^{41}Ca and ^{47}Ca, in literal terms have many-body structures in our model. Again, however, it might be expected that these many-body wave functions are dominated by the "single-particle" or "single-hole" terms. In Table III we list the experimental excitation energies for the lowest $3/2^+$ and $7/2^-$ states in each odd-mass nucleus in our study for which information is available in comparison with the results of the "df_t" interaction. From inspection of Table III we conclude that the observed energy separations of the $d_{3/2}$-like and $f_{7/2}$-like states throughout the range of A=33-47 nuclei are well reproduced, with the worst agreement occuring for ^{47}Ca.

We also show in Table III the observed and predicted energies for some "stretched" states in these nuclei, states in which experimentally observed properties and inspections of the model eigenfunctions both argue for structures in which a pair of $f_{7/2}$ nucleons coupled to J=7 is coupled ("weakly") to the ground state, either $3/2^+$ or $7/2^-$, of a neighboring nucleus. Experimental-theoretical comparisons for these states provide both an additional test of the $d_{3/2}$-$f_{7/2}$ single-particle-energy splitting and a test of whether the effective Hamiltonian is able to preserve a salient experimental feature of the region. The evidence is that the model results reproduce this feature systematically.

GROUND-STATE BINDING ENERGIES

The comparisons between experiment and theory presented in the preceding two sections have illustrated that the $d_{3/2}$-$f_{7/2}$ model space can produce a good facsimile of what is observed for the nominally simple "two-nucleon" and "single-nucleon" features of the A=33-48 region with a "reasonable" interaction. These features can be thought

Table III. Excitation energies of "$d_{3/2}$" and "$f_{7/2}$" "single particle" and "single hole" states in A = 33-47 nuclei, along with excitation energies of states nominally formed by coupling a "stretched" (J = 7) "$f_{7/2}$ pair" to a neighboring ground state. The conventions of the presentation are as explained in Table II.

	Energy (MeV)		2J	π	P
	Exp.	Theo.			
A = 35, 2T = 1	-32.704	-32.617	3	1	1
	3.163	3.158	7	1	-1
	8.844	8.230	17	1	1
A = 35, 2T = 3	-27.045	-26.913	3	1	1
	1.995	1.618	7	1	-1
A = 37, 2T = 1	-56.746	-57.042	3	1	1
	1.611	1.678	7	1	-1
	7.071	6.930	17	1	1
A = 37, 2T = 3	-51.595	-51.481	3	1	1
	3.103	3.088	7	1	-1
A = 37, 2T = 5	-41.238	-41.384	7	1	-1
	1.395	1.395	3	1	1
A = 39, 2T = 1	-81.790	-82.017	3	1	1
	2.814	2.442	7	1	-1
	7.780	7.381	17	1	1
A = 39, 2T = 3	-75.182	-75.557	7	1	-1
	1.517	1.278	3	1	1
	5.536	5.640	17	1	1
A = 41, 2T = 1	-105.795	-106.155	7	1	-1
	2.010	2.111	3	1	1
	5.220	5.528	17	1	1

A= 41, 2T= 3	-99.685	-100.259	3	1	1
	1.294	1.234	7	1	-1
A= 41, 2T= 5	-91.111	-91.538	7	1	-1
		1.321	3	1	1
A= 43, 2T= 1	-129.412	-130.215	7	1	-1
	0.152	0.625	3	1	1
	3.123	3.983	19	1	-1
A= 43, 2T= 3	-125.209	-125.586	7	1	-1
	0.990	1.822	3	1	1
	4.591	5.235	17	1	1
A= 43, 2T= 5	-116.862	-117.434	3	1	1
	0.738	1.096	7	1	-1
A= 45, 2T= 5	-143.755	-143.981	7	1	-1
		1.724	3	1	1
A= 45, 2T= 7	-133.023	-133.502	3	1	1
	1.080	1.626	7	1	-1
A= 47, 2T= 5	-169.840	-170.479	7	1	-1
	0.767	1.142	3	1	1
A= 47, 2T= 7	-161.428	-161.255	7	1	-1
	2.580	1.344	3	1	1
A= 47, 2T= 9	-148.253	-148.343	3	1	1
	1.970	2.426	7	1	-1

of as "relative" phenomena in that the focus is on the energy differences between states of "simple" character within a given nucleus. It is not evident that the total binding energies of multiparticle ensembles will emerge from a Hamiltonian whose features are dominated by these one-particle and two-particle phenomena. The ability of a model calculation to account simultaneously for the local spectral features of individual nuclei as well as systematic binding energy trends is indicative of an internal consistency in the relationship between model space, Hamiltonian and the dominant structure aspects of the nuclei under consideration.

In Table IV are presented comparisons of ground-state binding energies for the the isotopes of S, Cl, Ar, K, Ca, Sc and Ti. The experimental numbers have been obtained from Ref. 32 as described above, by subtracting the binding energy of ^{32}S and a Coulomb term constant for a given Z. The theoretical numbers are the results obtained with the "df_t" Hamiltonian. The key features of these results are the sytematic success with which the measured effects associated with the transition between the $d_{3/2}$ and $f_{7/2}$ shells are theoretically reproduced. The model is able to account for the effects of adding particles within a given orbit and, alternatively, in a new orbit, in a comprehensive way. The trends of binding energy as large neutron excesses are built up are also accounted for. The most evident deficiencies in the match between theory and experiment are the systematic overbinding in the A=42-46 region and the breaks in the systematic trends as the model space is exhausted for neutrons. The first of these problems may disappear upon a further iteration of the Hamiltonian. The latter, however, is probably intrinsic to the limitations of the space.

"MULTIPARTICLE-MULTIHOLE", OR "COLLECTIVE" STATES

The ultimate focus of this study is the nature of excited states, best known in the even-even heavy Ar and light Ca isotopes, which seem to manifest highly collective or "deformed" structures. In our approach such states would arise from configurations which can be considered as multiparticle-multihole excitations relative to the ground state configurations. Canonical examples are the first excited 0^+ states in ^{38}Ar and ^{42}Ca, which might be guessed to have $(d_{3/2})^{-4}$-$(f_{7/2})^{+2}$ and $(d_{3/2})^{-2}$-$(f_{7/2})^{+4}$ configurations, respectively. Another classic case in

Table IV. Binding energies for isotopes of Sulfur, Chlorine, Argon, Potassium, Calcium, Scandium and Titanium. The experimental values, from Reference 32, are given relative to ^{32}S, and have been corrected for the Coulomb energy of the last (Z-16) protons as described in the text. The theoretical values are those calculated with the "df_t" Hamiltonian.

	A	2J	P	Exp.	Theo.
S	33	3	1	-8.642	-8.617
	34	0	1	-20.059	-19.912
	35	3	1	-27.045	-26.913
	36	0	1	-36.934	-36.628
	37	7	-1	-41.238	-41.384
	38	0	1	-49.274 (0.012)	-49.262
	39	7	-1	-53.484 (0.200)	-53.565
	40	0	1	-61.077 (0.040)	-61.046
Cl	34	0	1	-20.059	0.148
	34	6	1	0.146	-20.060
	35	3	1	-32.704	-32.617
	36	4	1	-41.284	-41.379
	37	3	1	-51.595	-57.042
	38	4	-1	-57.703	-57.849
	39	3	1	-65.779	-66.109
	40	4	-1	-71.582 (0.500)	-71.879
	41	3	1	-79.512 (0.160)	-79.703
Ar	36	0	1	-47.958	-47.909
	37	3	1	-56.746	-57.042
	38	0	1	-68.584	-68.849
	39	7	-1	-75.182	-75.557
	40	0	1	-85.051	-85.441
	41	7	-1	-91.111	-91.538
	42	0	1	-100.579 (0.040)	-100.864
	43	7	-1	-106.209 (0.070)	-106.334
	44	0	1	-114.5590 (.020)	-115.084

	45	7	-1	-120.089 (0.060)	-119.874
	46	0	1	-128.159 (0.040)	-128.005
K	38	6	1	-68.714	-69.002
	39	3	1	-81.790	-82.017
	40	8	-1	-89.590	-89.956
	41	3	1	-99.685	-100.259
	42	4	-1	-107.221	-107.855
	43	3	1	-116.862	-117.434
	44	4	-1	-124.155 (0.040)	-124.509
	45	3	1	-133.023	-133.502
	46	4	-1	-139.903 (0.016)	-139.930
	47	3	1	-148.253	-148.343
Ca	40	0	1	-97.432	-97.713
	41	7	-1	-105.795	-106.155
	42	0	1	-117.276	-117.658
	43	7	-1	-125.209	-125.586
	44	0	1	-136.341	-136.640
	45	7	-1	-143.755	-143.981
	46	0	1	-154.151	-154.500
	47	7	-1	-161.428	-161.255
	48	0	1	-171.372	-171.260
Sc	42	0	1	-117.276	-117.658
	43	7	-1	-129.412	-130.215
	44				
	45	7	-1	-150.436	
	46	8	1	-159.196	-159.673
	47	7	-1	-169.840	-170.479
	48	12	1	-178.074	-178.279*
Ti	44	0	1	-145.711	-146.014
	45	7	-1	-155.240	
	46	0	1	-168.430	-168.855
	47	7	-1	-177.308	-177.674
	48	0	1	-188.934	-189.625

question is the first excited state in ^{40}Ca, conventionally assumed to have a "4 particle-4 hole" nature.

In Table V we present some results for some of these levels obtained from both the "df_t" and D_{av}" Hamiltonians. The two sets of results are similar overall as regards the degree of configuration mixing. The ground states exhibit an excess of 0.5-1.0 $f_{7/2}$ particles over the no-configuration-mixing limit, while the excited states have an excess of about 2 $f_{7/2}$ particles for A=38,42 and 44, and 3 or more for A=40.

The detailed changes in going from the "D_{av}" to the "df_t" Hamiltonian cannot be characterized simply. The "df_t" results for the lighter nuclei yield less configuration mixing in the ground states and more in the excited states. These distinctions disappear for the heavier masses, however.

We recall that the "df_t" results yielded better than 500 keV agreement with ground-state masses. Hence we see that the "D_{av}" ground states are systematically overbound by a few MeV out of 100 to 200 MeV. This quality of agreement between a model parameterized in terms of a very few spectral features is impressive testimony to the basic physical insight of the approach of Refs. 11 and 12. The improvements in ground state binding energies brought about by the iterations is not on its face very significant.

A more important consequence of the iteration to a better overall fit to experimental energies is observed for the energies of the excited states. The "D_{av}" excitations are quite a bit higher than experiment, with which the "df_t" results compare much more favorably.

CONCLUSIONS

The present study has comprehensively exploited the full capabilities of the $d_{3/2}$-$f_{7/2}$ model space. It has shown that the dominant one-particle and two-particle features of the low-lying levels of A=33-48 nuclei can be understood in terms of these degrees of freedom. The same parameterization of the $d_{3/2}$-$f_{7/2}$ model also yields a convincing accounting of ground-state binding energies and the excitation enrgies of low-lying collective states in this region, which cannot be understood in terms of models that incorporate a ^{40}Ca shell

Table V. Energies and occupation numbers from the "D_{av}" and the "df_t" Hamiltonians for selected positive-parity states.

A	2J	2T	#	Ham	Energy	$\langle d_{3/2} \rangle$	$\langle f_{7/2} \rangle$
38	0	2	1	"df_t"	−68.849	5.47	0.53
				"D_{av}"	−70.199	5.21	0.79
38	0	2	2	"df_t"	3.426	3.93	2.07
				"D_{av}"	4.405	4.16	1.84
40	0	0	1	"df_t"	−97.713	7.36	0.64
				"D_{av}"	−99.165	7.00	1.00
40	0	0	2	"df_t"	4.155	4.42	3.78
				"D_{av}"	5.868	5.00	3.00
42	0	2	1	"df_t"	−117.276	7.04	2.96
				"D_{av}"	−120.522	7.07	2.93
42	0	2	2	"df_t"	2.517	5.68	4.32
				"D_{av}"	4.923	5.76	4.24
44	0	4	1	"df_t"	−136.640	6.99	5.01
				"D_{av}"	−140.761	7.27	4.73
44	0	4	2	"df_t"	2.327	5.99	6.01
				"D_{av}"	5.330	5.88	6.12

closure. The results obtained so far in terms of binding energies and excitation spectra seem to justify further study of the E2 and particle-transfer properties of the model wave functions.

ACKNOWLEDGEMENTS

The work reported here is based upon research supported in part by the National Science Foundation under Grant No. PHY-8509736.

REFERENCES

* permanent address: National Tsing Hua University, Hsinchu, Taiwan.
1. I. Talmi and I. Unna, An. Rev. Nucl. Sci. 10, 353 (1960).
2. J. D. McCullen, B. Bayman, and L. Zamick, Phys. Rev. 134, B515 (1964).
3. J. Ginocchio and J. B. French, Phys. Lett. 7, 137 (1963).
4. P. W. M. Glaudemans, G. Wiechers and P. J. Brussaard, Nucl. Phys. 56, 548 (1964).
5. S. Goldstein and I. Talmi, Phys. Rev. 102, 589 (1956).
6. S. P. Pandya, Phys. Rev. 103, 956 (1956).
7. H. Nann, W. S. Chien, A. Saha, and B. H. Wildenthal, Phys. Rev. C15, 1959 (1977).
8. R. M. Del Vecchio, R. T. Kouzes and R. Sherr, Nucl. Phys. A265, 220 (1976).
9. H. Nann, W. S. Chien, A. Saha, and B. H. Wildenthal, Nucl. Phys. A292, 195 (1977). 1959 (1977).
10. H. Nann et al., Phys. Rev. Lett. 55, 578 (1985).
11. J. P. Schiffer and W. W. True, Rev. Mod. Phys. 48, 191 (1976).
12. W. W. Daehnick, Phys. Reports 96, 317 (1983).
13. P. Betz et al., Z. Phys. 271, 195 (1974); Z. Phys. 276, 295 (1976).
14. D. Cline, Physics of Medium Light Nuclei, Editrice Compositori, Bologna, Italy, p. 89 (1977).
15. S. T. Hsieh, R. B. M. Mooy and B. H. Wildenthal, Bull. Am. Phys. Soc. 30, 731 (1985).
16. F. C. Erne, Nucl. Phys. 84, 91 (1966).
17. A. E. L. Dieperink and P. J. Brussaard, Nucl. Phys. A106, 177 (1967).
18. P. Federman and S. Pittel, Phys. Rev. 186, 1106 (1969).
19. C. W. Towsley, D. Cline and R. W. Horoshko, Nucl. Phys. A204, 574 (1973).
20. M. Sakakura, A. Arima and T. Sebe, Phys. Lett. 61B, 335 (1976).

21. H. Hasper, Phys. Rev. C19, 1482 (1978).
22. I. P. Johnstone, Phys. Rev. C22, 2561 (1980).
23. S. T. Hsieh, M. C. Chang adn D. S. Chuu, Phys. Rev. C23, 521 (1981).
24. W. J. Gerace and A. M. Green, Nucl. Phys. A93, 110 (1967).
25. W. J. Gerace and A. M. Green, Nucl. Phys. A123, 241 (1969).
26. B. H. Flowers and L. D. Skouras, Nucl. Phys. A116, 529 (1968).
27. T. Erikson, Nucl. Phys. A205, 593 (1973).
28. D. Zwarts, Thesis, State University Utrecht, 1984 (unpublished).
29. W. Chung, Thesis, Michigan State University, 1976 (unpublished).
30. P. M. Endt and C. van der Leun, Nucl. Phys. A310, 1 (1978).
31. Table of Isotopes, Seventh Edition, edited by C. M. Lederer and V. S. Shirley, John Wiley and Sons, New York (1978).
32. A. H. Wapstra and G. Audi, Nucl. Phys. A432, 1 (1985).

RELATIVISTIC SHELL MODEL CALCULATIONS*

R. J. Furnstahl[†]
Physics Department and Nuclear Theory Center
Indiana University, Bloomington, IN 47405

ABSTRACT

Shell model calculations are discussed in the context of a relativistic model of nuclear structure based on renormalizable quantum field theories of mesons and baryons (quantum hadrodynamics). The relativistic Hartree approximation to the full field theory, with parameters determined from bulk properties of nuclear matter, predicts a shell structure in finite nuclei. Particle-hole excitations in finite nuclei are described in an RPA calculation based on this QHD ground state. The particle-hole interaction is prescribed by the Hartree ground state, with no additional parameters. Meson retardation is neglected in deriving the RPA equations, but it is found to have negligible effects on low-lying states. The full Dirac matrix structure is maintained throughout the calculation; no nonrelativistic reductions are made. Despite sensitive cancellations in the ground state calculation, reasonable excitation spectra are obtained for light nuclei. The effects of including charged mesons, problems with heavy nuclei, and prospects for improved and extended calculations are discussed.

I. INTRODUCTION

Nuclear structure physics has traditionally been based on the many-particle Schrödinger equation with nucleons interacting through static, two-body potentials. Historically, the shell model has provided a powerful framework for understanding this complicated many-body problem. In this approach, the nucleons are treated as independent particles moving in a self-consistent mean field arising from their interactions with all of the other nucleons. This nonrelativistic shell model picture is based in principle on Hartree-Fock theory.[1]

In recent years, there has been much interest in relativistic nuclear many-body theories based on baryons and mesons as the explicit dynamical degrees of freedom. We will refer to such theories as "quantum hadrodynamics" or QHD. A thorough discussion of QHD and its implications for nuclear physics is given in the review article by Serot and Walecka.[2] The general motivations for these theories are also given in that review.

Here we will present relativistic shell model calculations within the framework of quantum hadrodynamics. In the relativistic Hartree approximation to the full QHD theory as applied to the ground state of finite (closed-shell) nuclei, the observed spin-orbit splitting of single-particle levels and the resulting nuclear shell structure are predicted. In this discussion, we focus on particle-hole excitations in closed-shell nuclei in the Random Phase Approximation (RPA),

---

* Supported in part by NSF contracts PHY-81-07395 and PHY-48-30894 and DOE contract DE-AC02-81ER40047.
† Chester Davis Fellow.

using the single-particle wavefunctions and energies from the Hartree ground states as inputs.

Our calculational approach is to work consistently within QHD. Retardation in the meson propagators is neglected but otherwise the full Lorentz structure is maintained. Given a QHD model, the effective particle-hole interaction is completely specified by the approximation, with no additional parameters and no constants fit to excited state properties. The sensitive cancellations which are inherent in the ground state calculation mean that realistic excitation spectra are not guaranteed in this approach. The principal motivation for the present investigations is to test whether realistic spectra can be generated. By calculating consistently we can identify and examine the limitations of the approximations.

II. RELATIVISTIC HARTREE DESCRIPTION OF FINITE NUCLEI

Two models of quantum hadrodynamics are considered, as discussed by Serot and Walecka.[2] QHD-I is based on baryons and neutral vector and scalar mesons (σ, ω). The model QHD-II includes, in addition, charged vector (ρ) and pseudoscalar (π) fields in a renormalizable field theory which can be extended to include electromagnetic interactions. In principle, QHD-II provides a consistent framework for describing nuclear matter, finite nuclei, NN scattering, and the interaction of nuclei with hadronic and electromagnetic probes.

Exact solutions to the field equations of QHD-II are very complicated. As an approximate starting point, we use the relativistic Hartree approximation in which the meson field operators and sources are replaced by their expectation values.[8] In terms of Feynman diagrams, the Hartree approximation is achieved by self-consistently summing the tadpole contributions to the nucleon propagator.[2] In the present application, the density operators are normal ordered, so that only positive-energy nucleons contribute.

For the application of the Hartree approximation to QHD-II to closed-shell nuclei, we assume spherically symmetric ground states with total angular momentum zero. The ground states are assumed to have well-defined parity and charge. As a result, there are no ground state contributions from three vector fields, the pseudoscalar pion field, or the charged rho field. However, these constraints will not apply to the particle-hole interaction in the RPA.

The ground state wavefunction is described by the product of relativistic (four-component) single-nucleon wavefunctions. Each nucleon wavefunction satisfies a Dirac equation describing nucleons moving in condensed meson fields:

$$[i\vec{\alpha} \cdot \nabla + U_0(r) + \beta(M - U_s(r))] \psi_\alpha(\mathbf{x}) = E_\alpha \psi_\alpha(\mathbf{x}) \qquad (1)$$

where α labels the single-particle states. The nucleon mass is shifted by the Lorentz scalar potential $U_s(r) = g_s \phi_0(r)$. The condensed scalar field $\phi_0(r)$ satisfies the field equation:

$$(\nabla^2 - m_s^2)\phi_0(r) = -g_s \rho_s(r) \equiv -g_s \sum_\alpha^{occ.} \overline{\psi}_\alpha(\mathbf{x}) \psi_\alpha(\mathbf{x}) \qquad (2)$$

where g_s is the scalar meson coupling constant. Typically $U_s \approx 350$ MeV in the nuclear interior so the nucleon effective mass $M^* \equiv M - U_s$ is significantly

Meson	$J^\pi T$	Mass(MeV)	g^2	$C^2 \equiv g^2(M^2/m^2)$
σ	0^+0	520	109.626	357.47
ω	1^-0	783	190.431	273.87
ρ	1^-1	770	65.226	97.00

Table 1. Model parameters in the Hartree calculation of finite nuclei (as described in reference 3).

less than M. There is also a four-vector potential U_0 of comparable strength that is determined by the condensed ω, ρ, and Coulomb fields. Each satisfies an analogous field equation to (2). The nuclear ground state is determined by this system of coupled nonlinear differential equations, which must be solved self-consistently.

The masses of the nucleon and the vector mesons are taken from experiment, leaving four free parameters: the σ, ω, and ρ meson coupling constants and the mass of the scalar meson. These parameters are fit to properties of bulk nuclear matter and the rms charge radius in ^{40}Ca as described in reference 3. The resulting parameter set is given in Table 1. The same parameter set is used for all closed-shell nuclei and for the RPA calculations of excited states.

The outputs from the relativistic Hartree calculations of finite nuclei include charge densities, neutron densities, and rms radii of closed-shell nuclei. In Fig. 1, we show calculated Hartree charge densities from reference 3 for a variety of doubly-magic nuclei as compared to experiment and to nonrelativistic DDHF calculations. There is generally good agreement with experiment throughout the periodic table, from ^{16}O to ^{208}Pb.

As an example of the single-particle structure, Fig. 2 shows the Hartree single-particle levels for protons in ^{40}Ca. We find the welcome result that the nuclear shell structure is *predicted* with correct level orderings and spacings and major shell closures. The observed spin-orbit splitting of the levels is reproduced as a natural consequence of the motion of Dirac particles in condensed scalar and vector fields. Since the scalar and vector potentials add constructively to determine the spin-orbit interaction, its magnitude is expected to remain relatively stable with improved approximations.[2]

The same Hartree calculations, when combined with the empirical N-N scattering amplitudes and the impulse approximation, yield remarkably accurate results for medium-energy proton-nucleus cross sections and spin observables, with no further adjustment of parameters.[4] The success of the Hartree approximation in these contexts encourages us to extend this approach to the description of the excited state spectrum. Many of the techniques and insights

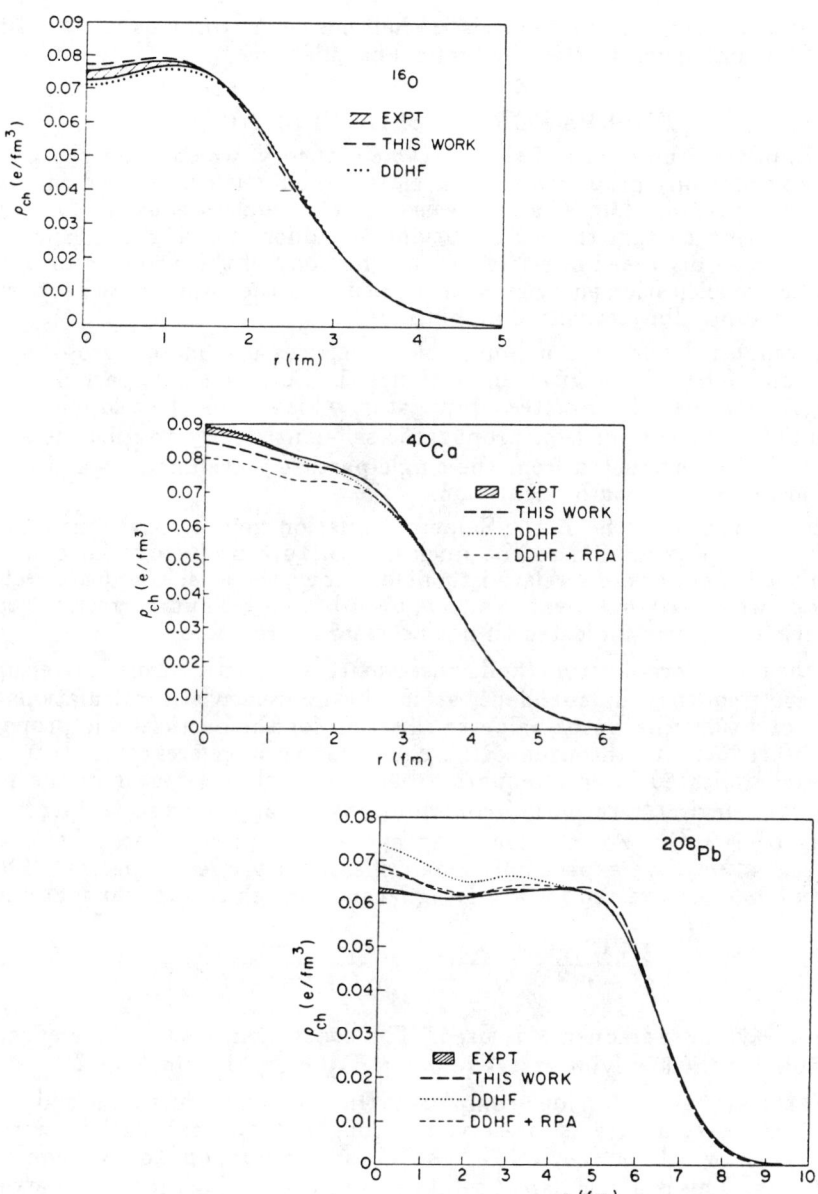

Fig. 1. Hartree charge densities for ^{16}O, ^{40}Ca and ^{208}Pb compared to experiment and to several nonrelativistic calculations (from reference 3). Relativistic Hartree results are indicated by the heavy dashed lines.

obtained from years of nonrelativistic shell model calculations can be directly applied to analogous Hartree shell model calculations.

III. RPA FOR CLOSED-SHELL NUCLEI

Within the framework of the QHD field theory, we consider the particle-hole (polarization) propagator for a closed shell nucleus in the one-meson-exchange approximation. That is, we use a Bethe-Salpeter equation for particle-hole scattering to sum the contributions of ladder and ring diagrams to the propagator, as discussed in reference 5. The poles of this propagator occur at the collective excitation energies of the nucleus and the residues are proportional to the corresponding transition amplitudes.

Feynman rules for the meson-nucleon vertices and meson propagators in QHD-I and QHD-II are given in reference 2. For the single-particle nucleon propagator we use the Hartree propagator, which sums the tadpole diagram contributions to the nucleon propagator self-consistently to all orders.[3] This propagator is constructed from the single-particle wavefunctions and energies found in the ground state calculation.

The solutions to the Bethe-Salpeter equation give a consistent and fully relativistic approximation to the nuclear structure problem. Since we work within the framework of a relativistic field theory, the level of approximation is manifest; we *know* what diagrams have been left out. Thus, corrections to this approach are clearly indicated (if not necessarily tractable!).

If the meson propagators (and consequently the particle-hole scattering kernel) were frequency independent, as in the nonrelativistic calculations with static potentials, the Bethe-Salpeter equation for the particle-hole propagator could be reduced to the usual RPA equations as in reference 5. Instead we have a complicated integral equation because of the q_0 factor in the propagator. We neglect retardation contributions by approximating $1/(q^2 - m^2)$ by $-1/(\mathbf{q}^2 + m^2)$. For the low-lying excited states considered here, where $E_{excited} \ll m_\pi$, we expect only small effects on the level spectra. This assumption has been tested for low-lying levels through the approximation:

$$\frac{1}{q^2 - m^2} \longrightarrow \frac{-1}{\mathbf{q}^2 + (m^2 - \langle q_0^2 \rangle)} \qquad (3)$$

where an averaged frequency is used. The conclusion is that these effects are negligible for the low-lying states (*e.g.* for $E_{ex} < 25$ MeV in ^{16}O).[6]

We stress that the full four-component Dirac matrix structure is maintained. This Lorentz structure is preserved throughout the calculation, retaining the physics which is inevitably lost in any reduction to two-component formalisms.[7] This is particularly useful when incorporating the nuclear structure wavefunctions into a consistent, relativistic theory of the interaction of nuclei with external probes (*e.g.* nucleon-nucleus scattering or electron scattering[8]).

Having neglected retardation, we obtain a simple summation of the diagrams. We obtain RPA equations precisely as in the standard nonrelativistic many-body theory.[2] The detailed derivation is given in reference 6. The matrix elements in these equations involve the four-component wavefunctions obtained from the Hartree ground state calculation, with the particle-hole interaction

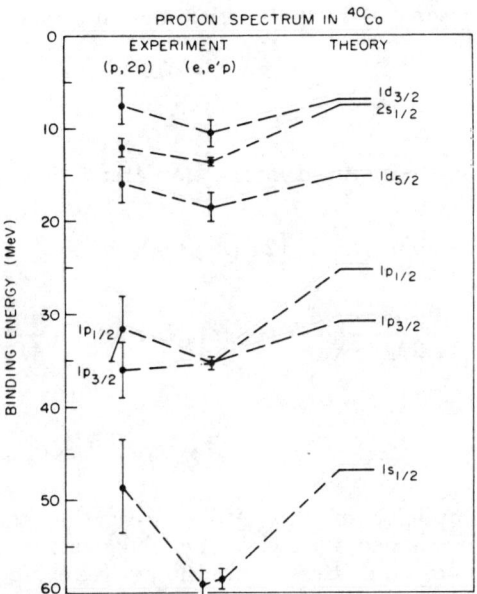

Fig. 2. Predicted Hartree proton spectrum in ^{40}Ca compared with representative experimental data (from reference 3).

determined from the QHD Feynman rules. In the QHD-II model, the particle-hole interaction $\mathcal{V}$ is the sum of contributions from the scalar (s), vector (ω), rho (ρ), and pi (π) mesons:

$$\mathcal{V}(1,2) = \left(\frac{-g_s^2}{4\pi}\right)\frac{e^{-m_s r_{12}}}{r_{12}} + \gamma_1^\lambda \gamma_{2\lambda}\left(\frac{g_\omega^2}{4\pi}\right)\frac{e^{-m_\omega r_{12}}}{r_{12}}$$
$$+ \gamma_1^\lambda \gamma_{2\lambda}\frac{\vec{\tau}_1 \cdot \vec{\tau}_2}{4}\left(\frac{g_\rho^2}{4\pi}\right)\frac{e^{-m_\rho r_{12}}}{r_{12}} + \gamma_1^5 \gamma_2^5 \,\vec{\tau}_1 \cdot \vec{\tau}_2 \left(\frac{g_\pi^2}{4\pi}\right)\frac{e^{-m_\pi r_{12}}}{r_{12}} \qquad (4)$$

Here $r_{12} \equiv |\mathbf{r}_1 - \mathbf{r}_2|$. Note that the spatial dependencies are simple Yukawa functions. Only the scalar and neutral vector contributions to $\mathcal{V}$ are considered for the QHD-I results. (Only pseudoscalar coupling for the pion is given here. Changes from the use of pseudovector coupling are discussed below.) The density and spin dependence of the particle-hole matrix elements is completely specified by the Lorentz structure of the interaction.

The particle-hole interaction $\mathcal{V}$ is deceptively simple. If the interaction were to be reduced for use with conventional two-component spinors, it would become an infinite series of terms with complicated radial and spin dependencies (*e.g.* tensor, spin-orbit, *etc.*). As an example, consider the contribution from the neutral vector meson (ω) to the particle-hole interaction:

$$\gamma_1^\lambda \gamma_{2\lambda} \left(\frac{g_\omega^2}{4\pi}\right) \frac{e^{-m_\omega r_{12}}}{r_{12}} . \tag{5}$$

The first few terms in the reduced interaction are:

$$\frac{g_\omega^2}{4\pi} m_\omega \left[Y(m_\omega r) + \frac{1}{12}(m_\omega/M)^2 \left[2Y(m_\omega r)\vec{\sigma}_1 \cdot \vec{\sigma}_2 - Z(m_\omega r) Y(m_\omega r) S_{12} \right] \right.$$

$$\left. + \frac{3}{2M^2 r}\left(\frac{d}{dr} Y(m_\omega r)\right) \mathbf{L} \cdot \mathbf{S} \right] + O\left(\frac{m_\omega^2}{M^2}\right) , \tag{6}$$

where

$$Y(z) = \frac{e^{-z}}{z} \quad \text{and} \quad Z(z) = \frac{3}{z^2} + \frac{3}{z} + 1 . \tag{7}$$

In practice, the simplicity of the QHD interaction makes the numerical evaluation of matrix elements easy. The Yukawa functions are expanded in spherical harmonics and Bessel functions, reducing the calculation of matrix elements to Slater integrals and standard angular momentum coupling coefficients and reduced matrix elements.

The nucleus is treated entirely consistently in this model. The parameters are fixed by experimental values or by the ground state calculation. The excited state energy levels and wave functions are calculated with the *same* parameters, using the Hartree single particle energies to form unperturbed levels and the Hartree spinors as the particle-hole basis. The particle-hole interaction is determined by the basic parameters of the theory: the meson masses and coupling constants, and by the Lorentz structure of the couplings to nucleons.

The incorporation of pions into QHD theories is a difficult and largely unsolved problem. The description of the pion in nuclear matter in QHD is discussed in detail by Matsui and Serot.[9] In the present work, we assume the free-particle mass and coupling constant of the pion (from scattering data) and study the magnitude of its effects on the excited states. We retain only the π–N coupling in the present model but both pseudoscalar and pseudovector couplings are considered.* They are found to yield significantly different results in the nuclear medium even though the free space descriptions are equivalent.

For $N = Z$ nuclei with no coulomb interaction included, the only significant difference between the two pion-nucleon couplings are factors of $(M^*(r_1)/M)$ $(M^*(r_2)/M)$ in the matrix elements in the pseudovector case.[6] (Other terms are smaller by a factor of $(\Delta\epsilon/2M) \ll 1$ where $\Delta\epsilon$ is the difference between the unperturbed p-h and RPA energies.) Since M^*/M varies from about 1/2 in the nuclear interior to 1 outside the nucleus, this can represent a substantial damping of the pion-nucleon coupling inside nuclei.

---

* Nonlinear couplings of the σ and π are *not* included and represent a correction to this approach.

Here we present results for the low-lying negative parity states in ^{16}O. For simplicity, the Coulomb interaction is neglected so we generate states of good isospin T. (Including Coulomb effects in the ground state does not alter the following qualitative discussion.[6]) The $1p_{1/2}$ and $1p_{3/2}$ states were chosen as the hole space. In the relativistic Hartree description of ^{16}O as described in reference 3, only the $1d_{5/2}$ and $2s_{1/2}$ single particle levels are bound. (There is a $d_{3/2}$ resonance at about 2.8 MeV.) Thus, to go beyond the smallest configuration space we must be able to deal with continuum states.

At present the continuum is discretized by imposing the boundary condition that the upper components of the single-particle wavefunctions vanish at a radius R. With R chosen to be at least several times the ground state radius, the bound state wave functions are negligibly affected and we generate an orthonormal basis which is approximately complete.[†] The advantage of this approach is that bound and continuum states are treated in essentially the same manner. Here we include configurations with unperturbed energies up to 50 MeV. This is sufficient to ensure stable energies for the levels of interest here. However, a more thorough treatment of the continuum is desirable for a quantitative description of excitations in the giant dipole region and above.

In figures 3 and 4, the energy level diagrams for the low-lying states in ^{16}O are shown. In each figure, the first column shows the lowest unperturbed Hartree particle-hole energies which are taken from the ground-state calculation of ^{16}O. These levels reflect the underlying nuclear shell structure in the Hartree ground state. The ordering and approximate location of the unperturbed levels agree (within 2 MeV at worst) with "experimental" particle-hole levels obtained from neighboring nuclei.[10]

The second column shows the QHD-I results (σ and ω only) and the third and fourth columns give the QHD-II levels with pseudoscalar (PS) and pseudovector (PV) pion-nucleon coupling, respectively. Selected experimental levels from reference 10 are given in column five. Since we can only expect to reproduce states which are primarily one-particle one-hole in nature, only these levels are included (as identified by one-nucleon transfer reactions[11]).

Both QHD-I and QHD-II PV reproduce the low-lying cluster of $T = 1$ levels at energies about 2 MeV too low. The QHD-II effective interaction with pseudovector coupling provides a superior reproduction of the relative ordering of the lowest four levels. Overall, given the simplicity of the inputs, the agreement with experiment by *both* models is impressive. We emphasize that even when the levels vary only slightly from their unperturbed energies, the contributions from the individual meson matrix elements are not small. The neutral meson matrix elements are typically 5–10 MeV or more. Furthermore, the matrix elements of mesons which do not contribute to the ground state can also be significant.

The differences between the pseudoscalar and pseudovector spectra are greatest in pion-like states ($J^\pi = 0^-, 2^-, 4^-$) where the pseudoscalar coupling is strongly attractive. For example, we find a 2^- state which remains close

---

† For ^{16}O we take R to be 12 fm. so that the first $d_{3/2}$ state occurs at the resonance energy.

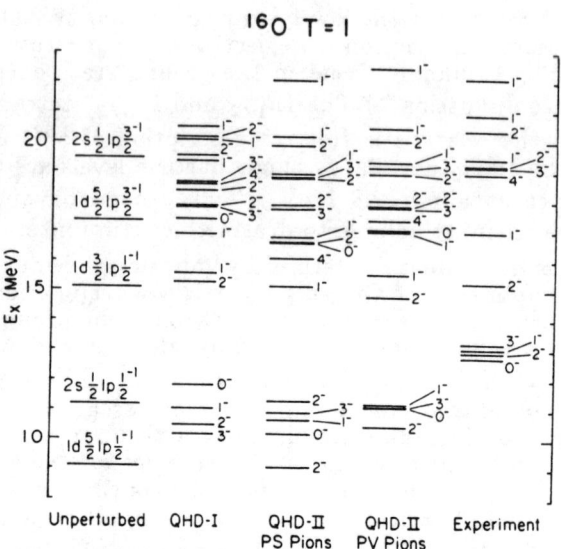

Fig. 3. Negative parity $T = 1$ RPA spectrum for ^{16}O. Only the $1s,d(1p)^{-1}$ unperturbed Hartree energies are shown. Experimental levels are from reference 10.

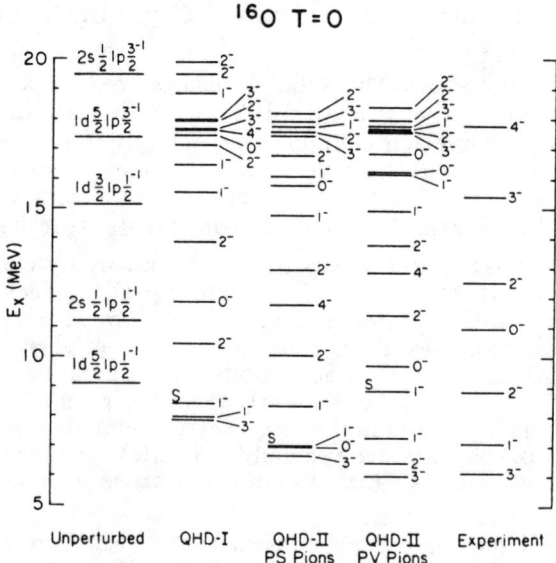

Fig. 4. Negative parity $T = 0$ RPA spectrum for ^{16}O. Only the $1s,d(1p)^{-1}$ unperturbed Hartree energies are shown. Experimental levels are from reference 10. Levels identified as spurious are indicated with an s.

to its unperturbed energy in the pseudovector case ($\sim$ 15 MeV), but with pseudoscalar coupling is pushed down below all other states. This reflects the damping of the pseudovector matrix elements in the nuclear medium.

If we now turn to the $T = 0$ spectrum, we have the complication of a spurious state. In describing the nucleus in a nuclear shell model, the translational invariance of the relativistic theory is broken. This leads to spurious "center-of-mass" components in the nuclear wave functions. In RPA calculations, the spurious components are most important in the $T = 0$, $J^\pi = 1^-$ states. At present we have no consistent method to insure that the spurious state does not mix with physical states. In fully self-consistent nonrelativistic RPA calculations, the center-of-mass excitation is orthogonal to the other $T = 0$, $J^\pi = 1^-$ states and appears at zero excitation energy. This result is not found in the present calculations.* However, the energy of the spurious state is lowered significantly when the model space is enlarged.

The $T = 0$ spectrum in QHD-I is generally consistent with experimental 1p-1h levels except that the lowest 3^- state is pushed down only 1 MeV from its unperturbed energy. Although the addition of the charged mesons pushes this state down to its experimental energy and increases its collectivity, the QHD-II interaction appears to be too attractive in the unnatural parity states, even with pseudovector coupling. The pseudoscalar pion certainly couples too strongly, resulting in a total disruption of the ordering of the lowest levels (including a 2^- state which is pushed down below 5 MeV).

Comparisons of the QHD-I and QHD-II spectra show that the 4^- levels are particularly sensitive to the charged mesons. Because of the simple structure of these levels in our model, we can easily determine that large rho meson matrix elements are acting to push these levels downward in energy.

In relativistic Hartree-Fock calculations in infinite nuclear matter, the inclusion of exchange terms in the ground state requires a much smaller value of the rho coupling constant to reproduce the empirical symmetry energy than in the Hartree approximation. At present, we do not have Hartree-Fock calculations for *finite nuclei*. However, we can indicate the prospects for an RPA calculation based on relativistic Hartree-Fock by repeating the Hartree calculation with the same wavefunctions and unperturbed energies but with a parameter set determined from a nuclear matter Hartree-Fock calculation.[2]

The results for some representative levels are shown in Figures 5 and 6. Columns 2 and 3 show levels with Hartree and Hartree-Fock parameters respectively, with the rho coupling constants indicated (the neutral meson parameters are comparable in magnitude). The $T = 1$ spectra show that the low-lying levels are essentially unchanged with the new parameters with the exception of the 4^- state. This state is pushed up almost two MeV to close to the experimental energy. The $T = 0$ spectra show more pronounced differences. The level ordering of the lowest levels changes to match the experimental ordering more closely because of the significant shifting of the 2^- state. The 4^- state is shifted dramatically, by almost 5 MeV, eliminating the most glaring discrepency with experiment. Finally, we note that the spurious state is very

---

* In figures 4 and 6, the levels identified as spurious are marked with an s.

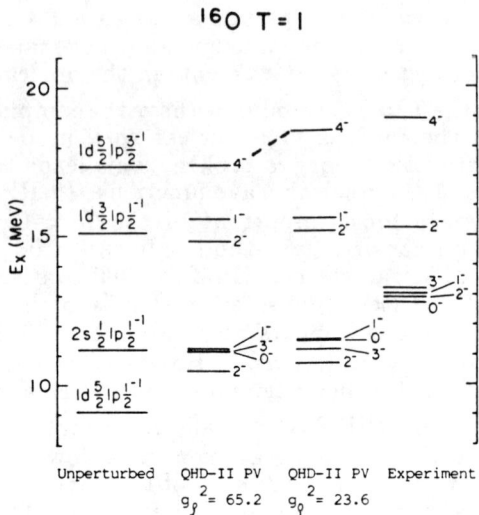

Fig. 5. Selected low-lying $T = 1$ levels from RPA calculations in ^{16}O with the QHD-II interaction and pseudovector pion-nucleon coupling. The second column uses Hartree parameters while the third uses Hartree-Fock parameters from reference 2.

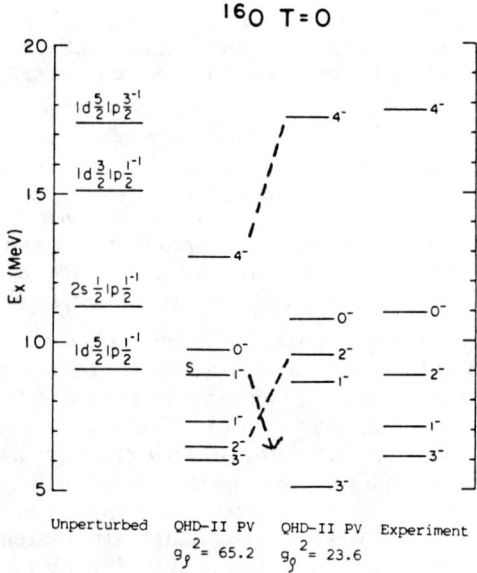

Fig. 6. Selected low-lying $T = 0$ levels from RPA calculations in ^{16}O with the QHD-II interaction and pseudovector pion-nucleon coupling. The second column uses Hartree parameters while the third uses Hartree-Fock parameters from reference 2.

sensitive to the parameter set change, and appears at almost zero energy in the calculation with Hartree-Fock parameters.

We must be cautious in drawing conclusions from these new calculations because they are not consistent. However, they do indicate that a Hartree-Fock description of the ground state could significantly improve the quantitative agreement with experiment.

If we repeat the ^{16}O calculations in another light nucleus. namely ^{40}Ca, we reach the same qualitative conclusions about the different QHD spectra and obtain roughly the same level of agreement with experiment.[6] However, we encounter difficulties in extending the calculations to heavy nuclei, and to ^{208}Pb in particular. In order to generate realistic RPA spectra for the low-lying levels, we must have a reasonable single-particle spectrum near the Fermi surface as given by the ground state calculation. This requisite is realized in the Hartree approximation for light nuclei, but is not satisfied for heavy nuclei such as lead.

In Fig. 7. we show the single-particle levels near the Fermi surface in the relativistic Hartree calculation of ^{208}Pb compared to experimental levels obtained from neighboring nuclei. We find that the unoccupied levels are somewhat underbound, and the level density of occupied levels is too small. These deficiencies preclude a meaningful calculation of RPA spectra in ^{208}Pb with the present approximations.

At least in part, the low level density can be attributed to the effective mass M^* being too small near the Fermi surface in the Hartree approximation. The situation will probably improve with a Hartree-Fock ground state because the effective mass will acquire a state-dependence which is missing in Hartree. In addition, nonrelativistic calculations of the single-particle level density in lead indicate the importance of coupling to particle-hole vibrations to increase the level density.[12] Thus an RPA calculation for the ground state may also be necessary to obtain a reasonable starting point for the excited state RPA calculations.

IV. SUMMARY AND CONCLUSIONS

We have examined particle-hole excited states in closed shell nuclei in the context of the shell model which arises naturally in the Hartree approximation to QHD. We have worked consistently within a relativistic framework with explicit mesonic degrees of freedom (QHD). The ground state and excited states are computed with the same set of parameters adjusted to fit static properties of bulk nuclear matter.

Particle-hole excited state energy levels and wavefunctions for light nuclei were calculated in the Random Phase Approximation (RPA) using the Hartree single-particle energies to form unperturbed levels and the Hartree spinors as the particle-hole basis. We make no nonrelativistic reductions of the interaction: matrix elements are calculated using four-component spinors. The particle-hole interaction is determined by the ground state parametrization of the meson masses and coupling constants, and by the Lorentz structure of the couplings to nucleons.

The principal result of these calculations is that despite the strong cancellations inherent in the QHD framework, experimental particle-hole

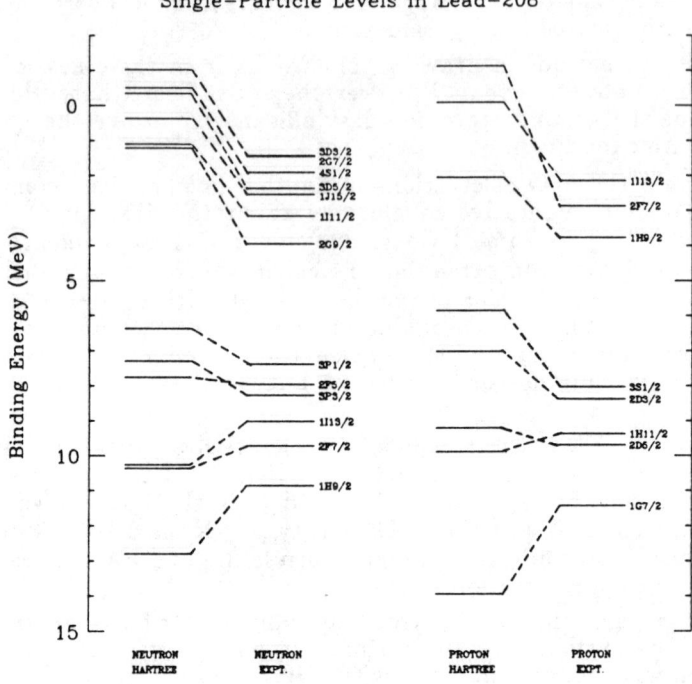

Fig. 7. Single-particle level spectrum in ^{208}Pb near the Fermi surface in the finite Hartree approximation. Both particle and hole states are included; the occupied levels are below 5 MeV binding energy. Experimental levels are from neighboring nuclei.

spectra in light nuclei are reasonably reproduced in both the QHD-I model (neutral mesons only) and in the QHD-II model with pseudovector pion coupling, with no adjustment of parameters. We find that meson retardation effects on the energy levels are generally negligible for the low-lying states (although this will not always be the case for other types of excitation).

Other conclusions we can draw are:
- In general, the pseudoscalar pion coupling is found to be too attractive in unnatural parity states. Use of pseudovector coupling significantly damps the relevant matrix elements and results in improved agreement with experiment.
- The large value of g_ρ used in the Hartree approximation is found to have disruptive effects on certain states. The parameters obtained from relativistic Hartree-Fock calculations in nuclear matter lead to more realistic spectra.
- An improved ground state, and a smaller single-particle level density near the Fermi surface in particular, is required to treat heavy nuclei with the

present approach. The first step is relativistic Hartree-Fock for finite nuclei and perhaps ground state RPA.

Since QHD provides a real many-body theory, we can, in principle, systematically calculate corrections to the present results. Important technical problems to be addressed include the elimination of spurious contributions and the development of an improved treatment of the continuum. By including more Feynman diagrams, a more realistic particle-hole interaction can be derived, including the effects of relativistic correlations and other higher-order effects. Finally, the QHD approach provides a consistent framework in which to describe the interactions of the nucleus with external electromagnetic and hadronic probes using the RPA wave functions generated here.

The author is pleased to thank Professors B. D. Serot and G. E. Walker for valuable discussions and comments.

REFERENCES

1. J. W. Negele, Rev. Mod. Phys. **54**, 913 (1982).
2. B. D. Serot and J. D. Walecka in: **Advances in Nuclear Physics**, Vol. 16, eds. J. W. Negele and E. Vogt (in press) and references cited in this article.
3. C. J. Horowitz and B. D. Serot, Nucl. Phys. **A368**, 503 (1981).
4. B. C. Clark, S. Hama, R. L. Mercer, L. Ray, and B. D. Serot, Phys. Rev. Lett. **50**, 1644 (1983);
J. A. McNeil, J. R. Shepard, and S. J. Wallace, Phys. Rev. Lett. **50**, 1439 (1983).
5. A. L. Fetter and J. D. Walecka, **Quantum Theory of Many-particle Systems**, (McGraw-Hill, New York, 1971).
6. R. J. Furnstahl, Phys. Lett. **152B**, 313 (1985);
R. J. Furnstahl, Ph.D. Thesis, Stanford University, (1985), unpublished.
7. E. D. Cooper, A. O. Gattone and M. H. Macfarlane, Phys. Lett. **130B**, 359 (1983).
8. B. D. Serot, Phys. Lett., **107B**, 263 (1981).
9. T. Matsui and B. D. Serot, Ann. Phys. (N.Y.) **144**, 107 (1982).
10. F. Ajzenberg-Selove, Nucl. Phys. **A375**, 1 (1982).
11. M. Waroquier, G. Wenes, and K. Heyde, Nucl. Phys. **A404**, 298 (1983).
12. S.-O. Bäckman, G. E. Brown and J. A. Niskanen, Phys. Rep. **124**, 1 (1985).

SESSION F

NON-NUCLEONIC DEGREES OF FREEDOM

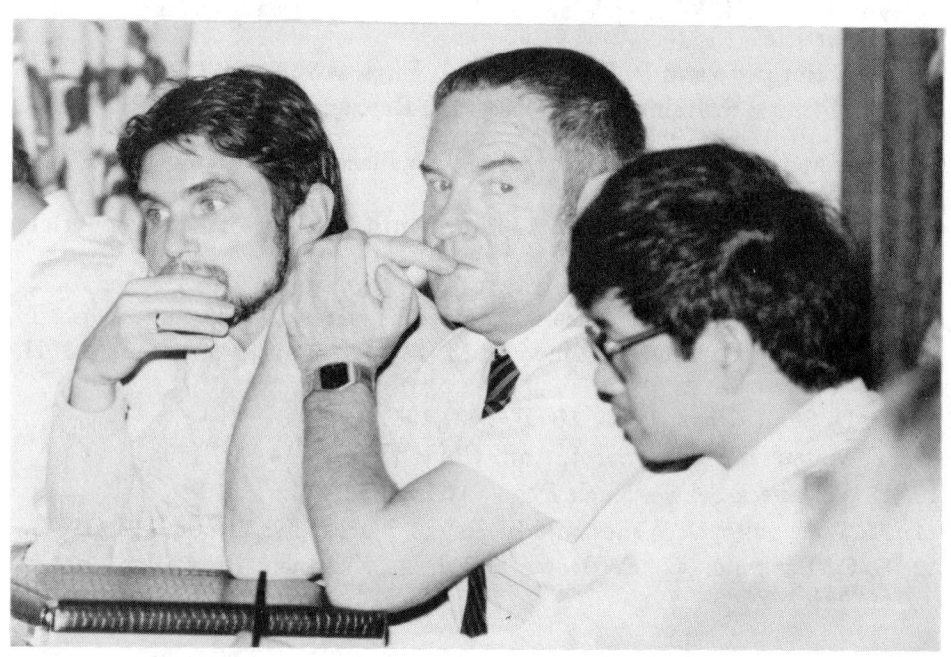

CHARGE EXCHANGE TO THE Δ-REGION

Clive Ellegaard[*]
Niels Bohr Institute, University of Copenhagen
Copenhagen, Denmark

ABSTRACT

Data on a variety of charge exchange reactions at intermediate energies (200 MeV/n - 1 GeV/n) are presented. The excitation of lowlying states will be briefly discussed to document the simplicity and selectivity of these reactions towards spin-isospin transitions. The bulk of the data on excitation of the Δ-resonance comes from the (^{3}He,t) reaction. The most striking feature of this systematic study is a constant shift for all nuclei of 35 MeV of the Δ-resonance from the position expected from quasi-free production.

Spectra with interesting features arising from the detailed structure of the projectile-ejectile system are shown for a variety of heavy-ion charge exchange reactions. Finally preliminary data for the (d,^{2}He) reaction are presented.

INTRODUCTION

The experiments presented here are from an American-French-Scandinavian collaboration at Laboratoire National Saturne. The program consists of a variety of charge exchange reactions at energies from 200 MeV per nucleon to 1 GeV per nucleon. The subject discussed here is Δ-production in nuclei by means of three different charge exchange reactions. They are the (^{3}He,t) reaction where the features are more or less established now, and two reactions that are just being explored now: heavy-ion charge exchange and the (d,^{2}He) reaction.

These reactions have the advantage that they have charged particles both as projectile and ejectile. Thus the instrument of analysis can be a magnet and is the spectrometer called SPES4. Its properties are well suited for this variety of reactions. One main point for all the reactions is the need to go to 0°. This is accomplished by having a magnet before the defining slit to sweep away the beam. Especially for the heavy-ion reactions mass and charge

---

[*]Results obtained in collaboration with:
M. Boivin, P. Radvanyi, J. Tinsley (Saturne), D. Bachelier, J.L. Boyard, T. Hennino, J.C. Jourdain, M. Roy-Stephan (Orsay), D. Contardo, M. Bedjidian, J.Y. Grossiord, A. Guichard, R. Haroutunian, J.R. Pizzi (Lyon), I. Bergqvist, A. Brockstedt, L. Carlén, P. Ekström, J. Lyttkens (Lund), C. Goodman (Indiana), and C. Gaarde, J.S. Larsen (Copenhagen).

Work supported in part by the Danish Natural Science Research Council.

identification is important. For this there is a 16 m flight path between scintillators at an intermediate focus and at the final focus, and two thick scintillators for charge identification. For the (d,2He) reaction - which maybe should be called the (d,2p) reaction - we are swamped by single protons from breakup of the deuteron. It is therefore essential that both the intermediate and final scintillators consist of 12 (13) separate sections so that a fast coincidence between pairs of them can be incorporated into the hardware trigger of the experiment.

The reactions are charge exchange reactions and in an energy range where the σ·τ interaction dominates; i.e. Gamow-Teller transitions are strongly favoured relative to the isobaric analog state. It is in the region where the great (p,n) work at Indiana[1] has shown the systematic occurrence of Giant Gamow-Teller resonances. We have followed this up with the (^{3}He,t) reaction, figure 1, and are beginning with the other reactions mentioned.

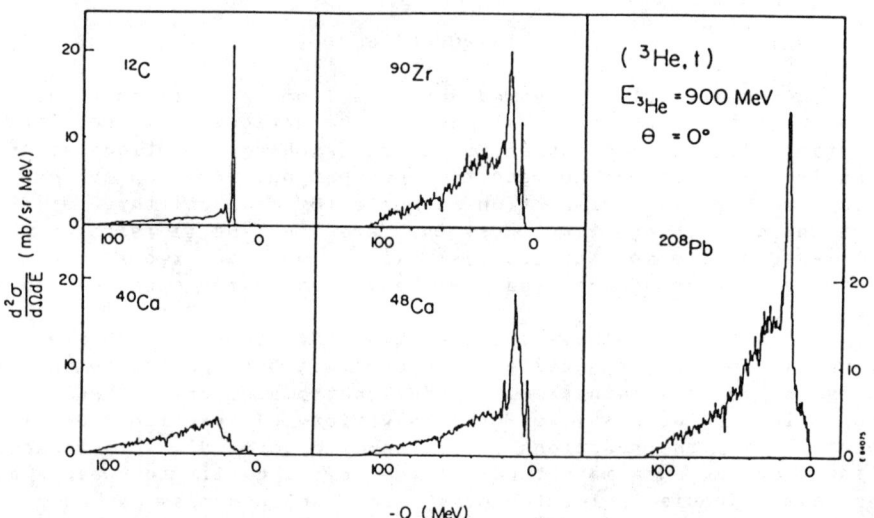

Fig. 1. Zero degree spectra of the (^{3}He,t) reaction at 900 MeV showing the systematic excitation of the Gamow-Teller resonance.

One outstanding problem is that a large fraction of the transition strength predicted[1] on the basis of a very general sumrule is missing; about 1/2 is missing. In spectra like those of figure 1 it is difficult to say how much strength is moved into a tail of the peak, but surely some must be through interaction with 2p-2h states. It is a way of saying: we have a problem defining the line

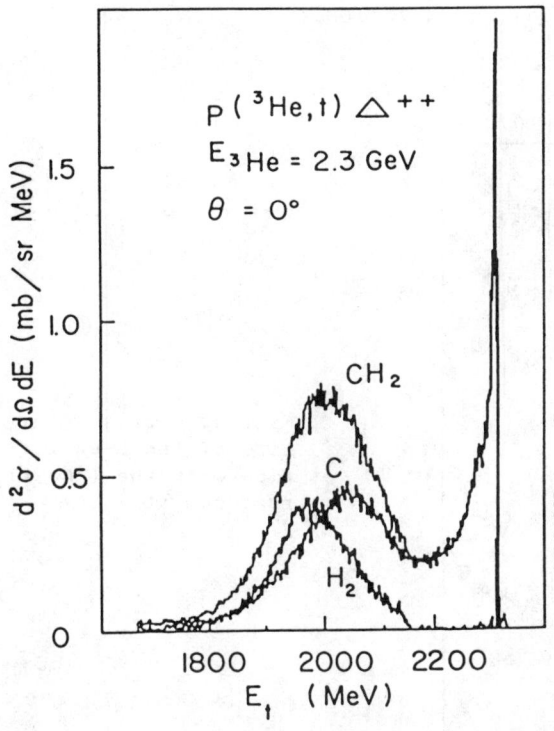

Fig. 2. (³He,t) spectrum on carbon at $E_{^3He}$=2.3 GeV, showing the Δ-excitation. Also shown is the Δ^{++}-excitation of the proton obtained by subtraction of the spectrum of carbon from CH_2.

shape of this resonance in the nuclear medium - or in this finite piece of nuclear matter. The first quantitative attempt at calculating it is by Bertch and Hamamoto[2], and they predict as much as half the strength in the region up to 50 MeV. We are working on disentangling this continuum part of the spectrum.

Another explanation for the missing strength is that it is moved to the Δ-region; and again some must be found there. The same interaction that creates the spin-isospin correlations in the nucleus will also excite a nucleon to the Δ-resonance - the first excited state of the nucleon. Thus these two regions would become admixed and lowlying strength would be moved up to ~300 MeV.

We are now studying the Δ-excitation in nuclei. We are using the same tools that we have learned about at low excitation energy, with the modifications necessary for going to this excitation. The main one is that in order to reach this excitation energy the momentum transfer is necessarily non-zero, and that is important. It means that there is a very small probability of exciting the Δ with $\ell=0$ transfer and thus directly studying the counter-part of the Gamow-Teller states. In any case it is interesting in itself to study the Δ excitation in nuclei, to measure the response of the nucleus to this excitation.

THE (^{3}He,t) REACTION

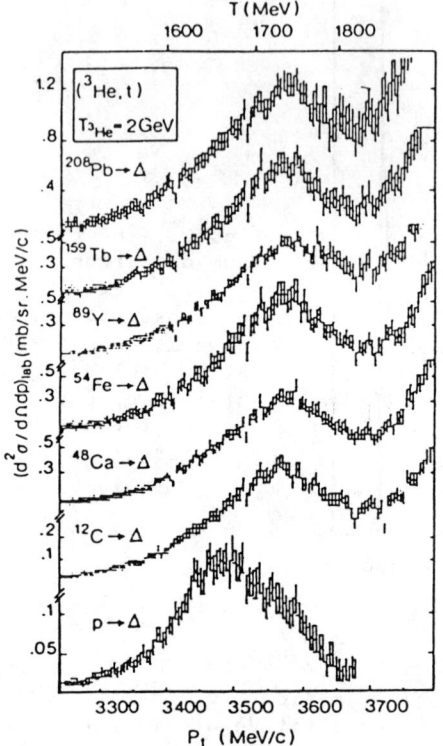

Fig. 3. Mass systematics of the Δ-excitation with the (^{3}He,t) reaction at 2 GeV.

Figure 2 shows spectra of the (^{3}He,t) reaction at 0° at 2.3 GeV. The figure shows the spectrum from ^{12}C with the sharp peak at low excitation energy that contains all the spin resonances, and around 300 MeV the Δ-resonance. In addition the figure indicates how the Δ-excitation on the proton is obtained. A distinct shift is observed in the energy of the Δ-excitation from the elementary excitation on the proton to that in a nucleus.

Figure 3 indicates that this is systematically so. All nuclei show the Δ-excitation at exactly the same energy and shifted 70 MeV relative to the proton. Part of this shift is of course kinematics from the light target, the proton, to the heavier nuclei. A shift in observed position is also induced by the ^{3}He or t formfactor - the fact that the probability that the triton can survive the momentum transfer becomes rapidly smaller the larger the transfer. The cross section is thus heavily weighted towards low excitation energy. This is true for both proton and heavier targets. These factors can be calculated[3], and fits to the proton data are shown in figure 4.

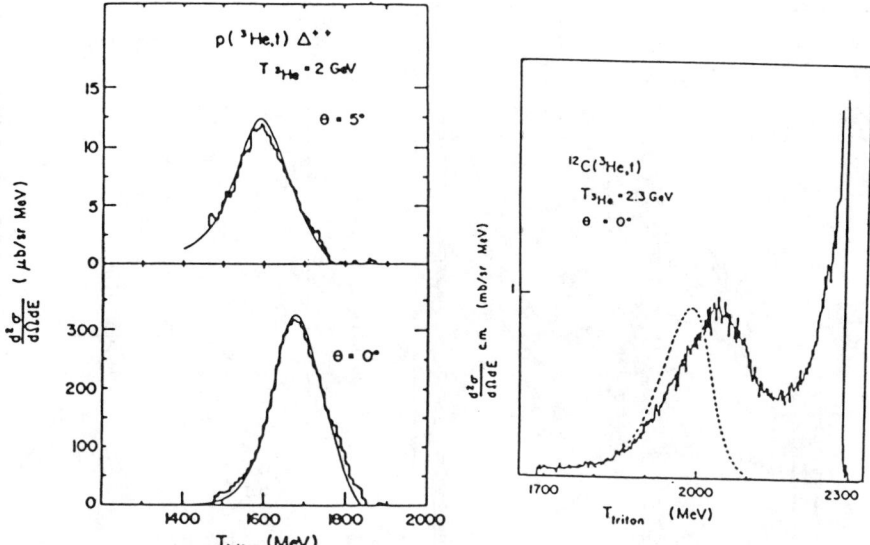

Fig. 4. Spectra at two angles for the p(^{3}He,t)Δ^{++} reaction. The curves are calculated cross sections in OPE approximation[3].

Fig. 5. The 0° spectrum for ^{12}C at 2.3 GeV shown together with the calculated cross section, arbitrarily normalized, for quasifree Δ-production.

With the same calculation for a nucleus the result of figure 5 is obtained. The nucleus is represented by nucleons in a harmonic oscillator, and the final Δ is in a plane wave state. This is what we would call quasifree production. The observed shift is reduced but we are still left with a 35 MeV shift.

Figure 6 shows how the spectra vary with angle. The Δ bump moves with kinematics as that of a proton, but retains its shift. It moves roughly parallel to the broad peak at low excitation where it may also be termed quasifree scattering.

Figure 7 shows that the angular distribution of the cross section for Δ production is the same for all nuclei and similar to the proton. Everything is consistent with quasifree production except the energy shift. That is the signal we have that the Δ is produced in a nucleus.

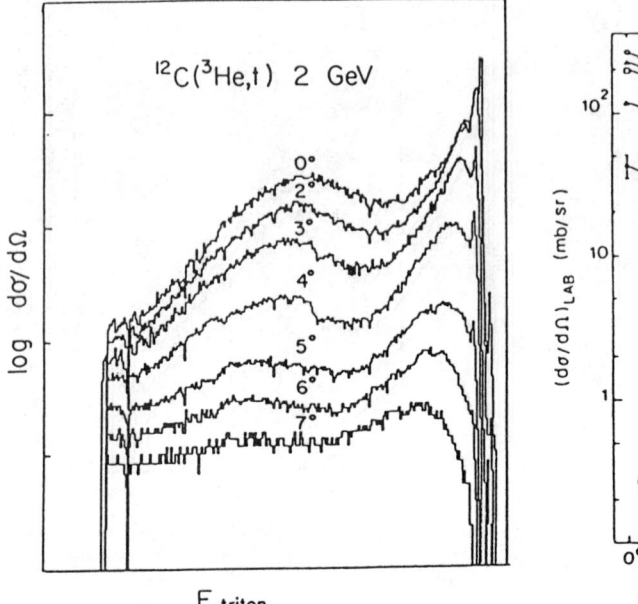
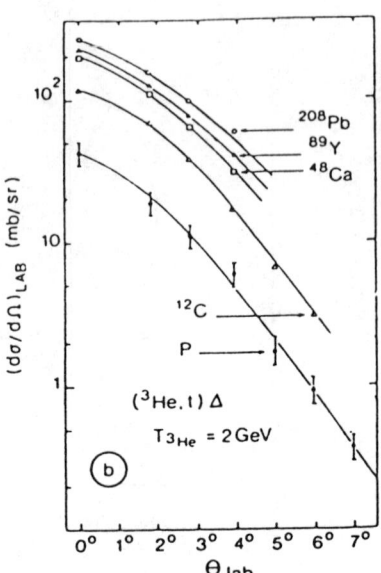

Fig. 6. Angular variation of energy spectra for the (^{3}He,t) reaction at 2 GeV.

Fig.7. Angular distributions of the cross sections for Δ peak for different nuclei and for the proton.

There have been several attempts to understand this shift. M. Ericsson[4] and Dmitriev and Suzuki[5] have calculated the response of nuclear matter to a σ τ probe leading to Δ formation. They obtain much larger shifts than observed. The calculations are for infinite nuclear matter whereas the reaction is surface dominated.

A suggestion put forward by Weise[6] is to look at the Δ production in π-nucleus scattering. Here a partial wave analysis shows that the observed Δ position is a function of transferred angular momentum. The (^{3}He,t) reaction mainly takes place via one pion exchange, and since the reaction is very peripheral, it puts a narrow window in the transferred ℓ-values. The reaction thus selects a specific position of the Δ. Faessler[7], on the background of calculated self energies of the Δ, makes the suggestion that the reaction takes place at a density of 0.3 times nuclear density. This is probably the same statement as the above. It remains to be seen whether these statements can be made quantitative in connection with the (^{3}He,t) reaction.

A recent calculation by Esbensen and Lee[8] takes its starting point in semi-infinite nuclear matter and thus gives the response of a surface to the probe. Figure 8 is a reproduction of their results.

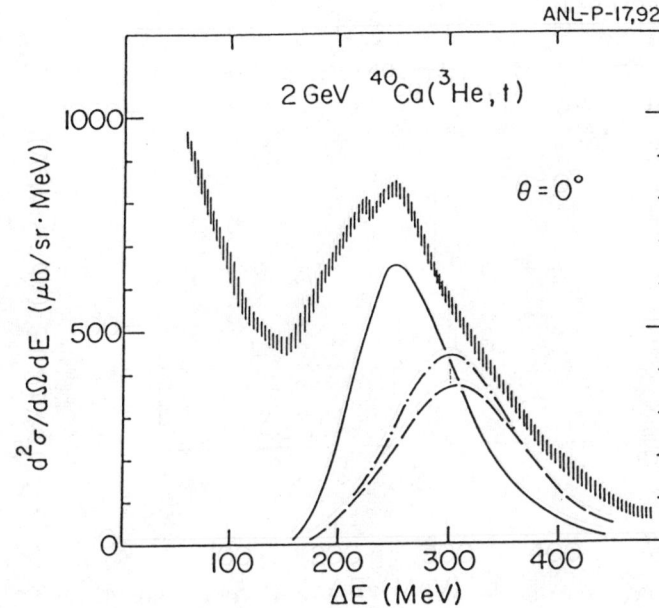

Fig. 8. A reproduction of the calculated results by Esbensen and Lee[8].

The dashed curve is the free response, i.e. without residual interactions. This should resemble figure 5. The dot-dashed curve shows the effect of the surface RPA, which does not change the response much. The full drawn curve shows the effect of including a Δ self energy. This is done in a way consistent with π-nucleus scattering data. A complex potential derived from π-nucleus scattering is translated into an effective mass of the Δ of 1204 MeV and an effective width of 200 MeV. In effect the surface response model is not really reproducing the shift, rather it is put into it. That the calculation is a surface calculation ensures that there is practically no difference between a calculation for ^{12}C and one for ^{208}Pb, which is consistent with the data.

HEAVY-ION CHARGE EXCHANGE

One way of probing different densities while exciting the Δ is to go to heavy-ion charge exchange reactions. One obvious choice would be a pair of reactions like:

$$^{12}C \rightarrow {}^{12}N, {}^{12}B$$

where it is known that almost all Gamow-Teller transition strength is collected in one state, the ground state, in the projectile-ejectile system. The result of such a reaction is shown in figure 9. There is strong excitation of the ground state region but practically no excitation of the Δ.

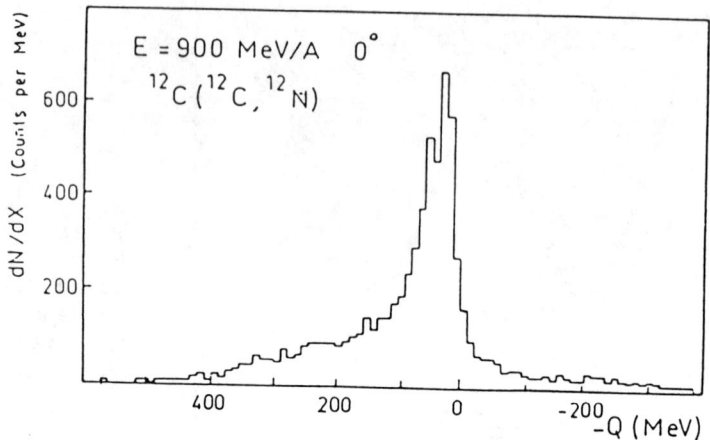

Fig. 9. Energy spectrum of the $^{12}C(^{12}C,^{12}N)^{12}B$ reaction at 0° and 0.9 GeV per nucleon.

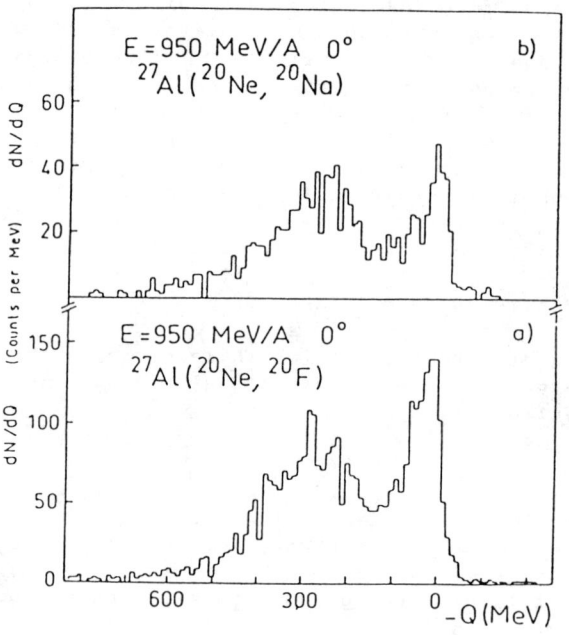

Fig. 10. Energy spectra of the $^{27}Al(^{20}Ne,^{20}Na)^{27}Mg$ and $^{27}Al(^{20}Ne,^{20}F)^{27}Si$ reactions at 0° and 0.95 GeV per nucleon

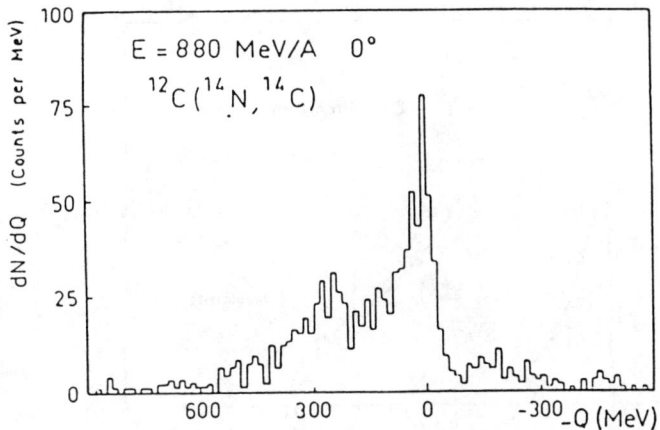

Fig. 11. Energy spectrum of the $^{12}C(^{14}N,^{14}C)^{12}N$ reaction at 0° and 0.88 GeV per nucleon.

In the system
$$^{20}Ne \rightarrow {}^{20}Na, {}^{20}F$$
the situation is very different with the Δ excitation dominating over the ground state region by a factor of 1.6.

In yet another case
$$^{14}N \rightarrow {}^{14}C$$
an intermediate situation is found. In this case, however, the reaction leading to ^{14}O shows no cross section whatsoever above the background.

The heavy ion reactions have turned out to be quite surprising. However all the above features may be explained by a detailed study of the projectile-ejectile system. Specifically: which particle-stable states can be populated in the ejectile and by which transitions. As mentioned above the main transition in the $^{12}C \rightarrow {}^{12}N$ reaction is the $\ell=0$ Gamow-Teller transition. It is expected that the formfactor for this larger system falls off more rapidly with momentum transfer than for the smaller (^{3}He,t) system. The observed spectrum of figure 9 is in accordance with this expectation. Then what makes the $^{20}Ne \rightarrow {}^{20}F, {}^{20}Na$ system so different? Information from other sources, mainly the $^{20}Ne(p,n)^{20}Na$ reaction[9], shows that besides a Gamow-Teller transition there is a very strong $\ell=1$ transition leading to a bound state in the ejectile. The $\ell=1$ transition has a maximum for a momentum transfer of ~0.6 fm^{-1} which is just the momentum transfer corresponding to excitation of the Δ at 0°. The reaction is thus matched for exciting the Δ which qualitatively explains the appearance of the spectra of figure 10.

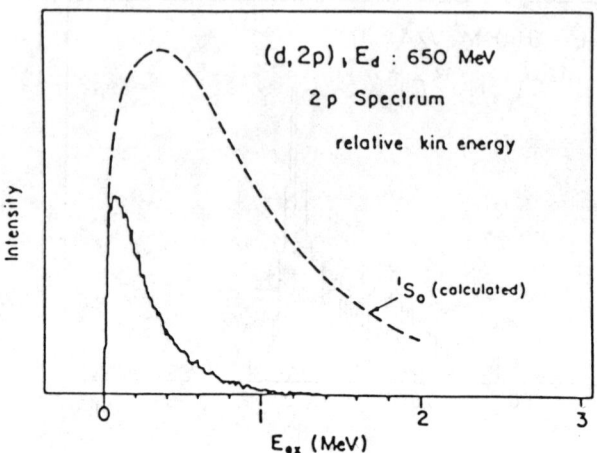

Fig. 12. Measured spectrum of the relativce kinetic energy of the two protons from the (d,2p) reaction shown together with the calculated spectrum for the $^1S_0$ state.

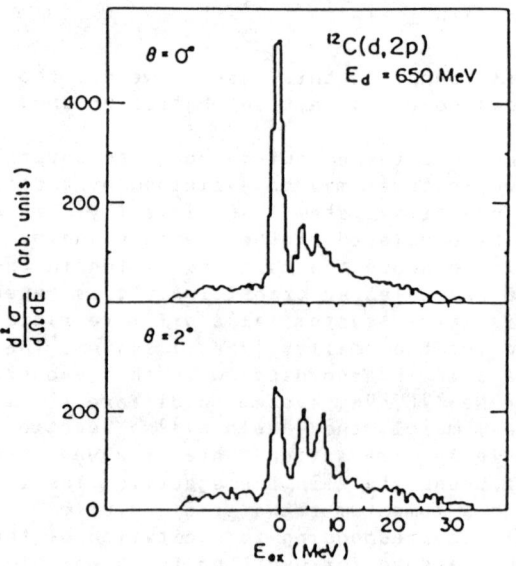

Fig. 13. Energy spectrum of the (d,^{2}He) reaction on carbon at 0° and 325 MeV per nucleon.

The absence of any transitions $^{14}N \to {}^{14}O$ is also a consequence of nuclear structure and β-decay probabilities. The only transition leading to a bound state is to the ground state of ^{14}O. The corresponding β-decay is very strongly hindered with the consequence that the transition cross section becomes very small. For transitions to ^{14}C there are additional bound states that may be reached by $\ell=0$ or $\ell=1$ transitions, yielding the spectrum of figure 11.

All the above effects are clearly demonstrating that we are dealing with a simple one-step process. The effect of nuclear structure on the formfactor is an added complication to these reactions, but, when understood, could be turned to an advantage.

THE (d,^{2}He) REACTION

The (d,^{2}He) reaction is also just being explored now. The ^{2}He is of course an unbound system of two protons. At the momenta of the present work the two protons are highly correlated and both can be detected in SPES4. The efficiency of the spectrometer has a Gaussian-like shape, and the total efficiency is 5-10% depending on the beam momentum. As mentioned before there is an overwhelming count rate of single protons, and a fast coincidence is essential for selecting the correlated protons.

What does it mean to detect a ^{2}He? In figure 12 is shown the calculated distribution of two protons in the $^1S_0$ state together with the measured distribution of the proton pairs accepted by SPES4. A p-state of the two protons would start at zero and still be very small at 1 MeV where they would no longer be detected in the spectrometer. Therefore the spectrometer is very selective to the $^1S_0$ state - the ^{2}He. That in turn makes the reaction from the ^3S state of the deuteron a specific spin transfer reaction.

Figure 13 shows a low excitation energy spectrum to illustrate how well the reaction works. It is the mirror spectrum of the (p,n) or the (^{3}He,t) and looks practically identical. To reach the Δ-region many momentum bytes are needed. Each reflects the Gaussian-like efficiency, and must be divided through by the corresponding efficiency curve. These are derived partly from a Monte Carlo calculation and partly from measurements of closely spaced momentum bytes. After dividing, chopping and assembling the final spectra appear as in figure 14.

From the reaction on carbon is again seen the sharp lowlying peak and the broad Δ peak. The figure also shows the reaction on the proton. Here is the first indication that the reaction is the inverse isospin reaction: The cross section for the $p(d,{}^2He)\Delta^0$ reaction is about 1/3 of that for the $p({}^3He,t)\Delta^{++}$ reaction, measured relative to carbon in both cases. This is what is expected from the isospin Clebsh Gordon coefficients. Otherwise, that the reaction is (n,p) like is mostly relevant for the lowlying states. Such reactions are very important for the solution of the missing strength question. An additional feature is that the deuteron beam is polarized. The deuteron with its spin one carries tensor polarization and thus with simple cross section measurements can yield information equivalent to double scattering experiments like

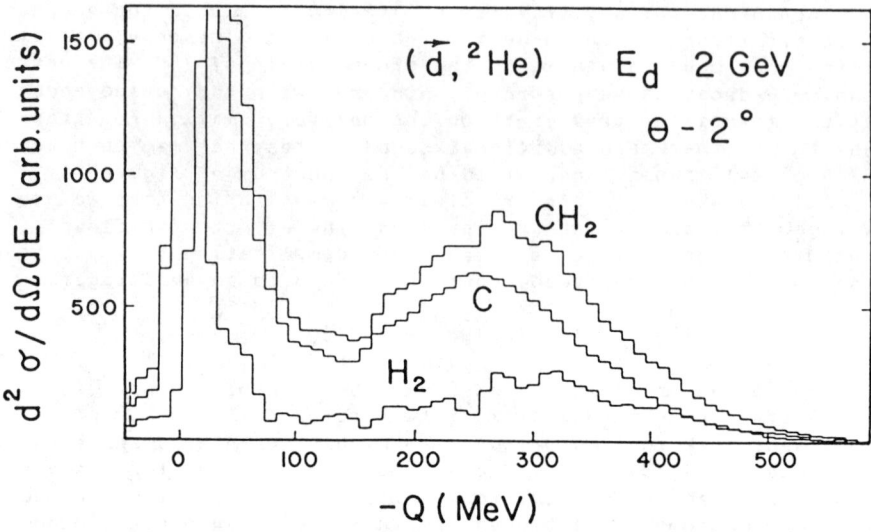

Fig. 14. (d,²He) spectrum on carbon at 0° and 1 GeV per nucleon showing the Δ-excitation. Also shown is the Δ⁰-ecxitation of the proton obtained by subtraction of the spectrum of carbon from CH_2.

($\vec{p}$,$\vec{n}$). For the Δ-region it should be possible to distinguish between σ·q and σxq excitations. These data have yet to be extracted.

REFERENCES

1. C. Goodman, Nucl. Phys. A374, 241c (1982), and references therein;
   C. Gaarde, Nucl. Phys. A396, 127c (1983) and references therein.
2. G. F. Bertch and I. Hamamoto, Phys. Rev. C26, 1323 (1982).
3. C. Ellegaard et al., Phys. Lett. 154B, 110 (1985).
4. G. Chanfray and M. Ericsson, Phys. Lett. 141B, 163 (1984).
5. V. F. Dmitriev and T. Suzuki, Nucl. Phys. A438, 697 (1985).
6. W. Weise, Prog. in Part. and Nucl. Phys. 11, 123 (1984).
7. W. H. Dichhoff, A. Faessler, J. Meyer-ter-Vehn and H. Müther, Phys. Rev. C23, 1154 (1981), and A. Faessler, private communication.
8. H. Esbensen and T. S. H. Lee, Argonne preprint ANL-P-17, 921 (1985).
9. R. P. DeVito et al., UICF Annual Report, p.32 (1982).

MODELS OF NUCLEON AND PION SUBSTRUCTURE

W. Weise

Department of Physics, SUNY,
Stony Brook, NY 11794
and
Institute of Theoretical Physics
University of Regensburg[1]
D-8400 Regensburg, W. Germany

INTRODUCTION

This paper discusses recent developments in understanding the structure of the nucleon and the pion in models based on the underlying Chiral Symmetry of QCD. As is well known, a very good description of a wide variety of nuclear physics phenomena at low and intermediate energies and momentum transfers is achieved in terms of nucleons and mesons, rather than quarks and gluons. In view of the relatively large electromagnetic hadron radii (the proton and pion r.m.s. charge radii are about 0.85 and 0.7 fm, respectively), this is obviously not a trivial observation which requires a deeper understanding.

The pion, introduced fifty years ago [1] as the basic generator of the nuclear force, has ever since played a fundamental role in nuclear physics. The presence of the pion in nuclei is a well established fact, e.g. from investigations of deuteron properties [2]. It is known that the pion exchange interaction, assuming *pointlike* pions and nucleons, gives a quantitatively correct description of the nucleon-nucleon force at distances larger than the pion Compton wavelength, $r \geq 1.4$ fm $= m_\pi^{-1}$. At shorter distances, two-pion and vector meson exchanges combine to give an extremely successful phenomenology.

Most remarkable conclusions are drawn from the analysis of deuteron electrodisintegration at small energy transfer but large momentum transfers up to $|\mathbf{q}| = 1$ GeV [3]. This process probes the spatial distribution of the meson exchange current down to small distances between nucleons. One would naively expect that a picture based on the exchange of pointlike pions breaks down and dissolves into a complicated quark-gluon problem at distances much smaller than 1 fm. This, however, is not the case: calculations of the d(e,e')pn differential cross section based on the exchange of pointlike pions between pointlike nucleons reproduce the data extremely well up to $|q^2| = 25\ fm^{-2}$ [4]. Short distance effects from ρ exchange and formfactors cancel out nearly completely. It is certainly remarkable that calculations involving only the pion work successfully down to distances where they are not supposed to work. It seems that quark and gluon degrees of freedom are

---

[1] permanent address

hidden in some way, at least down to length scales which are small compared to the QCD scale breaking parameter $\Lambda_{QCD}^{-1} \sim 1 fm$.

Recent developments guided by principles of non-perturbative QCD in the limit of a large number of colors N_c, suggest a description of low energy hadron physics in terms of meson (rather than quark-gluon) degrees of freedom. In this perspective, I will now first discuss aspects of pion structure and their implications, and then turn to chiral models of the nucleon and its electromagnetic properties.

THE PION AS A COLLECTIVE MODE

The very special features of the pion are related to the underlying chiral SU(2) x SU(2) symmetry of QCD which is exact in the limit of massless u and d quarks. There is strong evidence, both phenomenological and numerical [5], that chiral symmetry is dynamically broken. The pion is identified with the associated Goldstone boson. For massless u and d quarks, the pion mass is zero. The actual $m_\pi = 140$ MeV (a small number on hadronic scales) is understood in terms of non-zero but small current quark masses m_u and m_d of order 10 MeV. Current algebra gives the relation

$$m_\pi^2 f_\pi^2 = -(m_u + m_d) < \bar{q}q > \tag{1}$$

which connects, apart from the masses, two other fundamental parameters: the pion decay constant $f_\pi = 93$ MeV, and the quark condensate $< \bar{q}q > \stackrel{.}{=} -(240 \text{ MeV})^3$. The vacuum expectation value $< \bar{q}q >$ plays the role of an order parameter for the phase with broken chiral symmetry: it reflects the non-trivial structure of the vacuum in QCD. In fact, eq.(1) says that the basic pion properties (its mass and decay constant) are inseparably connected with the complexity of the non-perturbative QCD vacuum.

It has been proposed that the pion should not be viewed as a single bound pair of a valence quark and antiquark, but as a collective mode built on the QCD vacuum [6,7,8]. The notions developed in this connection have a far-reaching analogy with low-energy collective states of strongly interacting many-body systems.

To see how this works, consider the Nambu and Jona-Lasinio (NJL) model [9], which serves as a basic framework to illustrate the chiral symmetry breaking mechanism. The NJL Lagrangian is:

$$L_{NJL} = \bar{\psi} i \gamma_\mu \partial^\mu \psi + G[(\bar{\psi}\psi)^2 + (\bar{\psi} i \gamma_5 \vec{\tau} \psi)^2]. \tag{2}$$

Here the ψ fields describe u and d quarks (initially massless). Their effective interaction is proportional to a coupling strength G with the dimension of a squared length. The Lagrangian (2) is invariant under the chiral transformation

$$\psi' = exp[i\gamma_5 \vec{\tau} \cdot \vec{\theta}] \psi \tag{3}$$

of the quark fields. The NJL model can be thought of as a minimal effective Lagrangian, written in terms of strongly interacting Fermion fields, and representing basic properties of non-perturbative QCD in the long wavelength limit. This point has been investigated in ref.[10], where also the equivalence with widely used Boson representations (the non-linear σ model, or the Skyrme model) has been discussed.

The NJL model has a cutoff parameter Λ (say, $\Lambda = 1$ GeV). When the dimensionless coupling strength $\alpha = \Lambda^2 G$ reaches a critical value of order unity, the interaction produces a finite quark mass m, and the vacuum structure becomes non-trivial by the appearance of a quark condensate, i.e. a non-zero expectation value $<\bar{\psi}\psi>$. The self-consistent quark mass equation is of the form:

$$m = const.G \cdot tr <\bar{\psi}\psi>. \tag{4}$$

Once eq.(4) has non-trivial solutions, chiral symmetry is dynamically broken. At the same time, the interaction is strongly attractive in pseudoscalar-isovector quark-antiquark states, the ones which carry pion quantum numbers, in such a way that the mass of a pionic $q\bar{q}$ mode is moved down from $2m$ to zero. This is just what's required by the Goldstone theorem.

The NJL mechanism of dynamical chiral symmetry breaking is of a very general nature, quite independent of the detailed form of the chiral effective interaction. Several variants of this model have been discussed in the literature [7].

Results of QCD Monte Carlo calculations on the lattice [5] indicate a close connection between chiral symmetry breaking and confinement. We have investigated [11,12] a schematic model with these properties using the language of the Random Phase Approximation (RPA) familiar from the description of nuclear collective states. In this approach, the coupling constant G of the NJL model is replaced by a position dependent coupling strength $G(r)$ which is weak for small r and increases rapidly as r approaches values around $\Lambda_{QCD}^{-1} \cong 1 fm$, so that the self-consistent mass relation (4) now leads to a scalar potential $m(r)$ which confines quarks. This potential supports a set of localized valence quark-antiquark states

$$|(q\bar{q})_{\alpha\beta}> = b_\alpha^+ d_\beta^+ |0>, \tag{5}$$

where b_α^+ and d_β^+ create a quark and an antiquark, respectively, in orbits α and β. We can refer to these states as the "bag model" $q\bar{q}$ states.

When turning on the residual chiral quark-antiquark interaction in $q\bar{q}$ states with pion quantum numbers, the pion mode emerges as a coherent superposition of the bag model $q\bar{q}$ pairs:

$$|\pi> = \sum_i [X_i B_i^+ - Y_i B_i]|\tilde{0}>, \tag{6}$$

where $i = (\alpha\beta)$ and $B_i^+ = [b_\alpha^+ d_\beta^+]_{I=1}^{J=0^-}$. The vacuum $|\tilde{0}>$ has now a more complex structure than the one in eq.(4), in that it has $2q2\bar{q}$ etc. components already built in. The strength of these ground state correlations is measured by the Y-amplitudes in eq.(6). For originally massless u and d quarks, the pion mode $|\pi>$ must end up at zero mass in accordance with the Goldstone theorem. This is just a consequence of the particular combination, dictated by chiral symmetry, of the scalar-isoscalar $(\bar{\psi}\psi)^2$ part and the pseudoscalar-isovector $(\bar{\psi}i\gamma_5\vec{\tau}\psi)^2$ part of the interaction. Roughly speaking, the scalar term generates a locally varying effective mass (or confining potential) for the quarks, thereby breaking the chiral symmetry of the original Lagrangian, whereas the pseudoscalar term gives the right amount of attraction in the pionic $q\bar{q}$ channel so that the pion emerges as the triplet of zero mass Goldstone bosons. With finite current quark masses m_u and m_d, the RPA scheme leads naturally to the relation $m_\pi^2 = const.(m_u + m_d)$, consistent with the current algebra result, eq.(1).

As expected for a low-energy collective mode, the pion has a non-trivial structure in Fock space with strong admixtures of multi-$q\bar{q}$ components [12]:

$$|\pi> = \sqrt{Z}|q\bar{q}> + \sum_{n>1} a_n|nqn\bar{q}>, \qquad (7)$$

where n=3,5,7,... and $Z + \sum a_n^2 = 1$. The "spectroscopic factor" of a single valence $q\bar{q}$ pair is strongly reduced from unity. For model spaces with cutoffs around 1 GeV, the probability to find a single $q\bar{q}$ is $Z \dot{=} 0.3$ [12]. This Z-factor enters in the decay amplitudes for $\pi \to \mu\nu$ and $\pi^0 \to 2\gamma$. The pion weak decay constant is

$$f_\pi = \sqrt{3}\, m_\pi^{-1/2} \sum_i (X_i - Y_i)\phi_i(0), \qquad (8)$$

where $\phi_i(0)$ is proportional to the amplitude to find a $q\bar{q}$-pair in an orbit i at the origin. It is important to note that the combination $(X - Y)/\sqrt{m_\pi}$, and therefore f_π itself, is independent of m_π, consistent with basic current algebra requirements. One obtains

$$f_\pi = 0.4\sqrt{Z}\, C_v/R_\pi, \qquad (9)$$

where R_π is the pion r.m.s. radius. The factor 0.4 is of purely geometric origin, and $C_v \dot{=} 0.8$ is a vacuum correlation coefficient [12]. For a typical value $\sqrt{Z} = 0.6$ of the valence $q\bar{q}$ amplitude, one finds $f_\pi = 0.2/R_\pi$; the empirical f_π is obtained with $R_\pi = 0.4 fm$. We regard this R_π as an upper limit since components such as $|2q2\bar{q}>$, $|q\bar{q} + gluons>$ etc., not explicitly present in the Fock space expansion of the RPA wave function (6,7), will further reduce $\sqrt{Z}$, hence even smaller values of R_π are required to reproduce

f_π. The radius so obtained is considerably smaller than the measured pion charge radius [13],

$$< r_\pi^2 >^{1/2} = (0.66 \pm 0.02) fm. \tag{10}$$

We will now see how this can be understood.

QCD ORIENTED EFFECTIVE LAGRANGIANS AND THE VECTOR DOMINANCE MODEL

The small intrinsic size of the pion as a collective mode of the QCD vacuum suggests that a description of low energy dynamics in terms of point-like pions should work well. In fact, in the limit of large number of colors N_c, QCD effectively reduces to a non-linear local meson field theory [14,15]. For two massless flavors, i.e. with underlying $SU(2)_L \times SU(2)_R$ chiral symmetry, the theory in the low-energy, long wavelength limit is expressed in terms of the pion field alone. A minimal effective Lagrangian synthesizing these features is the Skyrme model[16]. It operates with the 2x2 unitary field

$$U(x) = exp[i\vec{\tau} \cdot \vec{\pi}(x)/f_\pi] \tag{11}$$

with $U^+ U = 1$. The Skyrme Lagrangian is

$$L = \frac{f_\pi^2}{4} tr[\partial_\mu U \partial^\mu U^+] + \frac{\epsilon^2}{4} tr[U^+ \partial_\mu U, U^+ \partial_\nu U]^2 + ..., \tag{12}$$

with possible extensions to higher orders to account for shorter wavelength properties.

Recent developments have pointed to interesting connections with the ρ meson and the Vector Dominance Model (VDM) which is known to work well as a low energy phenomenology of photon-hadron interactions. We give here a brief summary following Bando et al.[17]. The starting point is the observation that the non-linear σ model (the first term of the Lagrangian (12)) has a hidden $[SU(2)_V]_{local}$ gauge symmetry. The corresponding (composite) gauge boson is identified with the ρ meson. Its mass is generated by a Higgs mechanism, with a kinetic term properly added. The resulting Lagrangian is precisely of the form given by Weinberg [18]:

$$L = \frac{f_\pi^2}{4} tr[\partial_\mu U \partial^\mu U^+] - \frac{1}{4} \vec{F}_{\mu\nu} \cdot \vec{F}^{\mu\nu} + \frac{a}{2}(g^2 f_\pi^2) \vec{\rho}_\mu \cdot \vec{\rho}^\mu + \frac{a}{2} g \vec{\rho}_\mu \cdot (\vec{\pi} \times \partial^\mu \vec{\pi}) + ..., \tag{13}$$

where g is a coupling constant and a is an auxiliary parameter. Identifying

$$m_\rho^2 = ag^2 f_\pi^2, \qquad f_{\rho\pi\pi} = \frac{a}{2} g, \tag{14}$$

one has the celebrated KSFR relation $m_\rho^2 = 2f_\pi^2 f_{\rho\pi\pi}^2$ [19] for $a = 2$. In the long wavelength limit and for large m_ρ^2, the Skyrme quartic term can be shown [20] to be related to the ρ meson kinetic term, with

$$\epsilon = \frac{1}{\sqrt{8g}}. \tag{15}$$

Electromagnetic interactions are easily introduced into this scheme. The photon couples to the ρ meson as

$$L_{\gamma\rho} = -\frac{em_\rho^2}{g}\rho_3^\mu A_\mu, \tag{16}$$

and the direct $\gamma\pi\pi$ coupling vanishes,

$$L_{\gamma\pi\pi} = e(1 - \frac{a}{2})(\vec{\pi} \times \partial^\mu\vec{\pi})_3 A_\mu = 0, \tag{17}$$

for $a = 2$. This is just the Vector Dominance Model (VDM) picture, with the universality relation $g = f_{\rho\pi\pi}$. As an immediate consequence, the pion form factor $F_\pi(q^2)$ results from the process, Fig.1, in which the probing photon first converts into a ρ meson which then couples to the pion via the $\rho\pi\pi$ vertex in eq.(13). This yields the familiar VDM form:

$$F_\pi(q^2) = \frac{m_\rho^2}{m_\rho^2 - q^2 - im_\rho\Gamma_\rho(q^2)}. \tag{18}$$

Here the ρ meson width $\Gamma_\rho(q^2 = m_\rho^2) = 150\,MeV$ is obtained with $f_{\rho\pi\pi} = 5.9$. The resulting pion charge radius is

$$<r_\pi^2> = 6\frac{dF_\pi}{dq^2}\big|_{q^2=0} = \frac{6}{m_\rho^2} = (0.62\,fm)^2, \tag{19}$$

which is already remarkably close to the measured radius (10). Note that in this purely mesonic description with pointlike π and ρ, the apparent pion radius is determined entirely by the propagating ρ meson. Clearly, there is little space left for an intrinsic size of the pion. (By "intrinsic" we mean a size related to degrees of freedom *other* than those expressed in terms of *pointlike* π and ρ mesons.)

In the timelike region, the VDM reproduces $F_\pi(q^2)$ within 20 percent (see Fig.2b). This discrepancy, as well as the small deviation between the VDM result (19) and the measured $<r_\pi^2>$, can well be accounted for by phenomenological finite size corrections with an intrinsic pion radius

$R_\pi \doteq 1/3\,fm$, in accordance with the discussion in the previous section. The result is shown in Fig.2 [21].

THE BARYON SECTOR AND NUCLEON FORM FACTORS

Baryons emerge as topological solitons [16] of the effective Lagrangian (12). So far we have discussed primarily the role of the ρ meson which has a direct relation, via eq.(15), to the Skyrme fourth order term. The ω meson enters in the modified Skyrme model [22] by its coupling to the baryon (topological) current

$$B_\mu = \frac{\epsilon_{\mu\nu\lambda\rho}}{24\pi^2} tr[U^+(\partial^\nu U)U^+(\partial^\lambda U)U^+(\partial^\rho U)]. \tag{20}$$

The baryon number is identified with the toplogical charge (winding number) $B = \int d^3x\, B_0(x)$. Once the ω is introduced, the model has all the relevant features consistent with Vector Meson Dominance.

G.E. Brown, Mannque Rho and I have recently been investigating the implications of Vector Dominance and topology on nucleon electromagnetic form factors. Here I briefly summarize the results described in more detail in [21]. If $U(x)$ is a soliton configuration of the non-linearly coupled π fields, the universal constant g also determines the ρNN (or ρ-soliton) coupling, i.e. $g = f_{\rho\pi\pi} = f_{\rho NN}$. The $f_{\rho NN}$ appears in the phenomenological ρ-nucleon coupling Lagrangian

$$L_{\rho NN} = f_{\rho NN}[\overline{N}\gamma_\mu \frac{\vec{\tau}}{2} N \cdot \vec{\rho}^\mu + \frac{\kappa_\rho}{2M} \overline{N}\sigma_{\mu\nu}\frac{\vec{\tau}}{2} N \cdot \partial^\mu \vec{\rho}^\nu]. \tag{21}$$

As a consequence, we expect the isovector baryon form factor to be

$$G_V(q^2) = \frac{m_\rho^2}{m_\rho^2 - q^2} F_{soliton}(q^2), \tag{22}$$

where $F_{soliton}(q^2)$ describes the finite size of the soliton with a mean square radius $R_s^2 = 6 dF_{soliton}/dq^2|_{q^2=0}$. In an extended Skyrme model with 4th and 6th order terms constrained by Vector Dominance (e.g. eq.(15)), typical soliton radii are $R_s \doteq 0.6\,fm \doteq \sqrt{6}/m_\rho$. The overall radius is then $\sqrt{12}/m_\rho$, quite close to the measured proton charge radius

$$<r_p^2>^{1/2} = (0.86 \pm 0.01)fm. \tag{23}$$

Up to this point, one is tempted to conclude that a purely mesonic description of low energy phenomena works exceedingly well. However, one faces the problem of the large ρ-nucleon tensor - to - vector coupling ratio κ_ρ (see eq.(21)). Empirically [23],

$$\kappa_\rho = 6.6 \pm 0.6, \tag{24}$$

whereas strict Vector Dominance would identify κ_ρ with the isovector magnetic moment,

$$\kappa_V = 3.71. \tag{25}$$

This difference by about a factor 2 is a significant one. It is well established also from detailed analysis of isovector tensor correlations in nuclei [26], which tend to favour values of κ_ρ at the upper end of the errors given in eq.(24).

It has been suggested [21] that a description in terms of a chiral two phase model has definite advantages at this point. In this model, quarks move freely inside a sphere of radius R_0 which is surrounded by a soliton cloud. The baryon number is shared by the inside and outside domains in such a way that $B_{in}(R_0) + B_{out}(R_0) = 1$. For $R_0 \dot{=} 0.6 fm$ the chiral angle (defined by $\vec{\pi} = \hat{r}\theta(r)$) is $\theta(R_0) = \pi/2$, the quarks with $K = 0^+$ (where $\vec{K} = (\vec{\sigma} + \vec{\tau})/2$) sit at zero energy and the baryon number is fractioned half and half, i.e. $B_{in}(R_0) = B_{out}(R_0) = 1/2$. This baryon number fractioning is accompanied by a corresponding fractioning of spin[24] and charge. Thus a photon will now directly interact with half of the proton charge in the quark core. The other half is located in the soliton cloud which is probed via photon - vector meson conversion (see Fig. 3). The quarks, being at zero energy, have zero velocity and no magnetic coupling. Therefore the isovector Pauli form factor $F_2^V(q^2)$, the one related to the $\sigma_{\mu\nu} q^\nu$-term of the nucleon current, receives [B its contribution only from the meson cloud, with the factor of 1/2 due to fractionation discussed previously. Consequently, one has $\kappa_V \dot{=} \frac{1}{2}\kappa_\rho$.

This situation can be modelled [21] by combining the $0.6\,fm$ core represented by a form factor F_{core}, with an appropriate size for the vector meson - nucleon vertex. In fact, a monopole formfactor $F_{\rho NN}(q^2) = (\Lambda^2 - m_\rho^2)/(\Lambda^2 - q^2)$ with $\Lambda = \sqrt{2}m_\rho$ just simulates the 1/2 - 1/2 fractioning. For example, the isovector Dirac and Pauli form factors F_1^V and F_2^V become (with the fraction of charge $Z_{core} = 1/2$ carried by the core):

$$2F_1^V(q^2) = Z_{core} F_{core}(q^2) + \frac{m_\rho^2 F_{\rho NN}(q^2)}{m_\rho^2 - q^2}, \tag{26}$$

$$2F_2^V(q^2) = \kappa_\rho \frac{m_\rho^2 F_{\rho NN}(q^2)}{m_\rho^2 - q^2}. \tag{27}$$

Note that indeed $\kappa_V = \kappa_\rho F_{\rho NN}(q^2 = 0) = 1/2$ in this model. Corresponding expressions are obtained for the isoscalar form factors, with the role of the ρ meson now taken over by the ω meson, and the fraction of charge Z_{core} replaced by the fraction B_{core} of baryon number in the core ($B_{core} = 1/2$ in our example). Results so obtained for the proton and neutron charge form factors are shown in Figs.4,5. The core and cloud contributions to G_E^p are shown separately (Fig.4); note that the quark core dominates at large q^2, as it should. The neutron form factor (Fig.5) has contributions only from the cloud in this model.

SUMMARY AND CONCLUSIONS

We have first demonstrated in a schematic model how the pion emerges as a collective mode with small intrinsic size. For all purposes of long wavelength physics, the pion can be considered approximately pointlike. We have then investigated to what extent purely pionic non-linear theories work in explaining electromagnetic hadron form factors. The ρ meson is found to enter naturally and in accordance with the Vector Dominance Model. Pion and nucleon form factors are quite well described already at the level of the pure VDM with pointlike pions and ρ mesons, the nucleon being a soliton configuration of the non-linearly coupled pion fields in this theory.

However, the real challenge is to understand the large ρ-nucleon tensor to vector coupling ratio. A chiral topological two-phase picture, with a quark core of about $0.6\,fm$ radius surrounded by a soliton cloud and a half-and-half fractioning of charge and baryon number between core and cloud, leads to a consistent phenomenology including $\kappa_V = 1/2\,\kappa_\rho$. In this picture, one is beginning to probe the quark-gluon substructure for radii smaller than about $\sqrt{6}/m_\rho \doteq 0.6\,fm$.

One might question the uniqueness of this delineation between quark-gluon and meson degrees of freedom. In fact, bosonization theorems in (1+1) dimensions [25] imply that the bag radius does not play an observable role at low energy: fermion degrees of freedom can always be translated into mesonic ones; the bridge between the two representations is provided by topology. However, in (3+1) dimensions, complete bosonization requires infinitely many bosons with all spins. The large number of bosons might then prove prohibitive, at least more complicated than a description of shorter distance phenomena in terms of quarks and gluons. (This does not exclude the possibility that a non-linear effective meson Lagrangian with pions, vector mesons and a proper treatment of anomalies may lead to similar results).

The highly successful meson exchange phenomenology of the nucleon-nucleon interaction, truncated with J = 0 and J = 1 bosons, works well down to distances of about 0.5 fm. A description of low energy (non-strange) hadron dynamics based on Chiral Symmetry, with pointlike pions, vector mesons, and a small nucleon core, is consistent with this phenomenology and enjoys at the same time increasing theoretical justification.

I would like to thank Gerry Brown, Mannque Rho and Andreas Wirzba for many stimulating discussions and much helpful advice.

References

1. H. Yukawa, Proc. Phys. Math. Soc. Japan **17** (1935), 48.
2. D.O. Riska and G.E. Brown, Phys. Lett. **32B** (1970), 662; T.E.O. Ericson and M. Rosa-Clot,Nucl. Phys. **A405** (1983), 497.
3. M. Bernheim et al., Phys. Rev. Lett. **46** (1981), 402.
4. J.F. Mathiot, Nucl. Phys. **A412** (1984), 201.
5. J. Kogut et al., Phys. Rev. Lett. **48** (1982), 1140.
6. G.E. Brown, Nucl. Phys. **A358** (1981), 39c; T.H. Hansson and I. Zahed,*Stony Brook preprint*.
7. J. Finger, J. Mandula and J. Weyers, Phys. Lett. **69B** (1980), 367; S.L. Adler and A.C. Davis,Nucl. Phys. **B244** (1984), 469.
8. R. Brockmann, W. Weise and E. Werner, Phys.Lett. **122B** (1983), 201; W. Weise,Int. Rev. Nucl. Phys. **1** (1984), 57; W. Weise,Nucl. Phys. **A434** (1985), 685c.
9. Y. Nambu and G. Jona-Lasinio, Phys. Rev. **122** (1961), 345; Phys. Rev. **124** (1961), 246.
10. A. Dhar and S.R. Wadia, Phys. Rev. Lett. **52** (1984), 959.
11. V. Bernard, R. Brockmann, M. Schaden, W. Weise and E. Werner, Nucl. Phys. **A412** (1984), 349.
12. V. Bernard, R. Brockmann and W. Weise, Nucl. Phys. **A440** (1985), 605.
13. S.R. Amendolia et al., Phys. Lett. **138B** (1984), 454.
14. G. t'Hooft, Nucl. Phys. **B72** (1974), 461.
15. E. Witten, Nucl. Phys. **B160** (1979), 57.
16. T.H.R. Skyrme, Nucl. Phys. **31** (1962), 556; G. Adkins, C. Nappi and E. Witten,Nucl. Phys. **B228** (1983), 552.
17. M. Bando, T. Kugo, S. Uehara, K. Yamawaki and T. Yanagida, Phys. Rev. Lett. **54** (1985), 1215.
18. S. Weinberg, Phys. Rev. **166** (1968), 1568.
19. K. Kawarabayashi and M. Suzuki, Phys. Rev. Lett. **16** (1966), 255; Riazuddin and Fayyazuddin,Phys. Rev. **147** (1966), 1071.
20. M. Rho, *private communication*.
21. G.E. Brown, M. Rho and W. Weise, *preprint*.
22. G. Adkins and C. Nappi, Phys. Lett. **137B** (1984), 251; A. Jackson, A.D. Jackson, A.S. Goldhaber, G.E. Brown and L.C. Castillejo,Phys. Lett. **154B** (1985), 101.
23. G. Hoehler and E. Pietarinen, Nucl. Phys. **B95** (1976), 1.
24. A.J. Niemi, Phys. Rev. Lett. **54** (1985), 631.
25. V.A. Rubakov, Nucl. Phys. **B236** (1984), 109.
26. S.O. Backman, G.E. Brown and J. Niskanen, Phys. Reports **124** (1985), 1.

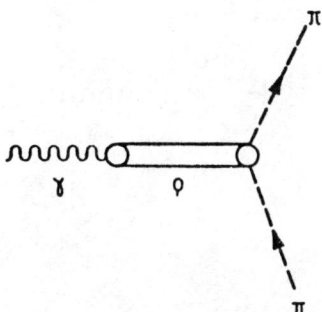

Figure 1: Pion form factor in the Vector Dominance Model

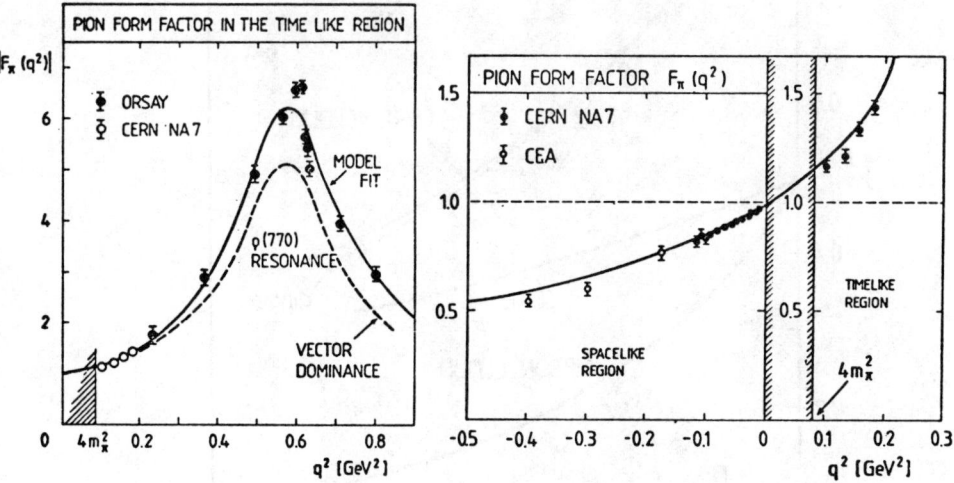

Figure 2: Pion form factor in the timelike and spacelike region. The dashed curve is obtained with the pure ρ-meson dominance model (with pointlike pion), eq.(18). The solid curve is the result of a model with an intrinsic pion radius of 0.3 fm [21].

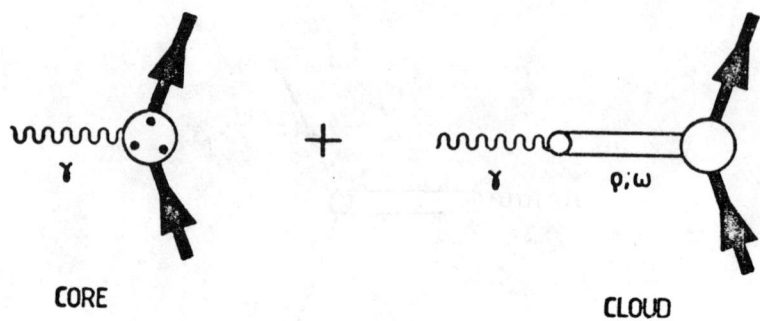

Figure 3: Nucleon electromagnetic form factor as described with a photon coupling to the quark core and to the meson cloud through an intermediate ρ and ω meson.

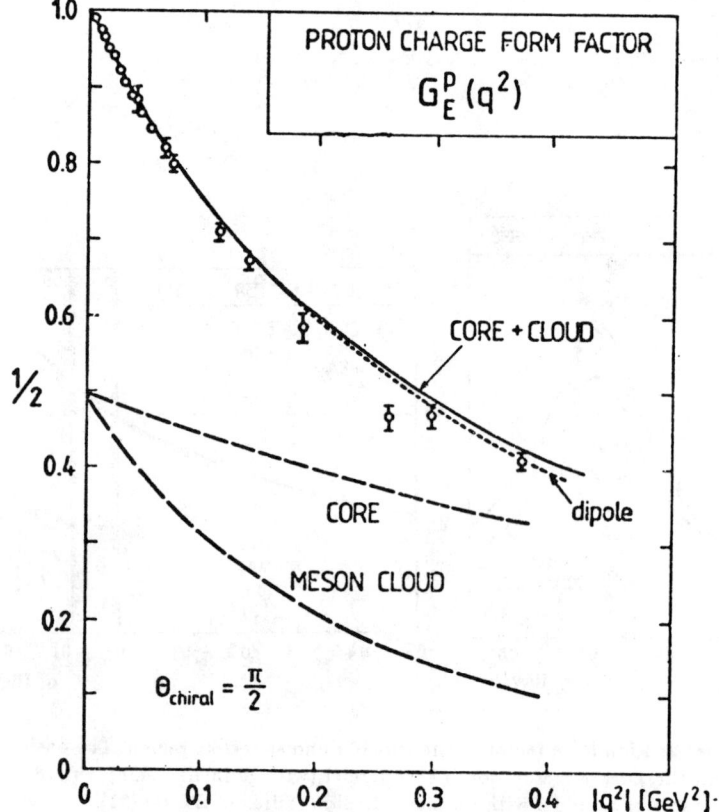

Figure 4: Proton charge form factor. The curves show the contributions of the quark core and the meson cloud, and the sum of both. See text and ref.[21].

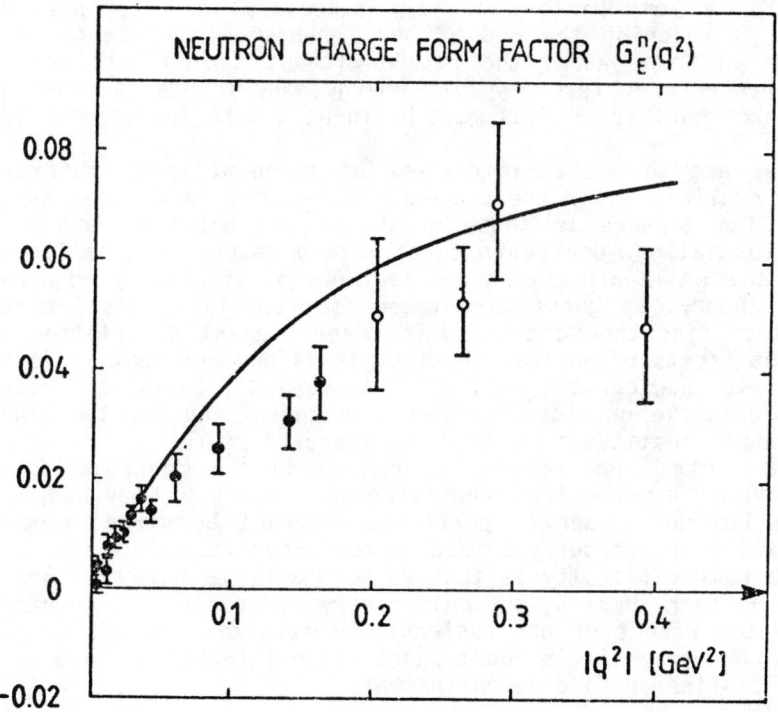

Figure 5: Neutron charge form factor. For details see ref.[21].

PROBING NON NUCLEONIC DEGREES OF FREEDOM WITH STRONG AND ELECTROMAGNETIC INTERACTIONS

Bernard Frois
Service de Physique Nucléaire - Haute Energie
CEN Saclay, 91191 Gif-sur-Yvette Cedex, France

INTRODUCTION

After a long history of very successful developments, nuclear physics is reaching the end of pure phenomenology. It is no longer possible to think about the future without taking into account the recent progress achieved by high energy experiments. A coherent description of nuclear physics must begin at a more fundamental level.

What are the relevant degrees of freedom? Is it sufficient to account explicitly for the presence of mesons? Do we need quarks and gluons? The answers to these questions are not yet known which is quite frustrating. One wants to achieve a description as fundamental as possible while preserving the features of simplicity required by a refined theory. We guess that there is probably no absolute answer. One has to find the most economical and elegant description for the different facets of nuclear physics. It is obvious that it is not the discovery of new constituents which is needed ; this has already been done by particle physics. One wants to understand how the binding of fundamental constituents creates nuclear structure.

The central problem is to understand the modification of the interactions between free constituents. It is a many-body problem where a limited number of particles interact by strong forces in a regime which cannot be described by perturbative techniques.

The main difficulty is that we have a large number of low energy data in nuclear physics, but only a very limited set of interpretable data on the effect of non nucleonic degrees of freedom. In particular, we know very little about short range interactions and the dynamics of confinement and deconfinement.

In this talk, I would like to examine our present view on non-nucleonic degrees of freedom with a few typical experimental results obtained recently both with hadronic and electromagnetic probes at intermediate energies.

It is the first generation of experimental data which has probed mesonic degrees of freedom with a spatial resolution of the order of 0.5 fm. This has made possible for example the measurement of the size of the pion-nucleon interaction region. This is very stimulating progress and we begin to have a coherent overview on the various reaction mechanisms which are induced by hadronic and electromagnetic probes.

CREATION OF PIONS AND DELTAS

A reaction induced by photons : $D(\gamma,pp)\pi^-$

This is a very simple reaction where the initial state is well defined. Since photons are not absorbed at the nuclear surface, pions and deltas are created throughout the nuclear volume. Argan et al.[1] have shown that this reaction for photon energies between 350 MeV and 500 MeV provides strong evidence that we understand the basic mechanisms of Δ creation in nuclei.

Three mechanisms contribute to this reaction :
I) pion photoproduction on a quasifree nucleon ;
II) creation of a Δ, by rescattering of the pion ;
III) simultaneous creation of a Δ and a pion, followed by the decay of the Δ in a proton and a pion which is reabsorbed by the spectator nucleon. This mechanism is a meson exchange.

These mechanisms are represented in Fig. 1 together with two sets of data which correspond to different kinematical conditions, chosen to enhance a specific mechanism, either pion rescattering (configuration A) or meson exchange (configuration B). The experimental data have been obtained by fixing the momentum p_2 of one proton and the mass Q of the system formed by the pion and the other nucleon. The method used to described these processes has been developed by Laget[2]. It is a diagrammatic method based on the expansions used in high energy physics. The total amplitude is developed in a series of elementary terms described by effective operators. These effective

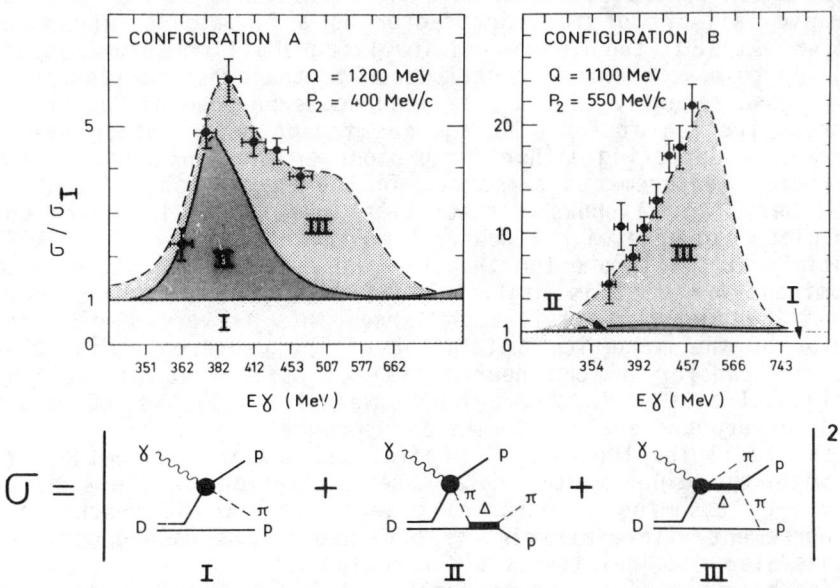

Fig. 1 $D(\gamma,pp)\pi^-$ cross section divided by the quasifree photopion production.

operators are once for all adjusted to reproduce the experimental pion photoproduction on the free nucleon. The $D(\gamma,pp)\pi^-$ data shown in Fig. 1 explore the singularities of mechanisms II (configuration A) and III (configuration B). For the first kinematical condition p_2 = 400 MeV/c and Q = 1200 MeV, when E_γ increases, the rescattering term has a dominant contribution up to E_γ = 380 MeV, then the meson exchange term becomes in its turn the dominant effect. The second kinematical condition (p_2 = 550 MeV/c and Q = 1100 MeV) singles out the meson exchange mechanism III ; the other mechanisms contribute for less than 5 % of the cross section.

The experimental results are in excellent agreement with the theoretical predictions[2]. One concludes that the basic mechanisms involving pions and Δ are well described by such a phenomenological diagrammatic expansion where the parameters are determined by the photopion production on the free nucleon. This experiment has been followed by studies of $D(\gamma,p\pi)$ [ref.[3]] and $D(\gamma,\pi^\pm)$ [ref.[4]] which are also well described by the same theoretical approach. This diagrammatic method can be used both for real and virtual pions, in particular for the reactions $NN \rightarrow D\pi$, $pD \rightarrow pD$, $pD \rightarrow T\pi$, $DD \rightarrow {}^3He\ n$ etc..

Effect of the delta in $\vec{d}$ + p elastic scattering at 180°CM

After this study of non nucleonic effects in a reaction induced by photons, the next step is to study similar effects induced by hadron scattering. The $\vec{d}$ + p elastic scattering at 180°C.M. is one of the simplest reaction because it is very selective. The diffraction process which dominates at forward angles vanishes at 180° because of the rapid fall-off of the form factor as a function of the momentum transfer. At 180° the mechanisms involving small momentum components are going to have a larger contribution to the cross section. Arvieux et al.[5] have recently used a polarized deuteron beam at Saturne. Both the cross section at 180°C.M. and the tensor polarization have been measured, by detecting either the protons or the deuterons emitted at 0°. Precise measurements exist now in the energy range from 0.3 GeV to 2.3 GeV. Fig. 2 shows a theoretical prediction including only a one-nucleon exchange with a deuteron wave function given by the Paris potential. At small momentum transfer the contribution of the S state is dominant while it is negligible at high momentum transfer where the D state contribution is much large. This is very similar to the behavior of the magnetic form factor of the deuteron[7]. Fig. 2 shows that the transfer of one neutron is not sufficient to explain the experimental data[5,6] in the region between T_p = 300 and 800 MeV. This region corresponds exactly to the Δ resonance.

In Fig. 3 the theoretical predictions are in much better agreement with the experimental data. The excitation of the Δ has been included by assuming a mechanism based on a two pion exchange. The best agreement with experiment is obtained by the dashed curve which includes also some relativistic corrections.

The behavior of the cross section at 180°C.M. is reasonably well explained by this two-meson exchange. However it is impossible to explain only with this mechanism the oscillations of t_{20} at 180°C.M

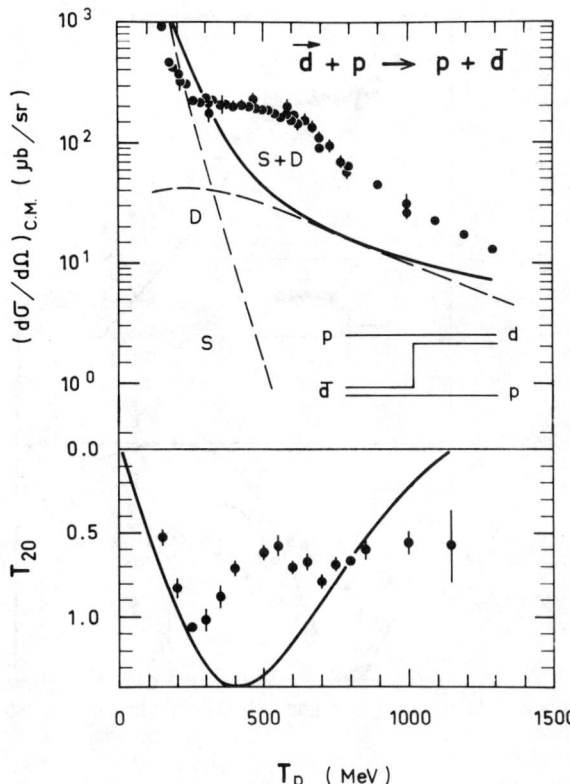

Fig. 2 Deuteron-proton elastic scattering cross section and tensor polarization T_{20} at 180°C.M. The theoretical prediction assumes only a one nucleon exchange.

as a function of energy. At low energy where non nucleonic effects are expected to be small, the agreement with theory is good but in the Δ resonance region it is very poor. At present the origin of the oscillations of t_{20} at 180°C.M. in pd elastic scattering is not understood. This is typical of polarisation data measured with intermediate energy hadronic probes. Their oscillations at backward angles are due to interference terms which will be reproduced only by a complete description of all the mechanisms, resonant and non-resonant.

To understand such complex reaction mechanisms one has first to extract the nuclear structure part by electron scattering, and then to study different hadronic reactions which enhance a particular mechanism.

The $\vec{p}p \to d\pi^+$ and $\vec{p}d \to T\pi^+$ reactions

Laget[8] has recently discussed such an approach for the $D(\pi^\pm,p)NN$ and the $D(\pi^\pm,\pi^\pm p)n$ reactions and also the $pp \to d\pi^+$ and $pd \to t\pi^+$ reactions. Fig. 4 shows the result of his calculations for angular distributions[9,10] of the $pp \to d\pi^+$ reaction at 578 MeV and the $pd \to T\pi^+$ reaction at 500 MeV. The solid curve represents the full calculation. The dashed curve is the same calculation without the creation of a Δ by a rescattering of the pion. This effect accounts for most of the cross section. The calculation depends on two parameters : the cut-off mass of the πNN form factor Λ_π and the ρ-nucleon coupling constant. The value of Λ_π which reproduces both hadronic and electromagnetic data is $\Lambda_\pi = 1.2$ GeV. The ρ-meson coupling constant which fits hadronic data is slightly higher than the one which fits electromagnetic data ($G_\rho^2/G_\pi^2 = 2.4$). The agreement with polarization data[10] is poor (Fig. 5). The general trend of the analyzing power of the

$\vec{p} + p \to d + \pi^+$ reaction is reproduced, but the backward data for the reaction $\vec{p} + D \to T + \pi^+$ have a completely different behavior. When the incident energy increases these calculations does not re produce the angular distributions of the pd → Tπ reaction above the Δ resonance for θ > 60°. Between 600 and 800 MeV the cross section has a sharp rise at backward angles that ones does not know how to explain. The analyzing power is also steadily increasing between 600 and 1100 MeV at an angle of 140°. This peak has been observed at Los Alamos by Kielczewska et al.[11] at 800 MeV. Bertini et al.[12] have shown recently that the position of this peak remains at $\theta_{CM} = 140°$ up to 1100 MeV. Kielczewska et al.[11] have estimated that this could be explained by a 3 % $\Delta°\Delta^+$ component in the ground state of the deuteron.

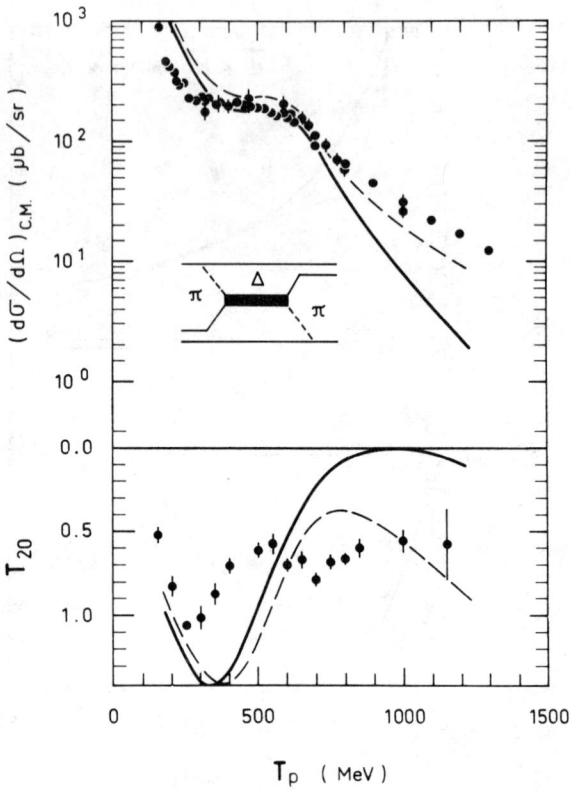

Fig. 3 Deuteron-proton elastic scattering cross section and tensor polarisation T_{20} at 180°C.M. The theoretical prediction take into account the excitation of the nucleon in a Δ by a two-pion exchange

The $\Delta°$ would be transferred while the Δ^+ would remain spectator. This does not appear now possible with the recent measurements of the magnetic form factor of the deuteron[7]. Sitarski and Lomon[13] have estimated that ΔΔ components must be smaller than 1 % to be compatible with these new measurements.

The $p + p \to d + \pi^+$ reaction is also difficult to interpret at energies higher than the Δ resonance. Recent data have been measured by Bertini et al.[14] at seven energies between T_p = 1.2 and 2.3 GeV. These data show a very rapid variation (Fig. 6). Showing a peak at $\sqrt{S_{\pi d}}$ = 2660 MeV with a width of 150 MeV. Possible interpretations are based on a one-pion exchange or a dibaryon resonance. The existence of such a resonance would be compatible with a $^1S_0$ six-quark states predicted by Lomon[15]. Further measurements are planned by Bertini et al.[14] to clarify this question.

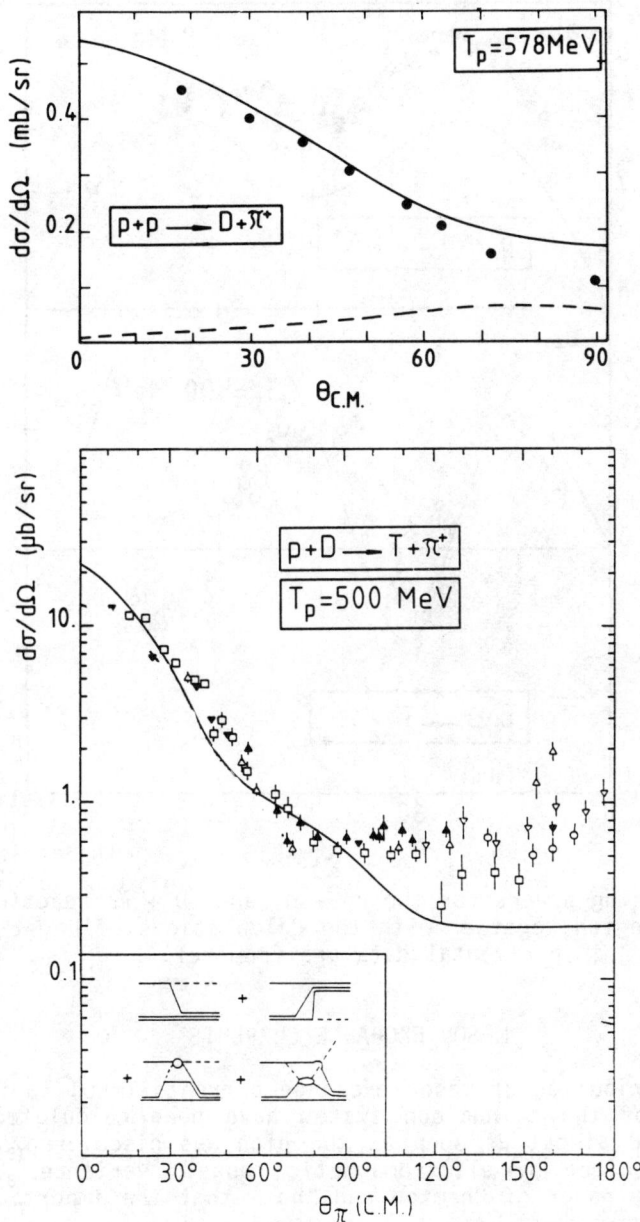

Fig. 4 Angular distributions of the reactions pp → dπ and pD → Tπ [refs.9,10]. The dashed curve[8] does not take into account the creation of a Δ in the pp → dπ reaction. The solid curve[8] is the full calculation.

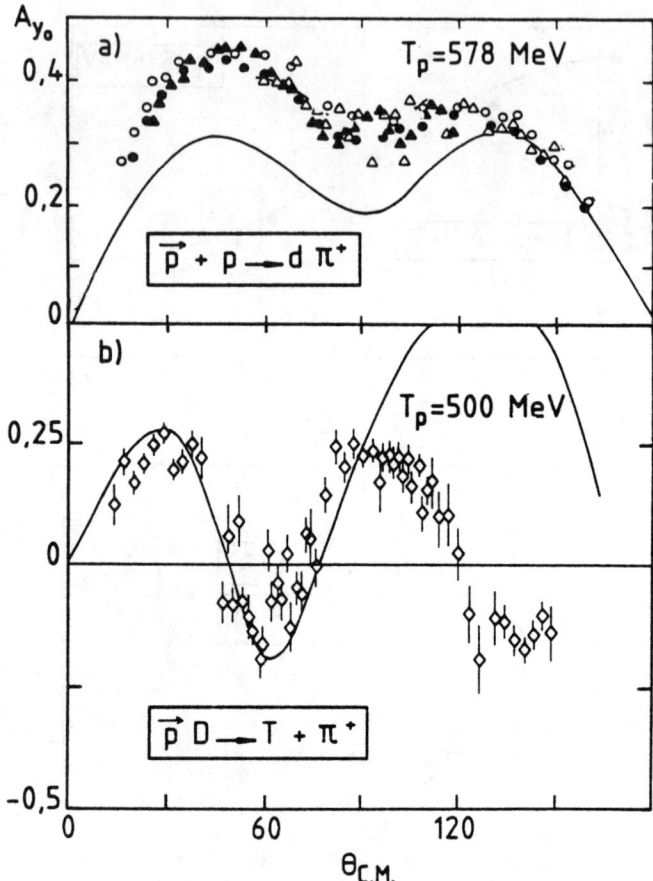

Fig. 5 Analyzing powers for the pp → dπ and pD → ππ reactions in the Δ resonance region together with the calculations of Laget[8]. The experimental data are from ref.[9].

MESON EXCHANGE CURRENTS

The contribution of meson exchange currents (MEC) to the magnetic moments of the trinucleon system have been calculated by Villars[16] in 1948 almost as soon as the pion was discovered. Such currents are required by electromagnetic gauge invariance. But it is only after the paper of Chemtob and Rho[21] that the importance of MEC has been recognized.

The first observation of meson exchange currents was only a 10 % increase[17] in the n + p → d + γ radiative capture of thermal neutrons. Riska and Brown[18] have shown in 1972 that this disagreeement was explained by the presence of the pion current. Shortly after, it was realized[19] that the inverse reaction e + d → n + p near threshold

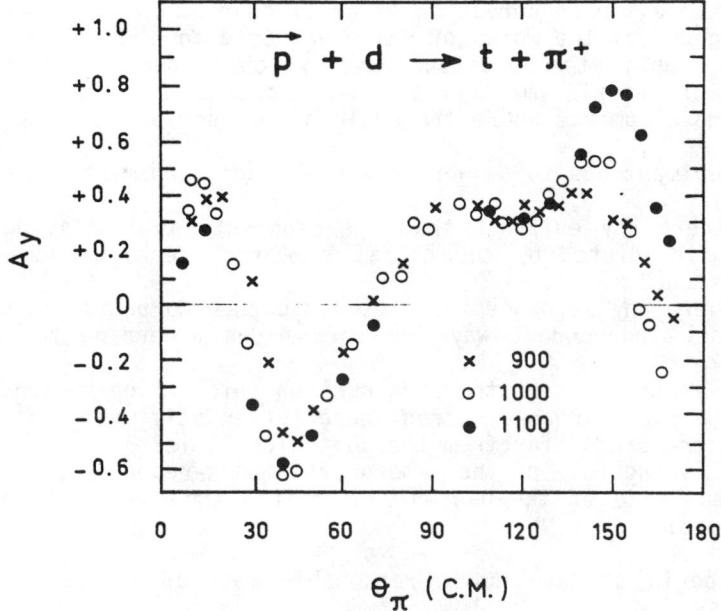

Fig. 6 Analysing power of the reaction pd → ππ for incident energies between 900 and 1100 MeV [ref.12].

might be much more sensitive to meson exchange currents. The interpretation of these two experiments showed in 1973 that, meson exchange currents can be isolated, and that, with the theoretical work developed around 1970 [ref.21], their description was within our reach. The following step was going to take about ten years of intense research. In the second part of this talk, I will present the progress achieved recently. I have chosen three examples which illustrate the frontier of our understanding, also discussed at this workshop from a theoretical point of view by Wolfram Weise.

Role of low energy theorems and chiral symmetry

The main difficulty of strong interactions is the absence of conservation laws similar to the electromagnetic current conservation. In order to reach a fundamental description of meson exchange currents it has been necessary to start by considering low energy pions. Such pions are usually called "soft pions". In this limit axial current would be exactly conserved if the pion mass was zero. This is discussed in detail in the book "Mesons in Nuclei" edited by Rho and Wilkinson[20]. If the pion mass was exactly zero, the conservation of axial current would lead to completely model independent predictions for soft pions. These are very powerful constraints for theoretical calculations, which are known as low energy theorems.

These constraints have a fundamental role since they are natural consequences of chiral symmetry. The presence of the pion is necessary to restore the conservation of axial charge. This role of the pion has emerged from the works of Nambu and Goldstone[22].

The pion is the Goldstone boson which is due to the spontaneous breaking of chiral symmetry. Its role is crucial in the recent theoretical developments where the nucleon is described as a Skyrmion[23].

Experiment has to answer to a number of fundamental questions :

1) Is there any evidence that the pion current is playing the central role predicted by the chiral symmetry breaking model of Nambu and Goldstone?
2) Is there any evidence that meson exchange currents can be derived in a model independent way for soft pions according to low energy theorems?
3) What is the size of the pion-nucleon interaction region?
4) Is the pion exchange current description still valid at very high momentum transfers, far from the soft pion limit?
5) What is the role of the ρ-meson and the Δ-resonance?
6) Is the size of the bag which confines the quarks in the nucleon an observable?

We begin to have now a reasonable idea of how to answer these questions.

The axial transition $^{16}O(0^+) \leftrightarrow ^{16}O(0^-)$

In 1978, Kubodera, Delorme and Rho[24] have shown that mesonic currents appear on the same footing as nucleonic currents, in the time component of axial transitions, and in the space component of magnetic isovector M1 transitions. The ideal experiments to test our understanding of mesonic currents are a study of such processes where the nucleonic contribution is strongly inhibited, either by selection rules or by destructive interferences.

The transition between the 0^+ ground state of ^{16}O and the 0^- state (E* = 120 keV) in ^{16}N is now the best example of the effect of meson exchange currents in an axial transition. This is a forbidden transition where the axial charge is modified by the presence of mesons in a spectacular way.

This transition has been recently measured to a remarkable accuracy due to ingenious experimental techniques. Garvey has summarized in 1984 at Osaka[25] the present status of these experiments, performed at Argonne and Montreal for β^- disintegration, and at Saclay for μ capture.

The transition probabilities are :

$$\Lambda_\beta = 0.486 \pm 0.020 \text{ s}^{-1} \quad [\text{réf.}26]$$

$$\Lambda_\mu = (1.560 \pm 0.094) \times 10^3 \text{ s}^{-1} \quad [\text{ref.}27]$$

which gives $\Lambda_\mu/\Lambda_\beta$ = 3.210 ± 0.234.

Extensive discussions of this transition can be also found in the papers of Guichon et al.[27] and in the paper of Towner and Khanna[28]. The experimental result is compared in Fig. 7 to different theoretical predictions[27-29]. Without mesonic contributions, these predictions are too high by a factor 3 or 4, and they are strikingly different. When meson exchange currents are included in the calculation, all the theoretical predictions, though they are based on different nuclear models, yield essentially the same result, in excellent agreement with experiment.

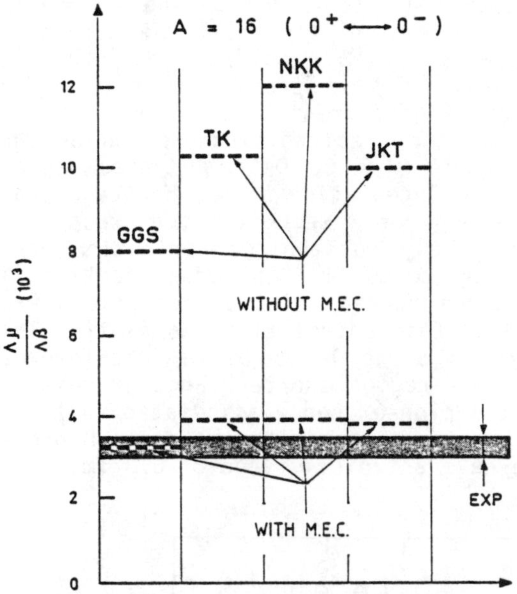

This was exactly the prediction of low energy theorems which are applicable because the momentum transfer is small, so the pions are really soft.

This result provides a striking verification of the model independence of the pion current at low energy for the time component of an axial transition. This result is confirmed by the very complete analysis of the famous $0^+ \leftrightarrow 1^+$ allowed transition in the triad A=12 [ref.[27]].

Fig. 7 Ratio of the muon capture rate of the β^- disintegration rate. The shaded area corresponds to the experimental uncertainty[25,27]. The dashed lines correspond to different theoretical predictions[27,29].

Electrodisintegration of the deuteron at threshold

In the previous example, as well as in the case of the radiative capture of thermal neutrons by hydrogen, one learns about meson exchange at small momentum transfers. This tells us very little about the spatial distribution of meson exchange currents and the size of their interaction region with the nucleon.

Electrons are the ideal probe to answer to the all the questions that we have raised about meson exchange currents. They are point particles which couple directly to quarks by exchange of a virtual photon. The virtual character of the photon allows to fix the energy transfer to study a specific transition, and by varying the momentum transfer to study the spatial distribution of charge and currents involved in nuclear transitions. The electromagnetic interaction is fully understood. Conservation of charge and current is a very powerful constraint to single out various processes. The virtual photons probe the entire nuclear volume without absorption. Such unique features have already been used extensively in nuclear physics to determine nucleon distributions and, in particle physics to determine quark distributions we will see that this also true for meson exchange currents.

The break up of the deuteron near threshold by electron scattering is at backward angle near 180° an M1 magnetic isovector transition. This is the second kind of transition in which mesonic currents appear to the same order as nucleonic current, because it involves the space component of the electromagnetic current. The nucleonic contribution to the cross section is vanishingly small due to a destructive interference between the $^3S_0 \to {}^1S_0$ and $^3D_1 \to {}^1S_0$ transitions. Thus, the remarkable feature of this process is that it is almost exclusively due to meson exchange currents.

Recent data[30] have mapped out the electrodisintegration of the deuteron for momentum transfers up to $Q^2 = 30$ fm^{-2}. A typical spectrum at 155° is shown in Fig. 8. The integration of the break-up peak at threshold over an excitation energy of 3 MeV gives the cross sections represented in Fig. 9. The purely nucleonic contribution vanishes at 12 fm^{-2} (dotted curve). Up to 16 fm^{-2} the cross section is due exclusively to the pion exchange current. These predictions are computed using the Paris nucleon-nucleon potential by Mathiot[31]. The inclusion of pion exchange currents alone leads to an interference minimum near $Q^2 = 25$ fm^{-2} ; in this region shorter range processes, namely the ρ-meson exchange and the pion-exchange associated to the Δ are needed to bring the theory into agreement with the data. Processes such as the ρ-meson exchange have a typical range of 0.3 fm.

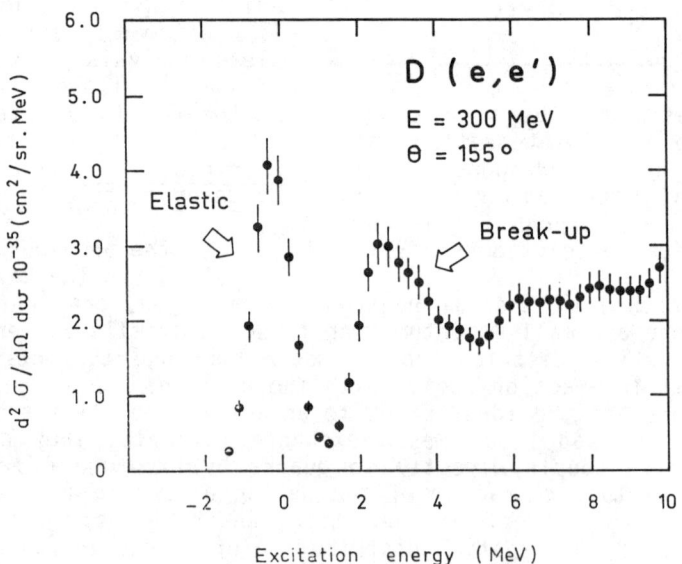

Fig. 8 Electrodisintegration of the deuteron. A typical experimental spectrum[30].

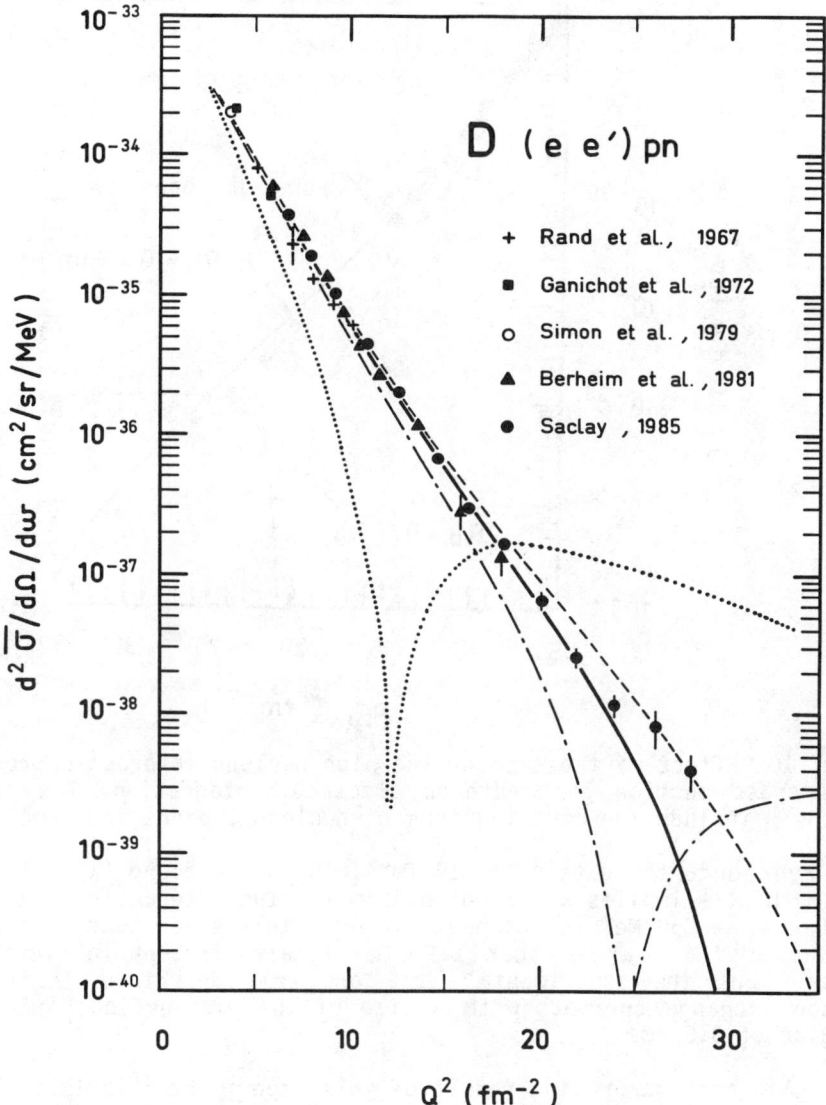

Fig. 9 Cross sections for the electrodisintegration of the deuteron at threshold[30]. The curves[31] represent the contributions of nucleons (dotted), nucleons + pions (dash-dotted), nucleons + pions + ρ-mesons (dashed), nucleons + pions + ρ-mesons + Δ (solid).

At the spatial scale probed by this experiment the internal structure of the nucleons and mesons plays a significant role. Fig. 10 shows the dramatic sensitivity to the size R_0 of the interaction region between mesons and nucleons which is accounted for by a πNN form factor. The monopole cut-off mass Λ_π which is adjusted in order

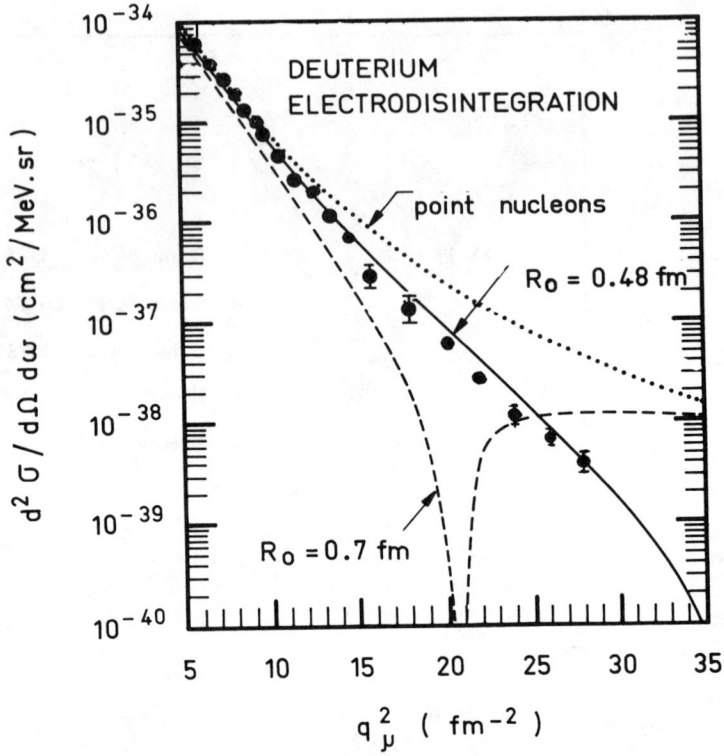

Fig. 10 Effect of the size of the pion nucleon interaction region on the cross section for deuteron electrodisintegration. The calculations[31] include the contributions of nucleons, pions and ρ-mesons.

to reproduce the data up to 18 fm^{-2}. The value found (Λ_π = 1.25 GeV) by Mathiot[31] implies a size of 0.5 fm for the interaction region. The value Λ_π = 850 MeV is incompatible with this experiment (R_0=0.7 fm). A value of Λ_π larger than 1.25 GeV is also incompatible with these data. Thus the experimental data are well described by including meson exchange currents with a size of the pion-nucleon interaction region of 0.5 fm.

The most surprising result of this experiment[30] is that there is not yet any evidence of a breakdown of mesonic theory even at such high momentum transfers beyond its expected limit of validity.

This experiment is also the first to give us some precise information on the shape of the nucleon form factor in the two-body current operator. The proper choice of form factor, either Sachs form factor G_E, Dirac F_1 or axial F_A, can be settled only in a fully relativistic description[33].

Fig. 11 shows that various predictions[31,32] with F_1 give a good agreement with experiment, while G_E leads to a complete disagreement. The difference between F_1 and F_A is too small to choose between the

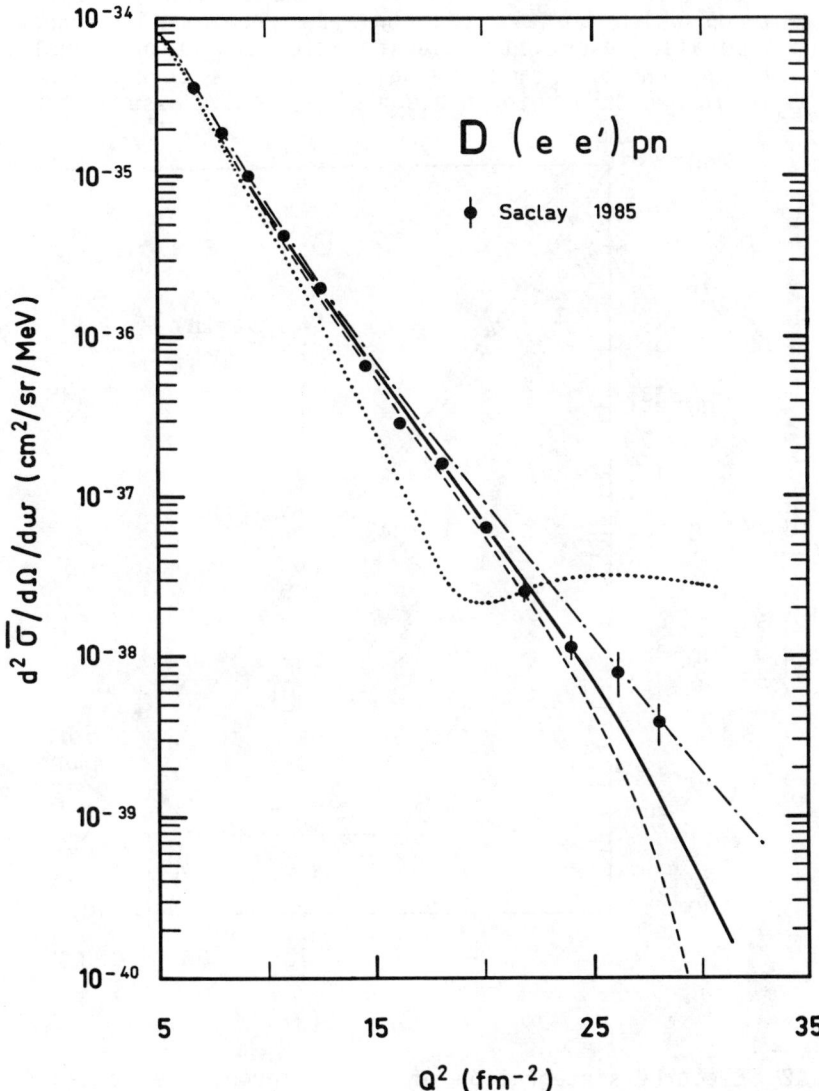

Fig. 11 Electrodisintegration of the deuteron. Comparison of the predictions with F_1 (dashed, dash-dotted, solid) and G_E (dotted)[30,32].

axial and the Dirac form factors, but the choice of G_E is incompatible with these data. It is interesting to note that it means that the pion is coupled to the nucleon, without contribution of the nucleon anomalous magnetic moment κ, since G_E and F_1 differ only by a term proportional to κ.

In this discussion the special role of the pion associated to the chiral symmetry breaking has not emerged. Fig. 12 shows that there is an alternative description of the electrodisintegration of the deuteron which answers exactly this problem. The experimental data can be also described by taking into account only nucleons and pions with a πNN form factor equal to 1. This corresponds to soft pion terms for which a point nucleon-pion term is assumed.

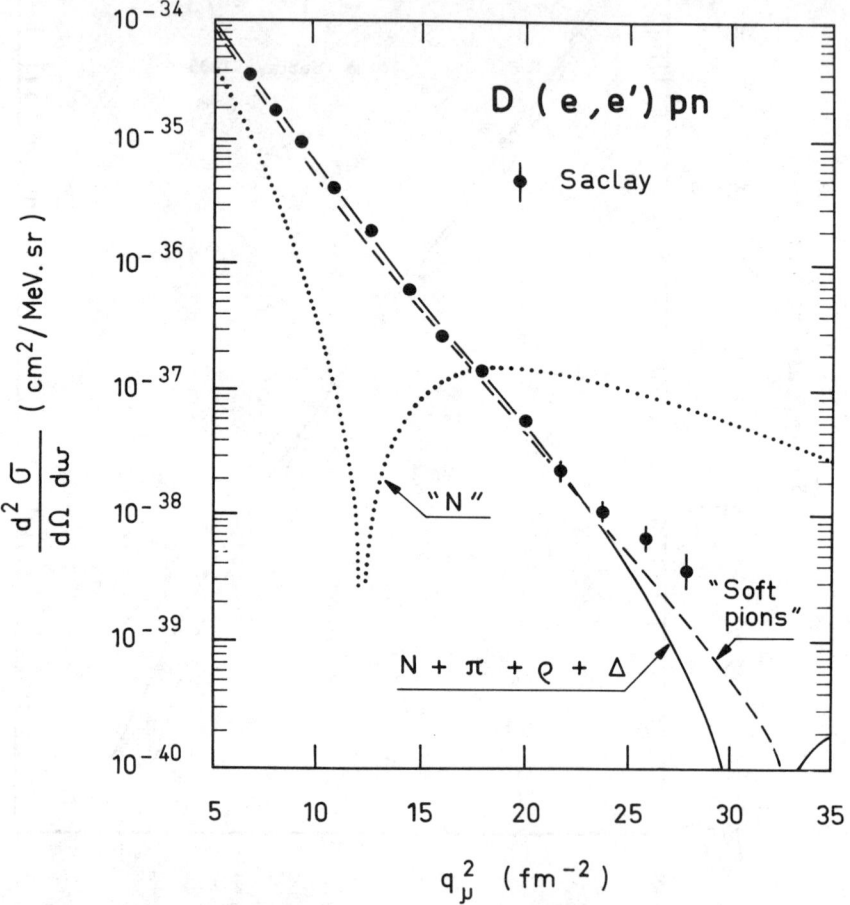

Fig. 12 Electrodisintegration of the deuteron. The dotted curve is the prediction with nucleons only. The dashed curve is the contribution of nucleons and pions with a point coupling. The solid curve is the full calculation with an meson nucleon interaction region of 0.5 fm.

This means that there are in fact two equivalent descriptions of the electrodisintegration of the deuteron :

- nucleons + pions with a point coupling ;
- nucleons + pions + ρ-mesons + Δ with form factors at each vertex corresponding to an interaction region of 0.5 fm.

There seems to be a very large cancellation between the effect of the ρ-meson and the Δ and the vertex form factors. The only term which survives to this cancellation is the pion exchange current. We have already observed that the anomalous magnetic moment of the nucleon does not contribute to the exchange current operator since it is absent in F_1. This is exactly what one expects from the Nambu-Goldstone picture of chiral symmetry breaking. Quarks are Dirac particles without anomalous moment and the absence of this term indicates that the only terms which are significant are due to a coupling of pions directly to the quarks. This explains why the apparent size of the πNN interaction region is smaller than the nucleon size (0.5 fm compared to 0.85 fm).

The equivalence of a description with a point interaction pion-nucleon and an extended region of radius 0.5 fm means that the size of the bag confining quarks is elusive. It might not be an observable. This is comparable to the smile of the cat in "Alice in wonderland" which fades away during his conversation with Alice. This picture proposed by Rho imerges naturally from the description of the nucleon as a Skyrmion[23]. This has been briefly discussed at this workshop by Weise.

The ground state form factors of ^{3}H and ^{3}He

The deuteron is an isoscalar system. It is impossible to unravel the isospin structure of the meson exchange currents. This missing information can be found in the comparison of the form factors of ^{3}H and ^{3}He.

A considerable theoretical progress has recently been made[34] by developing accurate numerical methods for describing the trinucleon ground state.

There remains a longstanding difficulty in the explanation of the charge form factor of ^{3}He. Large mesonic effects and three-body forces as well as quark degrees of freedom have been proposed to account for this discrepancy. In the case of ^{3}H, the isoscalar and isovector pieces of the meson exchange currents are predicted to cancel.

The experimental data for the charge form factor of ^{3}He are compared to the theoretical predictions of Hajduk, Sauer and Strueve[35]. The nucleonic contribution (dashed curve) is clearly not sufficient in the region of the second diffraction maximum. The contribution of Δ is directly included in the wave function by a coupled-channel method. The effect of Δ is small (dotted curve). The solid curve which is much closer to the experimental data includes both meson exchange currents and three-body force. The dominant mechanism which reconciles theory with experiment is the pion pair current (Z graph). However in the charge form factor it is an isoscalar process of relativistic order which cannot be derived from the same low energy theorems as the electrodisintegration of the deuteron which is an isovector magnetic process. In the case of an isoscalar pion current, we do not have model independent predictions for soft pions and there is no conservation law. So the best way to check this description is to make a similar comparison for the ^{3}H form factor which has been measured[36] recently at Saclay (Fig. 14). The prediction, which corresponds to the full calculation represented in Fig. 13 by a solid curve, is now in Fig. 14 the dashed curve. It is in striking disagreement with the experimental data. A much better agreement with the experimental data of both ^{3}H and ^{3}He is obtained in the quark constituent pair current model of Beyer, Giannini and Drechsel[37]. In this model the effect of quarks is to rescale the isovector and the isoscalar contributions in a different way. The main effect is due to the absence of anomalous magnetic moment. Sauer[38] has noticed that an

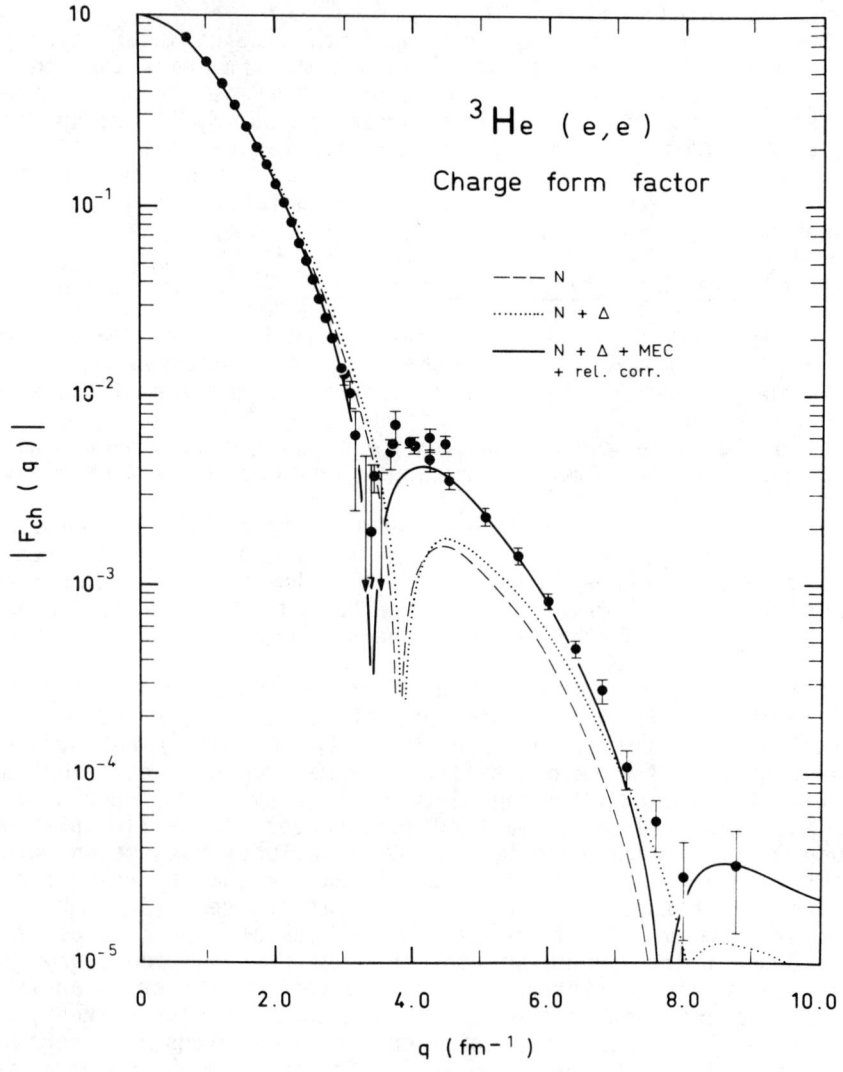

Fig. 13 The charge form factor of ^{3}He.

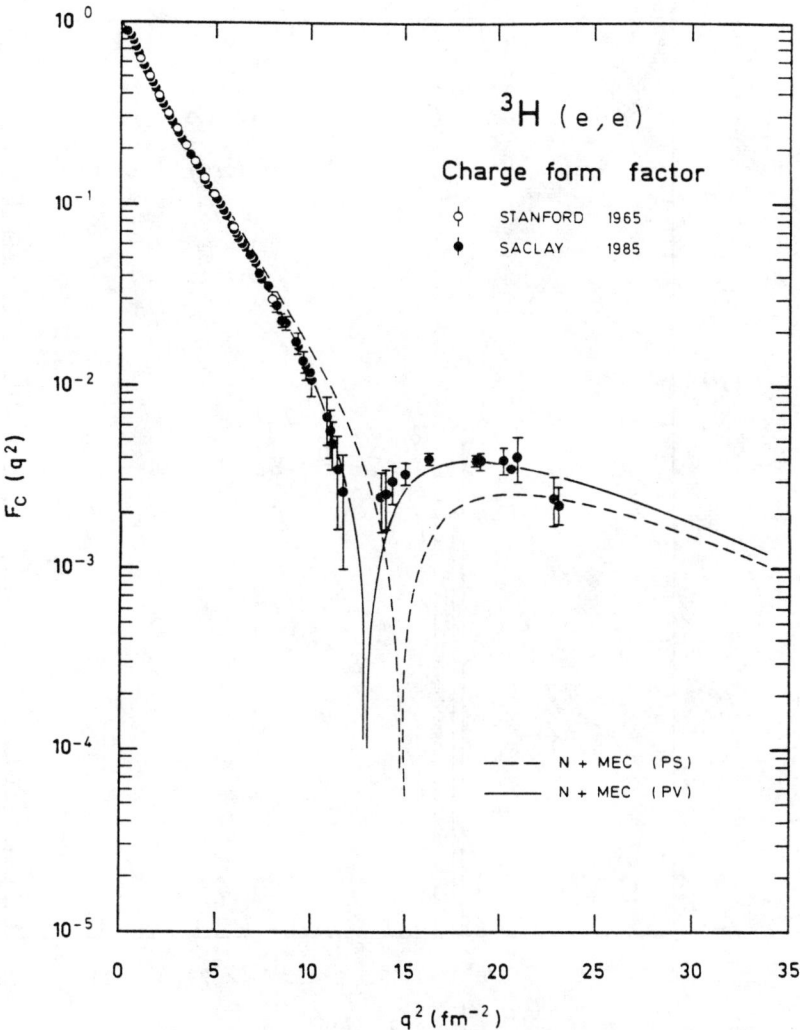

Fig. 14 The charge form factor of tritium[35,36].

almost similar agreement could be obtained by using a pseudo-vector pion nucleon coupling (in which there is no anomalous magnetic moment contribution). Previously the pseudoscalar coupling was used which led to the wrong prediction for the ^{3}H form factor. For ^{3}He the difference between pseudoscalar and pseudovector couplings is small (Fig. 15). This had not been noticed previously as a source of ambiguity for the charge form factor of tritium, though pseudo-vector coupling is required to give a correct description of pion photoproduction[39]. There remains a significant discrepancy for both nuclei which indicates that something is still missing.

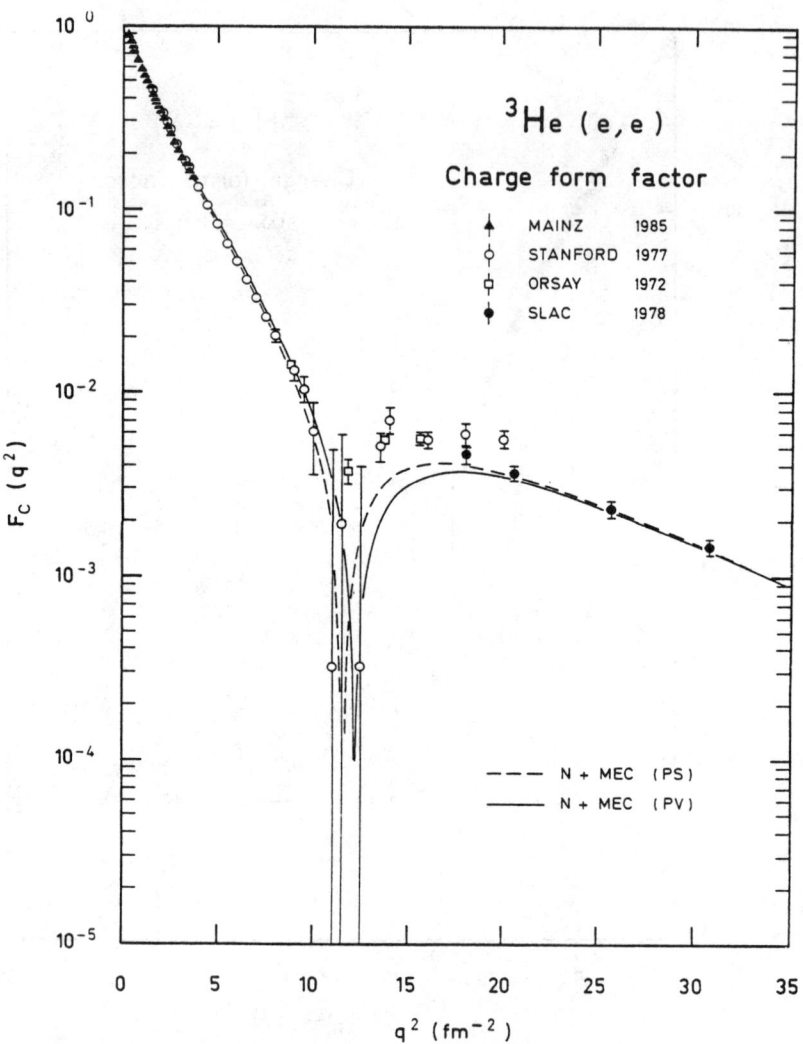

Fig. 15 The charge form factor of ^{3}He [rêfs.[35,40]].

A separation of the isospin components indicates that the isoscalar component predicted by Hajduk et al.[35] is in reasonable agreement with the experimental data. The source of disagreement is the isovector component of the charge form factor. It seems that the isoscalar pion exchange plays here also the same fundamental role, but we lack a theoretical guide for the moment.

For the magnetic form factors the agreement with theoretical predictions is very good. The isovector component is in perfect agreement with theoretical predictions while there is a small discrepancy for the isoscalar component.

CONCLUSIONS

The recent experimental data on subnucleonic degrees of freedom probed with hadronic and electroweak interactions are beginning to unravel fundamental nuclear processes.

The striking result is that there is no obvious breakdown of mesonic theory even at very high momentum transfer. The role of the pion appears to follow the picture of a boson directly coupled to quarks due to the chiral symmetry breaking at the boundary of the region where quarks are confined.

The size of the quark bag is elusive. The pion-nucleon interaction region has a size of 0.5 fm if the presence of the ρ-meson and the Δ are taken into account. If only the pion is considered, a point pion-nucleon coupling provides a good description of the electrodisintegration of the deuteron up to 30 fm^{-2}.

REFERENCES

1. P.E. Argan et al., Phys. Rev. Lett. 41, 86 (1978).
2. J-M. Laget, Phys. Reports 69, 1 (1981).
3. P.E. Argan et al., Nucl. Phys. A296, 373 (1978).
4. J.L. Faure et al., Nucl. Phys. A424, 383 (1984).
5. J Arvieux et al., Nucl. Phys. A431, 613 (1984) and references therein for all the compilation of unpolarized cross sections.
6. A. Boudard, Ph.D. Thesis, Report CEA-N-2386.
7. S. Auffret et al., Phys. Rev. Lett. 54, 649 (1985).
8. J.M. Laget, Proceedings of the second workshop on perspectives in nuclear physics at intermediate energies, Trieste, March 25-29 (1985) to be published by World Scientific, Singapore.
9. J. Hoftiezer et al., Phys. Lett. 100B, 462 (1981) ;
   E. Aprile et al., Nucl. Phys. A379, 369 (1982) ;
   E.G. Auld et al., Phys. Lett. 93B, 258 (1980) ;
   J.M. Cameron et al., ,Phys. Lett. 103B, 317 (1981).
10. E. Aslanides et al., Phys. Rev. Lett. 39, 1654 (1977).
11. D. Kielczewska et al., Few body problems in physics, vol. II, ed. by B. Zeitnitz (North-Holland, 1984).
12. R. Bertini et al., 6th Intern. Symp. on high energy spin physics, Marseille (1984).
13. P. Sitarski and E. Lomon, private communication.
14. R. Bertini et al., Phys. Lett. to be published.
15. E. Lomon, Nucl. Phys. A432, 139c (1985).
16. F. Villars, Helv. Phys. Acta 20, 476 (1947).
17. A.E. Cox et al., Nucl. Phys. 74, 497 (1965).
18. D.O. Riska and G.E. Brown, Phys. Lett. 38B, 193 (1972).
19. J. Hockert et al., Nucl. Phys. A217, 14 (1973) ;
    R.E. Rand et al., Phys. Rev. Lett. 18, 469 (1967).
20. "Mesons in nuclei", Edited by M. Rho and D. Wilkinson (North-Holland, 1979).
21. M. Chemtob and M. Rho, Nucl. Phys. A163, 1 (1971).

22. Y. Nambu, Phys. Rev. Lett. 4, 380 (1960) ;
    J. Goldstone et al., Phys. Rev. 127, 965 (1962) and ref. therein.
23. M. Rho, "Nuclear structure", Proceedings of the Niels Bohr Centennial, Copenhagen 1985, Edited by R. Broglia et al. (North-Holland, 1985).
24. K. Kubodera et al., Phys. Rev. Lett. 40, 755 (1978).
25. G.T. Garvey, "Nuclear spectroscopy and nuclear interactions", Proceedings of the International Symposium held in 1984 at Osaka, Edited by H. Ejiri and T. Fukuda (World Scientific Syngapore).
26. A.R. Heath and G.T. Garvey, Phys. Rev. C31, 2190 (1985).
    E.K. Warburton et al., to be published ;
    L. Lessard and L.A. Hamel, to be published ;
    C.A. Gagliardi et al., Phys. Rev. C28, 2423 (1983).
27. P.A.M. Guichon et al., Phys. Lett. 74B, 15 (1978) ;
    P.A.M. Guichon et al., Phys. Rev. C19, 987 (1979) ; Nucl. Phys. A382, 461 (1982).
28. I.S. Towner and F.C. Khanna, Nucl. Phys. A372, 331 (1981).
29. H.U. Jäger et al., Nucl. Phys. A404, 456 (1983) ;
    S. Nozawa et al., Phys. Lett. 140B, 11 (1984).
30. S. Auffret et al., Phys. Rev. Lett. 55, 1362 (1985).
31. J.F. Mathiot, Nucl. Phys. A412, 201 (1984).
32. W. Leidemann and H. Arenhövel, Nucl. Phys. A393, 385 (1983).
    D.O. Riska, Phys. Scripta 31, 471 and 107 (1985) ;
    A. Buchmann et al., Universität Mainz MKPH-T-85-2 ;
33. J. Friar and S. Fallieros, Phys. Rev. C13, 2571 (1976) ; C16, 908 (1977) ;
    J. Adam and E. Truhlik, Czech. J. of Phys. B34, 1157 (1984).
34. J.L. Friar et al., Ann. Rev. Nucl. Part. Sci. 34, 403 (1984) and references therein ;
    S. Ishikawa et al., Phys. Rev. Lett. 53, 1877 (1984) ;
    C.R. Chen et al., Phys. Rev. Lett. 55, 374 (1985).
35. C. Hasduk et al., Nucl. Phys. A405, 581 (1983).
36. F.P. Juster et al., Phys. Rev. Lett. 55, 2261 (1985) and references therein.
37. M. Beyer et al., Phys. Lett. 122B, 1 (1983).
38. P.U. Sauer, Trieste Workshop (1985).
39. J.L. Friar, Ann. Phys. (N.Y.) 104, 380 (1977).
40. R.G. Arnold et al., Phys. Rev. Lett. 40, 1429 (1978) ;
    J.S. McCarthy et al., Phys. Rev. C15, 1396 (1977) ;
    P.C. Dunn et al., Phys. Rev. C27, 71 (1983) ;
    C.R. Otterman et al., Nucl. Phys. A436, 688 (1985).

LOOKING FOR QUARK DEGREES OF FREEDOM

Ingo Sick
Dept.of Physics, University of Basel, Basel, Switzerland

1. INTRODUCTION

This session deals with non-nucleonic degrees of freedom of nuclei. In this talk, I want to concentrate on one particular aspect of this topic : the search for quark degrees of freedom. This theme mainly deals with future prospects rather than past achievements, and many of the ideas presented are somewhat speculative. The approach I want to discuss does, however, closely rely upon the things we have learned in the past when using electron scattering to look for mesonic degrees of freedom.

There are some obvious places where to look for quark degrees of freedom; let me cite a few. We can try to observe
1) the change of the nucleonic wave function due to the coupling to other channels,
2) the contribution of quarks to nuclear observables,
3) a change of the properties of the nucleon due to its internal degrees of freedom and the influence of the nuclear medium upon it, or
4) the presence of new configurations like 6-quark bags, hidden colour states, etc.

In this talk, I will say nothing on 4), since electron scattering provides very little information at present. In order to address this topic one has to do coincidence experiments with several GeV electrons. Such experiments will be feasible only once the CEBAF CW-accelerator is available. For the time being this most promising approach to look for quark degrees of freedom is not a practical one.

I want to address topics 1)-3), since some information can be obtained today using single-arm electron scattering experiments. Before discussing these points in more detail, I would like to ask the less specific question on the general rules one should try to follow when looking for degrees of freedom that are somewhat elusive.

2. LESSONS FROM SEARCH FOR MESONIC DEGREES OF FREEDOM

During the past ~ 15 years, a considerable effort has gone into the search for, and understanding of, mesonic degrees of freedom. From this experience, we can derive a few general guidelines.

a) Mesonic degrees of freedom, visible in electron scattering via the presence of meson exchange currents (MEC), in general are small. In selected observables, like the n-p radiative capture rate or the A=3 magnetic moments, they give an effect of $\sim 10\%$. In order to isolate a signal of $\sim 10\%$, one has to understand the nucleonic "noise" to much better than 10% accuracy. This at present is possible only for very light nuclei (A=2,3) where "exact" nucleonic wave functions are available. Alternatively, it is possible for observables that are very insensitive to nuclear structure, as, e.g. the inclusive cross sections at large energy loss.

Given the fact that quark degrees of freedom are likely to produce smaller effects than mesonic ones, initial concentration on the above cited observables seems wise.

b) We have learned in the past that MEC are much larger at large momentum transfer q. This is the case because of the following fact : One-body form factors typically depend on wave function overlaps in momentum space of the type $\int \psi(\vec{k}) \psi(\vec{k}+\vec{q}) d\vec{k}$. When the electron couples to exchange currents, the momentum transferred to the two nucleons involved is of order q/2, and the cross section depends on overlaps of the type $\int \psi(\vec{k}) \psi(\vec{k}+\vec{q}/2)$. At very large q, ($q \gg K_F$), $\psi(\vec{k}+\vec{q})$ becomes extremely small; $\psi(\vec{k}+\vec{q}/2)$ is much larger, and can lead to much larger cross sections despite the additional factors entering for processes involving 2 nucleons. At large q, we can enhance non-nucleonic over nucleonic degrees of freedom.

As an example for this feature, fig. 1 shows the ^{3}H magnetic form factor we measured[1] last year at Saclay. At q=0, the inclusion of mesonic degrees of freedom[2] changes the magnetic moment by $\sim 10\%$. At large q, the inclusion of MEC changes the cross section and form factor by an order of magnitude! While the purely nucleonic form factor (dashed curve) does not explain the data, the calculations including MEC (dotted and solid curves) are much closer to experiment. The difference between the two predictions including MEC, due to the treatment of effects of relativistic nature (the use of F_1 or G_e) is indicative of the size of the problems that remain to be understood.

The message of fig. 1 is clear : In order to enhance the signal one is looking for, one should study it at the distance scale appropriate for the phenomenon. At low q, long-range nucleonic properties dominate. At large q, the shorter-range MEC will be the most prominent ones.

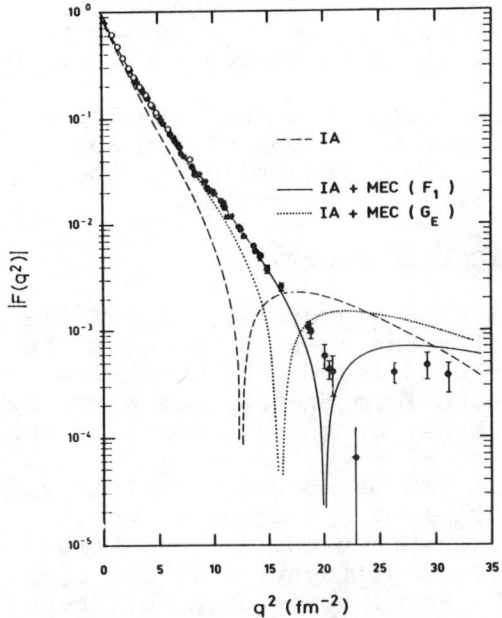

Fig.1. The tritium magnetic form factor[1] together with theoretical predictions[2] without (dashed) and with MEC contributions.

To isolate the effects of quark degrees of freedom upon the nuclear properties, we will have to look at nuclei with even better spatial resolution, that is at even larger momentum transfer.

c) MEC turn out to give much larger an effect in magnetic than in charge form factors. The reason for that is simple enough : For a charge form factor it matters little whether the charge is located in a nucleon, pion or quark. For a magnetic form factor, on the other hand, additional constituents lead to additional convection currents. Leightweight constituents in particular give larger contributions to these currents which are measurable via the total magnetization density of a nucleus. Accordingly, we would tend to search for quark effects in magnetic rather than charge form factors.

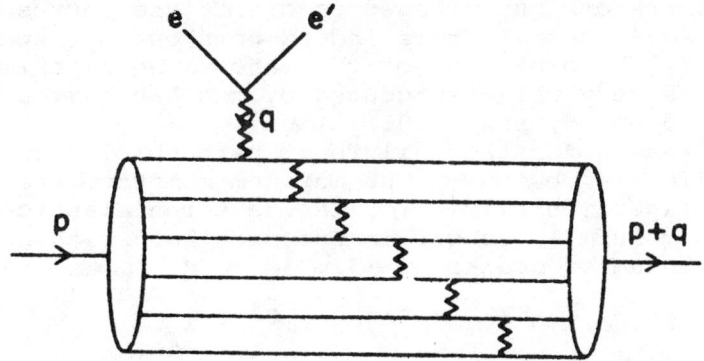

Fig.2. Democratic chain model for e-d scattering at very large q.

Of the three criteria a)-c) mentioned above, the most important one certainly is b), i.e. the isolation of phenomena via observables at the distance scale appropriate.

In the following, I will present a few examples of observables that, in accordance with some of the above considerations, have been or could be applied to the search for quark degrees of freedom in nuclei.

3. ASYMPTOTIC FORM FACTORS

At extremely large momentum transfers q, the fall-off of the elastic form factor can be used to count the number of constituents in the nucleus. The power-law that is expected to govern the form factor thus gives us information on whether nucleons, or quarks, dominate the observable.

In perturbative QCD, it can be shown[3], that at very large momentum transfer it is the diagram shown in fig. 2 that dominates. Suppose that the momentum transfer divided by the number of constituents N in the system is much larger than the Fermi momentum K_F of these constituents, $q/N \gg K_F$. In this case the nucleus can stay together only if the momentum transfer is transmitted to all constituents, such that every one has a momentum $\approx q/N$ after the scattering. It has been shown by Brodsky that in this case the elastic form factor will fall with the power

$$F(q) \propto q^{-2(N-1)} \qquad (1)$$

For the deuteron where the elastic form factor A(q) has been measured[4] up to very large q, N=6 for quark constituents leads us to expect a q^{-10} fall-off.

The interpretation of experimental data by Brodsky and collaborators has managed to create the impression that this q^{-10}-behaviour is indeed born out by experiment. This, however, is not the case. The experimental data can be very well reproduced by a q^{-n}-behaviour, but with an n=5.5±0.5, not n=10.

The reason for the failure of perturbative QCD to predict F(q) is obvious. The maximum momentum transfer achieved experimentally[4] for the deuteron elastic form factor is q=10fm^{-1}, so q/6 is not much larger than the Fermi momentum of quarks of $\sim$3fm^{-1}

$$q/N = 1.7 \text{ fm}^{-1} \not> 3 \text{ fm}^{-1} \simeq K_F \qquad (2)$$

Perturbative QCD-arguments are not applicable, and will never be with the nuclear form factors we could hope to measure in any foreseeable future.

A very simpleminded extension of eq.(1) is much more successful. (For a more sophisticated approach see ref.5). Assuming that at $q \sim 10 \text{fm}^{-1}$ we still see nucleons, we would expect a q^{-2}-behaviour of the body form factor of the deuteron. (In this case $q/2=5\text{fm}^{-1}$ is much larger than $K_F \sim 0.5\text{fm}^{-1}$). Given the composite nature of the nucleon, we have to add a q^{-4}-factor resulting from the q-dependence of the nucleon form factor. The q^{-6}-behaviour resulting from this primitive model (which ignores vertex form factors occurring in fig.2) agrees with the experimental observation of $q^{-5.5 \pm 0.5}$.

Above discussion shows that perturbative QCD is not appropriate to interpret elastic form factors that can be measured. It does not show which degrees of freedom really dominate the deuteron form factor at large q. To find out, we will need deuteron wave functions calculated in terms of the degrees of freedom of interest.

4. MEC AT SHORT RANGE

The work in electron scattering over the past 15 years has shown that there are a few show-case observables that exhibit the influence of MEC. In the A=3 magnetic form factors, or deuteron electrodisintegration at threshold, MEC give the dominating contribution to the cross section at large q. In these observables the nucleonic degrees of freedom are already strongly suppressed as compared to other ones. Going in these same observables to even higher momentum transfer could be hoped to get us into the regime where the internal structure of the mesons, due to quarks, become indispensable for an understanding.

A brief numerical estimate shows that we are not all that far away from the q-region of interest. For impulse approximation processes, the spatial resolution of electron scattering is $\sim 1.5/q$. For processes dominated by MEC, the resolution is closer to $\sim 3/q$, given the fact that in MEC processes the transfer is split to 2 nucleons, ultimately. The momentum transfer of 5-6fm^{-1} achieved today for A=2,3 corresponds to a resolution of 0.5fm, which is marginal, but not hopeless, to get a glimpse of quark degrees of freedom.

As an example, fig.3 shows the d(e,e') data[6] we recently extended to larger q at Saclay together with calculations[7] of J.F.Mathiot. The effect of MEC is large, and required to get the calculation into rough agreement with the data. The good agreement observed seems not to leave any room for additional degrees of freedom due to quarks. This, however, is not true upon closer inspec-

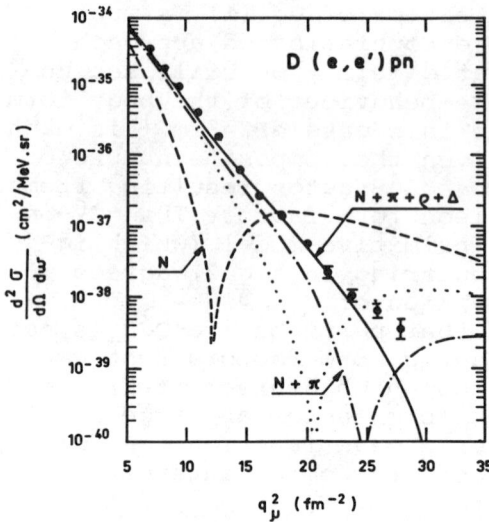

Fig.3. Deuteron electro-disintegration at threshold[6] together with theoretical predictions[7] without (N) and with (N+π+ρ+Δ) MEC. The dotted curve includes MEC with a πNN cutoff form factor with mass $\Lambda \sim 0.7$ GeV.

tion. The calculated results contain a "fudge factor" in the form of the πNN-vertex form factor. It has been parametrized with the usual dipole form, with a cut-off mass $\Lambda=1.2$ GeV fitted to the data. This parameter, which corresponds[7] to a size of the πNN interaction region of 0.48 fm, is too large, given the fact that the radius of the nucleon (pion) alone is already 0.85 (0.6) fm. Experimental data[8] from $p\bar{p} \to \bar{n}n$ give a much smaller value of Λ of $\Lambda \sim 0.7$ GeV, and this would lead to a large disagreement with the data (fig.3). There clearly is room for additional short-range effects that would change the cross section at large q.

The discussion shows that, even at 5 fm^{-1}, we already need to understand the internal structure of nucleons and mesons if we want to correctly predict MEC-dominated observables. In order to enhance these short-range effects even more, such as to clearly separate them from the mesonic effects dominating in the region of the diffraction feature in figs. 1,3, we should push the data to higher q. This can be done, since we were limited in the examples shown by accelerator energy only.

5. QUARK COMPONENTS OF NUCLEON WAVE FUNCTIONS

The internal structure of the nucleon lets us expect that in the nucleus we may find wave function components other than the one described by A nucleons. The most likely components perhaps are those where, during a short-range nucleon-nucleon collision, configurations more appropriately described by 6 quark bags are formed. Such configurations, of small size, are expected to contribute preferentially to the form factor at large q.

At present there are several calculations that describe nuclei in terms of hybrid models evoking, e.g. 6 quark components. For internucleon distances r_{NN} larger than some cut-off radius R, a purely nucleonic wave function is employed, for $r_{NN}<R$ a 6 quark component is assumed. By adjusting the 6q percentage within a sensible range, form factors at large q can be reproduced.

These exploratory calculations at present suffer from two main problems : the wave function is very crude, and MEC processes in general are not treated. These deficiencies make it difficulat to identify an unambiguous quark-signal.

The perhaps most realistic calculations are the ones performed in the framework of the resonating group approach. In this case, the wave function, initially parametrized in terms of quarks, can be split into a 2 nucleon-cluster component at long range, and a quark component at short range. Antisymmetrization of the wave function in terms of quarks is maintained. This type of approach has been highly successful in the past in the description of light nuclei (A=6,7) in terms of nucleons and, alternatively, nucleon clusters (d,^{3}He,^{4}He). It has managed to describe nuclear wave function in terms of two pictures that are not orthogonal, but complementary for the calculation of long/short-range properties.

Several resonating group calculations have recently been published[9]. In fig. 4 we show the prediction of Yamauchi et al. for the deuteron magnetic form factor B(q), an example chosen because MEC are expected to be small due to the isoscalar nature of the form factor. Fig.4 shows that in a resonating group calculation the effect of introducing a quark component in the wave function is fairly small in the region covered by existing data (q<6fm^{-1}). This smallness could be expected from the old experience with resonating group calculations, which showed that many observables can be represented in terms of either of the two types of constituents. At large q, where a diffraction minimum of B(q) is predicted, the genuine quark-effects start to become appreciable, and an experiment presently in progress at SLAC might provide data in the q-region of interest.

6. CHANGE OF NUCLEONS IN NUCLEI

Nucleons bound in nuclei are imbedded in a medium of rather large density; the average NN distance exceeds the

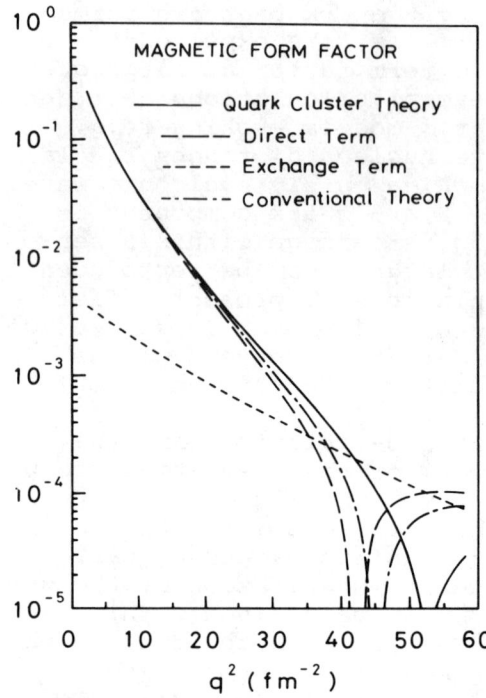

Fig. 4. Deuteron magnetic form factor.

nucleon diameter by only 20%. Accordingly, one may expect nucleons to change due to the environment.

Recently, one particular change has been discussed extensively in the literature : a general increase of the nucleon size in nuclei. Three types of considerations have led to propose such a change. The structure function of the bound nucleon as measured by the EMC experiment is smaller than the one of the free nucleon for X>0.3; an increase in nucleon size would explain[10] this observation, simply as a consequence of the uncertainty principle. The longitudinal sum rule measured in quasi-elastic electron-nucleus scattering is smaller than expected in the region $q=2 \div 3 fm^{-1}$; this could be explained if the bound-nucleon form factor were smaller than the free-nucleon one[11]. Calculations[12] of nuclear wave functions in terms of quarks suggest a general increase of the nucleon size. The various observations cited suggest a radius change of 10 - 20%.

A change of the nucleon due to the nuclear medium certainly is a plausible concept. The observations made, however, are far from unambiguous. Alternative explanations of the observed facts have been proposed. Here, we want do discuss a direct measurement of the bound nucleon size.

The general idea has been presented in a recent publication[13], so we do not need to go into great detail. Briefly, we consider electron-nucleus quasielastic scattering at large momentum transfer, and at energy loss ω lower than for the maximum of the quasielastic peak (i.e. at X>1). In this region the inelastic response function $\sigma(q,\omega)$ is dominated by elastic scattering off individual, moving, nucleons. The information on the nucleon form factor becomes accessible by exploiting the scaling property[14] of $\sigma(q,\omega)$ expected for quasielastic scattering,

$$\sigma(q,\omega)/\Sigma_A \sigma_{eN}(q) \cdot d\omega = F(y) \cdot dy \qquad (3)$$

with

$$y \cong (\omega^2 + 2m\omega - q^2)/2q \qquad (4)$$

In general, $\sigma(q,\omega)$ is a function of two independent variables. For quasielastic scattering, division through the (in-medium) electron nucleon cross section yields a function F(y) depending on a single variable y. This variable y corresponds to the nucleon momentum component parallel to $\vec{q}$, F(y) is the probability to find nucleons with momentum y before the scattering process.

Equations (3), (4) yield an F(y) independent of q provided one divides out a nucleon cross section with the q-dependence appropriate for nucleons in the medium. The q-independence of F(y) thus allows us to determine the q-dependence of the form factor of the bound nucleon. From the expression of the dipole form factor

$$G(q) = (1 + q^2 R^2/12)^{-2} \qquad (5)$$

which fits electron-nucleon scattering over a large q-range, the relation of q-dependence and nucleon size is evident.

This approach has been used to analyze the ^{3}He(e,e') data[15] we measured a number of years ago at SLAC. The data taken at energies between 3 and 15 GeV in the q-range 2÷12 fm^{-1}, show an impressive scaling behaviour as demonstrated by fig. 5. For y<0, cross sections varying over several decades define the same scaling function F(y). For y>0, where Δ-excitation and MEC play a role, no scaling is observed, as expected. The data in fig. 5 have been calculated using the free-nucleon form factor. If one changes the nucleon radius by, say, 10%, the scaling property is markedly degraded; the width of the F(y)-band is much larger (fig. 6). From this degradation of scaling a sensitive measure of the bound nuc-

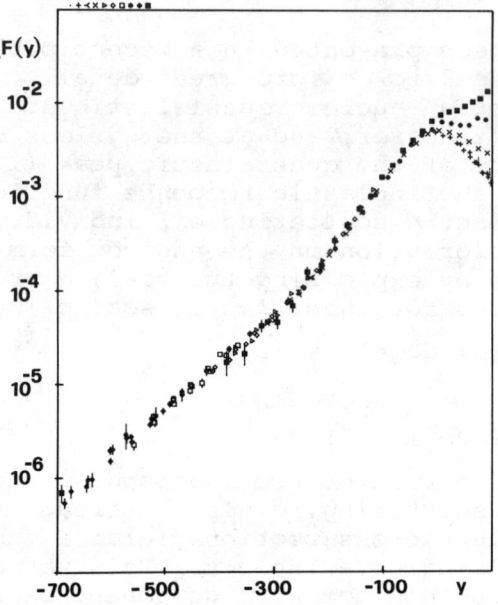

Fig. 5. Scaling function[4] for ^{3}He.

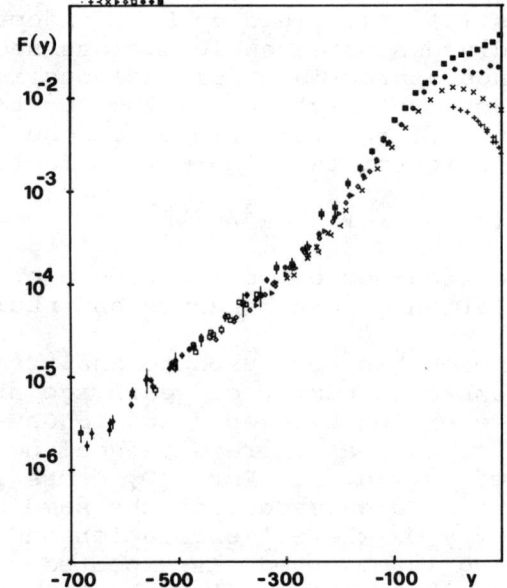

Fig. 6. Scaling function[4] for ^{3}He calculated with nucleon radius increased by 10%.

leon form factor and radius can be obtained. For ^{3}He, the nucleon-radius was found[13] to change by less than 3-6% due to the nuclear environment.

An experiment[16] we recently performed at SLAC, at the NPAS facility, allows us to extend this approach to heavy nuclei, 4≤A≤197. At present, the data on ^{56}Fe only have been analyzed. For iron, the range of q^2 covered by our experiment is 0.5-6 GeV^2/c^2, the X-region between 1 and 2.5. The data, when plotted in terms of F(y) show the same striking scaling behaviour as the one displayed in fig. 5 for ^{3}He. This scaling degrades rapidly as one changes the nucleon radius away from the free one.

In order to quantitatively analyze the goodness of this scaling, we have made a fit to the experimental F(y) points using a very flexible phenomenological parametrization (a sum of gaussians with 6 terms). This has been done for different values of the radius R (eq. 5). The χ^2 of these fits describes how well the points do define a unique F(y). Nonscaling, i.e. a q-dependence of F(y) leads to a large χ^2.

Fig. 7 shows the χ^2 as a function of changes of R relative to the free-nucleon radius. The lowest χ^2 is found near the free proton radius, and a rapid increase of χ^2 is observed for changes of R as low as 2%.

From this Fe(e,e') data, we conclude that the nucleon form factor and size change very little due to the nuclear medium.

I should add that the above experiment only concerns the average size of the nucleon. Multiquark configura-

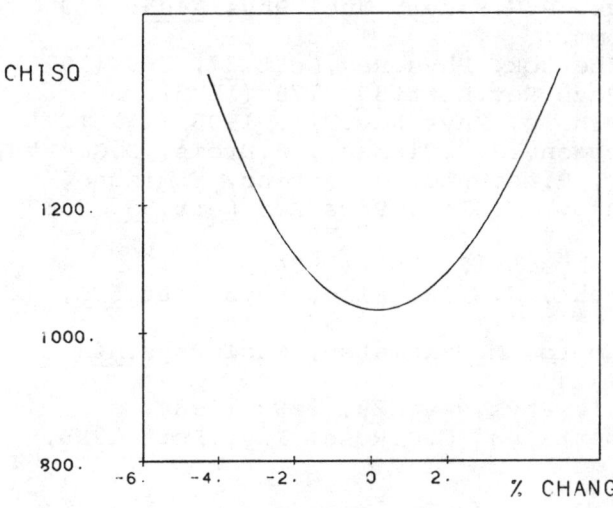

Fig. 7.
χ^2 for fit to scaling function of iron as function of nucleon radius change.

tions formed in e.g. NN collisions at short range, preferentially would couple to final states ≠2N. The energy transfer required to produce these states would shift strength from negative to a less negative y. Due to the rapid increase of $F(y)$ with increasing y, such strength might be difficult to detect via scaling-violation. A future search for scaling violations near X=2 will show whether such strength can indeed be detected in the data sample now available.

7. CONCLUSION

At present, we seem to have little firm evidence on the role quarks are playing in nuclei. Most nuclear properties still can be explained by evoking nucleonic and mesonic degrees of freedom only. The EMC experiment offers the first tantalizing evidence that quark degrees of freedom are relevant indeed.

In order to uncover the role played by quarks, we will have to go to higher momentum transfers that are more appropriate for the study of this short-range degree of freedom, and we will have to look at more exclusive channels that are more sensitive to specific wave function components. This can be done once a CW electron accelerator of several GeV energy is available.

REFERENCES

1. F.P.Juster, S.Auffret, J.M.Cavedon, J.C.Clemens, B.Frois, D.Goutte, M.Huet, P.Leconte, J.Martino, Y.Mizuno, X.H.Phan, S.Platchkov, S.Williamson, I.Sick, Phys,Rev.Lett., to be publ.
2. W.Strueve, C.Hajduk, P.U.Sauer, Nucl.Phys.$\underline{A405}$, 620 (1983) ans priv.comm.
3. S.J.Brodsky, B.T.Chertok, Phys.Rev,Lett.$\underline{37}$, 269 (1976).
4. R.Arnold et al., Phys.Rev.Lett.$\underline{35}$, 776 (1975).
5. C.Alabiso, G.Schierholz, Phys.Rev.$\underline{D11}$, 1905 (1975).
6. S.Auffret, J.M.Cavedon, J.C.Clemens, B.Frois, D.Goutte, M.Huet, F.P.Juster, P.Leconte, J.Martino, Y.Mizuno, X.H.Phan, S.Platchkov, I.Sick, Phys.Rev.Lett.$\underline{55}$, 9362 (1985).
7. J.F.Mathiot, Nucl.Phys.$\underline{A412}$, 201 (1984.
8. K.Bongardt, H.Pilkuhn, H.G.Schlaile, Phys.Lett.$\underline{52B}$, 271 (1974).
9. Y.Yamauchi, R.Yamamoto, M.Wakamatsu, Nucl.Phys.$\underline{A443}$, 628 (985);
   Y.E.Kim, M.Orlowski, Phys.Rev.$\underline{C29}$, 2299 (1984.
10. F.E.Close, R.G.Roberts and G.G.Ross, Phys.Lett.$\underline{129B}$, 346, (1983);

R.L.Jaffe, F.E.Close, R.G.Roberts and G.G.Ross, Phys. Lett.134B, 449 (1984).
11. J.V.Noble, Phys.Rev.Lett.46, 412 (1981).
12. L.S.Celenza, A.Rosenthal and C.M.Shakin, Phys.Rev. Lett.53, 892 (1984).
13. I.Sick, Phys.Lett.B157 13 (1985).
14. I.Sick, D.Day and J.S.McCarthy, Phys.Rev.Lett.45, 871 (1980).
15. D.Day et al., Phys.Rev.Lett.43, 1143 (1979).
16. D.Day, J.Jourdan, R.McKeown, Z.Meziani, R.Milner, D.Potterveld, R.Sealock, I.Sick, Z.Szalata, S.Thornton, to be publ.

STRANGENESS IN NUCLEI

R. Büttgen, K. Holinde[+] and B. Holzenkamp
Institut für Kernphysik der KFA Jülich,
D-5170 Jülich W. Germany

J. Speth[*]
Los Alamos National Laboratory
Los Alamos, NM 87545

ABSTRACT

We present further results of our general program, which is to construct meson-exchange potentials for hadronic systems involving strange particles. In this contribution we investigate the relationship between the free ΛN-interaction and the effective interactions inside of a nucleus. These polarization effects are taken into account within a generalized Brueckner G-matrix. Within this approximation we calculate the binding energy and effective mass of a Λ-particle in nuclear matter as well as the Landau-parameters for the ΛN-system.

INTRODUCTION

The most essential input to modern calculations of nuclear structure and nucleon-nucleus scattering is the nucleon-nucleon (NN) interaction. Because strong short-range correlations are introduced by a repulsive core in certain NN states the interactions used for these problems are necessarily effective ones which are qualitatively different from the free NN interaction. An understanding of the relationship between these effective interactions and their free counterparts has been one of the major objectives of nuclear physics over the past twenty years.

It is usually assumed that to a good first approximation the effective NN interaction (G) may be related to the free NN potential (V) via the Bethe-Goldstone (BG) equation

$$G(z) = V + V \frac{Q}{z-H_o+i\varepsilon} G(z) \qquad (1)$$

where the Pauli projection operator (Q) restricts the two-nucleon intermediate states to those unoccupied by the nuclear medium, H_o is a single-particle Hamiltonian that includes the kinetic energy

---

[+]also: Institut für Theoretische Kernphysik, Universität Bonn
[*]Permanent address KFA Jülich

and a single particle potential, and z represents the starting
energy of the two interacting nucleons. For positive z the
complex G-matrix is an appropriate effective interaction for
scattering. For the bound state problem z<0, the presence of the
Pauli operator in Eq. (1) ensures that the propagator has no
poles, and G(z) becomes the real reaction matrix used for nuclear
structure calculations. A review of the use of G-matrix
interactions in nuclear structure and nuclear reaction
calculations has been given by Love and Speth.[1] So far the medium
modifications in Eq. (1) enter only via the Pauli operator and the
single particle potential. It has only been, very recently that
so-called 'relativistic effects' that modify the free NN-
interaction V in Eq. (1), have been included in the calculation of
the G-matrix. These give rise to a strong density dependence in
the scalar-isoscalar channel of the G-matrix and reduce the
attraction in this channel considerably.[2,3] In order to solve
Eq. (1) the interaction V has to be given in an appropriate form
(on- and off-shell), which allows one to include all the medium
modifications mentioned above. Recently the Bonn group has
finished the construction of a meson-exchange model of the NN
interaction[4] that is based on suitable meson-nucleon-nucleon and
meson-nucleon-Δ isobar vertices W_α. The Hamiltonian is treated in
time-ordered (non-covariant) perturbation theory. This treatment
corresponds to standard many-body theory and thus leads to a
well-defined transition from the two- to the many-body problem.
The resulting (energy-dependent) quasipotential V contains, in
addition to single-meson exchange, explicit 2π-exchange
contributions consistent with results obtained from dispersion
theory. It was essential to include corresponding πρ-exchange
diagrams in order to obtain a quantitative fit to the NN data. It
has been demonstrated[4] that these contributions replace to a large
extent the fictitious σ-exchange used in former OBE-models. In
those meson-theoretical approaches one introduces form factors and
coupling parameters which are determined by fitting to the NN
phase shifts.

In this contribution we report on the application of the
Brueckner theory to the calculation of the effective
ΛN-interaction. In order to solve generalized Bethe-Goldstone
equations one needs the corresponding free ΛN-potential. We
recently started to calculate ΛN- and KN-potentials within the
meson-exchange picture.[5] Here one has to introduce new vertices
that are connected with strangeness exchange processes. In the
case of the KN-interaction the corresponding parameters can be
determined from the existing experimental phase shifts.
Unfortunately those data are not available for the ΛN-system.

THE FREE ΛN-INTERACTION

In comparing the free ΛN-interaction with the free
NN-interaction within the one-boson-exchange (OBE) model one
realizes that there are many similarities. However, because of
the strangeness exchange we have to distinguish two different
contributions: (1) The direct interaction (Fig. 1a) where, as in

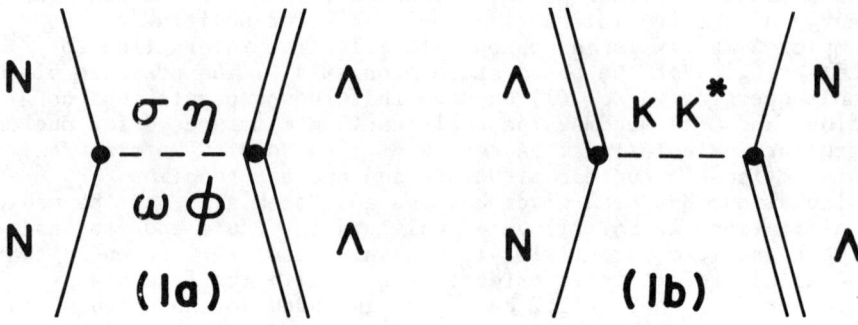

Fig. 1. One-boson exchange contributions: (1a) direct contribution (1b) strangeness-exchange processes in the ΛN system.

the NN-system, the σ,η,ω,ϕ-mesons are exchanged, and (2) the K and K^* meson that give rise to the exchange of strangeness (Fig. 1b). Our calculations of the free ΛN interaction[5] have been performed in the same framework as the NN-potential.[4] Several papers by the Nijmegen group[6] dealing with hyperon-nucleon scattering are available in the literature. However, they are based on OBE-potentials given in coordinate space, which require severe approximations for their derivation, as we will discuss shortly. In contrast, our procedure is to keep the original momentum-space expressions for the interaction. Moreover, the use of time-ordered pertubation theory provides a tractable three-dimensional formalism together with a reliable description of meson retardation effects, which have been shown to be of great importance in the NN system. In contrast, common 3-dimensional reductions of the Bethe-Salpeter equation, such as the Blankenbecler-Sugar choice taken in Ref. 6, provide a wrong description of meson retardation effects and thus lead to an unrealistically large size of higher order contributions. Specifically, the structure of the propagator in time-ordered perturbation theory due to lowest-order exchange is given by, (see Fig. 2)

$$g(z) = \frac{1}{2w}\left(\frac{1}{Z-E_1'-E_2-\omega} + \frac{1}{Z-E_1-E_2'-\omega}\right) \qquad (2)$$

Here Z is the starting energy. The result is essentially obtained by subtracting the energies involved in the intermediate state (dashed-dotted line in Fig. 2) from the starting energy.

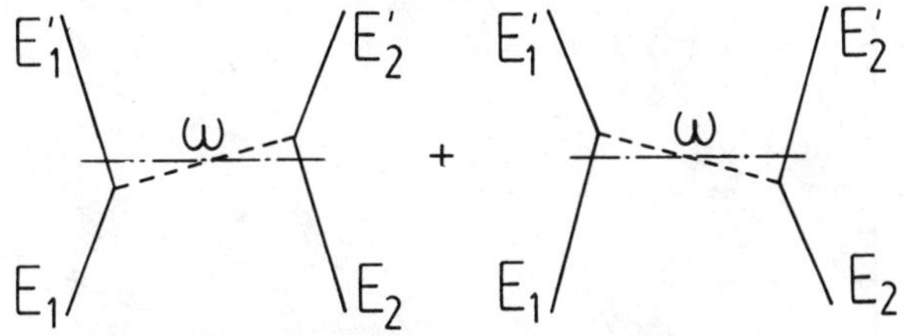

Fig. 2. Lowest order exchange in time-ordered perturbation theory.

Nelgecting recoil effect, that is, putting $Z = E_1'+E_2$ and E_1E_2', respectively, we obtain the conventional, static propagator $w^{-2} = 1/(\vec{k}^2+m_\alpha^2)$ where $\vec{k}$ is the momentum transfer and m_α the mass of the exchanged meson. The inclusion of recoil effects suppresses higher-order iterations of the processes of Fig. 2, because large intermediate momenta are involved. This is explicitly demonstrated in Fig. 4 for the higher order process shown in Fig. 3. In Fig. 4 we show the potential $V(p,q,z)$ due to this process, in the $^1S_0$ partial wave with and without inclusion of the recoil effect for $Z = 2 E_q$ and $q = 250$ MeV. Obviously, the recoil effects are non-negligible, suppressing the contribution by about 20 to 30%. The OBE contributions shown in Fig. 1 and the box diagrams of the form shown in Fig. 3 build up the

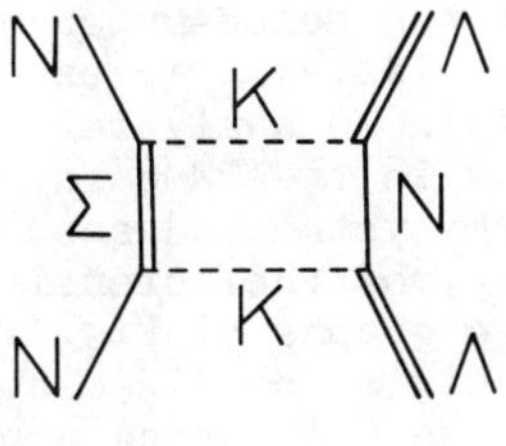

Fig. 3. Higher order box diagram contributing to ΛN scattering.

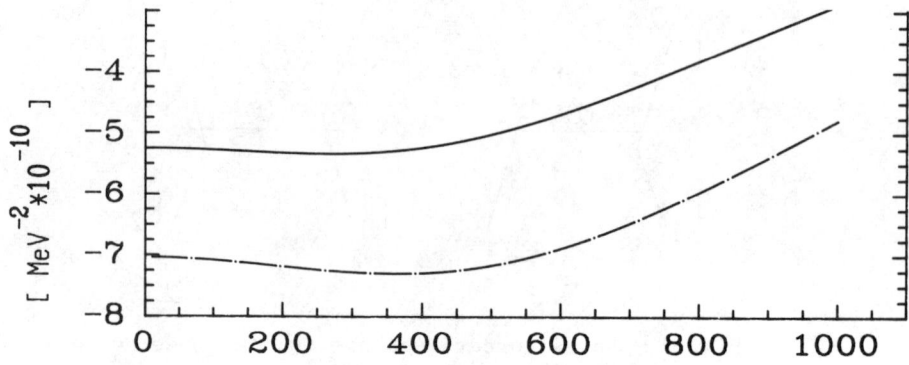

Fig. 4. Potential $V(p,q,z)$ due to the process in Fig. 3 in the $^1S_0$ partial wave, for $Z = 2E_q$ and $q = 250$ MeV, as a function of p. The solid line includes the retardation effects whereas they are omitted in the dashed-dotted curve.

quasipotential $V(z)$ and lead to the T-matrix by means of the unitarizing equation

$$T(z) = V(z) + V(z)\frac{1}{z-H_o+i\varepsilon}T(z) \qquad (3)$$

that determines the scattering phase shifts as a function of the lab energy. Using Eq. (3) we also investigated the effect of a relativistic treatment of the meson-hyperon vertices because the Nijmegen group made a nonrelativisitic expansion of those vertex functions in order to obtain an analytic Fourier transformation into coordinate space. As in the NN-case, this approximation has sizable effects for the ΛN system and should be avoided. As an example, we present in Fig. 5a the potential in the $^1S_0$ partial wave based on the exchange graphs of Fig. 1b. In Fig. 5b we show the corresponding phase shifts, obtained by solving the unitarizing equation for the scattering amplitude, Eq. (3). The results clearly demonstrate the inadequacy of the nonrelativistic approximation.

Finally, we want to report on an interesting observation concerning the strength of the tensor force in the ΛN system, that is primarily due to the diagrams in Fig. 1b. As Fig. 6a demonstrates, there is an extremely strong cancellation between the contributions of K- and K^*-exchange leading to a very small tensor force in the relevant momentum region. This interplay between K and K^*-exchange is analogous to π- and ρ-exchange in the

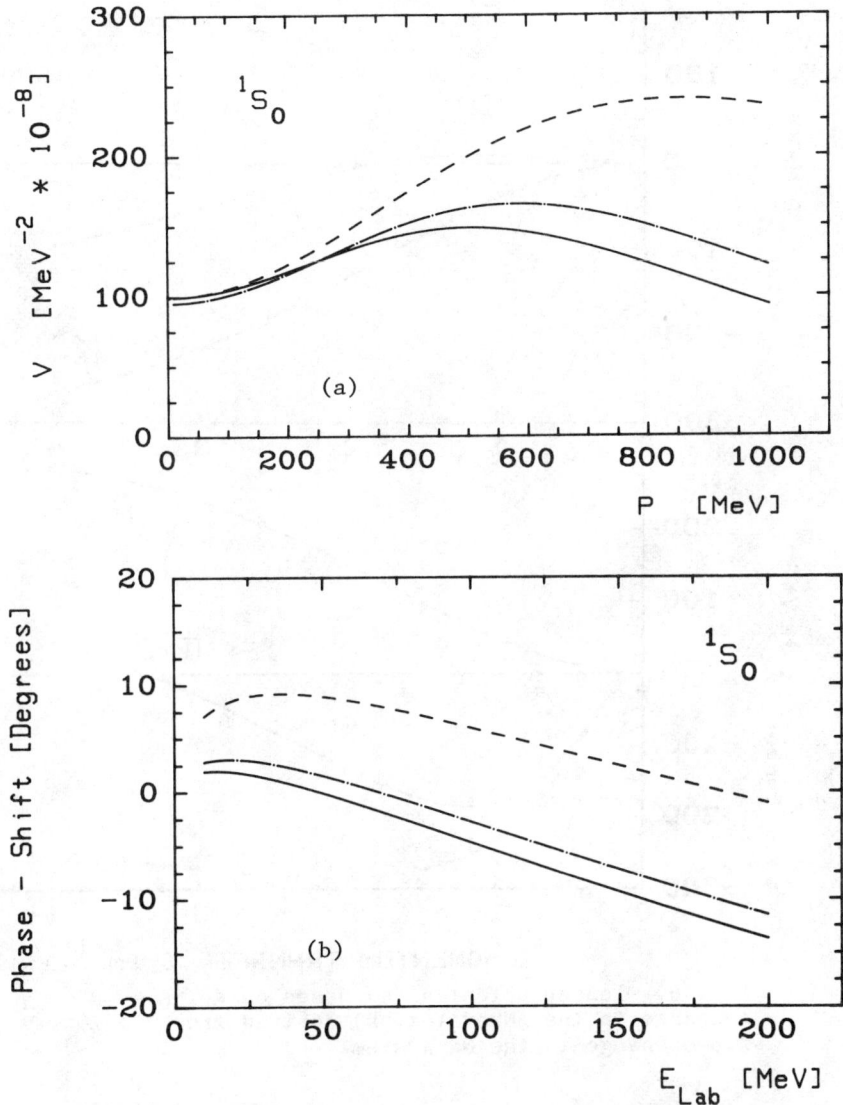

Fig. 5a) Potential $V(p,q,Z)$ due to the process in Fig. 1b in the $^1S_0$ partial wave, for $Z = 2E_q$ and $q = 250$ MeV, as a function of p. The solid line denotes the unapproximated result whereas, for the dashed-dotted curve, retardation effects have been omitted in the propagators. For the dashed curve, the nonrelativistic approximation at the vertices has been applied in addition. b) The corresponding result for the ΛN phase shift.

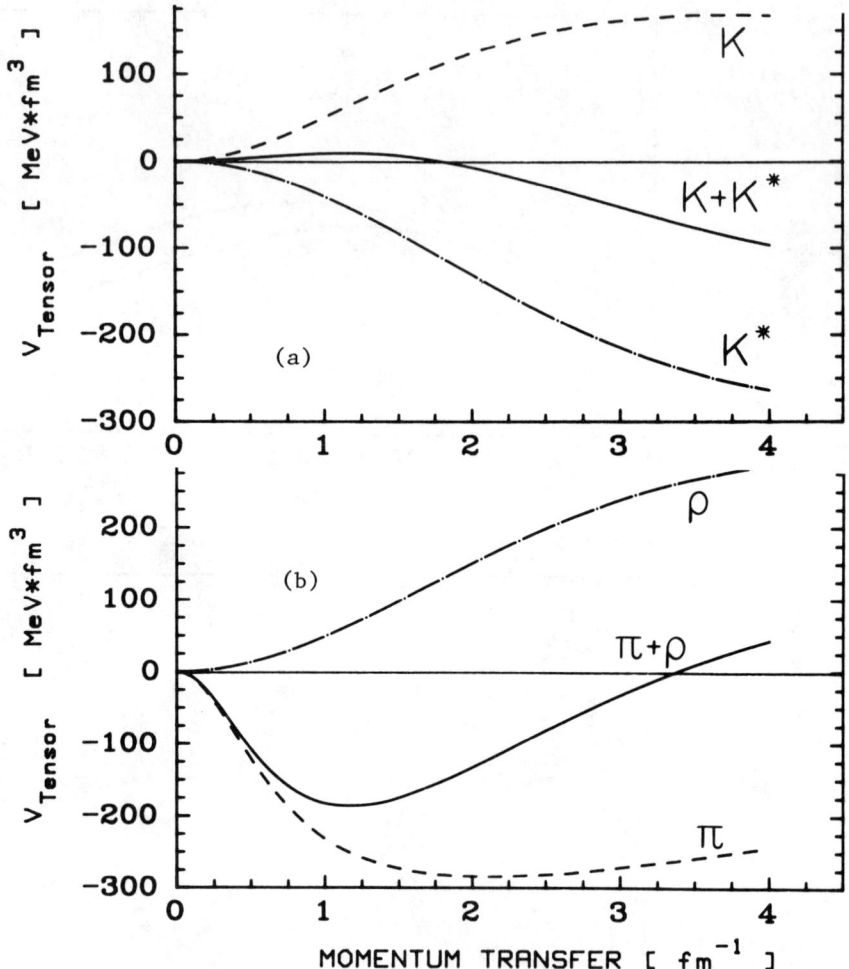

Fig. 6. Tensor potential a) based on K-, K*-exchange in the ΛN system; b) derived from π,ρ-exchange in the NN system.

NN system, see Fig. 6b. In fact, due to the larger mass of the kaon compared with the pion mass, the cancellation is much more nearly complete in the ΛN system. Consequently, the role of the tensor force in building up the binding energy of a Λ particle in nuclear matter should be very small.

G-MATRIX FOR THE Λ-N SYSTEM

As mentioned in the introduction, the effective interaction inside of a nucleus is related to the free potential via the Bethe-Goldstone equation (Eq. 1). In the following we report on results of such calculations for the ΛN-system. The G-matrix allows us to calculate, e.g., the binding energy of the Λ-particle in nuclear matter, the effective mass of the Λ, and the effective ΛN-interaction. A similar program has been carried out previously in Ref. 8 using the Nijmegen potential. In the present calculations we used the free ΛN-potential described in section 2 and emphasize especially the Landau parameters for the ΛN-system that are directly connected with the excited states of Λ-hypernuclei.

In analogy with the treatment of the free ΛN-interaction we introduce a direct (Fig. 1a) G-matrix, $G_{NΛ,NΛ}$, and an exchange (Fig. 1b) G-matrix, $G_{ΛN,NΛ}$. These G-matrices are defined by a coupled system of Bethe-Goldstone equations similar to the one shown in Eq. (1). In the ΛN case the Pauli-operator is restricted to nucleons only.

The single particle potential $U^Λ$ of a Λ in nuclear matter is calculated in the usual self-consistent way;

$$U^Λ(k_λ) = \sum_{n \in F} \left\{ \langle λn | G_{ΛN,ΛN}(E_λ+E_n) | Λn \rangle + \langle nΛ | G_{NΛ,ΛN}(E_Λ+E_n) | Λn \rangle \right\} \quad (4)$$

where the sum runs over all occupied nucleon states n. For the Fermi momentum $k_F=1.36$ [fm^{-1}] we obtain $m^*/m_Λ = 0.87$, $U_0^Λ = -28.5$ MeV. The parameters that have been used for the ΛN-interaction correspond to Nijmegen model D. We can compare our results with those in Ref. 8 where they obtained $m^*/m_Λ = 0.78$ and $U_0^Λ = -39.3$ MeV (Bandō et al.) and $U_0^Λ = -37.6$ MeV (Rosynek and Dabrowski). We believe that the differences are connected with the approximations made by the Nijmegen group which have been discussed in the previous section.

As in the NN-case we can obtain the operator structure of the ΛN effective interaction $G_{ΛN}$ directly from the G-matrix elements.[9] In the Landau limit, where one assumes that all particles in the initial and final states are on the Fermi sphere, the effective interaction $G_{ΛN}$ depends only on the (direct) momentum transfer q and the exchange momentum transfer Q. As the dependence on the latter is in general very weak, one obtains for the central part of the effective interaction the familiar form:

$$\tilde{G}_{ΛN}(q,k_F) = F(q,k_F) + g(q,k_F)\vec{\sigma}_1 \cdot \vec{\sigma}_2. \quad (5)$$

Here again we have to distinguish between the direct and the exchange contributions. In figures 8 and 9 we show the central components $F(q,k_F)$ and $g(q,k_F)$ in MeV-fm^3 as a function of the momentum transfer q in [fm^{-1}]. The Fermi momentum is $k_F=1.36$ fm^{-1}. The dashed curves indicate the 'bare potential'. The dashed-dotted curve is the result of a G-matrix calculation in

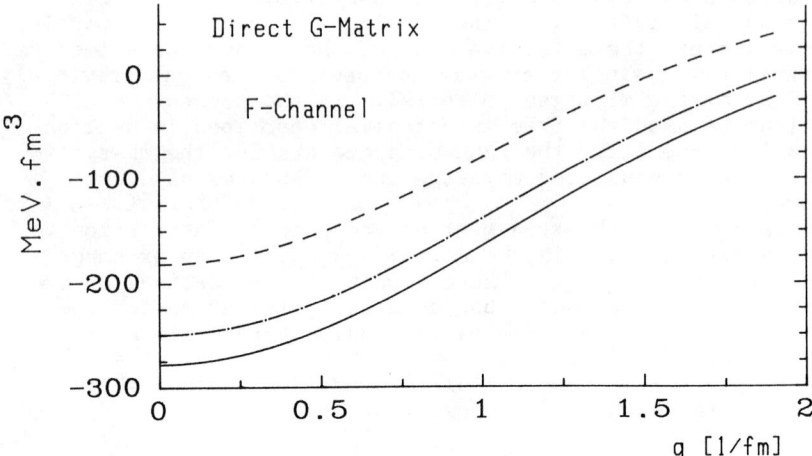

Fig. 7a) Spin independent, direct effective interaction in MeV fm³ as a function of the momentum transfer q. The dashed curve represents the free ΛN-interaction. The full and dashed-dotted lines are G-matrix results with and without the coupling between the direct and exchange channel, respectively.

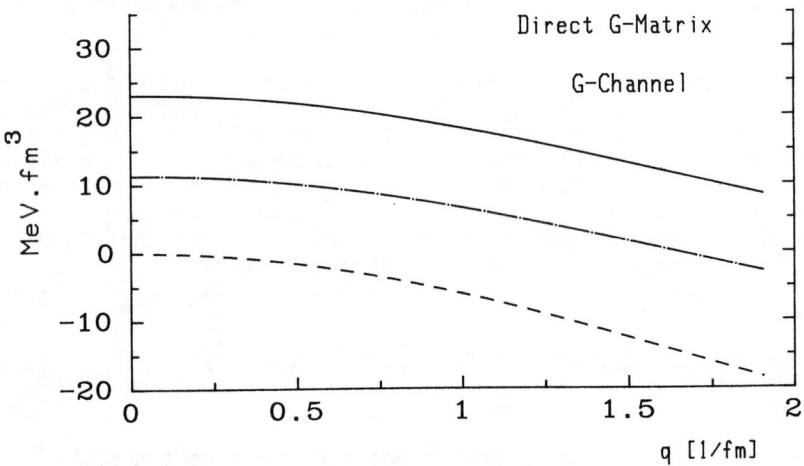

Fig. 7b) Same as in Fig. 7a for the spin-dependent direct effective interaction

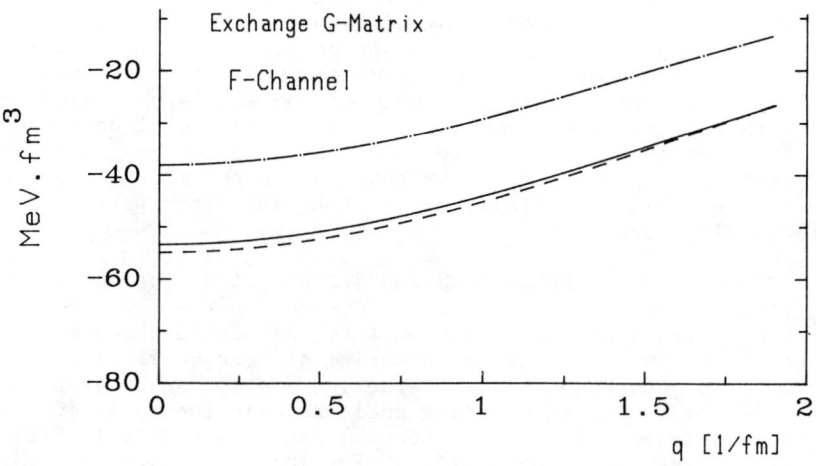

Fig. 8a) Same as in Fig. 7a, for the spin-independent exchange effective interaction.

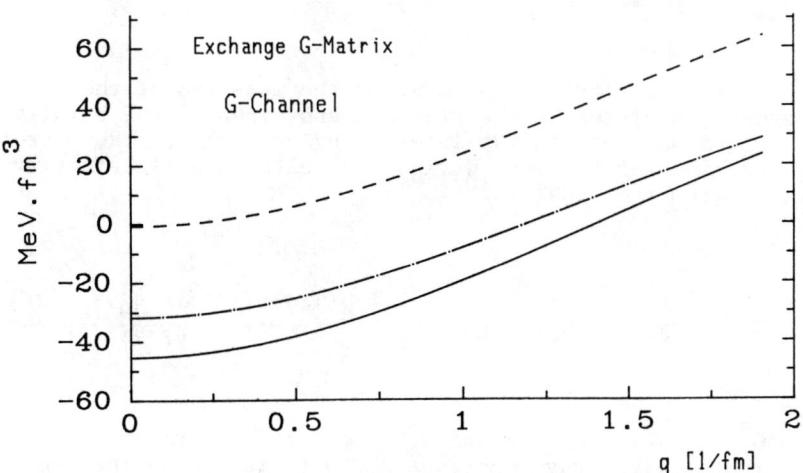

Fig. 8b) Same as in Fig. 7a, for the spin-dependent exchange effective interaction.

which we neglect the coupling between the direct-and exchange-G matrix. The full line corresponds to the results obtained from the solution of the coupled Bethe-Goldstone equations. The interaction in the direct spin-independent channel (Fig. 8a) is very attractive. However, from our experience in the NN-case[10] we expect large corrections in this channel due to the so-called induced interaction and relativistic effects that will both reduce the attraction appreciably. The strength of the force in all the other cases is weak and indicates that the energy of excited Λ-hypernuclei will essentially be given by the unperturbed single-particle energies.

CORRELATIONS IN HYPERNUCLEI

From conventional nuclear structure calculations one knows that the effective interaction in nuclei differs qualitatively from the free nucleon-nucleon interaction due to the large polarization effects of the surrounding nucleons. In the following, we compare correlated and uncorrelated NN- and ΛN-system's in free space as well as in nuclear matter. For this reason, we calculate the corresponding two-particle wavefunctions of the NN and ΛN system. For the free-scattering case, we obtained the partial waves $\psi_{LL'}^{JS}(r)$ from the R-matrix using the equation[10]:

$$\psi_{LL'}^{JS}(r) = j_L(q_0 r)\delta_{LL'} + P\int_0^\infty \frac{R_{LL'}^{JS}(Z|kq_0) j_{L'}(kr)}{Z - E(k)} k^2 dk. \quad (6)$$

Here q_0 is the starting momentum, k the momentum of the intermediate states, Z the starting energy and r the relative radial coordinate of the two interacting particles. The corresponding equation for two interacting particles in nuclear matter follows from the equation[10]:

$$\psi_{LL'}^{JS}(r) = j_L(q_0 r)\delta_{LL'} + \int_0^\infty \frac{Q(k) G_{LL'}^{JS}(Z|k q_0) j_{L'}(kr)}{Z - E(k)} k^2 dk \quad (7)$$

here $G_{LL'}^{JS}$ is the G-matrix and Q(k) the Pauli operator. The relative two-particle wavefunctions shown in the following are obtained by summing over the L':

$$\psi_L^{JS}(r) = \sum_{L'=J-S}^{J+S} \psi_{L'L}^{JS}(r). \quad (8)$$

Figures 9a and 10a show the two-particle wavefunction as a function of the relative distance between the two particles for the free scattering case of ΛN-system in the relative $^3S_1$ and $^1S_0$ states. For comparison, we also show in all the figures the uncorrelated plane wave (dashed curve). In both cases the forces are attractive and give rise to a moderate phase shift. One also

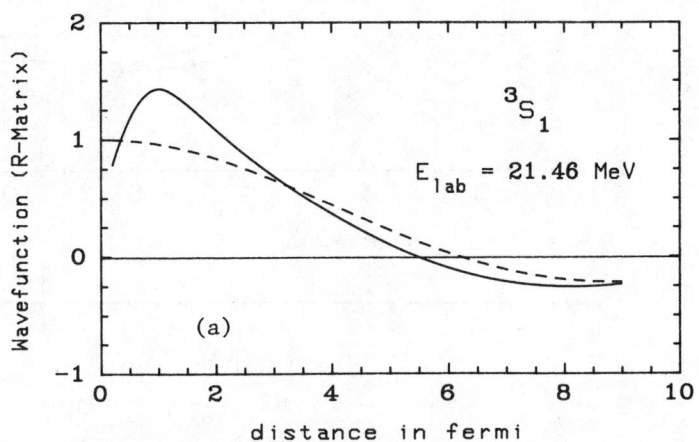

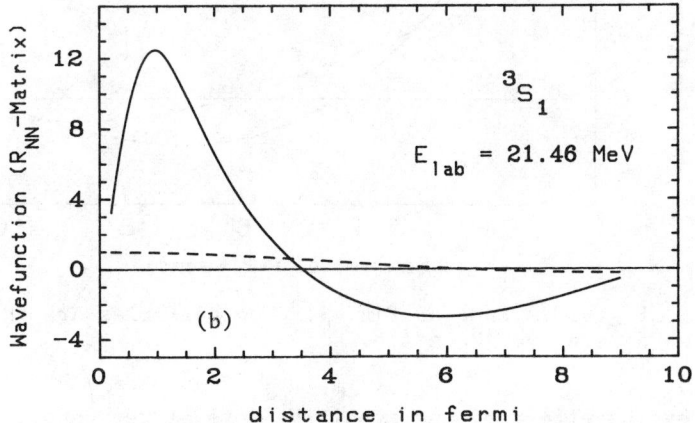

Fig. 9. Two-particle relative wave function $\psi_L^{JS}(r)$ of the ΛN (a) and NN (b) system in a relative $^3S_1$ state in free space (Eq. (6)). For comparison also the unperturbed wavefunction $j_0(q_0 r)$ (dashed line) is given.

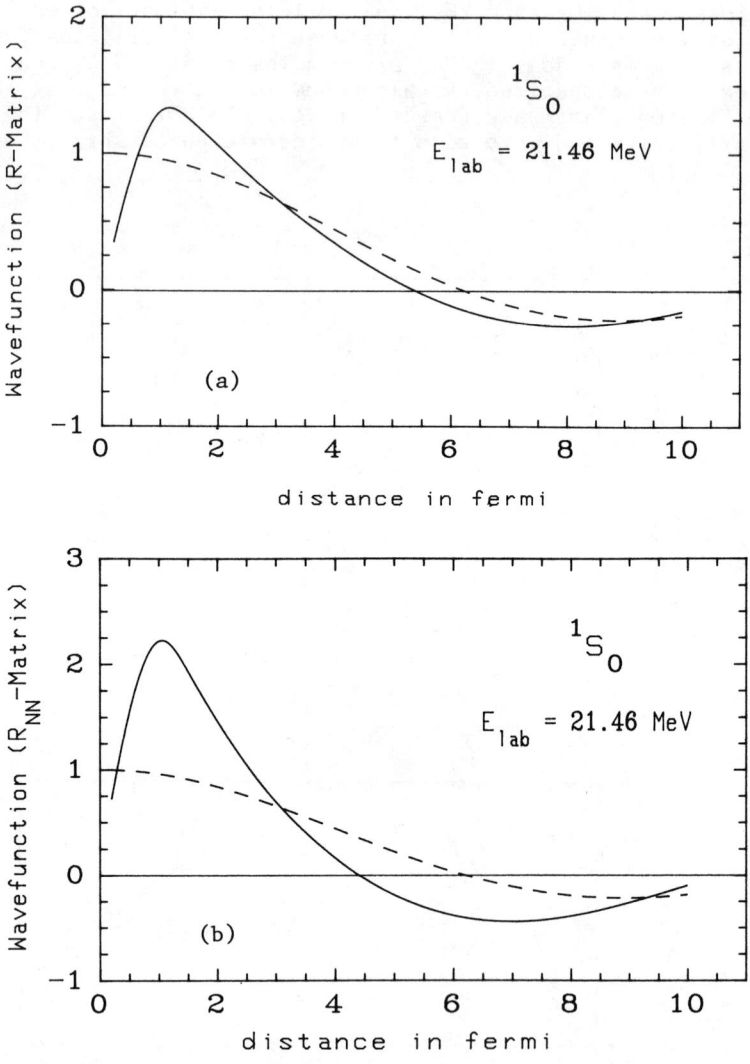

Fig. 10. Same as Fig. 9. The particles are in a relative $^1S_0$ state.

observes the effect of the repulsive core at very short distances that prevents the particles from overlapping. In figures 9b and 10b, corresponding results for the NN-system are shown. The huge difference between the two systems in the $^3S_1$ state is due to the difference in the tensor force. The tensor interaction in the free NN system is strong and attractive whereas in the ΛN system it is very weak (see the discussion in Section 2).

The correlated two-particle wavefunctions of the NN and ΛN system in nuclear matter are shown in figures 11 and 12. The starting momentum of q_0 = 100 MeV corresponds to the starting energy in the free case. In Figures 11a and 12a, we present results for the ΛN system that can be compared with the corresponding results for the NN system shown in Figures 11b and 12b. Here we observe that in nuclear matter the difference between the correlated two-particle wavefunctions of the NN and ΛN system are

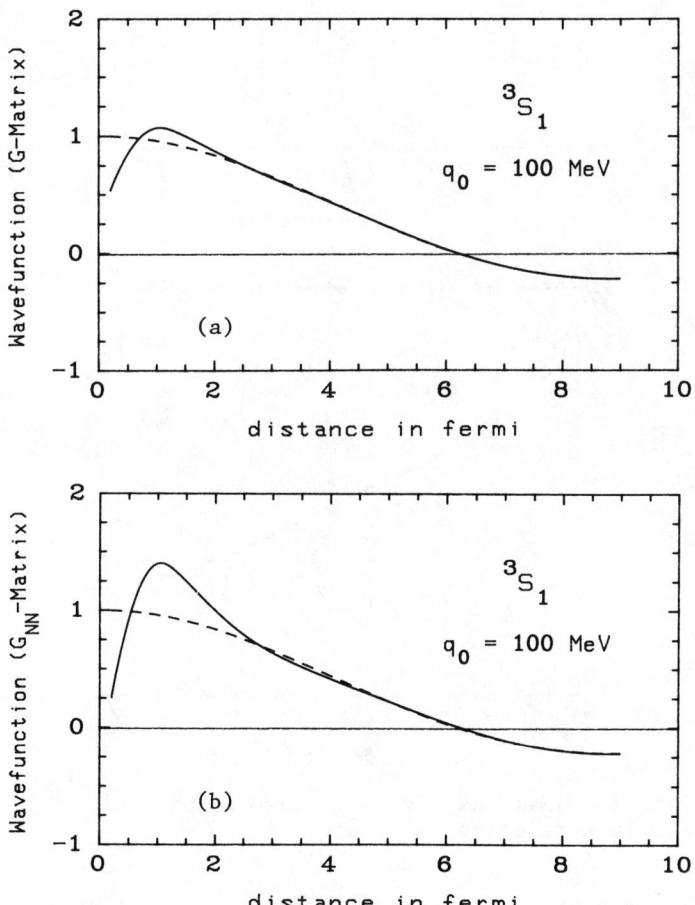

Fig. 11. Two-particle relative wave functions of the ΛN (a) and NN (b) system in nuclear matter (Eq.(7)). The particles are in a relative $^3S_1$ state. The dashed line denotes the unperturbed wave function $j_0(q_0 r)$.

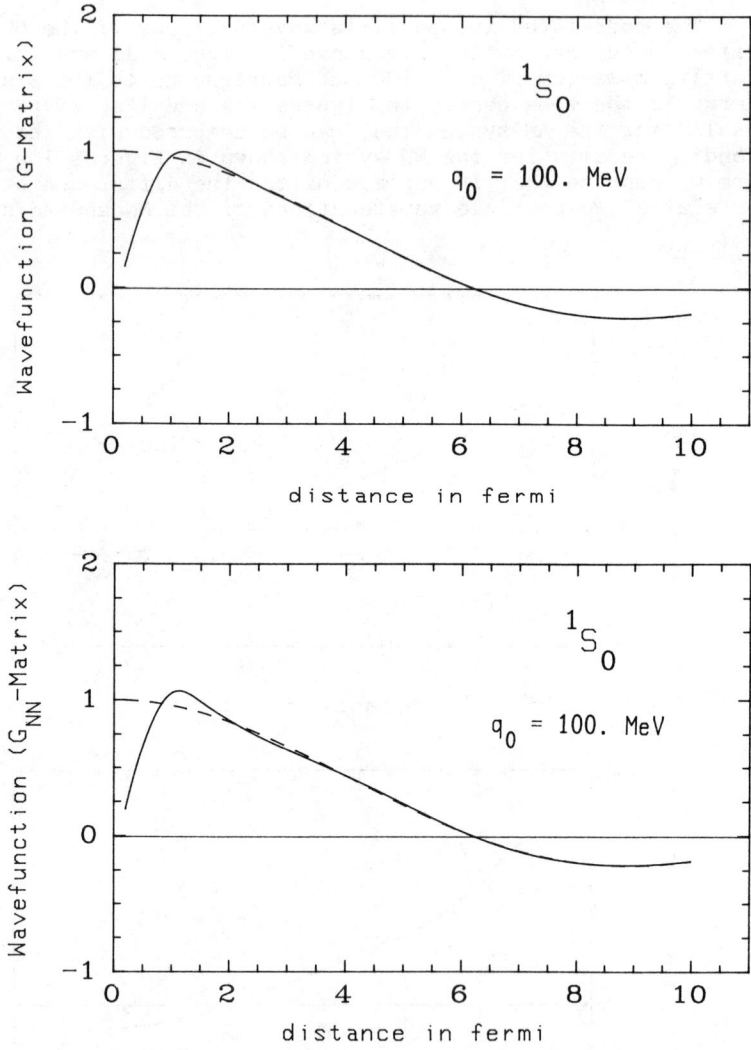

Fig. 12. Same as Fig. 11. The two particles are in a relative $^1S_0$ state.

much smaller compared with free scattering. In all cases, one notices the effect of an attractive interaction which is somewhat larger for the NN system. However, one also sees that at small distances the NN- and ΛN-system behave very similarly. The much stronger free NN interaction obviously gives rise to much larger polarization effects.

SUMMARY

The Bonn potential has been generalized to include strangeness exchange processes. This model has been applied to the ΛN interaction. All the calculations are performed in momentum space which allows one to include nonlocalities as well as relativistic effect. Using this ΛN interaction as a starting point, we have calculated generalized Brueckner G-matrices. From these G-matrices, we derived the effective ΛN interaction in nuclear matter. The present results indicate that the effective ΛN interaction is weak and that there probably exist no collective states in Λ-hypernuclei. We have also compared the two-particle correlation wavefunction of the NN- and ΛN-systems. There are large differences in free space, whereas in nuclear matter, the two systems behave quite similarly. The largest difference results from the tensor forces. In free space the NN tensor force is strong and attractive, whereas the tensor force in the ΛN system is weak. In nuclear matter the strong NN-tensor force is appreciably reduced due to polarization effects. Consequently, the large differences between the NN and ΛN system are reduced.

ACKNOWLEDGMENT

We thank Gerry Brown, Bill Gibbs, Ben Gibson, Mikkel Johnson and Frank Tabakin for valuable discussions. This work has been supported in part by NATO Grant RG 85/0093.

REFERENCES

1. W. G. Love and J. Speth, Comments Nucl. Part. Phys. 14, 185 (1985).
2. M. R. Anastasio, L. S. Celenzar, W. S. Pong, and C. M. Shakin, Phys. Rep. 100, 327 (1983).
3. K. Nakayama, S. Krewald, and J. Speth, Phys. Lett. B 145B, 310 (1984).
4. R. Machleidt, Lecture Notes in Physics 197, 352 (1984); Ch. Elster, K. Holinde, and R. Machleidt, preprint (to be published in Phys. Rep.).
5. R. Büttgen, diploma thesis, University of Bonn; Jül-Spez 314 (1985). B. Holzenkamp, diploma thesis, University of Bonn, Jül-Spez 315 (1986); R. Büttgen, K. Holinde and J. Speth, Phys. Lett. 163B, 305 (1985)
6. M. M. Nagels, Baryon-Baryon Scattering in a One-Boson Exchange Potential Model, Proefschrift (1975).
7. K. Holinde and H. Mundelius, Nucl. Phys. A364, 365 (1981).
8. H. Bandō et al., Progress of Theoretical Physics (Supplement) No. 81 (1985); J. Rosynek and J. Dabrowski, Phys. Rev. 20 C 1612 (1979)
9. K. Nakayama, S. Krewald, J. Speth, and W. G. Love, Nucl. Phys. A431, 419 (1984).
10. K. Nakayama, S. Krewald, and J. Speth, Phys. Lett. 145B, 310 (1984)
11. M. I. Haftel and F. Tabakin, Nucl. Phys. A158, 1 (1970).

SESSION G

OUTLOOK

SUMMARY AND OUTLOOK*

T. W. Donnelly
Massachusetts Institute of Technology, Cambridge, MA 02139

Various themes can be found running through this workshop. For example, its title suggests three points of focus in nuclear structure studies:
- High spin.
- High excitation.
- High momentum transfer.

The various detailed sessions at the Workshop provided another decomposition, this time into the categories
- Giant resonances.
- Stretched configurations.
- Transfer and knockout reactions.
- Shell model calculations.
- Non-nucleonic degrees of freedom.

Still another theme involves the interplay amongst the various classes of experiments, for instance as indicated in Fig. 1. On the hadronic side we have had a long and fruitful history of scattering, transfer, pickup, stripping,... reaction studies. On the electromagnetic side for single-arm (e, e') studies we also have had a long history (for example, the stretched particle-hole states were found using electron scattering in 1968). The advancements here are being driven by the new facilities, so that the coincidence $(e, e'x)$ reactions are becoming a fine tool as well. For me, the essence of the workshop was seeing the exploitation of all the different probes for a common purpose (*i.e.* to undertake the sudies implied in the Workshop title). The action was on the various links in Fig. 1, using densities from (e, e') studies as input to (h, h') analyses, for instance, as discussed by Kelly, or making use of the final-state similarities and differences in hadronic and electromagnetic initiated reactions to obtain spectroscopic information.

Furthermore, several themes involving theoretical issues could be seen to provide common threads throughout the Workshop:
- Relativity
  - Dirac phenomenology
  - Perturbation theory (as discussed in the opening talk by G. Brown)
- Shell model.
- $NN/\Lambda N$ interactions.

---

* This work was supported in part by funds provided by the U. S. Department of Energy (D.O.E.) under contract #DE-AC02-67ER03069.

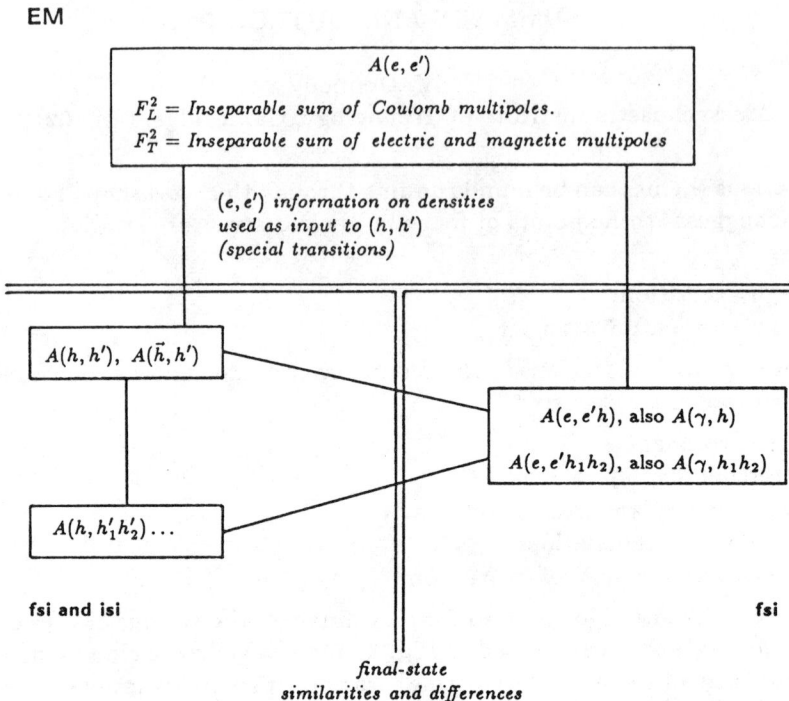

Fig. 1: Electromagnetic and hadronic interactions with nuclei (h = hadron).

- N, π substructures.
- Quarks.

Clearly this Workshop spanned a very broad range of topics in nuclear physics. In these discussions, I shall not attempt to summarize everything that was covered, but only to pick out a few things for further comment which caught my eye before concluding with a brief outlook towards some of the new initiatives for the future.

In the session on <u>Stretched Configurations</u> the central problem of the strength (*i.e.* the "missing" strength) was addresssed at the outset in the talk by Papanicolas who proposes that the quenching is related to the fractional occupancies in the relevant orbitals. Opposing points of view were also expressed; for example, later is discussing transfer and knockout reactions and the determination of absolute spectroscopic factors, Wagner concluded that the quenching of the 3s-proton contribution in the lead region is not the predicted effect of short-range correlations and tensor forces, but rather the result of configuration mixing between low-lying states. This interesting issue is apparently not yet resolved.

Concerning the relationships between the strengths found in (e,e') and (p,p') studies, Lindgren's presentation of results for the former where MEC effects were included for stretched configurations caught my eye. Perhaps now the electromagnetic and hadronic determinations of the quenching factors have been brought into line.

Anderson discussed the inter-relationships among (p,p'), (p,n) and (e,e') studies and highlighted the merits of (p,n) reactions as an isovector, non-normal-parity probe of stretched configurations. Of particular note were the strength differences seen for $0\hbar\omega$ excitations when compared with the more familiar $1\hbar\omega$ excitations. Nann discussed more complex two-nucleon transfers and focussed on studies of (d,α) and $(\vec{d},\alpha)$ reactions for fp-shell nuclei as a means of isolating specific J^π stretched configurations.

Taking stretched configurations to an extreme, Jacobs discussed (p,π) reactions and their selectivity in populating high-spin $2p-1h$ states progressing up to $19/2^-$ in ^{48}Ca and $37/2^+$ in ^{208}Pb! He at least of all the speakers satisfied all three Workshop topic criteria in showing how novel nuclear structure information could be found involving high spin, excitation and momentum transfer.

Turning to the more theoretical side, we had a stimulating session on current progress in Shell Model Calculations. I for one will be very interested to see predictions for form factors, cross sections, etc. made using these large-scale calculations. For example, Glaudemans discussed the importance of $2\hbar\omega$ configurations in light nuclei and displayed results with signficant $2p-1f$-shell admixtures in nominally $1p$-shell nuclei. Hints of these are already present from electron scattering at high momentum transfer (see Fig. 2). Glaudemans did sound a cautionary note, however, when he remarked that the calculated electromagnetic moments are found to be very sensitive to the choice of the effective interaction and how it is difficult to obtain a consistent description of both the predominant $1p$-shell states and the more complex intruder states.

The other speakers in this session helped to provide a picture of the state-of-the-art of shell model calculations. Wildenthal discussed many-$\hbar\omega$ calculations in the $d_{3/2}\ f_{7/2}$ model space and showed how simple "shell model" features as well as "deformed" features can emerge. Daehnick discussed the regularities and trends found in extracting residual interaction matrix elements. Schmid presented results using successively more complicated truncation schemes using projected HFB quasiparticle determinants as building blocks. Appropriately (it was near Halloween), he discussed MONSTER's, VAMPIR's and EXCITED VAMPIR's. Furnstahl brought in relativity in reporting progress in relativistic RPA calculations.

The session on Giant Resonances continued the theme of bringing different probes to bear on common problems in nuclear structure studies. Knöpfle showed some of the beautiful coincidence reaction $(^{28}Si(e,e'c)$, $c = p$ or $\alpha)$ data coming out of Mainz. These coupled with $(\alpha,\alpha'c)$ data provide evidence for large isovector $E2/E0$ contributions in the region of isoscalar giant

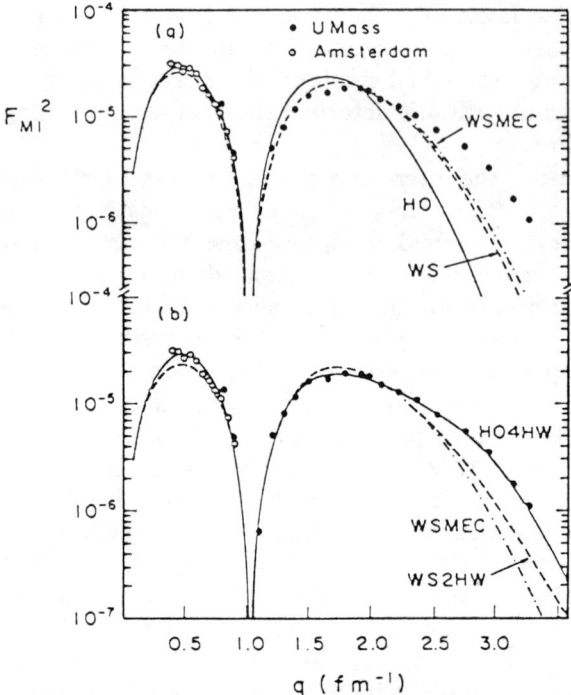

Fig. 2: Transverse form factor data and comparison with p-shell interpretations taken from R. S. Hicks et al., Phys. Rev. C26, 339 (1982). (a) Results of attempts to fit the data with a generalized p-shell model. HO, harmonic oscillator, $b = 1.73$ fm and $a = 0.85$; WS, Woods-Saxon, $R_0 = 1.17$ fm and $a = 48$; WSMEC, Woods-Saxon including one-pion exchange, $R_0 = 1.24$ fm and $a = 0.31$. All calculations reproduce the measured dipole magnetic moment. (b) Phonomenological evaluation of possible core polarization contributions. WSMEC, same as in (a); HO4HW, $4\hbar\omega$ harmonic oscillator fit, $b = 1.61$ fm; WS2HW, fit including $2\hbar\omega$ $(2p_{1/2}1p_{1/2}^{-1})_v$ matrix element calculated using Woods-Saxon wave functions. All curves include one-pion exchange current contributions evaluated for the 1p shell only.

$E2/E0$ resonances. Given the new generation of cw electron accelerators we are likely to see more examples where such inter-probe studies can be performed. Perhaps the main concern at present is: Where is the theory for such giant resonance studies? (See below for further comments on this point.)

Van der Woude discussed the nature of the decay of isoscalar giant resonances and how coincidence reactions may be used to get rid of experimental backgrounds (especially for 0° scattering of hadrons and for low-energy electron scattering experiments). Seestrom-Morris showed some of the current results for pion scattering, where the isospin selectivity for this probe can be used in identifying, for example, isovector quadupole strength in the giant resonance region.

In the session on Transfer and Knockout Reactions we heard in the talks by deWitt Huberts and Wagner how complementary information can be extracted using $(d,{}^3He)$ and $(e,e'p)$ reactions. Such detailed inter-comparisons have only been made possible with the recent availability of superb, relatively high resolution (100 – 150 keV) knockout data from NIKHEF. The different radial sensitivities of the two probes was graphically demonstrated by deWitt Huberts – to my mind such analyses highlight in the clearest terms why both hadronic and electromagnetic probes should be brought to bear on common problems in studies of nuclear structue.

The other speakers in this session discussed related subjects. Chant focussed mainly on knockout reactions using hadronic probes, $(p,2p)$, $(p,p'\alpha)$,..., but also touched upon hadronic/electromagnetic inter-comparisons, e.g. $(p,p'c)$ versus $(e,e'c)$, where $c=d$ or α. Gales discussed one- and two-nucleon transfer reactions at high incident energy for studies of high-spin inner and outer subshells (up to $1k_{17/2}$!). Ellegaard was one of the few speakers to refer to Δ excitation, in this case via charge exchange reactions at intermediate energies.

Macfarlane's comments at the end of the session appropriately summarized several main points: (1) Much of what we heard goes back to one basic relationship, $qR \sim \ell_t$, that is, high-q studies tell us about large transferred angular momentum (e.g. (p,π) reactions); (2) The spectroscopic studies were presented in terms of level occupancies, but "the principle quantum number is not an observable"; (3) There is no consistent overall reaction theory capable of handling electron scattering, 1-nucleon transfer, 2-nucleon transfer, ..., that is, "reaction theory is lagging behind experiment".

To take what we saw from NIKHEF in the $(e,e'p)$ reaction one step further, let me mention some of the results coming out of Bates on the same reaction but at higher energies (see Fig. 3). While the "$1p$" shell knockout is reminiscent of the "single-particle" structure seen at lower energies, the strength at higher missing energies is not at all understood (c.f. Macfarlane's point #3 above).

The final session on Non – Nucleonic Degrees of Freedom spanned a wide range of loosely related fundamental topics. Weise discussed the substructure of the nucleon and pion (the RPA pion) in terms of a model effective interaction which respects chiral symmetry, confines at large distances and is weak at short distances. Kisslinger discussed the electrodisintegration of the deuteron and the three-body charge form factors within the context of his hybrid quark hadron model and stressed the fact that, given the dominance of the impulse and pion pair currents in ${}^2H(e,e'p)$ which he finds, this is a good testing ground for a microscopic study of pion exchange effects, but not necessarily for short-range (e.g. q^6) phenomena. Frois and Sick showed some of the beautiful data coming out of the medium-to-high energy nuclear facilities at Saclay (ALS and Saturne) and SLAC (NPAS). The three-body ground-state form factors

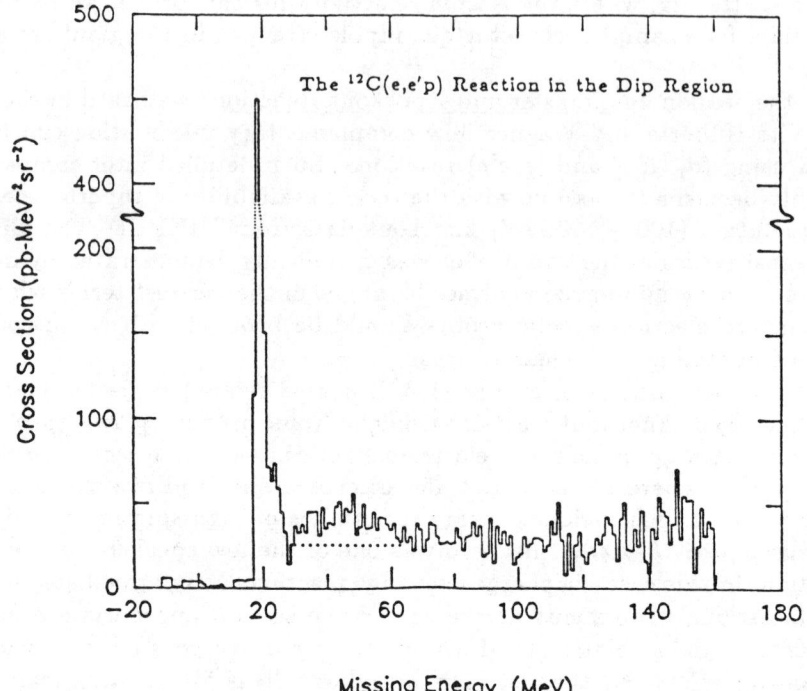

Fig. 3: $^{12}C(e, e'p)$ cross section taken from R. W. Lourie *et al.*, (Bates preprint submitted for publication). The sharp peak at about 20 MeV missing energy corresponds to $1p$ shell ejection. The strength above the horizontal dotted line is interpreted as "1s" shell knockout. No quantitative explanation exists for the remaining high missing energy strength.

(^3He and ^3H) have been a highlight of electron scattering experiments in the past few years. Frois showed the recent tritium results from Saclay together with state-of-the-art theory stressing the sensitivity to the choice of single-nucleon form factor (G_E or F_1). Sick continued with more discussion of the $A = 3$ systems and then showed some of the first radiatively-corrected SLAC data on y-scaling in complex nuclei. Finally, Speth discussed the recent work proceeding along the same lines followed in obtaining the Bonn NN potential, now extending these ideas to include strange particles, *e.g.* in KN or ΛN interactions.

During the course of the Workshop we heard about progress in topics extending from multi-nucleon heavy-ion transfer reactions to the EMC effect and quarks in nuclei. Providing any kind of comprehensive outlook spanning such a diffuse set of topics is not possible in a limited amount of time and so I have chosen to discuss only two specific future initiatives which are also currently of interest personally. Both are electromagnetic reaction studies which complement other work with hadronic probes and both depend on new experimental initiatives to be realized.

POLARIZATION DEGREES OF FREEDOM

With the availability of polarized beams and polarized targets (especially polarized targets internal to rings such as at IUCF or planned at Bates) the inter-probe comparisons can be carried to a finer level. The various classes of experiments summarized in Fig. 1 can now be extended as in Fig. 4. To take just one example of the new possibilities here, consider reactions $\vec{A}(e,e')$ and $\vec{A}(\vec{e},e')$ (together with the related electromagnetic process, $A(e,e'\gamma)$; see work already being done at Illinois). The new information available with polarized targets consists of combinations of the basic electromagnetic form factors in forms other than just the inseparable sums which occur in $A(e,e')$ studies. Thus, for a given nuclear transition one in effect has a "multipole meter". In using densities deduced from electron scattering for input to hadron scattering analyses, this addition knowledge will permit a wider class of transitions to be used.

Of course, the separated electromagnetic multipole information is interesting itself. For example, several "three-star" experiments naturally come to mind:

- $\vec{N}(\vec{e},e')N$ to extract G_{E_p} and especially G_{E_n}.
- $\vec{N}(\vec{e},e')\Delta$ to isolate the $C2/E2$ contributions from the dominant M1 multipole (to study the deformed Δ).
- $^2\vec{H}(e,e')^2H$ to separate the deuteron ground-state $C0$, $M1$ and $C2$ multipoles.

Some information of this last type (actually done as a tensor recoil polarization measurement) is already available (see Fig. 5); clearly carrying this to higher-q is of interest.

Even without having polarized targets there are new possibilities with polarized beams. Experiments of this sort have been performed with hadrons and yet we are just on the the threshold of being able to exploit these new possibilities with electrons. Specialized experiments of the $A(\vec{e},e')$ class to isolate parity-violating effects will soon be undertaken (a few have already). Furthermore, there are experiments which involve polarized electrons and are of interest even at the level of parity conservation. I would like to highlight just one of these as it is relevant to some of the Workshop discussions of giant resonance excitation.

Without having polarized electrons, the reaction $A(e,e'x)$ can be analyzed

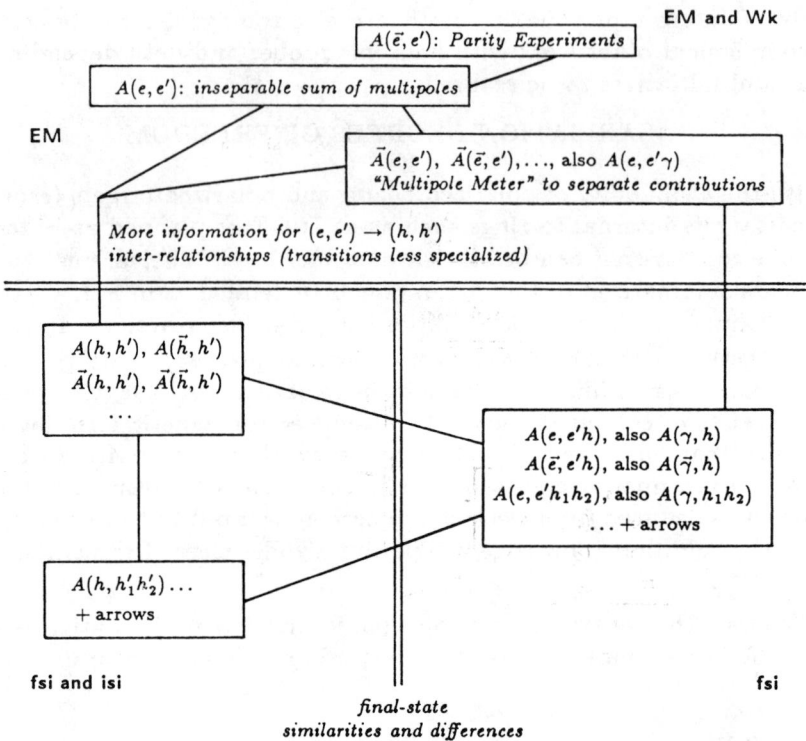

Fig. 4: Electroweak and hadronic interactions with nuclei including polarizations (h = hadron).

in terms of four basic response functions:

$$W_L \sim |L|^2$$

$$W_T \sim |T|^2$$

$$W_{TT} \sim \text{Re}(T^*_{(+1)} T_{(-)})$$

$$W_{TL} \sim \text{Re}(T^* L) \ ,$$

where "L" refers to longitudinal or charge projections of the electromagnetic current of the nuclear systems and "T" refers to transverse projections. The four responses can be separated by making a "Super Rosenbluth" decomposition, varying the kinematics in the reaction. Having polarized electrons adds

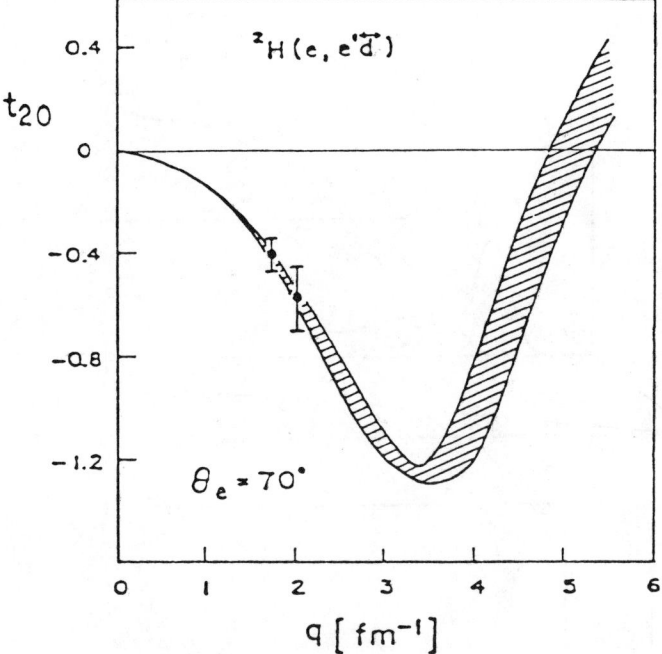

Fig. 5: Deuteron tensor polarization t_{20} including data from Bates (M. E. Schulze et al., Phys. Rev. Lett. **52**, 597 (1984)) and a shaded region indicating the spread obtained using representative meson-baryon-based calculations.

to these four a new fifth response function of the form

$$W_{TL'} \sim \text{Im}(T^*L) \ ,$$

which can be isolated by flipping the incident electron's helicity. This fifth response has some special properties: When a single doorway dominates the reaction we will have $L \sim |L|e^{i\delta}$ and $T \sim |T|e^{i\delta}$ with the same phase δ. The $T^*L = |T||L| = $ real and $W_{TL'}$ vanishes while W_{TL}, etc. are in general non-zero. Likewise, when final-state interactions are negligible, the phase shifts are negligible also and so again $W_{TL'} = 0$ (but not in general W_{TL}, etc.) Thus $W_{TL'}$ is a special type of interference response which, when non-zero, signals interesting effects are present. For instance, referring to Fig. 6, in the giant resonance region $W_{TL'}$ does not occur when only a given isolated resonance is present, but rather is sensitized to interferences between overlapping resonances or to interferences between a given resonance and the "direct" or "non-resonant" contributions. Work is in progress by Co' and Krewald at

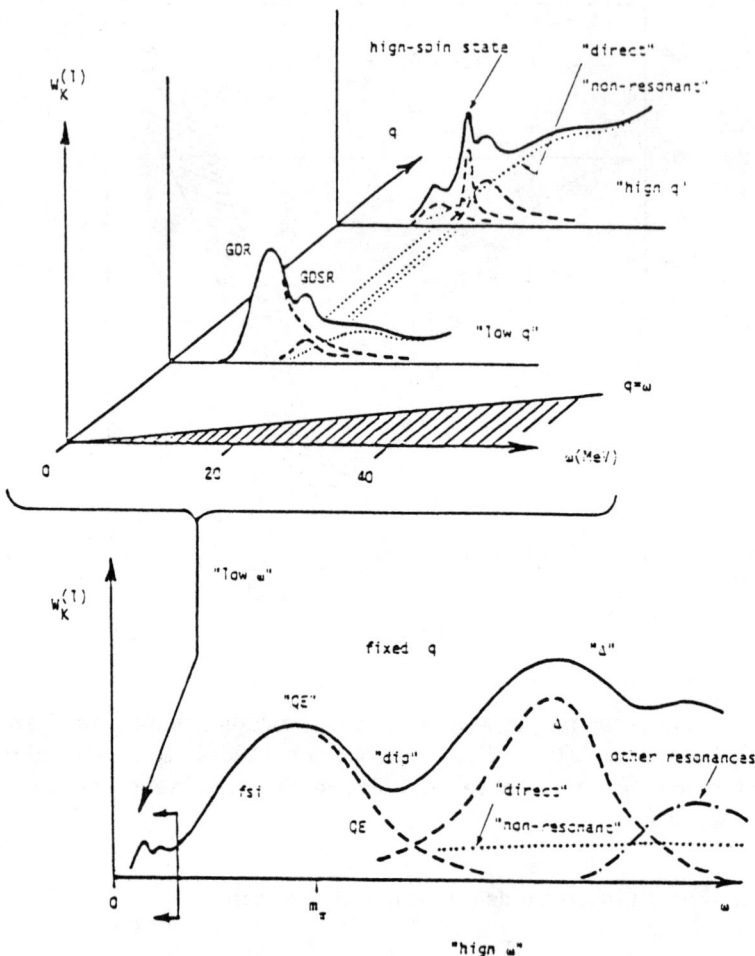

Fig. 6: Schematic representation of a typical electromagnetic response function in the $q\omega$-plane.

Jülich who are using a large scale continuum RPA code to explore all five response functions (see previous comment on reaction mechanism theory – here some significant progress is being made). Other kinematic regions remain to be studied (certainly the "Δ-region" should be very interesting, see Fig. 6) and so far no experiments have been undertaken.

PHOTO- AND ELECTRO-PRODUCTION OF KAONS
AND THE STUDY OF HYPERNUCLEI

Another area where hadronic and electromagnetic probes may play complementary roles is in the study of hypernuclei. To date most work has been done using the strangeness-exchange reaction $^A X(K^-, \pi^-)^A_\Lambda Y$ (and to a small

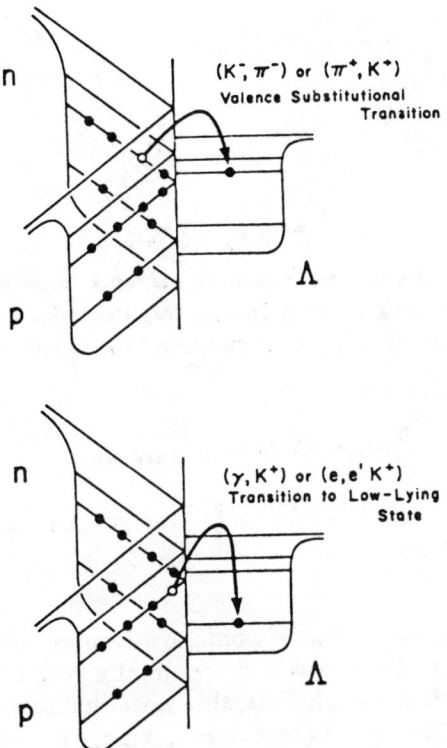

Fig. 7: Nucleus-to-hypernucleus transitions using hadronic or electromagnetic reactions.

extent $^A X(\pi^+, K^+)^A_\Lambda Y$), with similar reactions for Σ hyperons. A schematic picture of this process is shown in the upper part of Fig. 7. The K^- and π^- are both strongly interacting and so the reaction is predominantly surface-peaked and favors the valence sustitutional transitions as indicated. In contrast, the electromagnetic processes, $^A X(\gamma, K^+)^A_\Lambda Y$ and $^A X(e, e'K^+)^A_\Lambda Y$, involve only feeble interactions in the initial state and, because of the relative weakness of $K^+ N$ interactions compared to $K^- N$ interactions or πN interactions, very little asborption in the final state. Consequently, transitions such as indicated in the lower part of Fig. 7, involving a direct step to deep-lying hypernuclear configurations (including the ground state) can be studied. An example is given in Fig. 8; naturally in such cases a large angular momentum change occurs, for instance, we can consider

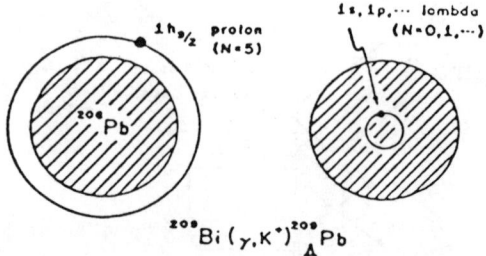

Fig. 8: Schematic representation of extreme nucleus-to-hypernucleus transitions proceeding directly from a high principal quantum number configuration (here $N = 5$) to deep-lying final states ($N = 0, 1, \ldots$).

$$[(1s_{1/2})_\Lambda (1h_{9/2})_p^{-1}]_{M4, E5}$$

$$[(1p_{3/2})_\Lambda (1h_{9/2})_p^{-1}]_{M3, E4, M5, E6}$$

$$\vdots$$

Just as with the nucleonic stretched configurations we have been hearing about in the Workshop, these Λp^{-1} states are optimally seen at intermediate to high values of momentum transfer. In fact, this is well-matched to conditions which should exist at high energy, high current, high duty factor nuclear physics facilities such as CEBAF.

In summary, returning to Fig. 1, we see at this point in time, as exemplified by the talks at this Workshop, that there is vigorous activity in all the areas with each of the types of probes and increasingly there are interprobe comparisons being undertaken to extract complementary information on nuclear structure. As I have tried to indicate in the present outlook using two specific new initiatives, I feel that we will see even more of this in the not-too-distant future.

Finally, to conclude, I would like to thank the organizers for another stimulating Workshop in this series and hope that the future holds more of the same hosted by IUCF which will meet the high standards set by this one.

CONFERENCE PARTICIPANTS

Jeannette Adams, University of Maryland, College Park, MD 20742, USA
Peter Alons, Indiana University, Bloomington, IN 47405, USA
G. Anagnostatos, Demokritos Athens, Attiki-Athens, 15310 Greece
Bryon Anderson, Kent State University, Kent, OH 44242, USA
Mohamad Asmar, Indiana University, Bloomington, IN 47405, USA
Norman Austern, University of Pittsburgh, Pittsburgh, PA 15260, USA
Salah Aziz, Indiana University, Bloomington, IN 47405, USA
Andrew Bacher, Indiana University, Bloomington, IN 47405, USA
Robert Bent, Indiana University, Bloomington, IN 47405, USA
Leslie Bland, Indiana University, Bloomington, IN 47405, USA
B. Alex Brown, Michigan State University, East Lansing, MI 48824, USA
Gerald Brown, SUNY, Stony Brook, NY 11794, USA
Jeremy Brown, Princeton University, Princeton, NJ 08544, USA
Barry Burks, Oak Ridge National Lab, Oak Ridge, TN 37831, USA
Roger Byrd, Indiana University, Bloomington, IN 47405, USA
John Calarco, University of New Hampshire, Durham, NH 03824, USA
Zong-Jian Cao, Indiana University, Bloomington, IN 47405, USA
Larry Cardman, University of Illinois, Urbana, IL 61801, USA
Thomas Carey, Los Alamos National Lab, Los Alamos, NM 87544, USA
James Carr, Florida State University, Tallahassee, FL 32306, USA
Nicholas Chant, University of Maryland, College Park, MD 20742, USA
Quan Chen, Indiana University, Bloomington, IN 47405, USA
Heinz Clement, Universitat Tubingen, D-7400 Tubingen, West Germany
Giampaolo Co', University of Illinois, Urbana, IL 61801, USA
Joseph Cohen, Indiana University, Bloomington, IN 47405, USA
David Cook, University of Minnesota, Minneapolis, MN 55455, USA
Gary Crawley, Michigan State University, East Lansing, MI 48824, USA
Vernon Cupps, II, Indiana University, Bloomington, IN 47405, USA
Wilfried Daehnick, Univ. of Pittsburgh, Pittsburgh, PA 15260, USA
John Dawson, Univ. of New Hampshire, Durham, New Hampshire 03824, USA
Peter de Witt Huberts, NIKHEF-K, 1009 AJ Amsterdam, The Netherlands
J.W.A. den Herder, NIKHEF-K, Amsterdam 1009 AJ, The Netherlands
C. Djalali, Michigan State University, East Lansing, MI 48824, USA
T.W. Donnelly, Massachusetts Inst. of Tech., Cambridge, MA 02139, USA
Clive Ellegaard, Niels Bohr Inst., DK-2100 Copenhagen, Denmark
Guy Emery, Indiana University, Bloomington, IN 47405, USA
Mirek Fatyga, Indiana University, Bloomington, IN 47405, USA
H. Terry Fortune, Univ. of Pennsylvania, Philadelphia, PA 19174, USA
Charles Foster, Indiana University, Bloomington, IN 47405, USA
Walter Fox, Indiana University, Bloomington, IN 47405, USA
Dennis Friesel, Indiana University, Bloomington, IN 47405, USA
Bernard Frois, CEN Saclay, Saclay, France
Richard Furnstahl, Indiana University, Bloomington, IN 47405, USA
Sydney Gales, Institut de Physique Nucleaire, F-91406 Orsay, France
Anibel Gattone, Indiana University, Bloomington, IN 47405, USA
Donald Geesaman, Argonne National Lab, Argonne, IL 60439, USA
James Gering, Indiana University, Bloomington, IN 47405, USA
Piet W.M. Glaudemans, Univ. of Utrecht, Utrecht, The Netherlands
Charles Glover, Oak Ridge National Lab, Oak Ridge, TN 37830, USA

Charles Goodman, Indiana University, Bloomington, IN 47405, USA
John Goodwin, Indiana University, Bloomington, IN 47405, USA
Peter Grabmayr, University of Tubingen, Tubingen 7400, West Germany
William Hersman, University of New Hampshire, Durham, NH 03824, USA
Ken Hicks, TRIUMF, Vancouver, B.C., V6T 2A3 Canada
Hsiao-Hua Hsu, Los Alamos National Lab, Los Alamos, NM 87545, USA
W-Y. Pauchy Hwang, Indiana University, Bloomington, IN 47405, USA
Will Jacobs, Indiana University, Bloomington, IN 47405, USA
Ronald Johnson, University of Surrey, United Kingdom
William Jones, Indiana University, Bloomington, IN 47405, USA
James Kelly, University of Maryland, College Park, MD 20742, USA
Leonard Kisslinger, Carnegie-Mellon Univ., Pittsburgh, PA 15213, USA
Karl-Tasso Knopfle, Max Planck Instititut, Heidelberg, West Germany
Elie Korkmaz, Indiana University, Bloomington, IN 47405, USA
Michael Kovash, University of Kentucky, Lexington, KY 40506, USA
Dieter Kurath, Argonne National Lab, Argonne, IL 60439, USA
Kris Kwiatkowski, Indiana University, Bloomington, IN 47405, USA
Helene Langevin-Joliot, Inst. de Physique Nucl., 91406 Orsay, France
Lawrence Lee, Univ. of Toronto, Toronto, Ontario, M5S 1A7 Canada
Richard Lindgren, U. of Mass./U. of Virginia, Amherst, MA 01003, USA
Jerry Lisantti, TRIUMF, Vancouver, B.C. V6T 2A3, Canada
Keh-Feh Liu, University of Kentucky, Lexington, KY 40506, USA
Huan Liu, Rutgers University, Piscataway, NJ 08854, USA
Tim Londergan, Indiana University, Bloomington, IN 47405, USA
W. Gary Love, University of Georgia, Athens, GA 30601, USA
Daniel Low, Indiana University, Bloomington, IN 47405, USA
Malcolm Macfarlane, Indiana University, Bloomington, IN 47405, USA
R. Machleidt, Virginia Polytech. Inst, Blacksburg, VA 24061, USA
Alan MacKellar, University of Kentucky, Lexington, KY 40506, USA
N. Matsuoka, Osaka University, Osaka 567, Japan
James McNeil, Drexel University, Philadelphia, PA 19104, USA
C. Andrew Miller, TRIUMF, Vancouver, B.C. V6T 2A3 Canada
Daniel Miller, Indiana University, Bloomington, IN 47405, USA
Gerald Miller, University of Washington, Seattle, WA 98195, USA
Johann Miranda, Indiana University, Bloomington, IN 47405, USA
Steven Moszkowski, UCLA, Los Angeles, CA 90024, USA
Hermann Nann, Indiana University, Bloomington, IN 47405, USA
Hideaki Ohsumi, Osaka University, Osaka, 560, Japan
Catherine Olmer, Indiana University, Bloomington, IN 47405, USA
Allena Opper, Indiana University, Bloomington, IN 47405, USA
Costas Papanicolas, University of Illinois, Urbana, IL 61801, USA
Peter Pella, Hendrix College, Conway, AK, 72032, USA
W. Karl Pitts, Indiana University, Bloomington, IN 47405, USA
Robert Pollock, Indiana University, Bloomington, IN 47405, USA
Charles Price, Indiana University, Bloomington, IN 47405, USA
Michael Price, Indiana University, Bloomington, IN 47405, USA
Edwin N.M. Quint, NIKHEF-K, Amsterdam 1009 AJ, The Netherlands
M. Radhakrishna, Indiana University, Bloomington, IN 47405, USA
Brian Raue, Indiana University, Bloomington, IN 47405, USA
George Rawitscher, University of Connecticut, Storrs, CT 06268, USA
Edward Redish, University of Maryland, College Park, MD 20742, USA
Karl Schmid, Universitat Tubingen, D-7400 Tubingen, West Germany

Peter Schwandt, Indiana University, Bloomington, IN 47405, USA
Susan Seestrom-Morris, Univ. of Minnesota, Minneapolis, MN 55455, USA
Brian Serot, Indiana University, Bloomington, IN 47405, USA
Qingbiao Shen, University of Kentucky, Lexington, KY 40506, USA
Edward Siciliano, University of Georgia, Athens, GA 30602, USA
Ingo Sick, Universitat Basel, CH-4056 Basel, Switzerland
Paul Singh, Indiana University, Bloomington, IN 47405, USA
James Sowinski, Indiana University, Bloomington, IN 47405, USA
Joseph Speth, KFA Julich, D-5170 Julich, West Germany
Edward Stephenson, Indiana University, Bloomington, IN 47405, USA
Terry Taddeucci, Indiana University, Bloomington, IN 47405, USA
Jeffrey Templon, Indiana University, Bloomington, IN 47405, USA
Thomas Throwe, Indiana University, Bloomington, IN 47405, USA
Adriaan van den Berg, Univ. of Utrecht, Utrecht, The Netherlands
Adriaan van der Woude, Univ. of Groningen, Groningen, The Netherlands
Steven Vigdor, Indiana University, Bloomington, IN 47405, USA
C.M. Vincent, University of Pittsburgh, Pittsburgh, PA 15260, USA
Victor Viola, Indiana University, Bloomington, IN 47405, USA
Gerhard Wagner, Universitat Tubingen, D-7400 Tubingen, West Germany
George Walker, Indiana University, Bloomington, IN 47405, USA
John Watson, Kent State University, Kent, OH 44242, USA
Wolfram Weise, University of Regensburg, 84 Regensburg, West Germany
Curtis Whiddon, Indiana University, Bloomington, IN 47405, USA
B. Hobson Wildenthal, Drexel University, Philadelphia, PA 19104, USA
Scott Wissink, Indiana University, Bloomington, IN 47405, USA
S.S.M. Wong, University of Toronto, Toronto, Ontario M5S 1A7, Canada
Stanley Yen, TRIUMF, Vancouver, B.C., Canada B6T 2A3
David Youngblood, Texas A & M Univ., College Station, TX 77843, USA

AIP Conference Proceedings

		L.C. Number	ISBN
No. 1	Feedback and Dynamic Control of Plasmas – 1970	70-141596	0-88318-100-2
No. 2	Particles and Fields – 1971 (Rochester)	71-184662	0-88318-101-0
No. 3	Thermal Expansion – 1971 (Corning)	72-76970	0-88318-102-9
No. 4	Superconductivity in d- and f-Band Metals (Rochester, 1971)	74-18879	0-88318-103-7
No. 5	Magnetism and Magnetic Materials – 1971 (2 parts) (Chicago)	59-2468	0-88318-104-5
No. 6	Particle Physics (Irvine, 1971)	72-81239	0-88318-105-3
No. 7	Exploring the History of Nuclear Physics – 1972	72-81883	0-88318-106-1
No. 8	Experimental Meson Spectroscopy –1972	72-88226	0-88318-107-X
No. 9	Cyclotrons – 1972 (Vancouver)	72-92798	0-88318-108-8
No. 10	Magnetism and Magnetic Materials – 1972	72-623469	0-88318-109-6
No. 11	Transport Phenomena – 1973 (Brown University Conference)	73-80682	0-88318-110-X
No. 12	Experiments on High Energy Particle Collisions – 1973 (Vanderbilt Conference)	73-81705	0-88318-111-8
No. 13	π-π Scattering – 1973 (Tallahassee Conference)	73-81704	0-88318-112-6
No. 14	Particles and Fields – 1973 (APS/DPF Berkeley)	73-91923	0-88318-113-4
No. 15	High Energy Collisions – 1973 (Stony Brook)	73-92324	0-88318-114-2
No. 16	Causality and Physical Theories (Wayne State University, 1973)	73-93420	0-88318-115-0
No. 17	Thermal Expansion – 1973 (Lake of the Ozarks)	73-94415	0-88318-116-9
No. 18	Magnetism and Magnetic Materials – 1973 (2 parts) (Boston)	59-2468	0-88318-117-7
No. 19	Physics and the Energy Problem – 1974 (APS Chicago)	73-94416	0-88318-118-5
No. 20	Tetrahedrally Bonded Amorphous Semiconductors (Yorktown Heights, 1974)	74-80145	0-88318-119-3
No. 21	Experimental Meson Spectroscopy – 1974 (Boston)	74-82628	0-88318-120-7
No. 22	Neutrinos – 1974 (Philadelphia)	74-82413	0-88318-121-5
No. 23	Particles and Fields – 1974 (APS/DPF Williamsburg)	74-27575	0-88318-122-3
No. 24	Magnetism and Magnetic Materials – 1974 (20th Annual Conference, San Francisco)	75-2647	0-88318-123-1

No. 25	Efficient Use of Energy (The APS Studies on the Technical Aspects of the More Efficient Use of Energy)	75-18227	0-88318-124-X
No. 26	High-Energy Physics and Nuclear Structure – 1975 (Santa Fe and Los Alamos)	75-26411	0-88318-125-8
No. 27	Topics in Statistical Mechanics and Biophysics: A Memorial to Julius L. Jackson (Wayne State University, 1975)	75-36309	0-88318-126-6
No. 28	Physics and Our World: A Symposium in Honor of Victor F. Weisskopf (M.I.T., 1974)	76-7207	0-88318-127-4
No. 29	Magnetism and Magnetic Materials – 1975 (21st Annual Conference, Philadelphia)	76-10931	0-88318-128-2
No. 30	Particle Searches and Discoveries – 1976 (Vanderbilt Conference)	76-19949	0-88318-129-0
No. 31	Structure and Excitations of Amorphous Solids (Williamsburg, VA, 1976)	76-22279	0-88318-130-4
No. 32	Materials Technology – 1976 (APS New York Meeting)	76-27967	0-88318-131-2
No. 33	Meson-Nuclear Physics – 1976 (Carnegie-Mellon Conference)	76-26811	0-88318-132-0
No. 34	Magnetism and Magnetic Materials – 1976 (Joint MMM-Intermag Conference, Pittsburgh)	76-47106	0-88318-133-9
No. 35	High Energy Physics with Polarized Beams and Targets (Argonne, 1976)	76-50181	0-88318-134-7
No. 36	Momentum Wave Functions – 1976 (Indiana University)	77-82145	0-88318-135-5
No. 37	Weak Interaction Physics – 1977 (Indiana University)	77-83344	0-88318-136-3
No. 38	Workshop on New Directions in Mossbauer Spectroscopy (Argonne, 1977)	77-90635	0-88318-137-1
No. 39	Physics Careers, Employment and Education (Penn State, 1977)	77-94053	0-88318-138-X
No. 40	Electrical Transport and Optical Properties of Inhomogeneous Media (Ohio State University, 1977)	78-54319	0-88318-139-8
No. 41	Nucleon-Nucleon Interactions – 1977 (Vancouver)	78-54249	0-88318-140-1
No. 42	Higher Energy Polarized Proton Beams (Ann Arbor, 1977)	78-55682	0-88318-141-X
No. 43	Particles and Fields – 1977 (APS/DPF, Argonne)	78-55683	0-88318-142-8
No. 44	Future Trends in Superconductive Electronics (Charlottesville, 1978)	77-9240	0-88318-143-6
No. 45	New Results in High Energy Physics – 1978 (Vanderbilt Conference)	78-67196	0-88318-144-4

No. 46	Topics in Nonlinear Dynamics (La Jolla Institute)	78-57870	0-88318-145-2
No. 47	Clustering Aspects of Nuclear Structure and Nuclear Reactions (Winnepeg, 1978)	78-64942	0-88318-146-0
No. 48	Current Trends in the Theory of Fields (Tallahassee, 1978)	78-72948	0-88318-147-9
No. 49	Cosmic Rays and Particle Physics – 1978 (Bartol Conference)	79-50489	0-88318-148-7
No. 50	Laser-Solid Interactions and Laser Processing – 1978 (Boston)	79-51564	0-88318-149-5
No. 51	High Energy Physics with Polarized Beams and Polarized Targets (Argonne, 1978)	79-64565	0-88318-150-9
No. 52	Long-Distance Neutrino Detection – 1978 (C.L. Cowan Memorial Symposium)	79-52078	0-88318-151-7
No. 53	Modulated Structures – 1979 (Kailua Kona, Hawaii)	79-53846	0-88318-152-5
No. 54	Meson-Nuclear Physics – 1979 (Houston)	79-53978	0-88318-153-3
No. 55	Quantum Chromodynamics (La Jolla, 1978)	79-54969	0-88318-154-1
No. 56	Particle Acceleration Mechanisms in Astrophysics (La Jolla, 1979)	79-55844	0-88318-155-X
No. 57	Nonlinear Dynamics and the Beam-Beam Interaction (Brookhaven, 1979)	79-57341	0-88318-156-8
No. 58	Inhomogeneous Superconductors – 1979 (Berkeley Springs, W.V.)	79-57620	0-88318-157-6
No. 59	Particles and Fields – 1979 (APS/DPF Montreal)	80-66631	0-88318-158-4
No. 60	History of the ZGS (Argonne, 1979)	80-67694	0-88318-159-2
No. 61	Aspects of the Kinetics and Dynamics of Surface Reactions (La Jolla Institute, 1979)	80-68004	0-88318-160-6
No. 62	High Energy e^+e^- Interactions (Vanderbilt, 1980)	80-53377	0-88318-161-4
No. 63	Supernovae Spectra (La Jolla, 1980)	80-70019	0-88318-162-2
No. 64	Laboratory EXAFS Facilities – 1980 (Univ. of Washington)	80-70579	0-88318-163-0
No. 65	Optics in Four Dimensions – 1980 (ICO, Ensenada)	80-70771	0-88318-164-9
No. 66	Physics in the Automotive Industry – 1980 (APS/AAPT Topical Conference)	80-70987	0-88318-165-7
No. 67	Experimental Meson Spectroscopy – 1980 (Sixth International Conference, Brookhaven)	80-71123	0-88318-166-5
No. 68	High Energy Physics – 1980 (XX International Conference, Madison)	81-65032	0-88318-167-3
No. 69	Polarization Phenomena in Nuclear Physics – 1980 (Fifth International Symposium, Santa Fe)	81-65107	0-88318-168-1

No. 70	Chemistry and Physics of Coal Utilization – 1980 (APS, Morgantown)	81-65106	0-88318-169-X
No. 71	Group Theory and its Applications in Physics – 1980 (Latin American School of Physics, Mexico City)	81-66132	0-88318-170-3
No. 72	Weak Interactions as a Probe of Unification (Virginia Polytechnic Institute – 1980)	81-67184	0-88318-171-1
No. 73	Tetrahedrally Bonded Amorphous Semiconductors (Carefree, Arizona, 1981)	81-67419	0-88318-172-X
No. 74	Perturbative Quantum Chromodynamics (Tallahassee, 1981)	81-70372	0-88318-173-8
No. 75	Low Energy X-Ray Diagnostics – 1981 (Monterey)	81-69841	0-88318-174-6
No. 76	Nonlinear Properties of Internal Waves (La Jolla Institute, 1981)	81-71062	0-88318-175-4
No. 77	Gamma Ray Transients and Related Astrophysical Phenomena (La Jolla Institute, 1981)	81-71543	0-88318-176-2
No. 78	Shock Waves in Condensed Mater – 1981 (Menlo Park)	82-70014	0-88318-177-0
No. 79	Pion Production and Absorption in Nuclei – 1981 (Indiana University Cyclotron Facility)	82-70678	0-88318-178-9
No. 80	Polarized Proton Ion Sources (Ann Arbor, 1981)	82-71025	0-88318-179-7
No. 81	Particles and Fields –1981: Testing the Standard Model (APS/DPF, Santa Cruz)	82-71156	0-88318-180-0
No. 82	Interpretation of Climate and Photochemical Models, Ozone and Temperature Measurements (La Jolla Institute, 1981)	82-71345	0-88318-181-9
No. 83	The Galactic Center (Cal. Inst. of Tech., 1982)	82-71635	0-88318-182-7
No. 84	Physics in the Steel Industry (APS/AISI, Lehigh University, 1981)	82-72033	0-88318-183-5
No. 85	Proton-Antiproton Collider Physics –1981 (Madison, Wisconsin)	82-72141	0-88318-184-3
No. 86	Momentum Wave Functions – 1982 (Adelaide, Australia)	82-72375	0-88318-185-1
No. 87	Physics of High Energy Particle Accelerators (Fermilab Summer School, 1981)	82-72421	0-88318-186-X
No. 88	Mathematical Methods in Hydrodynamics and Integrability in Dynamical Systems (La Jolla Institute, 1981)	82-72462	0-88318-187-8
No. 89	Neutron Scattering – 1981 (Argonne National Laboratory)	82-73094	0-88318-188-6
No. 90	Laser Techniques for Extreme Ultraviolt Spectroscopy (Boulder, 1982)	82-73205	0-88318-189-4

No. 91	Laser Acceleration of Particles (Los Alamos, 1982)	82-73361	0-88318-190-8
No. 92	The State of Particle Accelerators and High Energy Physics (Fermilab, 1981)	82-73861	0-88318-191-6
No. 93	Novel Results in Particle Physics (Vanderbilt, 1982)	82-73954	0-88318-192-4
No. 94	X-Ray and Atomic Inner-Shell Physics – 1982 (International Conference, U. of Oregon)	82-74075	0-88318-193-2
No. 95	High Energy Spin Physics – 1982 (Brookhaven National Laboratory)	83-70154	0-88318-194-0
No. 96	Science Underground (Los Alamos, 1982)	83-70377	0-88318-195-9
No. 97	The Interaction Between Medium Energy Nucleons in Nuclei – 1982 (Indiana University)	83-70649	0-88318-196-7
No. 98	Particles and Fields – 1982 (APS/DPF University of Maryland)	83-70807	0-88318-197-5
No. 99	Neutrino Mass and Gauge Structure of Weak Interactions (Telemark, 1982)	83-71072	0-88318-198-3
No. 100	Excimer Lasers – 1983 (OSA, Lake Tahoe, Nevada)	83-71437	0-88318-199-1
No. 101	Positron-Electron Pairs in Astrophysics (Goddard Space Flight Center, 1983)	83-71926	0-88318-200-9
No. 102	Intense Medium Energy Sources of Strangeness (UC-Sant Cruz, 1983)	83-72261	0-88318-201-7
No. 103	Quantum Fluids and Solids – 1983 (Sanibel Island, Florida)	83-72440	0-88318-202-5
No. 104	Physics, Technology and the Nuclear Arms Race (APS Baltimore –1983)	83-72533	0-88318-203-3
No. 105	Physics of High Energy Particle Accelerators (SLAC Summer School, 1982)	83-72986	0-88318-304-8
No. 106	Predictability of Fluid Motions (La Jolla Institute, 1983)	83-73641	0-88318-305-6
No. 107	Physics and Chemistry of Porous Media (Schlumberger-Doll Research, 1983)	83-73640	0-88318-306-4
No. 108	The Time Projection Chamber (TRIUMF, Vancouver, 1983)	83-83445	0-88318-307-2
No. 109	Random Walks and Their Applications in the Physical and Biological Sciences (NBS/La Jolla Institute, 1982)	84-70208	0-88318-308-0
No. 110	Hadron Substructure in Nuclear Physics (Indiana University, 1983)	84-70165	0-88318-309-9
No. 111	Production and Neutralization of Negative Ions and Beams (3rd Int'l Symposium, Brookhaven, 1983)	84-70379	0-88318-310-2

No. 112	Particles and Fields – 1983 (APS/DPF, Blacksburg, VA)	84-70378	0-88318-311-0
No. 113	Experimental Meson Spectroscopy – 1983 (Seventh International Conference, Brookhaven)	84-70910	0-88318-312-9
No. 114	Low Energy Tests of Conservation Laws in Particle Physics (Blacksburg, VA, 1983)	84-71157	0-88318-313-7
No. 115	High Energy Transients in Astrophysics (Santa Cruz, CA, 1983)	84-71205	0-88318-314-5
No. 116	Problems in Unification and Supergravity (La Jolla Institute, 1983)	84-71246	0-88318-315-3
No. 117	Polarized Proton Ion Sources (TRIUMF, Vancouver, 1983)	84-71235	0-88318-316-1
No. 118	Free Electron Generation of Extreme Ultraviolet Coherent Radiation (Brookhaven/OSA, 1983)	84-71539	0-88318-317-X
No. 119	Laser Techniques in the Extreme Ultraviolet (OSA, Boulder, Colorado, 1984)	84-72128	0-88318-318-8
No. 120	Optical Effects in Amorphous Semiconductors (Snowbird, Utah, 1984)	84-72419	0-88318-319-6
No. 121	High Energy e^+e^- Interactions (Vanderbilt, 1984)	84-72632	0-88318-320-X
No. 122	The Physics of VLSI (Xerox, Palo Alto, 1984)	84-72729	0-88318-321-8
No. 123	Intersections Between Particle and Nuclear Physics (Steamboat Springs, 1984)	84-72790	0-88318-322-6
No. 124	Neutron-Nucleus Collisions – A Probe of Nuclear Structure (Burr Oak State Park - 1984)	84-73216	0-88318-323-4
No. 125	Capture Gamma-Ray Spectroscopy and Related Topics – 1984 (Internat. Symposium, Knoxville)	84-73303	0-88318-324-2
No. 126	Solar Neutrinos and Neutrino Astronomy (Homestake, 1984)	84-63143	0-88318-325-0
No. 127	Physics of High Energy Particle Accelerators (BNL/SUNY Summer School, 1983)	85-70057	0-88318-326-9
No. 128	Nuclear Physics with Stored, Cooled Beams (McCormick's Creek State Park, Indiana, 1984)	85-71167	0-88318-327-7
No. 129	Radiofrequency Plasma Heating (Sixth Topical Conference, Callaway Gardens, GA, 1985)	85-48027	0-88318-328-5
No. 130	Laser Acceleration of Particles (Malibu, California, 1985)	85-48028	0-88318-329-3
No. 131	Workshop on Polarized ^{3}He Beams and Targets (Princeton, New Jersey, 1984)	85-48026	0-88318-330-7
No. 132	Hadron Spectroscopy–1985 (International Conference, Univ. of Maryland)	85-72537	0-88318-331-5

No. 133	Hadronic Probes and Nuclear Interactions (Arizona State University, 1985)	85-72638	0-88318-332-3
No. 134	The State of High Energy Physics (BNL/SUNY Summer School, 1983)	85-73170	0-88318-333-1
No. 135	Energy Sources: Conservation and Renewables (APS, Washington, DC, 1985)	85-73019	0-88318-334-X
No. 136	Atomic Theory Workshop on Relativistic and QED Effects in Heavy Atoms	85-73790	0-88318-335-8
No. 137	Polymer-Flow Interaction (La Jolla Institute, 1985)	85-73915	0-88318-336-6
No. 138	Frontiers in Electronic Materials and Processing (Houston, TX, 1985)	86-70108	0-88318-337-4
No. 139	High-Current, High-Brightness, and High-Duty Factor Ion Injectors (La Jolla Institute, 1985)	86-70245	0-88318-338-2
No. 140	Boron-Rich Solids (Albuquerque, NM, 1985)	86-70246	0-88318-339-0
No. 141	Gamma-Ray Bursts (Stanford, CA, 1984)	86-70761	0-88318-340-4

RAYMOND H. FOGLER LIBRARY
DATE DUE

**BOOKS ARE SUBJECT TO
RECALL AFTER TWO WEEKS**

APR 0 9 1987